PHYSICS OF CLIMATE

PHYSICS OF CLIMATE

José P. Peixoto
Professor of Physics

Geophysical Institute
University of Lisbon
Lisbon, Portugal

and

Abraham H. Oort

Geophysical Fluid Dynamics Laboratory/NOAA
U.S. Department of Commerce
Princeton, New Jersey 08542

Foreword by
Edward N. Lorenz

Professor Emeritus of Meteorology
MIT

Springer

Library of Congress Cataloging-in-Publication Data
Peixoto, José Pinto
    Physics of Climate / José P. Peixoto, Abraham H. Oort.
      p.   cm.
    Includes bibliographical references and index.
    ISBN 0-88318-711-6 (case). — ISBN 0-88318-712-4 (perfect)
    1. Climatology. 2. Dynamic meteorology. 3. Atmospheric
physics. I. Oort, Abraham H. II. Title.
QC981.P434 1991   91-11565
551.5—dc20   CIP

Printed on acid-free paper.

© 1992 Springer-Verlag New York, Inc.
AIP is an imprint of Springer-Verlag New York, Inc.

All rights reserved. This work may not be translated or copied in whole or in part without the written permission of the publisher (Springer-Verlag New York, Inc., 175 Fifth Avenue, New York, NY 10010, USA), except for brief excerpts in connection with reviews or scholarly analysis. Use in connection with any form of information storage and retrieval, electronic adaptation, computer software, or by similar or dissimilar methodology now known or hereafter developed is forbidden.
The use of general descriptive names, trade names, trademarks, etc., in this publication, even if the former are not especially identified, is not to be taken as a sign that such names, as understood by the Trade Marks and Merchandise Marks Act, may accordingly be used freely by anyone.

Printed and bound by Malloy, Lithographing, Inc., Ann Arbor, MI.
Printed in the United States of America.

10 9 8 7 6 5

ISBN 0-88318-712-4 Springer-Verlag New York Berlin Heidelberg SPIN 10714724

*Here watch the powerful machine of nature,
formed by ether and other elements.*

Noboru Nakamura

Artist

To the memory of
Professor Victor P. Starr
1909–1976

Unravelling a Ball of Yarn

It is to be hoped that by now the naive idea has been dispelled that problems like the atmospheric general circulation can be solved at one fell swoop through the lucky manipulation of the proper equations during a day's work of some genius in fluid mechanics. Solutions to problems involving systems of such complexity are not born full grown like Athena from the head of Zeus. Rather they evolve slowly, in stages, each of which requires a pause to examine data at great lengths in order to guarantee a sure footing and to properly choose the next step. The process is apt to be like the unwinding of a ball of yarn. If the right end is found, the unravelling may go forward quite systematically, but requires time—perhaps several decades, as was the case for the atmospheric circulation.

from

"Opinions and Impressions on Scientific and Related Subjects"

by Victor P. Starr

Contents

Foreword xvii
Preface xix
Acknowledgments xxi
List of symbols and definitions xxvii

1. Introduction 1
   1.1 Scope and background 1
   1.2 Layout of the book 4

2. Nature of the Problem 8
   2.1 Introduction 8
   2.2 Basic concepts of thermodynamic systems 9
       2.2.1 State of a system, extensive and intensive properties
       2.2.2 Classification of thermodynamic systems
       2.2.3 Forced and free behavior of open systems
       2.2.4 Random systems
   2.3 Components of the climate system 13
       2.3.1 Atmosphere
       2.3.2 Hydrosphere
       2.3.3 Cryosphere
       2.3.4 Lithosphere
       2.3.5 Biosphere
   2.4 The climate system 18
       2.4.1 The nature of the climate system
       2.4.2 The climate state
       2.4.3 Climate variability
   2.5 Feedback processes in the climate system 26
       2.5.1 Feedback concepts
       2.5.2 Applications to the climate system
       2.5.3 Some examples

3. Basic Equations for the Atmosphere and Oceans 32
   3.1 Equation of continuity 32
   3.2 Equations of motion 34
       3.2.1 Frictional effects
       3.2.2 Filtering of the basic equations for the atmosphere
       3.2.3 Filtering of the basic equations for the oceans

3.3　Vorticity equation 42
　　　　3.3.1　Some definitions of vorticity
　　　　3.3.2　General vorticity equation
　　　　3.3.3　Vorticity equation of the horizontal motion
　　3.4　Thermodynamic energy equation and some applications 46
　　　　3.4.1　First law of thermodynamics
　　　　3.4.2　Static stability
　　　　3.4.3　Potential vorticity
　　　　3.4.4　The thermodynamic energy equation and the local rate
　　　　　　　of change of temperature
　　3.5　Equation of state 51
　　　　3.5.1　Atmosphere
　　　　3.5.2　Oceans
　　　　3.5.3　Barotropy and baroclinicity
　　3.6　Equation of water vapor 58
　　3.7　Summary of the basic equations in Lagrangian and Eulerian form ... 58

4. Various Decompositions of the Circulation 61
　　4.1　Transient and stationary eddies 61
　　　　4.1.1　Time and horizontal resolutions of the circulation
　　　　4.1.2　Vertical resolution of the circulation
　　4.2　Spectral analysis of meteorological fields 65
　　　　4.2.1　Spectral analysis in space and time
　　　　4.2.2　Limitations of sampling
　　4.3　Empirical orthogonal function analysis 67

5. The Data 70
　　5.1　Observational networks 70
　　　　5.1.1　Atmospheric data
　　　　5.1.2　Oceanic data
　　　　5.1.3　Satellite data
　　　　5.1.4　International field projects
　　5.2　Data processing techniques 81
　　　　5.2.1　Atmospheric data
　　　　5.2.2　Oceanic data
　　　　5.2.3　Satellite data
　　5.3　Objective analysis methods 84
　　　　5.3.1　Atmospheric analyses
　　　　5.3.2　Reliability of the atmospheric analyses
　　　　5.3.3　Oceanic analyses
　　5.4　Other atmospheric data sets 88

6. Radiation Balance 91
　　6.1　Introduction 91
　　　　6.1.1　Nature of solar and terrestrial radiation
　　　　6.1.2　Global radiation balance
　　6.2　Physical radiation laws 95
　　　　6.2.1　Planck's law
　　　　6.2.2　Stefan-Boltzmann law
　　　　6.2.3　Wien displacement law
　　　　6.2.4　Kirchhoff's law
　　　　6.2.5　Beer-Bouger-Lambert law
　　6.3　Solar radiation 98
　　　　6.3.1　Solar spectrum and solar constant
　　　　6.3.2　Distribution of solar radiation at the top of the atmosphere
　　　　6.3.3　Aerosols
　　　　6.3.4　Absorption of solar radiation

		6.3.5 Scattering of solar radiation

       6.3.5 Scattering of solar radiation
       6.3.6 Effects of clouds on solar radiation
       6.3.7 Solar radiation at the earth's surface
6.4 Terrestrial radiation 104
       6.4.1 Introduction
       6.4.2 Absorption and emission spectra of atmospheric gases
       6.4.3 Rotational and vibrational bands
       6.4.4 Spectral lines—Lorentz formula
       6.4.5 Transmissivity functions
       6.4.6 Band models
       6.4.7 Nonhomogeneous paths
6.5 Radiative transfer 110
       6.5.1 Schwarzchild equation
       6.5.2 Radiative transfer equation
6.6 Radiation balance of the atmosphere 114
6.7 Radiation balance at the earth's surface 116
6.8 Observed radiation balance 117
       6.8.1 Radiation balance of the earth
       6.8.2 Global distribution of the radiation balance

7. Observed Mean State of the Atmosphere 131
7.1 Atmospheric mass and pressure 131
       7.1.1 Mass balance
       7.1.2 Distribution of mass in terms of pressure
7.2 Mean temperature structure of the atmosphere 137
       7.2.1 Global distribution of the temperature
       7.2.2 Vertical structure of the temperature
       7.2.3 Variability of the temperature
7.3 Mean geopotential height structure of the atmosphere 144
       7.3.1 Vertical structure of the geopotential
       7.3.2 Variability of the geopotential height
7.4 Mean atmospheric circulation 149
       7.4.1 Introduction
       7.4.2 Global distribution of the circulation
       7.4.3 Vertical structure of the circulation
       7.4.4 Variability of the circulation
7.5 Mean kinetic energy in the atmosphere 162
       7.5.1 Global distribution of the kinetic energy
       7.5.2 Vertical structure of the kinetic energy
7.6 Precipitation, evaporation, runoff, and cloudiness 165
       7.6.1 Precipitation
       7.6.2 Evaporation
       7.6.3 Surface runoff
       7.6.4 Cloudiness

8. Observed Mean State of the Oceans 176
8.1 Mean temperature structure of the oceans 176
       8.1.1 Global distribution of the temperature
       8.1.2 Vertical structure of the temperature
       8.1.3 Variability of the temperature
8.2 Mean salinity structure of the oceans 187
       8.2.1 Global distribution of the salinity
       8.2.2 Vertical structure of the salinity
8.3 Mean density structure of the oceans 190
       8.3.1 Global distribution of the density
       8.3.2 Vertical structure of the density

8.4 Mean ocean circulation 196
    8.4.1 Global distribution of the surface circulation
    8.4.2 Vertical structure of the circulation
8.5 Surface kinetic energy of the oceans 206

9. Observed Mean State of the Cryosphere 207
9.1 Role of the cryosphere in the climate 207
9.2 General features of the cryosphere 210
9.3 Ice sheets and glaciers 211
9.4 Sea ice 212
9.5 Snow 214
9.6 Permafrost 215

10. Exchange Processes Between the Earth's Surface and the Atmosphere 216
10.1 Introduction 216
10.2 Energy budget at the surface 217
    10.2.1 Energy fluxes at an ideal surface
    10.2.2 Energy budget of a layer
10.3 Development of the planetary boundary layer 222
    10.3.1 Some characteristic features of the planetary boundary layer
    10.3.2 Generation and maintenance of atmospheric turbulence
    10.3.3 Effects of stability
10.4 Exchange of momentum 226
    10.4.1 Eddy correlation approach
    10.4.2 Gradient-flux approach
    10.4.3 Mixing-length approach and wind profiles
    10.4.4 Bulk aerodynamic method
10.5 Transfer of mechanical energy into the oceans 231
10.6 Exchange of sensible heat 232
10.7 Exchange of water vapor, evaporation 233
    10.7.1 Eddy correlation, gradient-flux, and bulk transfer methods
    10.7.2 Energy balance method
    10.7.3 Combined approaches, Penman formula
10.8 Formation of atmospheric aerosol 240

11. Angular Momentum Cycle 241
11.1 Balance equations for angular momentum 241
    11.1.1 Introduction
    11.1.2 Angular momentum in the climatic system
    11.1.3 Angular momentum in the atmosphere
    11.1.4 Volume integrals
    11.1.5 Modes of transport
11.2 Observed cycle of angular momentum 255
    11.2.1 Angular momentum in the atmosphere
    11.2.2 Angular momentum exchange between the atmosphere and the underlying surface
    11.2.3 Cycle of angular momentum in the climatic system
    11.2.4 Angular momentum exchange between the oceans and the lithosphere

12. Water Cycle 270
12.1 Formulation of the hydrological cycle 270
    12.1.1 Introduction
    12.1.2 Water in the climatic system

12.2 Equations of hydrology 273
    12.2.1 Classic equation of hydrology
    12.2.2 Balance equation for water vapor
    12.2.3 Modes of water vapor transport
12.3 Observed atmospheric branch of the hydrological cycle 278
    12.3.1 Water vapor in the atmosphere
    12.3.2 Transport of water vapor
    12.3.3 Divergence of water vapor
12.4 Synthesis of the water balance 297
12.5 Hydrology of the polar regions 302
    12.5.1 Equations of hydrology including vapor, liquid, and solid water substance
    12.5.2 Observed water budget of the polar regions

13. Energetics 308
13.1 Basic forms of energy 308
13.2 Energy balance equations 310
    13.2.1 Introduction
    13.2.2 The climate equations
    13.2.3 Volume integrals
    13.2.4 Globally averaged climate equations
13.3 Observed energy balance 319
    13.3.1 Diabatic heating in the atmosphere
    13.3.2 Energy in the atmosphere
    13.3.3 Transport of atmospheric energy
    13.3.4 Energy in the oceans
    13.3.5 Transport of oceanic energy
    13.3.6 Synthesis of the energy balance
13.4 Energetics of the polar regions 353
    13.4.1 Formulation of the energy budget
    13.4.2 Observed energy budget in atmospheric polar caps
    13.4.3 Observed energy budget in ocean–land–ice polar caps
    13.4.4 Synthesis of the polar energetics

14. The Ocean–Atmosphere Heat Engine 365
14.1 Availability of energy in the atmosphere 365
14.2 Availability of energy in the ocean 370
14.3 Balance equations for kinetic and available potential energy 373
    14.3.1 Formal derivation of the balance equations
    14.3.2 Balance equations for mean and eddy kinetic energy in the atmosphere
    14.3.3 Balance equations for mean and eddy available potential energy in the atmosphere
14.4 Observed energy cycle in the atmosphere 379
    14.4.1 Spatial distributions of the energy and energy conversions
    14.4.2 The energy cycle
14.5 Maintenance and forcing of the zonal-mean state
    of the atmosphere 385
    14.5.1 Introduction
    14.5.2 Interactions between the eddies and the zonal-mean state
    14.5.3 Eliassen-Palm flux
    14.5.4 Modified momentum and energy equations
    14.5.5 Forcing of the mean meridional circulation
    14.5.6 Some examples of E–P flux diagrams
14.6 Observed energy cycle in the oceans 393
    14.6.1 Spatial distributions of the energy components
    14.6.2 The energy cycle

15. Entropy in the Climate System 401
15.1 Introduction 401
15.2 Balance equation of entropy 403
15.3 Observed entropy budget of the atmosphere 407
15.3.1 Global entropy budget
15.3.2 Regional entropy budgets

16. Interannual and Interdecadal Variability in the Climate System 412
16.1 Introduction 412
16.2 Quasibiennial oscillation 413
16.2.1 Observed features
16.2.2 Possible solar-QBO-climate connections
16.3 ENSO phenomenon 415
16.4 Regional teleconnections 426
16.5 Interdecadal fluctuations and trends 433
16.5.1 Anthropogenic influences
16.5.2 Atmospheric gases
16.5.3 Surface temperature
16.5.4 Upper air temperatures
16.6 Some special climatic phenomena 444

17. Mathematical Simulation of Climate 450
17.1 Introduction 450
17.2 Mathematical and physical structure of climate models 451
17.2.1 Basic parameters of a climate model
17.2.2 Model equations
17.2.3 Necessity of using numerical integrations
17.2.4 Parameterizations
17.2.5 Nature of the mathematical solutions
17.3 Hierarchy of climate models 463
17.4 General circulation models 464
17.4.1 Some general features
17.4.2 The development of general circulation modeling
17.4.3 Coupled ocean–atmosphere models
17.5 Statistical dynamical models 467
17.5.1 Energy balance models
17.5.2 Radiative–convective models
17.5.3 Two-dimensional statistical dynamical models
17.6 Uses and applications of models 473
17.6.1 Some general remarks
17.6.2 Data assimilation and network testing
17.6.3 Modeling of the hydrological cycle
17.6.4 Modeling of the ENSO phenomena
17.6.5 Modeling of the CO_2 effects
17.6.6 Effects of mountains and the simulation of an ice-age climate
17.6.7 Sensitivity to changes in astronomical parameters

Appendix A: Analysis in Terms of Fourier Components 481
A1. Introduction 481
A2. The Fourier spectrum 482

A3.　Multiplication and Parseval theorems 484
　　A4.　Spectral functions of the meteorological
　　　　 variables and equations 487
　　　　　A4.1.　*Linear quantities*
　　　　　A4.2.　*Quadratic quantities*
　　　　　A4.3.　*Meteorological equations*

Appendix B: Analysis in Terms of Empirical Orthogonal Functions (EOF's) 492
　　B1.　The problem 492
　　B2.　Solution of the problem 493
　　B3.　Variability of two-dimensional vector
　　　　 fields 496

References 497
Name Index 509
Subject Index 513

Foreword

The study of climate can assume many forms. Meteorologists have tended to think of the climate, and the changes that it continually undergoes, as special aspects of the weather. Oceanographers are likely to include ocean currents among the significant climatic features, and they may seek the roots of climatic changes in oceanic behavior. Geologists may attribute prehistoric climatic variations to changes in land forms and ultimately to the drifting of the continents.

Among meteorologists, climate has been an evolving concept. Early in the present century a local climate was often considered to be little more than the annual course of the long-term averages of temperature and precipitation. The existence of extensive regions of the globe with reasonably uniform local climates led to the concept of climatic zones; familiar geographical features such as rain forests, deserts, and tundra were to be found in particular zones.

By the middle of the century some meteorologists had extended the scope of climate to include not simply temperature and precipitation but virtually all atmospheric properties, at upper levels as well as near the earth's surface. To these investigators, climate consisted of the set of all long-term atmospheric statistics, and thus was almost synonymous with the general circulation of the atmosphere. Theoretical explanations of climate were no longer based mainly on the manner in which the atmosphere received and rid itself of heat and water, but also incorporated the role of the transports of energy, water, and momentum within the atmosphere, these transports having become essential elements of general-circulation theory.

Within more recent years the concept of a *climate system* has become firmly established. The basis for this view is the realization that the underlying ocean and land surfaces (and the ice, snow, lakes, rivers, and living things that are often found between these surfaces and the atmosphere) are not mere inert boundary conditions, to be taken for granted in seeking explanations for the atmosphere's behavior. On the contrary, they possess their own internal dynamics, and for them the atmosphere is one of the boundary conditions. Together with the atmosphere they form a larger system that may logically be studied as a single entity. From this perspective the atmospheric circulation problem is just one aspect of the climate problem. Needless to say, the study of the latter problem is no longer the prerogative of the meteorological community.

The authors of *Physics of Climate* and I have in common something that has shaped our scientific endeavors and, to some extent, our outlook upon life—we have all had Victor Starr as our mentor. Starr was a powerful theoretical meteorologist, but he is probably best remembered today for his extensive studies of the transports of angular momentum, energy, and water in the atmosphere, and his systematization of procedures for evaluating these transports from available meteorological data. His theoretical ideas had a great effect upon my subsequent work; his ideas regarding the measurement of transport processes are clearly discernible in the works of the authors of this volume.

Nearly twenty-five years ago, at the invitation of the World Meteorological Organization, I prepared a monograph for which I chose the title *The Nature and Theory of the General Circulation of the Atmosphere*. During the preparation of this volume I was aided by the opportunity to engage in almost daily conversations with Victor Starr, and the circulation that I wrote about was basically the circulation as Starr saw it. In the longest chapter I described the various exchanges between the atmosphere and its surroundings and the various fluxes within the atmosphere, and characterized these as the processes that maintained the circulation. The attitude toward climate being what it was at that time, I could probably, with a few modifications, have offered the volume as a monograph on the nature and theory of climate.

All this has changed. A modern treatment of the nature and theory of climate, as opposed to a descriptive account of specific climatic features, must deal with the entire climate system, whether or not the presentation scheme is to resemble the one that I chose. In *Physics of Climate* the authors have given us such a treatment.

Both authors are internationally known meteorologists. They have performed extensive evaluations of the processes that maintain the atmospheric circulation, using up-to-date observational data, and some of the results that they present in this volume are the fruits of their own previous work. They complement the atmospheric transport statistics with similar statistics for transports within the oceans. They devote a whole chapter to ice, snow, and permafrost.

The authors precede their detailed picture of the nature of the climate system with several chapters introducing the theory. Here the basic physical laws that govern the earth are expressed in the context of climatic variables. Their discussion is enhanced by the inclusion of mathematical expressions for the variables and mathematical equations representing the laws. However, there are numerous stretches of several pages where no mathematical symbols appear, even in the chapter on mathematical climate modeling, and the reader who shuns mathematical expositions will find that he can still follow the central ideas. The text is generously interspersed with figures—mainly charts and graphs.

So many relevant statistical properties of the climate system have by now been evaluated that it is unlikely that anyone would take the time to memorize them all. Gleaning them all from the original sources would be a lengthy undertaking. Having an extensive set together under one cover is one of the benefits that this volume can bestow on someone working in this area. At the other extreme, one who is overwhelmed by the size of the collection of climatic statistics will find numerous interesting little items. For example, the authors note that about 80% of the world's fresh water is in frozen form, they present a simple description of how winds can produce upwelling and downwelling in the oceans, and they include a graph showing the variations of the surface elevation of the Great Salt Lake in Utah over the past century

Edward N. Lorenz
Professor Emeritus of Meteorology
Massachusetts Institute of Technology
Cambridge, MA

September 1991

Preface

In some sense, writing a book may be compared with creating a piece of art. As the sculptor who may start with a vague image, an empty board, a bag of clay, and some sculptor's carving tools, the writer may start with nothing but some general ideas of what he or she wants to write about and to include in the book. But then, in a seemingly miraculous way, gradually—one by one—the ideas start to take shape, to grow on their own and to develop into more concrete and better defined structures. Handwritten notes, outlines, sketches, and maps begin to replace the original ideas and start to provide the needed framework. We found that the process of building a book is a constantly evolving one—one adds new sections and modifies, transplants, or discards other older sections. Again, it resembles the sculptural process of adding pieces of clay to the developing model where they are needed and to cut away the parts that are superfluous or out of place.

The creative process and interplay with the observational and theoretical material on the earth's climate system available to us has been an exciting and rewarding task. It also was a fluid process of interchange between two very different minds, with very different makeup and backgrounds. But, as mentioned in Edward Lorenz's Foreword, there was one unifying factor that brought us together almost three decades ago and has bound us together as friends and colleagues ever since. This was our common background of studies at the Department of Meteorology at MIT under an inspiring teacher, Victor P. Starr, to whose memory we dedicate this book. In fact, about two decades ago Starr suggested to us the idea of writing a book on the observational foundations of the atmospheric general circulation. However, we found out rather quickly that the available material was incomplete and insufficient and that the time was obviously not yet ripe to undertake this task. Numerical models were in their infancy, and the fields of meteorology, oceanography, climatology, and hydrology were still rather disconnected.

Now after two further decades have elapsed, much comprehensive material on the global atmosphere and oceans has become available. Our joint articles and those of our colleagues and students on the observed angular momentum, water, and energy cycles in the atmosphere and oceans, together with the course notes of the classes we have taught at the University of Lisbon and Princeton University have provided the basic material for this book.

In this book we hope to offer a broad perspective on the earth's climate system. It provides a synthesis of our present observational knowledge and the physical understanding we have gained on how the climatic engine works. More than 50% of the book is devoted to the atmosphere, nearly 20% to the oceans, about 5% to the cryosphere, solid earth, and biosphere, and the remaining 25% to more general discussions. Thus it is clear that the atmosphere and oceans have the major emphasis in our book. In a different breakdown, the discussions of the physical principles and background take up about 40% of the book and the observations and diagnostic discussions about 60%. We decided not to include any theories and tentative explanations of why the climate system works as it does. The book is written, we hope, in simple, easily understandable terms in order to make the field accessible to a broad range of interested students and scientists from other disciplines.

In addition to being a textbook this volume represents a report on the research which has occupied us over a period of many years. However, the large number of topics covered include some in which we had no previous expertise. For this and other reasons, some sections of the book are not as skillfully phrased as they could have been. Therefore, we shall welcome suggestions for improvements.

We foresee that our book will be of interest and useful to a rather broad audience of scientists, including meteorologists, oceanographers, geographers, climatologists, and other geophysicists, as well as physicists, mathematicians, ecologists, and engineers with an interest in climate. The level of treatment is geared mainly to graduate students and researchers, but the material should also be understandable for motivated undergraduate students since we have kept the mathematics to a modest level and concentrated on the physical understanding. Furthermore, some chapters of the text can be used as a graduate level course on the general circulation of the atmosphere, or as a supplement to advanced courses in the atmospheric and oceanic sciences.

Acknowledgments

We want to acknowledge many of our colleagues at the Geophysical Fluid Dynamics Laboratory (GFDL), Atmospheric and Environmental Research, Inc. (AER), National Center for Atmospheric Research (NCAR), European Centre for Medium Range Weather Forecasts (ECMWF), Climate Analysis Center (CAC), and Cooperative Institute for Research and Environmental Sciences (CIRES), who reviewed the various chapters of our book, and gave valuable contributions with their constructive criticisms and suggestions that have enriched the book. We are especially grateful to Drs. Syukuro Manabe and Ngar-Cheung Lau from GFDL, Drs. Richard D. Rosen and David A. Salstein from AER, Prof. Lev Gandin from the University of Maryland, Prof. Barry Saltzman from Yale University, and Prof. Edward N. Lorenz from MIT who were willing to set aside other pressing commitments and contributed with their insight, guidance, and deep knowledge to the broadening and improvement of the book.

The publication of this book would have not been possible without the encouragement and continuous support of the directors of GFDL, Drs. Joseph Smagorinsky and Jerry Mahlman, who have so generously made available the GFDL resources for this undertaking. We want to thank Mr. Roy Jenne from NCAR, Mr. Sydney Levitus from GFDL, and Dr. Chester F. Ropelewski from CAC for sharing some of their climatological data used in this book. We are also indebted to the various supporting agencies, the National Oceanic and Atmospheric Administration (NOAA), National Science Foundation (NSF), and the Portuguese National Council of Research (INIC), for providing the essential financial support for our joint research.

The many staff members of the Administrative, Computer Systems, Computer Operational, Scientific Illustration, and other Technical Support Groups at GFDL and, in particular, Mr. James S. Byrne, Mr. John N. Connor, Mrs. Gail T. Haller, Mrs. Catherine Raphael, Mr. Melvin Rosenstein, and Mr. Philip G. Tunison are gratefully acknowledged for their highly competent help in facilitating our research and in preparing the manuscript. We thank Mr. James G. Welsh for providing us with his remarkable computational expertise during so many years of our research. Mrs. Joyce Kennedy deserves special recognition for her efficiency and cheerfulness in typing the almost endless series of revisions of the manuscript. Finally, we want to thank Dr. Noboru Nakamura at GFDL for creating the frontispiece, the editors, Dr. Thomas von Foerster, Mr. Michael Hennelly, and Mr. Andrew Prince at the American Institute of Physics for their able assistance and advice in the organization and preparation of the manuscript for publication.

Grateful acknowledgment is made to the following publishers and authors for permission to reprint previously published material.

ACADEMIC PRESS:

Mitchell, J. M., 1976: "An overview of climatic variability and its causal mechanisms." Quaternary Res. **6**, 481–493.
Oort, A. H., 1985: "Balance conditions in the earth's climate system." Adv. Geophys. A **28**, 75–98.
Oort, A. H. and J. P. Peixoto, 1983: "Global angular momentum and energy balance requirements from observations." Adv. Geophys. **25**, 355–490.
Wallace, J. M. and P. V. Hobbs, 1977: *Atmospheric Science: An Introductory Survey*. Academic, New York, 467 pp.

ALLEN AND UNWIN:

Tolmazin, D., 1985: *Elements of Dynamic Oceanography*. Allen and Unwin, Winchester, MA, 181 pp.

AMERICAN GEOPHYSICAL UNION:

Cayan, D. R., J. V. Gardner, J. M. Landwehr, J. Namias, and D. H. Peterson, 1989: In *Aspects of Climate Variability in the Pacific and the Western Americas* (D. H. Peterson, editor), Introduction. Geophys. Mon. **55**, American Geophysical Union, Washington, D.C., pp. xiii–xvi.
Ellis, J. S., T. H. Vonder Haar, S. Levitus, and A. H. Oort, 1978: "The annual variation in the global heat balance of the earth." J. Geophys. Res. **83**, 1958–1962.
Keeling, C. D., R. B. Bacastow, A. F. Carter, S. C. Piper, T. P. Whorf, M. Heimann, W. G. Mook, and H. Roeloffzen, 1989: "A three-dimensional model of atmospheric CO_2 transport based on observed winds: I. Analysis of observational data." In *Aspects of Climate Variability in the Pacific and the Western Americas* (D. H. Peterson, editor), Geophys. Mon. **55**, American Geophysical Union, Washington, D.C., pp. 165–236.
Manabe, S. and J. L. Holloway, Jr., 1975: "The seasonal variation of the hydrological cycle as simulated by a global model of the atmosphere." J. Geophys. Res. **80**, 1617–1649.
Nakamura, N. and A. H. Oort, 1988: "Atmospheric heat budgets of the polar regions." J. Geophys. Res. **93**, 9510–9524.
Oort, A. H., S. C. Ascher, S. Levitus, and J. P. Peixoto, 1989: "New estimates of the available potential energy in the world ocean." J. Geophys. Res. **94**, 3187–3200.
Peixoto, J. P., A. H. Oort, M. de Almeida, and A. Tomé, 1991: "Entropy budget of the atmosphere." J. Geophys. Res. **96**, 10981–10988.

AMERICAN METEOROLOGICAL SOCIETY:

Abramopoulos, F., C. Rosenzweig, and B. Choudhury, 1988: "Improved ground hydrology calculations for global climate models (GCMs): Soil water movement and evapotranspiration." J. Climate **1**, 921–941.

Hastenrath, S. 1982: "On meridional heat transports in the World Ocean." J. Phys. Oceanogr. **12**, 922–927.

Jones, P. D., 1988: "Hemispheric surface air temperature variations: Recent trends and an update to 1987." J. Climate **1**, 654–660.

Julian, P. R. and R. M. Chervin, 1978: "A study of the Southern Oscillation and Walker circulation phenomenon." Mon. Weather Rev. **106**, 1433–1451.

Lau, N.-C. and A. H. Oort, 1981: "A comparative study of observed Northern Hemisphere circulation statistics based on GFDL and NMC analyses. Part I: The time-mean fields." Mon. Weather Rev. **109**, 1380–1403.

Levitus, S. and A. H. Oort, 1977: "Global analysis of oceanographic data." Bull. Am. Meteor. Soc. **58**, 1270–1284.

Manabe, S. and R. F. Strickler, 1964: "Thermal equilibrium of the atmosphere with a convective adjustment." J. Atmos. Sci. **21**, 361–385.

Manabe, S. and R. T. Wetherald, 1967: "Thermal equilibrium of the atmosphere with a given distribution of relative humidity." J. Atmos. Sci. **24**, 241–259.

Naujokat, B., 1986: "An update of the observed quasibiennial oscillation of the stratospheric winds over the tropics." J. Atmos. Sci. **43**, 1873–1877.

Oort, A. H., 1989: "Angular momentum cycle in the atmosphere–ocean–solid earth system." Bull. Am. Meteor. Soc. **70**, 1231–1242.

Pan, Y.-H. and A. H. Oort, 1983: "Global climate variations connected with sea surface temperature anomalies in the eastern equatorial Pacific Ocean for the 1958–1973 period." Mon. Weather Rev. **111**, 1244–1258.

Ropelewski, C. F. and M. S. Halpert, 1987: "Global and regional scale precipitation patterns associated with the El Niño/Southern oscillation." Mon. Weather Rev. **115**, 1606–1626.

Salstein, D. A. and R. D. Rosen, 1986: "Earth rotation as a proxy for interannual variability in atmospheric circulation, 1860–present." J. Climate Appl. Meteor. **25**, 1870–1877.

Trenberth, K. E. and D. J. Shea, 1987: "On the evolution of the Southern Oscillation." Mon. Weather Rev. **115**, 3078–3096.

van Loon, H. and K. Labitzke, 1988: "Association between the 11-year solar cycle, the QBO, and the atmosphere. Part II: Surface and 700 mb in the Northern Hemisphere in winter." J. Climate **1**, 905–920.

Wallace, J. M. and D. S. Gutzler, 1981: "Teleconnections in the geopotential height field during the Northern Hemisphere winter." Mon. Weather Rev. **109**, 784–812.

Weare, B. C., A. R. Navato, and R. E. Newell, 1976: "Empirical orthogonal analysis of Pacific sea surface temperatures." J. Phys. Ocean **6**, 671–678.

CAMBRIDGE UNIVERSITY PRESS:

Lamb, P. J. and R. A. Peppler, 1991: West Africa. In *Teleconnections Linking Worldwide Climate Anomalies: Scientific Basis and Societal Impact* (M. H. Glantz, R. W. Katz, and N. Nicholls, editors), Cambridge University, New York.

Untersteiner, N., 1984: "The cryosphere." In *The Global Climate* (J. T. Houghton, editor), Cambridge University, New York, pp. 121–140

CANADIAN JOURNAL OF FISHERIES AND AQUATIC SCIENCES:

Mysak, L. A., 1986: "El Niño, interannual variability, and fisheries in the Northeast Pacific Ocean." Can. J. Fish. Aquat. Sci. **43**, 464–497.

ELSEVIER SCIENCE PUBLISHERS:

Baumgartner, A. and E. Reichel, 1975: *The World Water Balance*. Elsevier, Amsterdam, 179 pp.

Manabe, S., K. Bryan, and M. J. Spelman, 1979: "A global ocean–atmosphere climate model with seasonal variation for future studies of climate sensitivity." Dyn. Atm. Oceans **3**, 393–426.

GIDROMETEOIZDAT:

Berlyand, T. G. and L. A. Strokina, 1980: "Global distribution of total cloudiness." Gidrometeoizdat, Leningrad, 71 pp.

SYDNEY LEVITUS:

Levitus, S., 1982: Climatological Atlas of the World Ocean. NOAA Professional Paper No. 13, U.S. Government Printing Office, Washington, D.C., 163 pp.

MARINE TECHNOLOGY SOCIETY:

Wyrtki, K., 1982: "The Southern oscillation, ocean–atmosphere interaction, and El Niño." Marine Technol. Soc. J. **16**, 3–10.

McGRAW-HILL:

Gast, P. R., 1965: "Solar electromagnetic radiation." In *Handbook of Geophysics and Space Environments*, Air Force Cambridge Research Laboratory, U.S. Air Force, pp. 16.1–16.9.

THE MIT PRESS:

Houghton, H. G., 1985: *Physical Meteorology*. MIT, Cambridge, MA, 442 pp.

MUNSKGAARD INTERNATIONAL PUBLISHERS:

Vinnichenko, N. K., 1970: "The kinetic energy spectrum in the free atmosphere—1 second to 5 years." Tellus **22**, 158–166.

OXFORD UNIVERSITY PRESS:

Goody, R. M., 1964: *Atmospheric Radiation: I. Theoretical Basis*. Clarendon, Oxford, 436 pp.

PERGAMON PRESS:

Labitzke, K. and H. van Loon, 1988: "Associations between the 11-year solar cycle, the QBO, and the atmosphere. Part I: The troposphere and stratosphere in the northern hemisphere in winter." J. Atmos. Terr. Phys. **50**, 197–206.

Ropelewski, C. F., 1989: "Monitoring large-scale cryosphere/atmosphere interactions." Adv. Space Res. **9**, 213–218.

REIDEL PUBLISHING COMPANY:

Budyko, M. I., 1986: *The Evolution of the Biosphere*. Reidel, Dordrecht, 423 pp.

Peixoto, J. P. and A. H. Oort, 1983: "The atmospheric branch of the hydrological cycle and climate." In *Variations of the Global Water Budget*. Reidel, London, pp. 5–65.

ROYAL METEOROLOGICAL SOCIETY:

Gardiner, B. G., 1989: "The Antarctic ozone hole." Weather **44**, 291–298.

Newell, R. E., D. G. Vincent, T. G. Dopplick, D. Ferruza, and J. W. Kidson, 1970: "The energy balance of the global atmosphere." In *The Global Circulation of the Atmosphere* (G. A. Corby, editor), Royal Meteorological Society, London, pp. 42–90.

SPRINGER-VERLAG:

Oort, A. H., Y.-H. Pan, R. W. Reynolds, and C. F. Ropelewski, 1987: "Historical trends in the surface temperature over the oceans based on the COADS." Climate Dyn. **2**, 29–38.

List of Symbols and Definitions

Only the principal symbols are listed. A horizontal, two-dimensional vector is indicated by sans serif type, a three-dimensional vector by bold face type, and a three-dimensional tensor by "serif gothic" type.

a_λ	absorptivity for radiation at wavelength $\lambda = I_{\lambda a}/I_\lambda$
c	(1) speed of light ($= 2.998 \times 10^8$ m s^{-1}); (2) condensation rate per unit mass; (3) specific heat of the surface layer (Chapters 6 and 10, Table 10.1)
c^*	conductive capacity of the surface layer
c	three-dimensional wind velocity $= (u, v, w)$
c$_A$	three-dimensional velocity vector in non-rotating absolute frame
c_I	specific heat of snow/ice
c_L	specific heat of land surface
c_o	specific heat of ocean water (4187 J kg^{-1} K^{-1} $\approx c_{po} \approx c_{vo}$)
c_p	atmospheric specific heat at constant pressure ($= 1004$ J kg^{-1} K^{-1})
c_s	speed of sound ($\simeq 300$ m s^{-1})
c_v	atmospheric specific heat at constant volume ($= 717$ J kg^{-1} K^{-1})
curl	curl operator
d	day
d	sun–earth distance (Chapter 6)

d_m	mean sun–earth distance
dA	area element
div	divergence operator
dm	mass element
ds	surface element
$d_e s$	transfer of entropy across boundaries of open system
$d_i s$	entropy produced within a system
dV	volume element
dyn	1 dyne $= 10^{-5}$ N (1 dyne cm$^{-2} = 10^{-1}$ Pa)
e	(1) evaporation rate per unit mass; (2) water vapor pressure (Chapter 10)
e_s	saturation water vapor pressure
f	(1) Coriolis parameter $= 2\Omega \sin \phi$ ($= 1.03 \times 10^{-4}$ s^{-1} at 45° latitude); (2) frequency; (3) feedback of a system (Chapter 2); (4) ratio real and potential evaporation $= E/E_0$ (Chapter 10)
f'	$= 2\Omega \cos \phi$
f_N	Nyquist frequency $= \frac{1}{2} \Delta t$
g	acceleration due to gravity ($= 9.81$ m s^{-2} at sea level; $= 9.75$ m s^{-2} at 20 km altitude)
g	apparent gravity vector
gpm	geopotential meter (1 gpm $\approx$ 1 m)
grad	gradient vector
h	hour
h	(1) Planck constant ($= 6.626 \times 10^{-34}$ J s); (2) hour angle of sun (Chapter 6); (3) enthalpy of oceans $= (p/\rho + I)$ (Chapter 14)
$h(x)$	depth of ocean bottom
h_r	enthalpy reference state ocean
k	(1) von Karman constant ($= 0.4$); (2) Boltzmann constant ($= 1.380 \times 10^{-23}$ J K^{-1}); (3) wavenumber $= 2\pi R \cos \phi / L$ (Appendix A)
k_λ, k_ν	extinction coefficient for radiation at wavelength λ, frequency ν
$k_{\lambda a}$	absorption coefficient for radiation at wavelength λ
$k_{\lambda s}$	scattering coefficient for radiation at wavelength λ
l	mixing length

m_a	mass of the atmosphere ($\simeq 5.136 \times 10^{18}$ kg)
m_d	molecular weight of dry air ($= 28.9 \times 10^{-3}$ kg mol^{-1})
m_e	mass of the earth ($\simeq 5.98 \times 10^{24}$ kg)
m_v	molecular weight of water vapor ($= 18.0 \times 10^{-3}$ kg mol^{-1})
mb	millibar (1 mb $= 10^2$ Pa $= 1$ hPa $\approx 10^3$ dyne cm^{-2})
n	wavenumber
$\mathbf{n}$	unit vector (directed outward of volume)
n_i	number of moles for dissolved component i
p	pressure
p_D	bottom pressure at depth D in ocean
p_{Ei}, p_{Wi}	pressure at east, west sides of ith mountain range
p_r	reference pressure
p_0	surface pressure
p_{00}	reference level pressure (≈ 1000 mb)
p_{SL}	pressure at mean sea level (Chapter 7)
q	specific humidity (in g/kg moist air)
q_s	specific humidity of saturated air
r	correlation coefficient
$\mathbf{r}$	radius vector
$\mathbf{r}_r$	position vector of a point in rotating system
r_λ	reflectivity for radiation at wavelength $\lambda = I_{\lambda r}/I_\lambda$
s	second
s	(1) entropy per unit mass; (2) path length
$s(q)$	sources and sinks of water vapor
s^*	static stability in oceans $\left(= -\dfrac{1}{\rho}\dfrac{\partial \rho}{\partial z} - \dfrac{g}{c_s^2} \right)$
sr	steradian
t	time
u	(1) eastward wind component; (2) optical depth or optical path length $= \int k_\lambda \rho \, ds$; (3) wind velocity
u_*	friction velocity $= \sqrt{-\tau_0/\rho}$
u_{ag}, v_{ag}	ageostrophic eastward, northward component of the wind

u_g, v_g	geostrophic eastward, northward component of the wind
v	northward wind component
v_l, v_s	meridional velocity of liquid, solid water suspended in air
$[\bar{v}]_{\text{ind}}$	indirectly computed $[\bar{v}]$ from angular momentum balance
V	horizontal wind vector $= (u,v)$
w	upward wind component
w_E	Ekman vertical velocity
yr	year
z	(1) geometric height; (2) geopotential height $\left(\equiv \dfrac{1}{9.80}\int_0^z g\,dz \text{ in MKS system} \simeq \text{geometric height}\right)$ (3) zenith angle (Chapter 6)
z_0	(1) surface height; (2) surface roughness length (Chapter 10)
z_{SA}	height of pressure level in standard atmosphere
A	(1) area of integration; (2) albedo; (3) arbitrary parameter; (4) constant Wien's law (Chapter 6)
A_{sfc}	surface albedo
AU	astronomical unit, mean earth–sun distance $(= 1.4960 \times 10^{11}$ m$)$
$\mathcal{A}$	symbol for atmosphere
B	(1) Bowen ratio $= F^{\uparrow}_{SH}/F^{\uparrow}_{LH}$; (2) rate of production of turbulent kinetic energy by buoyancy forces $= -g\,\overline{\rho' w'}$
B_λ, B_ν	black body radiation at wavelength λ, frequency ν
$\mathcal{B}$	symbol for biosphere
C	continental torque
$C(A, B)$	rate of energy conversion from form A into form B
C_D	bulk drag coefficient ($\simeq 0.0013$ over ocean)
C_H	heat transfer coefficient (bulk)
C_W	water vapor transfer coefficient (bulk)
COADS	Comprehensive Ocean-Atmosphere Data Set
$\mathcal{C}$	symbol for cryosphere; continental torque (Chapter 11)

D	(1) turbulent diffusion rate of water vapor into a unit volume $= -\alpha \text{ div } \mathbf{J}_q^D$; (2) rate of viscous dissipation (Chapter 10)
D_A	dissipation rate of kinetic energy in atmosphere $= -\int \alpha \, \tau: \text{grad } \mathbf{c} \, dm$
D_{oc}	mechanical energy input into oceans $= \int (\omega z - \tau_0 \cdot \mathbf{c}) ds$
DJF	December–February
$D(K)$	rate of dissipation of kinetic energy
E	(1) rate of evaporation (includes sublimation); (2) total energy
E–P flux	Eliassen–Palm flux (Chapter 14)
E_a	evaporation power of air
E_1	rate of evaporation at the surface
E_0	potential evaporation or potential evapotranspiration
E_s	rate of sublimation at surface
EBM	energy balance model (Chapter 17)
ENSO	El Niño–Southern Oscillation phenomenon
EOF	empirical orthogonal function
EU	Eurasian Oscillation
$\mathscr{E}$	symbol for environment
$F^\uparrow, F^\downarrow$	upward, downward fluxes of radiant energy
$\mathbf{F}$	(1) frictional force $= (F_x, F_y, F_z) = (F_\lambda, F_\phi, F_z) = -\alpha \text{ div } \tau$ (2) Eliassen–Palm flux vector (F^ϕ, F^p) (Chapter 14)
F_{BA}	downward flux of energy at bottom of atmosphere
F_{BO}	upward (geothermal) heat flux at ocean bottom
$F_G^\downarrow$	downward heat flux into ground
F_l	meridional flux of water across wall in liquid phase (Chapter 12)
$F_{LH}^\uparrow$	upward flux of latent heat at surface
$\mathbf{F}_{LH}$	horizontal flux of latent heat
$F_{LW}, F_{LW}^\downarrow, F_{LW}^\uparrow$	longwave radiation flux, downward flux, upward flux
F_M	latent heat flux involved in melting and freezing
F_{net}	$= F^\downarrow - F^\uparrow$
F_o	net flow of ocean water across latitudinal wall
$\mathbf{F}_{rad}$	radiation flux at top of atmosphere

$F_{\text{rad}}^{\text{sfc}}$	downward radiation flux at surface
F_{riv}	northward transport of water in rivers across latitudinal wall
F_s	(1) freezing rate of water; (2) meridional flux of water across wall in solid phase (Chapter 12)
$F_{\text{SH}}^{\uparrow}$	upward flux of sensible heat at surface
F_{SW}	shortwave radiation flux (irradiance)
$F_{\text{SW}}^{\downarrow}, F_{\text{SW}}^{\uparrow}$	downward flux, upward flux of shortwave radiation
F_{TA}	net energy flux at top of atmosphere
F_{wall}	flux across latitudinal wall
F_v	meridional flux of water across wall in vapor phase
$\mathscr{F}$	$= \mathbf{F}_{\text{rad}} + \mathbf{F}_{\text{LH}} + \mathbf{J}_H^D + \tau \cdot \mathbf{c}$
G	gain or transfer function of a system (Chapter 2)
$G(P)$	generation rate of P
G_{LH}	net latent heat flux across boundaries
G_{rad}	net radiative flux across boundaries
G_{SH}	net sensible heat flux across boundaries
GCM	general circulation model
H	(1) total heating $= \int \dot{Q}\, dm + \int_{\text{sfc}} F_{\text{SH}}\, ds$; (2) maximum depth ocean basin; (3) hour angle at sunrise and sunset (Chapter 6); (4) feedback factor (Chapter 2)
Hz	hertz
$\mathscr{H}$	symbol for hydrosphere
I	internal energy
I_a	moment of inertia of atmosphere
I_e	moment of inertia of solid earth ($\simeq 8.04 \times 10^{37}$ kg m^2)
I_λ	intensity at wavelength λ (power per unit area per unit solid angle per unit wavelength per unit time)
ITCZ	intertropical convergence zone
J	joule
$\mathbf{J}$	angular momentum transport density vector $= (\mathscr{J}_\lambda, \mathscr{J}_\phi, \mathscr{J}_p)$
$\mathbf{J}_E$	total energy transport density vector $= (\mathscr{J}_{E\lambda}, \mathscr{J}_{E\phi}, \mathscr{J}_{Ep})$
$\mathbf{J}_H$	enthalpy transport density vector $= (\mathscr{J}_{H\lambda}, \mathscr{J}_{H\phi}, \mathscr{J}_{Hp})$
$\mathbf{J}_H^D$	heat flux due to molecular and turbulent eddy diffusion
$\mathbf{J}_K$	mechanical energy transport density vector $= (\mathscr{J}_{K\lambda}, \mathscr{J}_{K\phi}, \mathscr{J}_{Kp})$
$\mathbf{J}_q$	water vapor transport density vector $= (\mathscr{J}_{q\lambda}, \mathscr{J}_{q\phi}, \mathscr{J}_{qp})$

$\mathbf{J}_q^D$	water vapor flux due to molecular and turbulent eddy diffusion
JJA	June–August
K	degree Kelvin
K	(1) (total) kinetic energy; (2) thermal conductivity of surface layer (Chapter 10); (3) large-scale eddy diffusion coefficient for momentum (Chapter 17)
K^*	thermal diffusivity $= K/\rho c$ (Chapter 10)
K_A	large-scale eddy diffusion coefficient for heat (Chapter 17)
K_E	eddy kinetic energy
K_H	eddy diffusion coefficient for enthalpy;
K_h	horizontal kinetic energy $= (u^2 + v^2)/2$
K_M	(1) eddy diffusion coefficient for momentum; (2) mean kinetic energy
K_{SE}	stationary eddy kinetic energy
K_t	kinetic energy in turbulent motions
K_{TE}	transient eddy kinetic energy
K_W	eddy diffusion coefficient for water vapor
L	latent heat of evaporation, melting or sublimation
L_e	latent heat of evaporation ($= 2501$ J g^{-1})
L_m	latent heat of melting ($= 334$ J g^{-1})
L_s	(1) latent heat of sublimation ($= 2835$ J g^{-1}) (2) wavelength
LH	latent heat energy
$\mathscr{L}$	symbol for lithosphere
M	(1) mass of a given volume (2) absolute angular momentum (Chapter 12)
$\mathbf{M}$	angular momentum vector
$\mathbf{M}_E$	Ekman mass transport vector $= (M_{Ex}, M_{Ey})$
M_r	relative angular momentum
M_s	melting rate of snow and ice
M_Ω	Ω-angular momentum
MBT	mechanical bathythermograph (ocean data)
$\mathbf{N}$	baroclinicity vector (Chapter 3)
N	Brunt-Väisälä frequency
NAO	North Atlantic Oscillation
NMC	National Meteorological Center (Washington, D.C.)

NPC	north polar cap
NPO	North Pacific Oscillation
NWP	numerical weather prediction
$\mathcal{O}$	symbol for oceans
P	(1) precipitation rate in liquid or solid form; (2) total potential energy (Chapter 12); (3) available potential energy (Chapter 14)
Pa	pascal ($= 10$ dyne cm^{-2})
P_l	rate of precipitation in liquid form
P_E	eddy available potential energy
P_M	mean available potential energy
P^O, P^A	available potential energy in the oceans, atmosphere
P_s	rate of precipitation in solid form
P_{SE}	stationary eddy available potential energy
P_{TE}	transient eddy available potential energy
PNA	Pacific–North American oscillation
$\mathcal{P}$	pressure torque
$\mathcal{P}_L$	pressure torque over land
$\mathbf{Q}$	net horizontal flux vector of water in vapor phase $= (Q_\lambda, Q_\phi)$
$\mathbf{Q}_c$	net horizontal flux vector of water in condensed phase $= (Q_{c\lambda}, Q_{c\phi})$
Q	diabatic heating rate
Q_f	rate of diabatic heating by friction
Q_h	rate of diabatic heating by energy fluxes across boundaries
Q_0	integrated daily value of solar radiation at top of atmosphere
QBO	quasibiennial oscillation
R	mean radius of the earth ($\simeq 6.371 \times 10^6$ m)
R_d	gas constant for dry air ($= 287$ J kg^{-1}K^{-1})
R_e	Reynolds number $= uL/\nu$
R_i	Richardson number
R_o	(1) Rossby number $= u/fL$; (2) surface runoff
R_u	underground runoff
RCM	radiative-convective model (Chapter 17)

S	(1) salinity (in $°/_{oo}$);
	(2) solar constant ($\simeq 1360$ W m^{-2});
	(3) line intensity (Chapter 6);
	(4) rate of production of turbulent kinetic energy
	$\quad = -\rho \, \overline{\mathbf{v}'w'} \cdot \partial \overline{\mathbf{v}}/\partial z$ (Chapter 10);
	(5) total entropy of the system (Chapter 15)
$S(W_l), S(W_s), S(W_v)$	source of water in liquid, solid and vapor phase
S_A	rate of energy storage in atmosphere
S_I, S_L, S_O	rate of heat storage in snow and ice, land, oceans
S_{LHI}	rate of latent heat storage in snow and ice
S_v	Sverdrup = 10^6 m^3 s^{-1}
SDM	statistical dynamical model (Chapter 17)
SPC	south polar cap
SST	sea surface temperature
$\mathscr{S}$	symbol for climate system
T	(1) temperature;
	(2) period;
	(3) rate of transfer of kinetic energy to turbulent eddies from other eddies
T^*	reference or equivalent temperature (Chapter 15)
T_A	northward energy transport in atmosphere across latitudinal wall
T_e	(1) equivalent temperature = $T(1 + Lq/c_p T)$;
	(2) radiative equilibrium temperature
T_{oc}	northward energy transport in oceans across latitudinal wall
T_I	transport of sensible plus latent energy into oceanic polar cap in form of snow and ice
T_s	sea surface temperature
T_{SA}	temperature in standard atmosphere
T_V	virtual temperature = $T(1 + 0.61q)$
T_{sfc}	surface temperature
$\mathscr{T}$	friction torque
$\mathscr{T}_L, \mathscr{T}_O$	friction torque over land, ocean
U	relative humidity = e/e_s
V	volume of integration
V_F, V_S	portion of output signal that is fed back, input signal
W	precipitable water in vertical column (in vapor form)

W_c	precipitable water in vertical column (in condensed form)
W_i, W_k	energy states of an atom or molecule
W_1, W_s, W_v	precipitable water in liquid, solid, vapor phase
XBT	expendable bathythermograph (ocean data)
Z	zenith angle of sun
ZPS triangle	zenith–pole–sun triangle
α	(1) specific volume $= 1/\rho$; (2) half width of spectral line
α_w	molecular diffusivity for water vapor
β	variation of Coriolis parameter with latitude $= df/Rd\phi$ ($= 1.62 \times 10^{-11}$ m^{-1}s^{-1} at 45° latitude)
γ	(1) environmental lapse rate $= -\partial T/\partial z$; (2) Bowen constant ($= 0.61$ mb °C^{-1}; Chapter 10)
γ_d	dry adiabatic lapse rate $= g/c_p$
δ	declination of the sun
δm	mass element
δV	volume element
$\delta \rho$	stability measure ocean $= \rho(z + \zeta) - \rho_{\theta L}(z)$
ε	(1) infrared emissivity surface; (2) quantum of energy
$\tilde{\varepsilon}$	weighted band emissivity
ε_λ	emissivity at wavelength $\lambda = I_\lambda/B(\lambda, T)$
ζ	(1) vertical component of the vorticity (2) displacement in z-direction
η	(1) absolute vorticity; (2) height of sea level (Chapter 11); (3) efficiency heat engine (Chapter 14)
κ	$= R_d/c_p (= 0.286)$
λ	(1) longitude; (2) wavelength
μ	dynamic coefficient of viscosity
μ_i	chemical potential of element i
μm	micrometer
ν	(1) kinematic coefficient of viscosity $= \mu/\rho$; (2) frequency $= c/\lambda$
ρ	density
ρ_1, ρ_s	density of liquid, solid water suspended in air

$\rho_{\theta L}$	local potential density
σ	(1) Stefan-Boltzmann constant $(= 5.6701 \times 10^{-8}$ W m^{-2} K$^{-4})$; (2) measure of ocean density (in kg m^{-3}) $= \rho(T,S,p) - 1000$; (3) rate of generation of entropy; (4) sigma coordinate $= p/p_0$ (Chapter 17)
$\sigma_{dis}, \sigma_{LH}, \sigma_{rad}, \sigma_{SH}$	rate of entropy generation associated with kinetic energy dissipation, latent heat release, radiational heating, sensible heating
σ_t	measure of ocean density (in kg m^{-3}) $= \rho(T, S, p_0) - 1000$, where p_0 is pressure at ocean surface
τ	(1) surface stress; (2) transmissivity (Chapter 6); (3) time period
τ^*	diffuse transmissivity
$\boldsymbol{\tau}, \boldsymbol{\tau}_0$	surface stress vector (positive if momentum is transferred upward)
$\boldsymbol{\tau}$	3-dimensional stress tensor $= \rho \, \overline{\mathbf{c'c'}}$
τ_λ	transmissivity at wavelength $\lambda = I_{\lambda\tau}/I_\lambda$
ϕ	(1) latitude; (2) geopotential $= \phi_N - \Omega^2 r^2$
ϕ_N	earth's gravitational potential
ψ	streamfunction for mass
ψ_E	streamfunction for energy
ψ_M	streamfunction for angular momentum
ψ_q	streamfunction for water vapor
ω	(1) vertical velocity in (x, y, p, t) system $= dp/dt$; (2) solid angle (element $d\omega = \cos \phi \, d\phi \, d\lambda$)
ω_c	vertical velocity in (x, y, p, t) system for condensed water in air
Γ	mean stability parameter $= -(\kappa\theta/pT)(\partial\tilde{\theta}/\partial p)^{-1}$ $= (\gamma_d/\tilde{T})(\gamma_D - \tilde{\gamma})^{-1}$
Γ_a	circulation of velocity field around boundary contour
Δ	slope of saturation vapor pressure versus temperature curve $= \partial e_s/\partial T$
θ	(1) potential temperature; (2) angle between incoming solar beam and vertical
Λ	availability $= (\Phi + I) - (\Phi + I)_r$
Π	product
Σ	summation

Φ	potential energy
Ω	angular velocity of the earth (7.292×10^{-5} rad s^{-1})
$\boldsymbol{\Omega}$	angular velocity vector

Special Symbols Used in Appendices A and B

a_k, b_k, c_k	Fourier cosine, sine, complex coefficients
$\{\mathbf{e}_i\}$	set of orthogonal base vectors
$\mathbf{e}_i^T$	transpose vector $\mathbf{e}_i$
f_n	data vector
$f(\lambda)$	function in λ-space ⎫
$F(k)$	function in wave number (k)-space ⎬ Fourier pair
$F_1(k), F_2(k)$	real, imaginary parts of $F(k)$
$\widehat{F}(k)$	complex conjugate of $F(k)$
$\mathscr{F}(A)$	Fourier transform of A
$f(\lambda)*g(\lambda)$	convolution of $f(\lambda)$ and $g(\lambda) = \int_{-\infty}^{\infty} f(\lambda - \zeta) g(\zeta) d\zeta$
ε_k	phase angle for wave number k
$\theta_f(j), \theta_g(j)$	phase angles
$\Phi_{fg}(j)$	covariance spectrum (co-spectrum) of $f(\lambda)$ and $g(\lambda) = \widehat{F}(j) G(j)$
$\Phi_{ff}(j)$	variance spectrum of $F(\lambda) = \widehat{F}(j) F(j)$

Mathematical Operators

$\overline{A} = (t_2 - t_1)^{-1} \int_{t_1}^{t_2} A \, dt$	time average of A
$A' = A - \overline{A}$	departure from time average of A
$[A] = (2\pi)^{-1} \int_0^{2\pi} A \, d\lambda$	zonal average of A
$A^* = A - [A]$	departure from zonal average of A
$\langle A \rangle$	(1) mass-weighted "vertical" average of A
	$= (p_0 - p_t)^{-1} \int_{p_t}^{p_0} A \, dp;$
	(2) ensemble average (Chapter 2)
$A'' = A - \langle A \rangle$	departure from vertical average

$\{A\}$	areal average of A over a region, polar cap, hemisphere or the globe
$\tilde{A}$	global mean of A over a constant z, p, θ or ρ surface (Chapter 14)
A^*	(1) departure from areal average $= A - \{A\}$ (2) $A - \tilde{A}$ (Chapter 14)
$A \cup B$	union of sets A and B
$A \cap B$	intersection of sets A and B
$\mathbf{A} \times \mathbf{B}$	vector product of vectors $\mathbf{A}$ and $\mathbf{B}$
$\mathbf{A} \cdot \mathbf{B}$	inner product of vectors $\mathbf{A}$ and $\mathbf{B}$
$\mathbf{a} : \mathbf{b}$	inner product of tensors $\mathbf{a}$ and $\mathbf{b}$

$$\operatorname{curl} \mathbf{A} = \left(\frac{\partial A_z}{\partial y} - \frac{\partial A_y}{\partial z}, -\frac{\partial A_z}{\partial x} + \frac{\partial A_x}{\partial z}, \frac{\partial A_y}{\partial x} - \frac{\partial A_x}{\partial y} \right)$$

$$\operatorname{div} \mathbf{A} = \frac{\partial A_x}{\partial x} + \frac{\partial A_y}{\partial y} + \frac{\partial A_z}{\partial z}$$

$$\operatorname{grad} A = \left(\frac{\partial A}{\partial x}, \frac{\partial A}{\partial y}, \frac{\partial A}{\partial z} \right)$$

$\nabla^2 F =$ Laplacian operator of $F =$ div grad F

Example of Nomenclature

$$\langle [\overline{vA}] \rangle = \langle [\overline{v'A'}] \rangle + \langle [\overline{v}^*\overline{A}^*] \rangle + \langle [\overline{v}][\overline{A}] \rangle$$

$\langle [\overline{vA}] \rangle$	vertical average of northward (meridional) transport of A due to all motions
$\langle [\overline{v'A'}] \rangle$	vertical average of northward transport of A resulting from transient eddies
$\langle [\overline{v}^*\overline{A}^*] \rangle$	vertical average of northward transport of A resulting from stationary eddies
$\langle [\overline{v}][\overline{A}] \rangle$	vertical average of northward transport of A resulting from mean meridional circulations

CHAPTER 1

Introduction

Historically, the study of the atmosphere and oceans, our human environment, together with the study of our broader celestial environment of the sun, moon, and planets, provided the initial impetus to develop the field of physics. In more recent times astronomy, with its new discoveries, has expanded the scope of physics into its modern breadth. However, the earth sciences of meteorology and oceanography have had comparatively little influence on modern developments, and have often been regarded as old, antiquated, and "solved" branches of physics with little intellectual challenge.

Lately a change in view has taken place, and many physicists have become aware of the new and, in our view, exciting developments in the earth sciences. For example, the study of simple nonlinear systems first developed by Lorenz (1963, 1969) for the study of climate has started a boom in this field among applied mathematicians and physicists.

Since no single comprehensive book appears to have been published so far on the physics of climate, our goal has been to create such a book dealing with the earth's climate as a physical system. The emphasis in this monograph is on a general but concise description of what we consider to be the most important features of the climate system consisting of the atmosphere, oceans, and cryosphere (snow and ice fields) as its principal components.

1.1 SCOPE AND BACKGROUND

We will regard the climate in a very broad sense in terms of the mean physical state of the climatic system. The climate can then be defined as a set of averaged quantities completed with higher moment statistics (such as variances, covariances, correlations, etc.) that characterize the structure and behavior of the atmosphere, hydrosphere, and cryosphere over a period of time. This definition of climate includes the more narrow traditional concept of climate based on the mean atmospheric conditions at the earth's surface.

We know that climate has undergone many changes in the past and that it will continue to change in the future. In other words, the climate is always evolving and it must be regarded as a living entity. Thus we should avoid the misleading concept of the constant nature of climate.

Throughout the text an effort will be made to focus on the various components of the climate system in an integrated way rather than to consider them independently and to stress the importance of physics for understanding the complex interactions between them. Since the atmosphere is the part of the climate system in which we live, the primary focus of attention will be on this component of the climate system. Moreover, the rise of climatology as a science is closely related to the progress and rapid development of meteorology as well as to the increase in our capacity to obtain more and better observations of the atmosphere.

We, of course, accept that the processes occurring in the climate system obey the laws of physics and that these laws must be applied in an appropriate way. Since these laws are expressed in mathematical terms, a basic knowledge of mathematics is also needed. Further, since changes in atmospheric composition, such as the increase in carbon dioxide and the depletion of ozone, are observed facts it is now evident that chemistry is becoming increasingly important. Thus the understanding of climate requires some knowledge of the three basic sciences of physics, mathematics, and chemistry.

It is true that most of the climatic features can be described and interpreted without referring to physics or mathematics. This has been the traditional approach used in descriptive climatology. However, we think that the concepts in climatology should be built within a physics framework in order to get the proper meaning and interpretation of the results. In fact, studies of the climatic phenomena and processes in the light of physics during the last few decades have greatly improved our understanding of the atmosphere, oceans, and cryosphere.

We must keep in mind that climatology is an observational science so that the growth in understanding strongly depends on the improvement of the measurements and observing systems. The recent improvements come from an expansion and better quality of the classic observing networks and the application of new technologies (radar, laser, lidar, infrared radiometers, scatterometers, acoustic tomography, etc.). For example, satellite-based observations have transformed our perspective of the earth's climate, leading us to regard it both as an integrated global system and as a three-dimensional entity. It can be expected that these improvements will continue in the near future.

The progress in both physical insight and in the observations is reflected in the rapid development of comprehensive mathematical models. These models tend not only to improve our insight into the processes that make up the climate but may also enable us to predict future climate conditions.

The observations provide the basis for monitoring the climate and are essential to test any theory of climate. An understanding of the climate can be gained from diagnostic studies based on the observations analyzed in the light of the physical laws as well as from mathematical models. In our book we will be concerned with the trilogy formed by the monitoring of climate, diagnostic studies, and mathematical modeling, but our main focus will be on the diagnostic studies. As indicated in Fig. 1.1, this trilogy must be decisive for the development of a theory of climate. Such a theory would contain a broad understanding and the potential to predict climatic change.

This last possibility to predict the climate would have strong socio-economic implications. As we know in the past, climate has been a determining factor in, e.g., the patterns of food production and the distribution of populations. It is also clear that climatic fluctuations, such as the pronounced droughts in the Sahel region, episodes of El Niño, and failing monsoon rains, have had tremendous social and economic consequences, causing hardships on millions of people, and that such fluctuations have exerted profound stress on the biosphere in general. Furthermore, the evidence is growing that humans have been influencing the climate through their activities, so

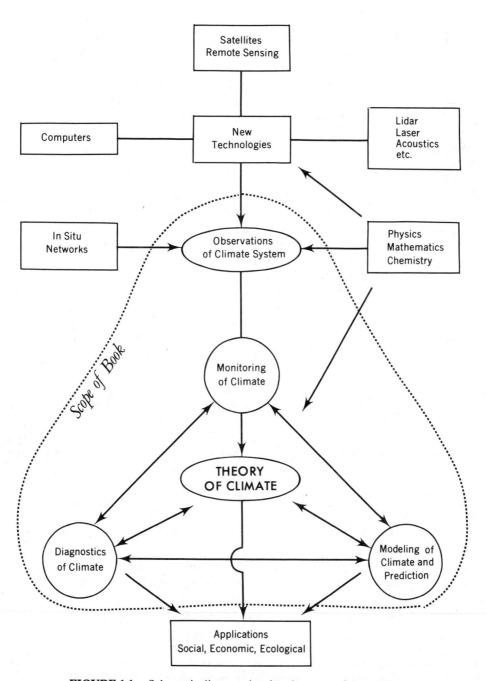

FIGURE 1.1. Schematic diagram showing the scope of the book.

that it is of utmost importance to be able to know in advance the implications of any new actions, which is only possible through a better theory and understanding of climate and climate change.

1.2 LAYOUT OF THE BOOK

The material presented is based largely on direct, *in situ* observations complemented with some satellite observations, and is chosen to best convey to the reader our knowledge and understanding of the physical processes that maintain the present climate. We concentrate on the discussion of the large-scale, long-term mean climate over the globe, its annual cycle, and its fluctuations with time scales up to several decades.

Following the schematic outline of the book shown in Fig. 1.2 we first describe in Chap. 2 the state and behavior of the total climate system composed of the atmosphere, oceans, cryosphere, lithosphere, and biosphere, each one with a different range of time and space scales. We further discuss why the climate system is such a highly interactive and complex system with many feedbacks. We will see that changes in one part of a subsystem may eventually affect all other parts of the climate system and that feedback processes from the slower subsystems, such as the oceans and glaciers, can initiate fluctuations with a long time scale in the faster subsystems, such as the atmosphere. In fact, this feedback process can lead to what we may call climatic cycles and climate trends. For example, during El Niño the predictability of the behavior of the atmosphere may increase substantially.

The basic equations, as presented and explained in Chap. 3 and the averaging schemes used to synthesize the various properties of the general circulations in the atmosphere and oceans as given in Chap. 4, lay the theoretical basis for the later chapters. The observational foundations are given in Chap. 5 where the original observations, the reduction of the observations into representative statistics, and various objective analysis techniques are discussed. Chapter 5 is supplemented by a set of Appendices dealing with various techniques of data reduction and analysis. We may note here that only since World War II has the systematic study of the atmosphere and oceans become feasible on a global scale, mainly through the increase of *in situ* observations. During the last one or two decades, global observing networks have become available in which satellites play a central role not only as remote observing platforms in space for homogeneous monitoring of the atmosphere and the earth's surface at regular time intervals, but also as collectors of more standard *in situ* observations. Besides the atmospheric data, satellites provide us, for the first time, with direct measurements of the fluxes of radiant energy entering and leaving the atmosphere.

The radiation laws necessary to understand the transfer of radiant energy (solar and terrestrial) through the atmosphere and the emission of long-wave radiation by the various components of the climate system are presented in Chap. 6. This chapter also contains a discussion of the observed planetary radiation balance. This balance is, of course, of utmost importance because radiation forms the basic driving factor for practically all processes occurring in the climate system and for all atmospheric and oceanic motions. Essentially all the energy that enters the earth comes from the sun.

The observed mean states of the three principal climatic components, i.e., the atmosphere, oceans, and cryosphere, are given in Chaps. 7–9. Together with the later

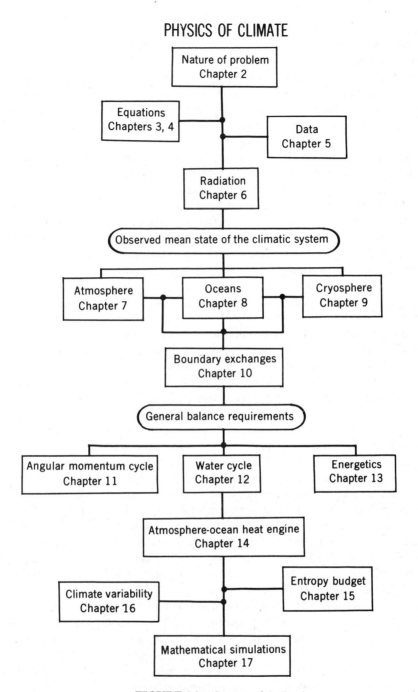

FIGURE 1.2. Layout of the book.

chapters on the balance equations they form the main bulk of this book. Next, the exchange processes at the earth's surface are formulated in Chap. 10. Because of the lack of direct observations and the uncertainties in the transfer formulas used, this is a relatively weak spot in our understanding of the climate system.

A more physical understanding of the climate system can be gained from the material given in Chaps. 11–14 on the angular momentum, water, and energy cycles.

These cycles clearly show the relative importance of the atmosphere, oceans, cryosphere, and solid earth in maintaining the various balances and what processes transfer angular momentum, water, and energy from one region of the globe to another or between the different media. The energy generation, conversion, and dissipation rates in the atmosphere and oceans, and estimates of the efficiency of the atmosphere and oceans regarded as heat engines are discussed in Chap. 14. Chapter 15 addresses the question of how entropy, a measure of the disorder in the climate system, is generated and exported to outer space.

Some statistics on climatic fluctuations and trends observed in the atmosphere ranging from intermonthly to interdecadal time scales are presented in Chap. 16.

In view of the paramount importance of mathematical climate models for simulating, understanding, and possibly predicting climate, we give in Chap. 17 an evaluation of the foundations, structure, and status of the various models now in use. In general, these models are based on physically sound formulations of the different processes that occur and make up the climate. We stress the importance of simple models for clarifying the mechanisms involved in the various atmospheric and oceanic processes leading to a better understanding of the complex interactions between the various components of the climate system. However, our main emphasis will be on the large-scale general circulation models (GCM's) of the atmosphere and ocean, since they lead to the most realistic climate simulations. GCM's have been playing an important role in investigating the various physical factors that shape the climate, and in assessing the possible impact of man's activities on the climate.

As a final note, we may add that the development of the electronic computer and its application to numerical weather forecasting (e.g., Smagorinsky, 1983) has enabled meteorologists to give, for the first time, quantitative forecasts based on the basic laws of physics. As mentioned earlier, similar numerical models with high spatial and temporal resolutions now provide reasonably accurate simulations of the average climatic conditions in the atmosphere and oceans, their seasonal climatology, their regional structure, and, to some extent, their natural year-to-year variability.

In a sense, this development comes in the nick of time. Man's impact on the local environment is evident all around us and, as shown in Chap. 16, even globally there is strong evidence of important changes due to human activity. We may mention the measured steady rate of increase in atmospheric CO_2 since the beginning of the industrial revolution and the first claims that the predicted rise in tropospheric temperature is already detectable above the "climate noise." General circulation models are giving us tentative answers as to what to expect for regional changes in temperature and precipitation patterns as a function of CO_2 concentration. In the near future, such predictions may affect national and international policy decisions regarding, e.g., the use of coal or thermonuclear fuel as alternative energy sources.

The selection of topics included in this book represents, of course, our personal choice and is determined by what we consider to be most important as well as by our areas of expertise. Thus we have excluded such topics as

(1) The classification of climate types (see, e.g., Trewartha, 1968).

(2) Atmospheric and oceanic tides and other diurnal variations, because they are not thought to directly affect the large-scale circulations.

(3) Gravity waves, in spite of their apparent importance in determining some of the vertical structure in the atmospheric winds.

(4) Climatic effects of the atmosphere above the lower stratosphere (i.e., above about 20-km height), because the data are sparse and also because it contains less than 5% of the total atmospheric mass.

(5) Climatic variations longer than several decades, because of the lack of direct observations and because our primary focus is on the present-day climate. There are many excellent books that deal with the paleoclimatic records (e.g., Lamb, 1972; Lamb, 1977; Budyko, 1982; and Crowley and North, 1991).

Some other outstanding works have been published related to certain other aspects of the earth's climate. Among those we should refer to the monograph *The Nature and Theory of the General Circulation of the Atmosphere* by Edward N. Lorenz (1967), now considered to be a classic, and in dealing with yet larger systems, such as the atmospheres of Jupiter and the Sun, we should mention the work *Physics of Negative Viscosity Phenomena* by the late Victor P. Starr (1968).

CHAPTER 2

Nature of the Problem

2.1 INTRODUCTION

In order to understand the mechanisms and the physical processes responsible for the climate it is necessary to first have a clear picture of the characteristic features of the structure and behavior of the climate.

Since the atmosphere is a thermo-hydrodynamical system, it can be characterized by its composition, its thermodynamical state as specified by the thermodynamic variables, and its mechanical state (motions). A complete description of the state of the atmosphere should also include other variables, such as cloudiness, precipitation, and diabatic heating which affect the large-scale behavior of the atmosphere. Traditionally, the most important elements of the climate are considered to be the temperature and precipitation. It is on the basis of these elements that the climates are usually classified. By and large their geographical distribution shows warm and moist climates in the low latitudes, warm and much drier climates in the subtropics, temperate and moist climates in mid to high latitudes, and, finally, cold and dry climates in the polar and subpolar regions. However, we know that this distribution does not give a complete description of the local and regional climates. We have to also consider the land–sea contrast and the moderating influence of the oceans on the temperature; the effects of mountains on precipitation, cloudiness, and temperature; the influence of the ice fields on the temperature; and other similar influences.

It is well known that the climate is modulated by both external and internal factors. The external factors may be grouped into (a) general factors such as the solar radiation, the sphericity of the earth, the earth's motion around the sun and its rotation, the existence of continents and oceans; and (b) regional and local factors, such as distance to the sea, topography, nature of the underlying surface, vegetation cover, and proximity to lakes. Internal factors deal with the intrinsic properties of the atmosphere, such as the atmospheric composition, various instabilities, and the general circulation.

The atmosphere ($\mathscr{A}$), as a thermodynamic system, cannot be considered separately from its neighboring systems (see Fig. 2.1). The adjoint systems are the hydrosphere ($\mathscr{H}$), including the oceans ($\mathscr{O}$), lakes, and rivers, the cryosphere ($\mathscr{C}$) formed by the snow and ice masses of the earth, the underlying lithosphere ($\mathscr{L}$), and the marine and terrestrial biosystems ($\mathscr{B}$). Although these natural systems are very different in their composition, physical properties, structure, and behavior, they are all linked together by fluxes of mass, energy, and momentum forming a world-wide system, the so-called climatic system. As we will see, the total climatic system ($\mathscr{S}$) is extremely complex due to the nonlinear interactions of its components.

At this point, it is important to make clear the distinction between weather and climate. Weather is concerned with detailed instantaneous states of the atmosphere and with the day-to-day evolution of individual synoptic systems. The atmosphere is characterized by relatively rapid random fluctuations in time and space so that the weather, identified as the complete state of the atmosphere at a given instant, is continuously changing. The climate, on the other hand, can be considered as the "averaged weather," completed with some measures of the variability of its elements and with information on the occurrence of extreme events. Thus we may note that the same variables that are relevant in the weather and in other branches of meteorology are also those that are important in the characterization of climate. However, what distinguishes the problem of climate from the problem of weather is the neglect of details of the daily fluctuations in the state of the atmosphere. Instead, we include in climate the various statistics produced by considering an ensemble or a sequence of instantaneous states, so that the climate is independent and free from the statistical fluctuations that would characterize any individual realization. Thus we see that the thermo-hydrodynamical conservation laws for mass, momentum, and energy that form the physical foundation for studying the instantaneous behavior of the atmosphere, i.e., the weather, are essentially the same as those required for studying the physics of climate.

To facilitate the later analyses it seems convenient to first review some general concepts about systems.

2.2 BASIC CONCEPTS OF THERMODYNAMIC SYSTEMS

We will define a system, within the framework of thermodynamics, as an arbitrary geometric portion of the universe with fixed or movable boundaries (walls) which may contain matter, energy, or both (Tisza, 1966). The whole universe will be divided into two parts: the system and the surroundings or the environment.

2.2.1 State of a system, extensive and intensive properties

The state of a system is specified by a set of physical additive or extensive properties represented by the variables $X_1, X_2, ..., X_N$ necessary for a description of the system. These variables are proportional to the size of the system. As examples we have the volume, the masses of individual components, the internal energy, and the entropy. A composite system is a union of spatially adjoint or nonoverlapping simple systems separated by conceptual or real partitions (walls) (see Fig. 2.2). The simple systems that form the composite system are called subsystems. The amount of the quantity X_i for the global system is then the sum over all subsystems a of X_i^a, i.e., $X_i = \Sigma_a X_i^a$. The set of all X_i specifies the state of the composite system.

The amount of X_i transferred from system a to an adjacent system b during a given interval of time is denoted by $X_i^{a,b}$. Such a transfer leads to a change of the state of the system and the system is said to undergo a thermodynamic *process* (see Fig. 2.2).

PHYSICS OF CLIMATE

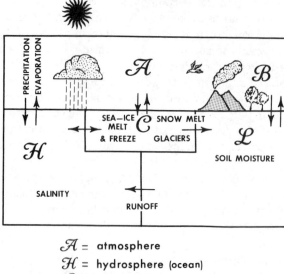

FIGURE 2.1. Schematic diagram of the total climate system and its subsystems, highlighting some aspects of the hydrological cycle.

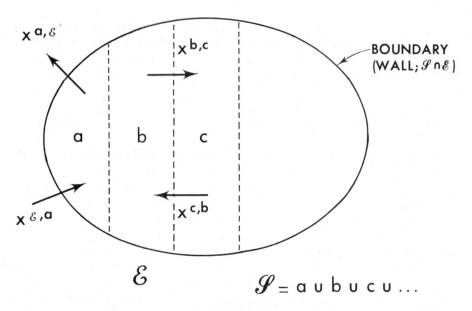

FIGURE 2.2. Schematic diagram of a thermodynamic system $\mathscr{S}$ with the subsystems $a,b,c,...$ and their interactions with each other and the environment $\mathscr{E}$.

When X_i is conserved it obeys a continuity equation, so that the net increase ΔX_i^a of X_i for the subsystem a is such that

$$\Delta X_i^a + \sum_b X_i^{a,b} = 0, \qquad (2.1)$$

where the summation is taken over all subsystems b involved in the transfer of X_i.

When $X_i^{a,b} \neq 0$ we will say that the systems a and b are interactive or coupled by means of X_i exchanges. If $X_i^{a,b} = 0$, the systems a and b are uncoupled and the wall $a \cap b$ is said to be restrictive for X_i.

The state of a system may also be specified in terms of intensive properties P_i which are local in character, independent of the size and total mass of the system, and are defined at a given point and at a given instant. The intensive properties may vary both in space and time and can be regarded as defining a field. As examples of intensive properties we have temperature, pressure, forces, and velocities.

The size of a system may be characterized by a scale factor such as its volume or its mass. We will define the density of a property X_i as the ratio of the amount of X_i and the chosen scale factor. Where this scale factor is mass M, the densities are called specific quantities ($x_i = X_i/M$). The densities are also regarded as intensive properties.

In general, the intensive and extensive variables are paired (conjugate) in the energy, which means that the product of P_i and X_i has the dimension of energy.

2.2.2 Classification of thermodynamic systems

A transfer of an extensive property across the boundary of a system, $X_i^{a,b}$, will induce variations inside the system a in both its extensive and intensive properties. For example, if we identify X_i with energy, the entropy, pressure, and temperature in system a will change if energy is transferred across the boundary $a \cap b$. When the boundary is restrictive for all quantities X_i so that the transfers vanish, the system is said to be an *isolated* system. In a more limited sense an isolated system is one that does not exchange energy with its surroundings. When a boundary is restrictive only for matter (i.e., the boundary is an impermeable wall) the system is said to be *closed*.

A system in which the transfers of matter and energy are allowed is called an *open* system. Most natural systems, such as the atmosphere, oceans, and biosphere, belong to this group. Open systems are sustained by a continuous supply and removal of matter and energy. They may attain a steady-state condition in which the properties are invariant when averaged over a given time interval. However, the instantaneous values of the open system may undergo oscillations due to the existence of net fluxes across the boundaries.

Open systems can be classified into three main categories: *decaying*, *cyclic*, and *randomly fluctuating systems*. Decaying or dissipative systems consume their own mass or energy, or both (e.g., river runoff during a dry season). Cyclic systems follow an imposed regular oscillatory behavior (e.g., systems forced by the diurnal or annual cycles). Randomly fluctuating systems change in an irregular way with fluctuations unpredictable in time and unpredictable as to their size (e.g., turbulent whirls in the atmosphere).

Cascading systems are important in nature. A cascading system is a chain of open subsystems which are dynamically linked by a cascade of matter or energy so that the output of mass or energy from one subsystem becomes the input for the next subsystem. Many of the processes observed in nature can be described in terms of cascading systems, such as the hydrological cycle and the cycle of incoming solar radiation. The input can be partially stored in one of the subsystems, which may then act as a regulator, controlling the amounts of mass or energy available as output for the next subsystem. However, the regulator may also be external to the system.

2.2.3 Forced and free behavior of open systems

The characteristics and physical behavior of the various subsystems in a composite system may differ substantially in space and time. It is then convenient to separately consider one subsystem or a combination of subsystems with similar (but not identical) behaviors (e.g., the oceans and the atmosphere). These form an *internal* system that is embedded in an *external* system consisting of the remaining subsystems (e.g., ice masses and land) and the *environment*. In general, we assume that the time scale characterizing the behavior of the internal system is much shorter than that of the external system, so that the external system can be considered to be in a steady state. Through the boundary conditions the external system may influence the behavior of the internal system leading to a forced adjustment of the internal properties of the system, the so-called *external forcing*.

To analyze the stability of the internal system, we will consider a small but finite perturbation within one of the subsystems. The initial perturbation may die away (stable case) or grow (unstable case). In the unstable case, the perturbation will result in a completely different new state of the internal system, but when the perturbation is caused by an almost periodic external forcing, it may lead to an oscillatory response of the internal system.

We know that under certain conditions a system may undergo fluctuations of various time scales. These variations may be caused by changes in the external forcing, called *forced* variations (e.g., diurnal and seasonal variations), or they may be independent of any external influences and due to internal feedbacks and instabilities. It is clear that the latter variations, called *free* variations, are associated with nonlinearities inside the system which may create the conditions for the growth of small instabilities (e.g., generating cyclonic disturbances). Negative feedbacks will have a stabilizing influence, whereas positive feedbacks can lead to pronounced instabilities independent of any external forcings (see Sec. 2.5).

2.2.4 Random systems

Let us consider an open randomly fluctuating system subject to steady boundary conditions, and let us assume that this system behaves in a "turbulent regime," going through many different physical states. The system may seem hopelessly complicated, but the complexity itself provides a basis for the successful analysis through a statistical approach. Since we are not concerned with the detailed behavior of each individual state we can apply statistical arguments, as long as the number of states is sufficiently large, which is usually the case.

It is important to keep in mind that when we want to describe a situation from a statistical point of view (using probabilities) it is always necessary to consider an assembly of events. In the present case we will call it an *ensemble*, using the approach of statistical mechanics. An ensemble consists of a large number N (in principle $N \rightarrow \infty$) of identically constructed systems each of which is in a state that is independent of the state of the other members in the ensemble. The ensemble can be formed by an infinite number of mental reproductions or replicas of the state of the same system. An ensemble is primarily a conceptual aid to get a better understanding of the physical behavior of a system.

The probability of occurrence of a certain event (state) is then defined with respect to this particular ensemble and is given by the fraction of members in the ensemble that are characterized by the occurrence of this specific event (state). The properties of the ensemble are described by a probability distribution.

Since the state variables are random functions of time, we can use two types of statistical averages (Reif, 1965). The first of these is the ensemble average of a state variable $y(t)$ *at a given instant of time*, where the average is taken over all members of the ensemble. The ensemble average, denoted by $\langle y(t) \rangle$, is given by

$$\langle y(t) \rangle = \frac{1}{N} \sum_{k=1}^{N} y^k(t), \qquad (2.2)$$

where $y^k(t)$ is the value of $y(t)$ in the kth member of the ensemble and N is the very large number of members in the ensemble.

The second average of interest is the time average of $y(t)$ for *a given member of the ensemble* over some long-time interval 2τ (where $\tau \to \infty$). The time average is denoted by an overbar, so that for the kth member of the ensemble it is given by

$$\overline{y^k(t)} = \frac{1}{2\tau} \int_{-\tau}^{\tau} y^k(t+t') \, dt'. \qquad (2.3)$$

In more pictorial terms, if the member states are arranged in the vertical and time in the horizontal direction, we can say that the ensemble average is taken vertically for a given time t, whereas the time average is taken horizontally for a given member k. We note that the time average and ensemble average operators commute:

$$\langle \bar{y} \rangle \equiv \overline{\langle y \rangle}.$$

Now consider a situation which is statistically stationary with respect to y. This means that there is no preferred origin in time for the statistical description of y so that the same ensemble results when all member functions $y^k(t)$ of the ensemble are shifted by an arbitrary amount in time. For such stationary ensembles there is a connection between ensemble and time averages, if we assume that $y^k(t)$ for each member of the ensemble in the course of a sufficiently long time passes through all different states accessible to it. This is the *ergodic* assumption. Under this assumption the time average is equivalent to the ensemble average. Hence, in such a stationary ensemble the time average of y taken over some very long-time intervals must be independent of time t. Furthermore, the ergodic assumption implies that the time average must be the same for essentially all systems of the ensemble: $\overline{y^k(t)} = \bar{y}$ independent of k. Similarly in such a stationary ensemble the ensemble average of y must be independent of time: $\langle y(t) \rangle = \langle y \rangle$.

In conclusion, for a stationary ergodic ensemble,

$$\langle y \rangle = \bar{y}. \qquad (2.4)$$

In other words, ensemble averages that are averages at a fixed time over all members in an ensemble can be replaced by time averages over a single member.

2.3 COMPONENTS OF THE CLIMATE SYSTEM

We will discuss in more detail the various components of the climate system, such as the atmosphere, hydrosphere, cryosphere, and lithosphere.

2.3.1 Atmosphere

The earth's atmosphere is a comparatively thin film of a gaseous mixture which is distributed almost uniformly over the surface of the earth. In the vertical direction, more than 99% of the mass of the atmosphere is found below an altitude of only 30 km. In comparison, the horizontal dimensions of the atmosphere may be represented by the distance between the north and south poles, and is on the order of 20 000 km. If proportions were preserved, the thickness of the atmosphere would be represented on an ordinary office globe by scarcely more than the thickness of a coat of paint. However, in spite of its small relative mass and thickness, the atmosphere constitutes the central component of the climatic system. It shows an impressive amount of detail and great variability of its properties both in time and space.

The atmosphere can be divided into several layers which differ in composition, temperature, stability, and energetics (see Fig. 2.3). Starting from the surface, the main layers are the troposphere, stratosphere, mesosphere, and thermosphere, separated by conceptual partitions called pauses (e.g., tropopause). The composition of the atmosphere up to the mesopause is practically uniform with regard to the concentrations of nitrogen, oxygen, and other inert gases. Among the variable components, water vapor is predominant in the lower troposphere and ozone in the middle stratosphere. Carbon dioxide is well mixed below the mesopause. The composition of the atmosphere is further complicated by the presence of various substances in suspension, e.g., liquid and solid water (clouds), dust particles, sulfate aerosols, and volcanic ash. The concentrations of these aerosols also vary in time and space.

The response time of the atmosphere to an imposed change is much shorter than that of any other component of the climatic system. By response or relaxation time we mean the time it takes for a system to re-equilibrate to a new state after a small perturbation has been applied to its boundary conditions or forcing. The response time of the atmosphere is on the order of days to weeks and is due to its relatively large compressibility and low specific heat and density. These properties make the atmosphere more fluid and more unstable. The troposphere shows a large-scale general circulation with eddy motions in midlatitudes such as the weather systems and random, turbulent motions mainly in the planetary boundary layer and near the jet streams. Because of gravity, the atmosphere is stratified with the densest layers at the surface, and the atmosphere is in a state of almost hydrostatic equilibrium in the vertical.

The atmosphere is set into motion primarily through differential heating by the sun. Thus the study of atmospheric motion is, basically, a problem in convection under the influence of rotation. It is a complex process because the motions of the atmosphere are influenced by many factors besides the rotation of the earth, such as inhomogeneous thermodynamic and mechanical surface conditions. However, when we disregard the irregular details of the flow, we find a pronounced tendency for the motions of the atmosphere to be organized on a global scale.

Historically, the first important discovery of the interrelationships between the behavior of the atmosphere in one area of the globe and that in other areas was made by the early navigators who noted the existence of an extensive and persistent easterly flow of air (i.e., the trade winds) at sea level in low latitudes. Later the discovery of the meandering westerly currents which encircle the globe throughout most of the depth of the atmosphere has given substance to the concept of the general circulation as a planetary phenomenon, the behavior of which is governed by broad-scale conditions rather than by the local vagaries of the weather.

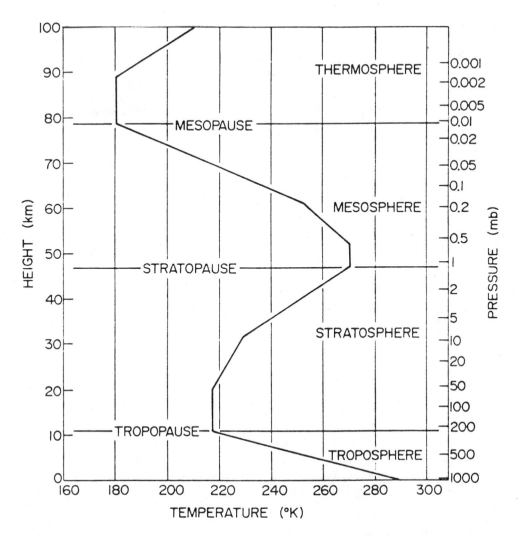

FIGURE 2.3. Idealized vertical temperature profile according to the U.S. Standard Atmosphere (1976). Also shown are the names commonly used for the various layers and pauses in the atmosphere from Wallace and Hobbs (1977).

To demonstrate the great variability of the processes occurring in the atmosphere and to show the relative importance of the different scales of motion we present a spectrum of the kinetic energy with periods between seconds and several years in Fig. 2.4. Most of the kinetic energy is concentrated in the low frequencies, namely, at 10^0, around 10^{-1}, and between 10^{-2} and 10^{-3} day^{-1}. The first and third peaks are associated with the diurnal and annual cycles, whereas the second maximum (days to weeks) is associated with large-scale transient disturbances that occur in midlatitudes along the polar front. The relative maximum around 10^3 day^{-1} is due to small-scale turbulent motions which, combined with molecular friction, are included in the internal energy. Thus we will not regard them as part of the kinetic energy of the circulation, in spite of their importance in the boundary layers of the atmosphere and oceans.

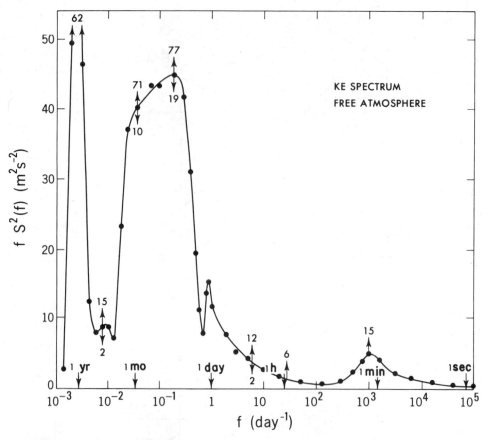

FIGURE 2.4. Spectrum of atmospheric kinetic energy between 10^{-5} and 10^3 days adapted from Vinnichenko (1970). The kinetic energy spectrum is tentative and somewhat schematic since it is patched together using limited data from a few stations only. [The abscissa axis is in units of $\log f$ and the ordinate axis in units of $fS^2(f)$, where f is the frequency and $S^2(f)$ is the explained variance. This representation is sometimes used if the frequency range is very large (necessitating a $\log f$ scale) and one, nevertheless, wants to preserve the area under the curve to be equal to the total variance $\int fS^2(f)d\ln f = \int S^2(f)df$.]

2.3.2 Hydrosphere

The hydrosphere consists of all water in the liquid phase distributed on the earth. It includes the oceans, interior seas, lakes, rivers, and subterranean waters. By far the most important for climatic studies are the oceans. They cover approximately two-thirds of the earth's surface so that most of the solar radiation reaching the globe falls on the oceans and is absorbed by them. Because of their large mass and specific heat the oceans constitute an enormous reservoir to store energy. Energy absorbed by the ocean results in a relatively small change of the surface temperature as compared to the change that would occur over land. Due to their thermal inertia the oceans act as buffers and regulators for the temperature. Since the oceans are more dense than the atmosphere, they also have a larger mechanical inertia and a more pronounced stratification. The upper part of the ocean is the most active. It contains a surface mixed layer with a thickness on the order of 100 m.

The oceans show much slower circulations than the atmosphere. They form large quasihorizontal circulation gyres with the familiar ocean currents and slow thermohaline overturnings (i.e., overturnings that are due to density variations associated with changes in temperature and salinity). On a smaller scale the circulation also shows eddies, but turbulence is much less pronounced than in the atmosphere. The response or relaxation time for the ocean varies within a wide range that extends from weeks to months in the upper mixed layer to seasons in the thermocline (found at several hundred meters depth), to centuries or millenia in the deep ocean. The ocean currents transport part of the heat stored in the oceans from the intertropical regions where there is an excess of heat due to the more intense incident solar radiation toward colder midlatitude and polar regions.

The atmosphere and oceans are strongly coupled. Air–sea interactions occur on many scales in space and time through the exchange of energy, matter, and momentum at the atmosphere–ocean interface as can be seen, e.g., from air masses modifications, such as from maritime to continental air. The exchange of water vapor through evaporation into the atmosphere supplies the water vapor and part of the energy for the hydrological cycle leading to condensation, precipitation, and runoff. On the other hand, precipitation strongly influences the distribution of ocean salinity.

There are internal interactions in the atmosphere and oceans mainly when and where the gradients of their intensive properties (e.g., temperature and salinity) are large.

The lakes, rivers, and subterranean waters are essential elements of the terrestrial branch of the hydrological cycle and thereby are an important factor in the global climate. They also influence the climate on a regional or local scale. For example, rivers are an important factor in the ocean salinity near the coasts.

2.3.3 Cryosphere

The cryosphere comprises the large masses of snow and ice of the earth's surface. It includes the extended ice fields of Greenland and Antarctica, other continental glaciers and snow fields, sea ice, and permafrost. The cryosphere represents the largest reservoir of fresh water on the earth, but its importance to the climatic system results mainly from its high reflectivity (albedo) for solar radiation and its low thermal conductivity. Since continental snow cover and sea ice change seasonally they lead to large intra-annual and sometimes interannual variations in the energy budgets of the continental regions and of the upper mixed layer of the ocean. In addition to seasonal variations of the cryosphere, large changes may occur over much longer periods of time. Due to the high reflectivity of snow and ice for solar radiation (see Table 6.1) and the low thermal diffusivity of sea ice compared to that of stirred water (see Table 10.1), the snow and ice fields act at high latitudes as insulators for the underlying land and waters, preventing them from losing heat to the atmosphere. The strong cooling of the atmosphere near the earth's surface stabilizes the atmosphere against convection and contributes to the occurrence of a colder local climate.

The large continental ice sheets do not vary rapidly enough to influence the climate on a seasonal or interannual basis. However, they play a major role in climatic changes on much longer time scales up to tens of thousands of years, such as the glacial and interglacial periods that have occurred during the Pleistocene. A glaciation will lower sea level considerably, possibly on the order of 100 m or more, thus affecting the shape and boundaries of the continents. Owing to their large mass and compactness, the ice sheets develop a dynamics of their own with very slow motions. Occasionally ice sheets over the oceans may break up, forming icebergs. Mountain glaciers move slowly downward under the action of gravity and may spread or disappear in the course of centuries depending on local accumulation of snow and on the temperature.

2.3.4 Lithosphere

The lithosphere includes the continents whose topography affects air motions, and the ocean floor. Excluding the upper active layer, in which temperature and water content can vary in response to atmospheric and oceanic phenomena, the lithosphere has the longest response time of all components of the climatic system. On the time scales considered in this book, the lithosphere may be regarded as an almost permanent feature of the climatic system.

There is a strong interaction of the lithosphere with the atmosphere through the transfer of mass, angular momentum, and sensible heat, as well as through the dissipation of kinetic energy by friction in the atmospheric boundary layer. The transfer of mass occurs mainly in the form of water vapor, rain and snow, and, to a lesser extent, in the form of other particles and dust. Volcanoes throw matter and energy from the lithosphere into the atmosphere, thereby increasing the turbidity of the air. The added particulate matter, as well as the ejected sulfur-bearing gases that may condense in the stratosphere, together forming what is called aerosol, may have an important effect on the radiation balance of the atmosphere and therefore on the earth's climate (e.g., Mass and Portman, 1989). There is also a large-scale transfer of angular momentum between the lithosphere and the oceans presumably through the action of torques between the oceans and the continents.

The soil moisture of the active layer in the continental lithosphere has a marked influence on the local energy balance at the surface affecting the rate of evaporation, the surface albedo, and the thermal conductivity of the soil.

2.3.5 Biosphere

The biosphere comprises the terrestrial vegetation, the continental fauna, and the flora and fauna of the oceans. The vegetation alters the surface roughness, surface albedo, evaporation, runoff, and field capacity of the soil. Furthermore, the biosphere influences the carbon dioxide balance in the atmosphere and oceans through photosynthesis and respiration. On the whole, the biosphere is sensitive to changes in the atmospheric climate and it is through the signature of these changes in fossils, tree rings, pollen, etc., during past ages that we obtain information on paleoclimates of the earth.

At this point we may mention the human interaction with the climatic system through such activities as agriculture, urbanization, industry, pollution, etc.

2.4 THE CLIMATE SYSTEM

In order to get a better understanding of the nature of the problems involved with the climate system we will start by discussing the time scales involved and what we mean by the state of the climate system and its variability.

2.4.1 The nature of the climate system

As we have seen the climate system ($\mathscr{S}$) is a composite system consisting of five major interactive adjoint components: the atmosphere ($\mathscr{A}$), the hydrosphere ($\mathscr{H}$) with the oceans ($\mathscr{O}$), the cryosphere ($\mathscr{C}$), the lithosphere ($\mathscr{L}$), and the biosphere ($\mathscr{B}$), i.e.,

$$\mathscr{S} \equiv \mathscr{A} \cup \mathscr{H} \cup \mathscr{C} \cup \mathscr{L} \cup \mathscr{B}. \tag{2.5}$$

As shown schematically before in Fig. 2.2, all the subsystems are open and nonisolated. The global climate system $\mathscr{S}$, as a whole, is assumed to be a nonisolated system for energy but a closed system for the exchange of matter with outer space. The atmosphere, hydrosphere, cryosphere, and biosphere act as a cascading system linked by complex physical processes involving fluxes of energy, momentum, and matter across the boundaries and generating numerous feedback mechanisms.

The components of the climatic system are heterogeneous thermo-hydrodynamical systems, which can be characterized by their chemical composition and their thermodynamic and mechanical states. The thermodynamic states are specified, in general, by certain intensive variables (e.g., temperature, pressure, specific humidity, specific energy, density, and salinity) whereas the mechanical state is defined by other intensive variables that characterize the motions (e.g., forces and velocities).

The estimated time scales (which are proportional to the response time) for various components of the climatic system vary widely from one subsystem to another and even within the same subsystem. The time scale for the atmospheric boundary layer varies from minutes to hours; for the free atmosphere from weeks to possibly months; for the upper mixed layer in the oceans it ranges from weeks to years; for the deep ocean waters from decades to millenia; for sea ice from weeks to decades; for inland waters and vegetation from months to centuries; for glaciers the time scale is on the order of centuries; for ice sheets on the order of millenia and beyond; and for tectonic phenomena on the order of tens of millions of years.

Due to the complexity of the internal climatic systems and on the basis of the different response times, it is convenient to consider a hierarchy of internal systems first taking the systems with the shortest response times so that all other components are considered to be part of the external system. For example, for time scales of hours to weeks the atmosphere can be regarded as the sole internal component of the climatic system ($\mathscr{S} \equiv \mathscr{A}$) with the oceans, ice masses, land surfaces, and biosphere treated as the external forcings or boundary conditions. For time scales of months to centuries, the climatic internal system must include the atmosphere and the oceans ($\mathscr{S} \equiv \mathscr{A} \cup \mathscr{O}$) as well as snow cover, sea ice, and the biosphere. For the study of the variability of climate beyond centuries the entire cryosphere and the biosphere must also be included in the internal system ($\mathscr{S} \equiv \mathscr{A} \cup \mathscr{O} \cup \mathscr{C} \cup \mathscr{B}$) considering the influence of $\mathscr{L}$ as an external forcing.

A schematic picture of the time scales for the various components of the climate system is presented in Fig. 2.5. The figure also shows some of the external processes that may cause fluctuations in the climate system. Thus the whole climate system must be regarded as continuously evolving with parts of the system leading and others lagging in time. The highly nonlinear interactions between the subsystems tend to occur on many time and space scales. Therefore, the subsystems of the climate system are not always in equilibrium with each other, and not even in internal equilibrium. The subsystems also have feedback loops that we will describe later.

The climatic system is subject to two main external forcings that condition its global behavior, namely, solar radiation and the action of gravity. Among these exter-

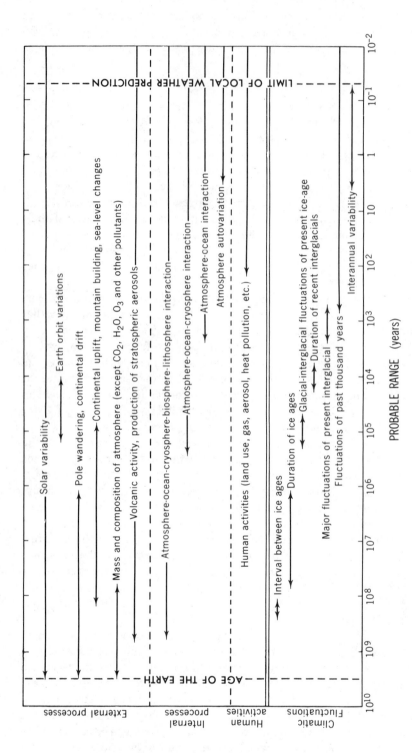

FIGURE 2.5. Characteristic time scales for the various components of the climate system and for certain external forcing factors between 10^{-2} and 10^{10} yr (adapted from National Academy of Sciences, 1975; Bergman et al., 1981).

nal forcings we must consider the solar radiation as the primary factor since it provides almost all the energy that drives the climate system. The solar radiation reaching the top of the atmosphere is partially transferred, partially transformed into other forms of energy that are eventually dissipated by the general circulations of the atmosphere and oceans, and partially used in chemical and biological processes. Within the climate system energy occurs in a variety of forms, such as heat, potential energy, kinetic energy, chemical energy, and short-wave solar and long-wave terrestrial radiation. Of all the forms of energy, we may disregard the electric and magnetic energies since they are only of importance in the very high atmosphere.

The short-wave radiation is unequally distributed over the various parts of the climate system due to the sphericity of the earth, the orbital motion, and the tilt of the earth's axis. More radiation reaches and is absorbed in the tropical regions than at polar latitudes. Taken over the globe as a whole, observations show that the system loses about the same amount of energy through infrared radiation as it gains from the incoming solar radiation. However, small currently unmeasurable imbalances could occur for both short and long periods (Saltzman, 1977).

Due to the observed range of temperature between the equator and the poles, the decrease of emitted terrestrial radiation with latitude is much less pronounced than the decrease in absorbed solar radiation, leading to a net excess of energy in the tropics and a net deficit poleward of 40° latitude. This source and sink distribution provides the basic impetus for almost all thermodynamic, in general irreversible, processes occurring inside the climate system, including the general circulations of the atmosphere and oceans.

As a final comment, it may be of interest to note that the entropy of the incoming solar energy is much lower than the entropy exported by the system through long-wave radiation. The reason for this difference is that the solar radiation originates with a temperature on the order of 6000 K, whereas the terrestrial radiation is emitted at a temperature of about 250 K. The change of entropy Δs is given by

$$\Delta s = \Delta Q/T,$$

where ΔQ is the heat transferred at the temperature T. Since the earth is, on the average, in radiative equilibrium at the top of the atmosphere, the generation of entropy for all internal processes in the climate system is 20 to 30 times larger than the amount of imported entropy, as we will show later in Chap. 15. Furthermore, in view of the importance of the frictional, diffusive, and other irreversible processes, the climate system must be regarded as a highly dissipative system.

2.4.2 The climate state

Even under steady external forcing the internal system is always subject to random fluctuations in time and space. We may then consider a large ensemble of climate states corresponding to the same external forcing and apply the ideas given in Sec. 2.2.4 and define climate in terms of the ensemble of internal states and in terms of a probability distribution. For fixed external forcings we assume the uniqueness of the limiting set of statistics and accept the ergodic hypothesis so that we can replace the ensemble averages by time averages. Then we can define a climate state as a set of averages over the ensemble completed with higher moment statistics, such as variances, correlations, etc., together with a description of the state of the external system (Leith, 1978).

If we consider the traditional case of the atmosphere as the internal system we can define the climate in terms of the atmospheric state together with the mean conditions of the oceans, cryosphere, land, and other external forcings. Thus for the atmosphere the averaging interval of time must, at least, exceed the average life span of the synoptic weather systems in the atmosphere. We can then define climatic states for a month, a season, a year, a decade, and so on. The traditional 30-year averaging interval (determined by the International Meteorological Organization) to define the climate through its mean values and the higher moments is a particular case for the atmosphere and is still a useful concept.

For a different set of the complete external conditions we may obtain a different climate state of the internal system so we can define a climate change as the difference between two climate states of the same kind, such as the difference between the climate states of two typical August months, two typical decades, etc. This difference should include the differences between the averages and moments of higher order. A climate anomaly will then be defined as a departure of a particular climate state for a given interval of time from the ensemble of the equivalent states.

In our earlier discussion of weather and climate, we mentioned that the physical laws that govern their evolution are basically the same. However, the applications of the equations to these problems are different. In climate studies it is necessary to consider not only the internal effects but also the complex interactions between the atmosphere and its external system. For weather prediction, the atmosphere behaves almost inertially, so that slowly acting boundary conditions can be ignored. For example, the fluctuations of the sea surface temperature and snow and ice cover can be disregarded in weather forecasts for up to one or two weeks. Nevertheless, these changes gradually affect the lower atmosphere, and become important when we want to study the climate.

Starting with a given initial state, the solutions of the equations that govern the dynamics of a nonlinear system, such as the atmosphere, result for an infinite interval of time in a set of infinitely long-term statistics. For different initial conditions the limiting solutions may or may not be unique (see Fig. 2.6). If all initial states ultimately lead to the same set of statistical properties, the system is *ergodic* or *transitive*. If instead there are two or more different sets of statistical properties, where some initial states lead to one set while the other initial states lead to another, the system is called *intransitive*. If there are different sets of statistics that a transitive system may assume in its evolution from different initial states through a long, but finite, period of time, the system is called *almost intransitive* (Lorenz, 1969; Saltzman, 1983). In the transitive case, the equilibrium climate statistics are both stable and unique, whereas in the almost-intransitive case the system in the course of its evolution will show finite periods during which distinctly different climatic regimes prevail. This may be due to internal feedbacks or instabilities involving the different components of the climatic system. The glacial and interglacial periods in the earth's history may be manifestations of an almost intransitive system. However, the present geological and historical information is not yet sufficient to conclusively decide what type of solution applies to the climate. Of course, the issues of uniqueness and stability of the solutions are very important in the analysis of the determinism of climate (Lorenz, 1969).

2.4.3 Climate variability

The terrestrial climate has varied significantly and continuously on time scales ranging from years to glacial periods and to the age of the earth. The variability of climate can be expressed in terms of two basic modes: the *forced* variations which are the response of the climatic system to changes in the external forcing and the *free*

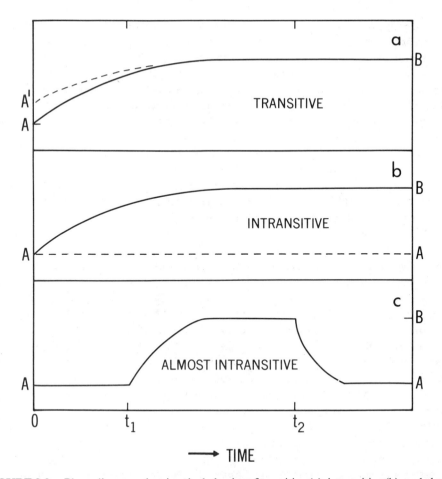

FIGURE 2.6. Phase diagrams showing the behavior of transitive (a), intransitive (b), and almost intransitive (c) climate systems starting from an initial state A. In a transitive system two different initial states A and A' evolve into the same equilibrium state B. An intransitive system will have two or more alternative equilibrium states A and B for the same boundary conditions. An almost intransitive system may behave as if it were intransitive for a given period of time, shifting (e.g., at time t_1) to an alternative climate state B where it may remain until a time t_2. Then after returning to climate state A it may remain there or undergo further shifts.

variations due to internal instabilities and feedbacks, leading to nonlinear interactions among the various components of the climatic system.

The changes in the purely external factors that affect the climatic system, but are not influenced by the climatic variables themselves, constitute what may be called the external causes of climatic changes, whereas those changes that are related to nonlinear interactions among the various physical processes in the internal system are called internal causes. The distinction between the two classes of causes is not always very clear.

The external causes comprise variations in both astronomical and terrestrial forcings. The astronomical factors would include changes (a) in the intensity of solar irradiance; (b) in the orbital parameters of the earth (eccentricity of the orbit, axial precession and obliquity of the ecliptic); and (c) in the rate of rotation of the earth.

Among the terrestrial forcings we must consider: (a) variations in atmospheric composition (mixing ratios of carbon dioxide and ozone, aerosol loading, etc.) due to volcanic eruptions and human activity; (b) variations of the land surface due to land use (deforestation, desertification, etc.); (c) long-term changes of tectonic factors such as continental drift, mountain-building processes, polar wandering, etc. Some other possible terrestrial and astronomical forcing mechanisms have been suggested, such as changes in solar output, the collision of the earth with interplanetary matter, changes in volcanic activity, and changes in the geothermal flux.

The internal causes are associated with many positive and negative feedback mechanisms and other strong interactions between the atmosphere, oceans, and cryosphere. These processes can lead to instabilities or oscillations of the system which can either operate independently or introduce strong modifications on the external forcings. Let us show by some examples what we mean by the difference between an externally forced variation and an internally free change. The seasonal or diurnal variations of the climate are clearly related to the external astronomical forcing. But there are day-to-day weather variations that take place independently of any changes in external forcing. These irregular fluctuations with time scales of several days to a week may be connected with the passage of migratory atmospheric perturbations (highs and lows on the weather map) or with the passage of a frontal system. They are to be considered free, because they result from the internal baroclinic instability of the zonal current (see Secs. 3.5.3 and 13.3.3.5) which depends only on the critical value of the latitudinal gradient of the temperature.

To further illustrate the large range in variability in time for the atmosphere we present here a spectrum for the temperature (internal energy) near the earth's surface. Figure 2.7 shows an idealized variance spectrum of the atmospheric temperature during its past history as evaluated by Mitchell (1976). The analysis of the spectrum shows several spikes and broader peaks. The spikes are astronomically dictated, strictly periodic components of climate variation, such as the diurnal and annual variations and their harmonics, whereas the broad peaks represent variations that are, according to Mitchell, either quasiperiodic or aperiodic—however, with a preferred time scale of energization. Many of these broad peaks cannot be directly explained by the known external forcings. They indicate the existence of a strong free variability within the system.

The peak at three to seven days is associated with the synoptic disturbances mainly at midlatitudes. The slightly raised region of the spectrum at 100–400 years is associated with the "little ice age" that began near the early 17th Century with rapid expansion of the mountain glaciers in Europe. The peak near 2500 years is perhaps due to the cooling observed after the "climatic optimum," about 5000 years ago, which predominated during the great ancient civilizations. The next three peaks are perhaps related to deterministic astronomical variations of the orbital parameters of the earth, which are supposed to be responsible for the ice ages (Milankovitch, 1941): (a) the eccentricity of the orbit of the earth with a cycle of about 100 000 years, (b) the axial precession with a cycle of around 22 000 years, and (c) the change in the obliquity of the ecliptic or the axial tilt with a period of about 41 000 years. Finally, the peaks near 45 and 350 million years may, according to Mitchell (1976), be related to glaciations due to orogenic and tectonic effects and to continental drift.

For a linear system the externally forced variations would lead to a simple relationship of cause and effect: if the forcing is an oscillatory process the response of the system would have exactly the same frequency. As we have seen, this is not always the case, since the internal climatic system is inherently unstable and never reaches the equilibrium state.

In conclusion, climate variability results from complex interactions of forced and free variations because the climate system is a dissipative, highly nonlinear sys-

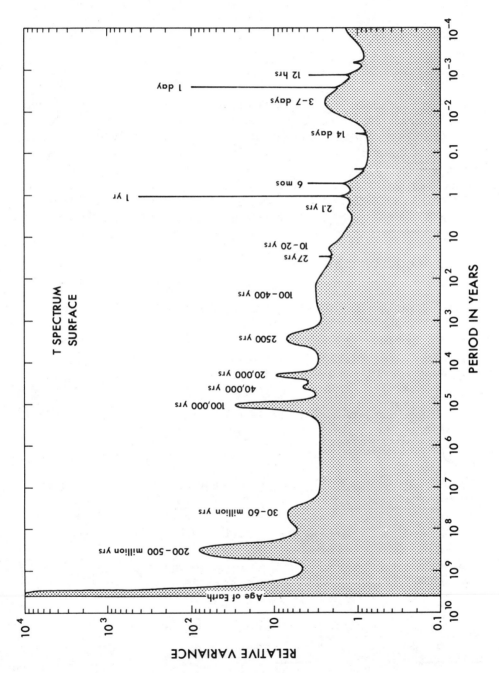

FIGURE 2.7. Idealized, schematic spectrum of atmospheric temperature between 10^{-4} and 10^{10} yr adapted from Mitchell (1976).

tem with many sources of instabilities. The interactive and often nonlinear nature of the instabilities and the feedback mechanisms of the climatic system make it very difficult to obtain a straightforward interpretation of cause and effect.

2.5 FEEDBACK PROCESSES IN THE CLIMATE SYSTEM

Of particular importance in open systems such as the components of the climatic system is feedback, a concept often used in electrical engineering. The feedback mechanisms act as internal controls of the system and result from a special coupling or mutual adjustment among two or more subsystems. Part of the output returns to serve as an input again, so that the net response of the system is altered; the feedback mechanism may act either to amplify the final output (positive feedback) or to dampen it (negative feedback). There is a large number of such mechanisms operating within the various components of the climatic system and between the subsystems.

2.5.1 Feedback concepts

Because of the importance and the increasing use of the feedback concept in numerical simulations, we will give here an elementary formulation of it, using an approach similar to that used in electrical engineering (Smith, 1987). As shown in Fig. 2.8, the basic idea is to use a portion of the output signal to modify the input signal. We will define the gain or transfer function of a system, G, as the ratio of the output signal V_2 to the input signal V_1:

$$G = V_2/V_1. \tag{2.6}$$

If V_F denotes the portion of the output signal that is fed back we can define a feedback factor H:

$$H = V_F/V_2. \tag{2.7}$$

Therefore, the input signal $V_1 = V_S + V_F$, where V_S is the initial input signal. Then we can write [see Fig. 2.8(a)]

$$V_2 = GV_1 = G(V_S + V_F) = G(V_S + HV_2) = GV_S + GHV_2. \tag{2.8}$$

Let us now relate the signal input V_S to the final output V_2 by introducing an effective transfer function G_F which includes the feedback [Fig. 2.8(b)] so that

$$G_F = V_2/V_S. \tag{2.9}$$

Solving for V_2 from Eq. (2.8) and substituting V_2 into Eq. (2.9), we obtain for the gain with feedback the following expression:

$$G_F = \frac{G}{1 - GH} = \frac{G}{1 - f}, \tag{2.10}$$

where $f = GH$ is the feedback of the system. Thus

$$V_2 = \frac{G}{1 - GH} V_S = \frac{G}{1 - f} V_S. \tag{2.11}$$

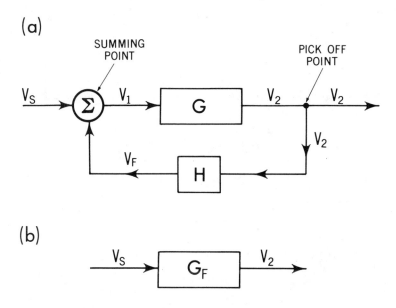

FIGURE 2.8. Schematic diagrams of a simple feedback loop often used in electrical engineering, where V_s is the initial input signal and V_2 the output signal. G, H, and G_F represent the gain, feedback factor, and effective gain of the system, respectively.

If the system has several feedback mechanisms linked in series (a cascading system) the final transfer function will be the product of the individual gains:

$$G_F = \Pi_i G_{Fi}.$$

since, e.g.,

$$V_4/V_S = (V_2/V_S)\,(V_3/V_2)\,(V_4/V_3).$$

If the feedback processes are in parallel, i.e., independent of each other, the response is linear and the feedbacks are additive so that the final gain is given by

$$G_F = \frac{G}{1 - G\Sigma_i H_i} = \frac{G}{1 - \Sigma_i f_i}. \qquad (2.12)$$

Let us again consider the single feedback case given in Eq. (2.10). In the case of zero feedback, $f = 0$ and $G_F = G$. For a negative feedback, $f < 0$ and $0 < G_F < G$. As f becomes larger and larger negative, G_F tends asymptotically to zero. For positive feedbacks with $0 < f < 1$, we find $G_F > G$. However, as f approaches unity, G_F becomes infinite. For $f > 1$, G_F would become negative, which is a physically unrealistic case.

2.5.2 Applications to the climate system

Let us apply the previous concepts to the climate system. Under equilibrium conditions, the net radiation (solar minus terrestrial) at the top of the atmosphere F_{TA} is zero. The mean surface temperature of the earth T_{sfc} can then be regarded as one of the responses of the climatic system to the imposed external forcings, taking into consideration all possible internal feedbacks.

Any external perturbations due, e.g., to changes in solar output, atmospheric water vapor, clouds, carbon dioxide, volcanic eruptions, etc., will induce an imbalance in the net radiation at the top of the atmosphere, ΔF_{TA}. If the system is damped, the variables of the climate system, including T_{sfc}, will change in order to adjust to the new equilibrium state. Thus we will regard ΔF_{TA} as the signal input V_S for the climate system and ΔT_{sfc} as the output V_2 (see Fig. 2.9). We can then write

$$\Delta T_{\text{sfc}} = G_F \Delta F_{\text{TA}}, \tag{2.13}$$

where, in this context, G_F is a sensitivity factor that is sometimes denoted as the gain. Basically, it is a transfer function as we discussed above.

The transfer function G_F incorporates all feedback processes that take place in response to the externally imposed ΔF_{TA}. For the feedback mechanisms the input signal is now ΔT_{sfc} and each of the processes is characterized by a transfer function, H_i. Assuming that the feedback processes are independent (see Fig. 2.9) we can write for the final input

$$\Delta F'_{\text{TA}} = \Delta F_{\text{TA}} + \left(\sum_i H_i\right) \Delta T_{\text{sfc}}. \tag{2.14}$$

Added to the external forcing ΔF_{TA} is a net effect given by the second term on the right-hand side of Eq. (2.14) resulting from internal feedback processes, acting by themselves. Then we can write

$$\Delta T_{\text{sfc}} = \left(\frac{G}{1 - G\Sigma_i H_i}\right) \Delta F_{\text{TA}} = \left(\frac{G}{1 - \Sigma_i f_i}\right) \Delta F_{\text{TA}}, \tag{2.15}$$

where G is the gain without feedback and f_i represents the feedbacks of the different processes.

Without feedbacks ($\Sigma_i f_i = 0$) the temperature response, denoted by ΔT^*_{sfc}, would, of course, be very different from ΔT_{sfc}. The ratio of the gain with and without feedbacks is then given by

$$\frac{G_F}{G} = \frac{\Delta T_{\text{sfc}}}{\Delta T^*_{\text{sfc}}} = \frac{1}{1 - \Sigma f_i}. \tag{2.16}$$

We should note that these analyses have some limitations because they do not permit interactions among the various processes or nonlinear responses.

Since

$$\frac{\partial G_F}{\partial f_i} = \frac{G}{(1 - \Sigma f_i)^2}, \tag{2.17}$$

we see that G_F is sensitive to small fluctuations in f_i which means that f_i must be evaluated with high precision. We see that the approach discussed here, simple as it may be, permits us to get some insight into the complex feedback mechanisms in the climate system.

In general, the transfer functions can be expressed in terms of partial derivatives. In the present case we can assume that F_{TA} depends on external variables x_e, internal variables x_i, and, explicitly, on the surface temperature T_{sfc}. Furthermore, $x_i = x_i(T_{\text{sfc}})$ so that we can write the following expression for the total variation:

$$\Delta F'_{\text{TA}} = \sum_e \frac{\partial F'_{\text{TA}}}{\partial x_e} \Delta x_e + \left(\sum_i \frac{\partial F'_{\text{TA}}}{\partial x_i} \frac{\partial x_i}{\partial T_{\text{sfc}}} + \frac{\partial F'_{\text{TA}}}{\partial T_{\text{sfc}}}\right) \Delta T_{\text{sfc}}. \tag{2.18}$$

Comparing this equation with Eq. (2.14) we see that

$$\sum_i \frac{\partial F'_{\text{TA}}}{\partial x_i} \frac{\partial x_i}{\partial T_{\text{sfc}}} = \sum_i H_i$$

represents the contributions of the internal feedbacks.

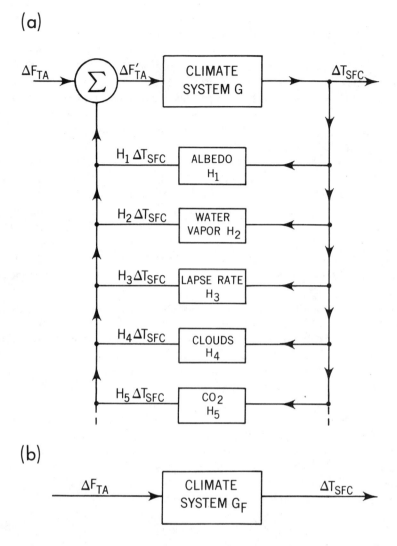

FIGURE 2.9. Example of a set of possible feedback loops in the climate system, where the input is an imbalance in the net radiation at the top of the atmosphere $\Delta \mathbf{F}_{TA}$ and the output is the change in surface temperature of the earth. The symbols G, H_1, H_2,... and G_F represent the gain, feedback factors, and effective gain of the climate system, respectively.

This type of analysis or some other variations of it (Schlesinger and Mitchell, 1987) are being used frequently in assessing the behavior of numerical climate models.

2.5.3 Some examples

The reflectivity (albedo) for solar radiation is a very important factor in the energy balance. The high albedo values of snow and ice make them a dominant factor in the climate mainly in the polar regions. The extent of ice and snow depends largely on the near-surface temperature of the air. If the temperature decreases for some reason the amount of snow and ice will generally increase or last longer, which will lead to an increase of the planetary albedo. Then more solar radiation will be reflected and less energy will be available to heat the atmosphere, and the temperature of the atmosphere–snow/ice system will further decrease.

On the other hand, suppose that the snow or ice cover decreases in extent, then because of the decreased albedo less solar radiation will be reflected and the temperature will increase leading to a further decrease of snow or ice cover. These snow/ice–albedo–temperature interactions are examples of positive feedbacks (see Fig. 2.9).

Changes in vegetation cover can also cause variations in the surface albedo leading to important local feedback effects, such as exemplified by progressive desertification (Charney *et al.*, 1977).

As another example of a positive feedback mechanism we can mention the water vapor–greenhouse effect. An increase in surface temperature, in the absence of other changes, will cause the evaporation at the earth's surface and the amount of water vapor in the atmosphere to increase as well. Since water vapor is a strong absorber of long-wave radiation more terrestrial radiation will be trapped, heating the lower atmosphere and leading to a further increase of the temperature. If on the other hand, the temperature becomes lower due to some other reasons (e.g., ice–albedo feedback), the amount of water vapor decreases and the greenhouse effect becomes less effective.

Another way of expressing this same feedback mechanism is to accept that the time-mean relative humidity at a particular altitude tends to remain almost constant within a relatively large range of temperatures in the lower atmosphere. However, while the relative humidity remains practically the same, the absolute humidity increases rapidly with temperature. Thus an increase of temperature at constant relative humidity increases the amount of water vapor in the air leading through the absorption of long-wave radiation to a further increase in the temperature. We should note that the name water vapor–greenhouse effect is actually a misnomer since heating in the usual greenhouse is due to the reduction of convection, whereas in the case of water vapor the heating is due to the trapping of infrared radiation. Other gases, such as carbon dioxide, can also contribute to the greenhouse effect.

As an example of a negative internal feedback, we can consider the temperature–long-wave radiation coupling in the atmosphere. If the temperature increases, the atmosphere will generally lose more long-wave radiation to space, thus reducing the temperature and attenuating the initial perturbation.

Sometimes the cloudiness–temperature interaction is given as an example of a simple feedback system. However, clouds can lead to many different feedback processes because they are both excellent absorbers of infrared radiation and effective reflectors of solar radiation. These two opposing effects make the clouds an essential but very complex modulating factor in the radiation balance of the earth. The amount of infrared radiation emitted to space depends on the temperature of the cooler cloud tops and is generally lower than the amount of radiation emitted by the warmer clear atmosphere and warmer earth's surface. Thus the net amount of outgoing terrestrial radiation to space is reduced by the presence of clouds and some radiation is trapped, augmenting the water vapor–greenhouse effect. It seems that the influence of the high reflectivity for solar radiation predominates for low and middle clouds leading to a cooling with increased cloudiness, whereas high cirrus clouds are more transparent to short-wave than to long-wave radiation leading to a reinforced greenhouse effect. It is clear that the final outcome of the various possible cloud feedbacks is difficult to assess because it depends not only on changes in the amount but also on changes in the type of clouds, the cloud heights, the cloud liquid and ice water content, and the size of the cloud particles (e.g., Ramanathan *et al.*, 1989, and Cess *et al.*, 1989).

The extreme complexity of the multiple feedback interactions in the climatic system also becomes apparent in the case of air–sea exchange. In this case, sea surface temperature anomalies tend to strongly affect the thermal structure of the lower atmosphere and eventually, through the atmospheric general circulation, they also affect the surface wind stresses. These anomalous wind stresses form the feedback mechanism from the atmosphere back to the oceans by generating changes in the general circulation in the oceans that, in turn, modify the sea surface temperature anomalies, closing the loop.

As we have seen there are in nature many positive and negative feedback processes. However, it must be noted that a positive feedback process cannot proceed indefinitely because it would lead to runaway situations that have not been observed on earth but may have happened in the case of Venus. Therefore, a compensation between positive and negative feedback processes must prevail in the mean. There is some geological evidence (Crowley, 1983) for catastrophic changes in the climatic state (e.g., at the end of the Cretaceous and during the sudden glaciations of the Pleistocene) that could involve some runaway process and in which a change to a new and different climatic state occurred.

CHAPTER 3

Basic Equations for the Atmosphere and Oceans

Our discussion will be organized around the governing equations of the atmosphere taken as one of the internal subsystems of the climatic system. These equations express the principles of conservation of mass, momentum (Newton's second law of motion), and energy following the approach advocated by Starr (1951). We will assume that the atmosphere behaves as a homogeneous gaseous system that obeys the ideal gas law when unsaturated. We will further assume conservation of water substance in the various phases for the entire climatic system. Thus we must also include the laws governing evaporation, condensation, and the conversion of cloud droplets into precipitation (raindrops and snow crystals). The radiation laws expressing short-wave absorption, reflection, scattering, and infrared radiative transfer will be discussed extensively in Chap. 6. We will consider here the fluid components of the climatic system as a continuum and a thermo-hydrodynamical system. Usually a local Cartesian coordinate system (x, y, z, t) is used with x parallel to the latitude circle (positive to the east), y parallel to the meridian circle (positive to the north), and z in the vertical direction (positive upward), and t denoting time.

3.1 EQUATION OF CONTINUITY

The principle of conservation of mass for a given element of mass $\delta m = \rho \delta V$ is given by $d(\rho \delta V)/dt = 0$, which leads to the equation of continuity. In fact, if we expand the previous derivative we obtain

$$-\frac{1}{\rho}\frac{d\rho}{dt} = \frac{1}{\delta V}\frac{d}{dt}\delta V,$$

where the right-hand side represents the relative rate of expansion of the volume element. This expansion rate is just the three-dimensional divergence of the wind velocity vector

$$-\frac{1}{\rho}\frac{d\rho}{dt} = \operatorname{div} \mathbf{c}, \qquad (3.1)$$

where **c** is the three-dimensional wind velocity with components u (positive if eastward), v (northward), and w (upward), and ρ is the density.

Another form of the continuity equation is given by

$$\frac{d\alpha}{dt} = \alpha \text{ div } \mathbf{c}, \tag{3.2}$$

where $\alpha = \rho^{-1}$ is the specific volume.

The time derivatives in these equations are material or substantial derivatives, expressing the rate of change seen by an observer moving with the flow. Since the density is a function of space and time we can expand the total derivative $d\rho/dt$ as follows:

$$\frac{d\rho}{dt} = \frac{\partial \rho}{\partial t} + u\frac{\partial \rho}{\partial x} + v\frac{\partial \rho}{\partial y} + w\frac{\partial \rho}{\partial z} = \frac{\partial \rho}{\partial t} + \mathbf{c}\cdot\text{grad } \rho,$$

where $\partial \rho/\partial t$ is the local, Eulerian time derivative, and $\mathbf{c}\cdot\text{grad } \rho$ represents the advection of mass. Thus the equation of continuity can be written in a local form as

$$\frac{\partial \rho}{\partial t} = -\text{div } \rho\mathbf{c}, \tag{3.3}$$

since $\text{div } \rho\mathbf{c} = \rho \text{ div } \mathbf{c} + \mathbf{c}\cdot\text{grad } \rho$.

Under hydrostatic equilibrium, the vertical pressure gradient dp/dz balances the gravity force ρg, or $dp = -\rho g\, dz$. As the atmosphere is almost always in hydrostatic equilibrium (with the exception of small-scale phenomena, such as cumulus convection), the geometrical (x, y, z, t) coordinate system can be replaced by the (x, y, p, t) system, where the pressure p is used to specify the position in the vertical rather than the geometrical height z. In this new system the continuity equation can be written, noting that $\delta m = -\delta x \delta y \delta p / g$, as

$$\frac{d}{dt}\delta m = 0$$

or

$$\frac{1}{\delta m}\frac{d}{dt}\delta m = \frac{1}{\delta x \delta y}\frac{d}{dt}(\delta x \delta y) + \frac{1}{\delta p}\frac{d}{dt}(\delta p) = 0$$

or

$$\text{div } \mathbf{v} + \frac{\partial \omega}{\partial p} = 0, \tag{3.4}$$

where **v** is the horizontal component of the wind vector and $\omega = dp/dt$. The quantity ω can be thought of as the component of velocity along the p axis. In fact, since

$$\frac{dp}{dt} = \frac{\partial p}{\partial t} + \mathbf{v}\cdot\text{grad } p + w\frac{\partial p}{\partial z} \tag{3.5}$$

and since $\partial p/\partial t$ and $\mathbf{v}\cdot\text{grad } p$ are, in general, much smaller than the last term we find that

$$\omega \approx w\frac{\partial p}{\partial z}$$

or assuming quasihydrostatic equilibrium

$$\omega \approx -\rho g w.$$

It must be pointed out that in this form of the continuity equation we are dealing with the two-dimensional horizontal divergence, and that the equation does not explicitly contain the density and time derivatives.

Using the longitude λ and latitude ϕ instead of x and y we may define a new (λ, ϕ, p, t) system in which the continuity equation becomes

$$\frac{\partial u}{R \cos \phi \partial \lambda} + \frac{\partial v \cos \phi}{R \cos \phi \partial \phi} + \frac{\partial \omega}{\partial p} = 0, \qquad (3.6)$$

where R is the mean radius of the earth.

3.2 EQUATIONS OF MOTION

The equations of motion are an expression of Newton's second law, i.e., the law of conservation of momentum. They state that, in an absolute nonaccelerating coordinate system, the mass in a given volume can gain or lose momentum in three ways. The first way is through body forces acting upon the mass, such as gravity, the second through surface forces acting at the boundaries, such as pressure forces, and, finally through the exchange of mass possessing different amounts of momentum across the boundaries of the volume. The effects of friction are included. Sometimes it is convenient to express the friction force $\mathbf{F}$ as a shear stress acting at the boundaries of the volume. These stresses are important where the atmosphere and oceans are in contact with the lithosphere or in contact with each other.

The body force for a unit mass can be expressed as the gradient of the earth's gravitational potential $-\operatorname{grad} \phi_N$. The net force per unit volume due to changes of pressure is $-\operatorname{grad} p$ so that the force per unit mass is given by $-(1/\rho)\operatorname{grad} p$.

Now the equations of motion can be written per unit mass as

$$\frac{d\mathbf{c}_A}{dt} = -\frac{(\operatorname{grad} p)}{\rho} - \operatorname{grad} \phi_N + \mathbf{F}, \qquad (3.7)$$

where $\mathbf{c}_A$ is the three-dimensional velocity vector measured in a nonrotating, absolute reference frame and $\mathbf{F}$ is the friction force. For a frame of reference that rotates with the earth with an angular velocity Ω, the equations have to be adjusted. Let the subscript A refer to quantities measured with respect to the fixed or absolute system and the subscript r refers to quantities measured relative to the rotating frame. Thus a point with a fixed position in the rotating system $\mathbf{r}_r$ has a velocity $\Omega \times \mathbf{r}_r$. When the point is moving with respect to the rotating frame its velocity relative to the fixed frame is now given by

$$\frac{d\mathbf{r}_A}{dt} = \frac{d\mathbf{r}_r}{dt} + \Omega \times \mathbf{r}_r$$

or

$$\mathbf{c}_A = \mathbf{c} + \Omega \times \mathbf{r}_r.$$

Applying this operation twice we obtain for the acceleration

$$\frac{d\mathbf{c}_A}{dt} = \frac{d}{dt}(\mathbf{c} + \Omega \times \mathbf{r}_r) + \Omega \times (\mathbf{c} + \Omega \times \mathbf{r}_r)$$

or

$$\frac{d\mathbf{c}_A}{dt} = \frac{d\mathbf{c}}{dt} + 2\Omega \times \mathbf{c} + \Omega \times (\Omega \times \mathbf{r}_r).$$

Thus the acceleration in the inertial frame $d\mathbf{c}_A/dt$ is expressed as the sum of the acceleration observed in the rotating frame (the so-called apparent acceleration) $d\mathbf{c}/dt$, the Coriolis acceleration $2\Omega \times \mathbf{c}$, and the centripetal acceleration $\Omega \times (\Omega \times \mathbf{r}_r)$, which acts inward at right angles to the axis of rotation. This last term can be written in the form

$$\Omega \times (\Omega \times \mathbf{r}_r) = \Omega \times (\Omega \times \mathbf{r}) = -\Omega^2 \mathbf{r},$$

where $\mathbf{r}$ is the radius vector of the latitude circle ($|\mathbf{r}| = R \cos \phi$), and it can also be written in the form of the gradient of a scalar quantity:

$$\Omega^2 \mathbf{r} = \text{grad} \tfrac{1}{2} (\Omega \times \mathbf{r})^2.$$

The equations of motion (3.7) in the relative framework can now be written in the form

$$\frac{d\mathbf{c}}{dt} + 2\Omega \times \mathbf{c} = -(\text{grad } p)/\rho - \text{grad } \phi + \mathbf{F},$$

where ϕ is the geopotential defined by

$$\phi = \phi_N - \frac{1}{2} \Omega^2 R^2 \cos^2 \phi,$$

i.e., the sum of the gravitational potential and the centripetal potential.

Finally, introducing the apparent gravity vector $\mathbf{g} = -\text{grad } \phi$, we find

$$\frac{d\mathbf{c}}{dt} = -2\Omega \times \mathbf{c} - (\text{grad } p)/\rho + \mathbf{g} + \mathbf{F}. \tag{3.8}$$

The word "apparent" gravity implies that the centripetal force does not appear explicitly in the equations of motion, but is absorbed in the geopotential.

In writing the Coriolis term $-2\Omega \times \mathbf{c}$, on the right-hand side, Eq. (3.8) becomes formally equivalent with Eq. (3.7). However, it includes a new apparent force which, as the centripetal force, arises from the motion of the air with respect to the ground. This force is the well-known Coriolis force, given in vector form by $-2\Omega \times \mathbf{c}$. It acts normal to the velocity vector and is everywhere at right angles to the earth's rotation axis, or, in other words, parallel to the plane of the equator. In general, the horizontal components of the Coriolis force have a greater significance than the vertical component because the latter is usually negligible compared with gravity and the vertical component of the pressure gradient force. The horizontal components in the λ and ϕ directions are given by $fv - f'w$ and $-fu$, respectively, where $f = 2\Omega \sin \phi$, the so-called Coriolis parameter, and $f' = 2\Omega \cos \phi$. The value of f at 45° latitude is $1.03 \times 10^{-4} \text{ s}^{-1}$.

The equations of motion (3.8) can be rewritten for the three individual components in a spherical coordinate system (λ, ϕ, z, t) since we may assume, for our purposes, that the geoid can be approximated by a sphere. For a derivation see, e.g., Holton (1972):

$$\frac{du}{dt} = \frac{\tan \phi}{R} uv - \frac{uw}{R} + fv - f'w - \frac{1}{\rho} \frac{\partial p}{R \cos \phi \partial \lambda} + F_\lambda, \tag{3.9a}$$

$$\frac{dv}{dt} = -\frac{\tan \phi}{R} u^2 - \frac{vw}{R} - fu - \frac{1}{\rho} \frac{\partial p}{R \partial \phi} + F_\phi, \tag{3.9b}$$

$$\frac{dw}{dt} = \frac{u^2}{R} + \frac{v^2}{R} + f'u - \frac{1}{\rho} \frac{\partial p}{\partial z} - g + F_z, \tag{3.9c}$$

where $u = R \cos \phi \, d\lambda/dt$, $v = R d\phi/dt$, and $w = dz/dt$.

The first two terms on the right-hand side of Eqs. (3.9a)–(3.9c) are of the form ξ_i/R and show the influence of the geometry of the earth on the motion field. It is interesting to point out that these terms, when multiplied by the corresponding velocity components, do not "perform work" (generate kinetic energy) since $u\xi_1/R + v\xi_2/R + w\xi_3/R = 0$. Also, the Coriolis terms involving f and f' satisfy a similar invariance relationship, i.e., they do not generate kinetic energy.

3.2.1 Frictional effects

The friction force, which was symbolically written as $\mathbf{F} \equiv (F_x, F_y, F_z) = (F_\lambda, F_\phi, F_z)$, is equal to the divergence of a stress tensor τ:

$$\mathbf{F} = -\alpha \, \text{div} \, \tau. \tag{3.10}$$

The stress tensor can be written as a time covariance of the microscopic velocity fluctuations $\tau = \overline{\rho \mathbf{c'c'}}$, where the bar indicates a time average (say for the period of an hour) and the prime a departure from the time average. The method of notation for the various components of the stress tensor τ is illustrated in Fig. 3.1. Not only molecular-scale processes but also those scales of motion that are not explicitly resolved in calculating the advection of momentum are lumped into the stress; it is a kind of residual term. The friction force can be neglected except close to the earth's surface, the ocean bottom, and in regions of strong wind shear near the jet streams. The "eddy friction" due to cumulus momentum mixing and breaking gravity waves may also be important in other parts of the atmosphere.

The zonal (eastward) component of the friction force consists of three parts:

$$F_x = -\frac{1}{\rho}\left(\frac{\partial \tau_{xx}}{\partial x} + \frac{\partial \tau_{yx}}{\partial y} + \frac{\partial \tau_{zx}}{\partial z}\right).$$

Near the earth's surface the strongest wind shears are in the vertical direction, so that one can neglect the first two terms in F_x. Further, a flux-gradient relationship $\tau_{zx} = (-\mu/\rho)(\partial \bar{u}/\partial z)$ (Newton's law of viscosity) holds very well for τ_{zx} near the surface. Thus

$$F_x = -\frac{1}{\rho}\frac{\partial \tau_{zx}}{\partial z} = -\frac{1}{\rho}\frac{\partial \rho \overline{u'w'}}{\partial z} \approx -\frac{\partial \overline{u'w'}}{\partial z}$$

$$= \frac{\partial}{\partial z}\left(\frac{\mu}{\rho}\frac{\partial \bar{u}}{\partial z}\right) = \frac{\partial}{\partial z}\left(v\frac{\partial \bar{u}}{\partial z}\right), \tag{3.11}$$

where μ = dynamic coefficient of viscosity and $v = \mu/\rho$ = kinematic coefficient of viscosity $\approx 1.2 \times 10^{-5}$ m^2 s^{-1}.

Since small-scale turbulent eddies in the atmospheric boundary layer are much more efficient than viscous effects for transferring momentum, the kinematic coefficient of viscosity has to be replaced by a much larger coefficient of eddy viscosity K_M in order to obtain realistic results:

$$F_x = \frac{\partial}{\partial z}\left(K_M \frac{\partial \bar{u}}{\partial z}\right). \tag{3.12}$$

For example, to include all effects of eddies up to scales of roughly 10 m an eddy coefficient of $K_M \approx 10^{-1}$ m^2 s^{-1} must be used (see also Sec. 10.4). Expressions similar to Eqs. (3.11) and (3.12) can be established for F_y and F_z.

For the oceans, the characteristic values of the molecular and eddy kinematic viscosities are 1.9×10^{-6} m^2 s^{-1} and $0.0002 - 0.75$ m^2 s^{-1}, respectively. Whereas the molecular viscosity is constant with depth and depends only slightly on temperature, the eddy viscosity changes with the density stratification and with the other characteristics of the shear flow.

3.2.2 Filtering of the basic equations for the atmosphere

Equations (3.9a)–(3.9c) can be specialized, or filtered, in order to exclude certain phenomena, such as sound waves, that are not thought to be important for the large-scale climate in the atmosphere and oceans. Further, because the vertical extent of the atmosphere is small compared with its horizontal dimensions, there is the tendency for the circulations to be predominantly horizontal and in quasihydrostatic equilibrium. These assumptions can be incorporated into Eqs. (3.9a)–(3.9c) by selecting the dominant terms using scale analysis (Charney, 1948). Scale analysis is basically a technique for estimating the orders of magnitude of the various terms in the governing equations for a particular class of motions. It is based on dimensional analysis and on an adequate choice of the characteristic values of, e.g., the length, depth, and time scales for the fluctuations. These characteristic values are used as basic units to measure the various terms in the governing equations.

For the typical scales of motion in the atmosphere we assume the following characteristic values as being representative of the observed values:

horizontal length scale	$L \approx 10^6$ m;
depth scale	$H \approx 10^4$ m;
horizontal velocity scale	$u \approx 10$ m s^{-1};
vertical velocity scale	$w \approx 10^{-2}$ m s^{-1};
horizontal pressure scale	$\Delta p \approx 10$ mb $= 10^3$ Pa;
time scale	$L/u \approx 10^5$ s.

Since $\Omega = 7.29 \times 10^{-5}$ s^{-1}, the Coriolis parameter $f = 2\Omega \sin \phi$ is approximately 10^{-4} s^{-1} in middle latitudes, and we have

$$\left(\frac{du}{dt}, \frac{dv}{dt}\right) \sim \frac{u^2}{L} = 10^{-4} \text{ m s}^{-2};$$

$$\left(\frac{dw}{dt}\right) \sim \frac{uw}{L} = 10^{-7} \text{ m s}^{-2};$$

$$(fu, fv) \sim 10^{-3} \text{ m s}^{-2};$$

$$(f'w) \sim 10^{-6} \text{ m s}^{-2};$$

$$\left(\frac{uv}{R}, \ldots\right) \sim 10^{-5} \text{ m s}^{-2};$$

$$\left(\frac{\partial p}{\rho R \partial \phi}, \ldots\right) \sim 10^{-3} \text{ m s}^{-2};$$

$$\left(\frac{1}{\rho}\frac{\partial p}{\partial z}\right) \sim 10 \text{ m s}^{-2};$$

$$(F_\lambda, F_\phi) \sim 10^{-4} \text{ to } 10^{-5} \text{ m s}^{-2};$$

$$(F_z) \sim 10^{-6} \text{ to } 10^{-7} \text{ m s}^{-2};$$

and

$$(g) \sim 10 \text{ m s}^{-2}.$$

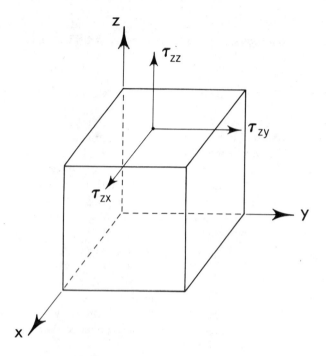

FIGURE 3.1. Schematic diagram showing the three stress components τ_{zx}, τ_{zy}, and τ_{zz} on the $z =$ constant plane in the x, y, and z directions, respectively. The friction force is given by the divergence of the three-dimensional stress tensor $\mathbf{F} = -\alpha \, \text{div} \, \tau$.

Frictional effects can be generally neglected for synoptic-scale motions above the planetary boundary layer.

The filtering of the third equation of motion (3.9c) leads to a quasibalance between the vertical pressure gradient force and gravity:

$$\frac{\partial p}{\partial z} \approx -\rho g, \tag{3.13}$$

which is the condition of hydrostatic equilibrium. As mentioned earlier, it is convenient, at times, to use pressure as the vertical coordinate leading to the so-called (x, y, p, t) system since pressure and height are related through Eq. (3.13).

In small-scale turbulence, cumulus convection, and meso-scale phenomena, the vertical velocities may be of the same order of magnitude as the horizontal velocities, and hydrostatic equilibrium conditions are not observed. However, the departures from hydrostatic equilibrium are of short duration and confined to small regions.

If we keep only the terms of the order 10^{-3} m s^{-2} in the horizontal equations of motion (3.9a) and (3.9b), we find an approximate balance between the horizontal pressure gradient and the Coriolis force, the so-called geostrophic balance:

$$fv_g = \frac{1}{\rho} \frac{\partial p}{R \cos \phi \, \partial \lambda}, \tag{3.14a}$$

$$fu_g = -\frac{1}{\rho} \frac{\partial p}{R \partial \phi}, \tag{3.14b}$$

or

$$\mathbf{V}_g = \frac{1}{\rho f} \mathbf{k} \times \text{grad} \, p, \tag{3.15}$$

where u_g and v_g indicate the horizontal components of the geostrophic wind $\mathbf{v}_g$. The geostrophic solutions show that the winds tend to blow parallel to the isobars with high pressure on the right in the Northern Hemisphere and on the left in the Southern Hemisphere.

In the (x,y,p) system the same filtering would lead to

$$fv_g = \frac{g\partial z}{R\cos\phi\partial\lambda}, \qquad (3.16a)$$

$$fu_g = -\frac{g\partial z}{R\partial\phi}, \qquad (3.16b)$$

or

$$\mathbf{v}_g = \frac{1}{f}\mathbf{k}\times \operatorname{grad} gz. \qquad (3.17)$$

The geostrophic approximation applies when the acceleration is much smaller than the Coriolis force. The nondimensional quantity $R_0 \approx (dv/dt)/fu \approx u/fL$, the so-called Rossby number, sets a criterion for the validity of the geostrophic approximation. In the atmosphere, $R_0 \approx 10^{-1}$, whereas for the oceans, $R_0 \approx 10^{-3}$, implying a much stronger geostrophic constraint for the ocean circulations.

In Eqs. (3.9a) and (3.9b), u and v can be regarded as the sum of the geostrophic components u_g and v_g and the corresponding ageostrophic components u_{ag} and v_{ag}, i.e., $u = u_g + u_{ag}$ and $v = v_g + v_{ag}$. The horizontal divergence of mass is largely due to the ageostrophic component of the flow. Of course, the geostrophic conditions are not valid near the equator where f becomes very small. Also, close to the earth's surface frictional effects have to be included leading to a cross-isobaric flow and a reduction in intensity of the wind. As in the case of the hydrostatic equation, the geostrophic approximation is not valid for small-scale circulations.

If we use a filter of 10^{-4} m s^{-2} the horizontal equations (3.9a) and (3.9b) can be written

$$\frac{du}{dt} = fv - \frac{1}{\rho}\frac{\partial p}{R\cos\phi\partial\lambda} + F_\lambda, \qquad (3.18a)$$

$$\frac{dv}{dt} = -fu - \frac{1}{\rho}\frac{\partial p}{R\partial\phi} + F_\phi. \qquad (3.18b)$$

If we want to study global-scale circulations with length scales on the order of the radius of the earth, such as the quasistationary waves or the mean meridional circulations in the Hadley and Ferrel cells (see Sec. 7.4.3), we have to also include some of the metric terms so that Eqs. (3.9a) and (3.9b) become

$$\frac{du}{dt} = \frac{\tan\phi}{R}uv + fv - \frac{1}{\rho}\frac{\partial p}{R\cos\phi\partial\lambda} + F_\lambda, \qquad (3.18c)$$

$$\frac{dv}{dt} = -\frac{\tan\phi}{R}u^2 - fu - \frac{1}{\rho}\frac{\partial p}{R\partial\phi} + F_\phi. \qquad (3.18d)$$

These equations are more general than the geostrophic equations because they include the accelerations and friction. They therefore become prognostic equations (i.e., they contain time derivatives), whereas the geostrophic equations are only diagnostic equations (i.e., they do not contain time derivatives).

Assuming hydrostatic equilibrium, $\partial p/\partial z = -\rho g$, and using the mathematical expressions for derivatives, such as

$$\left(\frac{\partial p}{\partial x}\right)_z = -\left(\frac{\partial z}{\partial x}\right)_p \left(\frac{\partial z}{\partial p}\right)^{-1} = \rho g \left(\frac{\partial z}{\partial x}\right)_p, \quad (3.19)$$

we can rewrite the horizontal equations of motion (3.18c) and (3.18d) in the (λ,ϕ,p,t) system in the following form:

$$\frac{du}{dt} = \frac{\tan \phi}{R} uv + fv - \frac{g\partial z}{R \cos \phi \partial \lambda} + F_\lambda, \quad (3.20a)$$

$$\frac{dv}{dt} = -\frac{\tan \phi}{R} u^2 - fu - \frac{g\partial z}{R\partial \phi} + F_\phi, \quad (3.20b)$$

where

$$\frac{d}{dt} = \frac{\partial}{\partial t} + u\frac{\partial}{R \cos \phi \partial \lambda} + v\frac{\partial}{R\partial \phi} + \omega \frac{\partial}{\partial p} \quad (3.21)$$

and $\omega = dp/dt$.

3.2.3 Filtering of the basic equations for the oceans

For large-scale oceanic circulations we have the following characteristic values:

horizontal length scale	$L \approx 10^6$ m;
depth scale	$H \approx 4 \times 10^3$ m;
horizontal velocity scale	$u \approx 10^{-1}$ m s^{-1};
vertical velocity scale	$w \approx 10^{-4}$ m s^{-1};
horizontal pressure scale	$\Delta p \approx 10$ mb;
time scale	$\approx 10^7$ s.

The Rossby number for the oceans is then

$$R_0 \approx u/fL = 10^{-3}.$$

Therefore, the advective terms can be neglected except possibly in strong boundary currents, such as the Gulf Stream and Kuroshio, and in the equatorial undercurrents.

For the oceans, we can use a similar procedure as in the atmosphere to filter the equations of motion. The results are formally the same as Eqs. (3.18a) and (3.18b):

$$\frac{du}{dt} = fv - \frac{1}{\rho}\frac{\partial p}{R \cos \phi \partial \lambda} + F_\lambda,$$

$$\frac{dv}{dt} = -fu - \frac{1}{\rho}\frac{\partial p}{R\partial \phi} + F_\phi,$$

where

$$F_\lambda \approx -\frac{1}{\rho}\frac{\partial \tau_{zx}}{\partial z} \approx \frac{\partial}{\partial z}\left(K_M \frac{\partial u}{\partial z}\right)$$

and

$$F_\phi \approx -\frac{1}{\rho}\frac{\partial \tau_{zy}}{\partial z} \approx \frac{\partial}{\partial z}\left(K_M \frac{\partial v}{\partial z}\right).$$

In oceanography, it is more common to define the stress with respect to the oceans rather than with respect to the atmosphere or, in other words, to use the opposite sign for the stress. However, to avoid confusion we will keep the same sign convention as for the atmosphere, and use a positive sign for τ_{zx} and τ_{zy} when momentum is transferred from the ocean to the atmosphere.

Hydrostatic equilibrium is also observed:

$$\frac{\partial p}{\partial z} = -\rho g,$$

where ρ is almost constant.

The steady-state equations for the oceans can be reduced to

$$-fv = -\frac{1}{\rho} \frac{1}{R \cos \phi \partial \lambda} \frac{\partial p}{} + \frac{1}{\rho} \frac{\partial(-\tau_{zx})}{\partial z},$$

$$fu = -\frac{1}{\rho} \frac{\partial p}{R \partial \phi} + \frac{1}{\rho} \frac{\partial(-\tau_{zy})}{\partial z}, \qquad (3.22)$$

since the advection terms are very small.

The velocity can then be separated into two parts, one part driven by the pressure gradient (the geostrophic components u_g and v_g) and the other part driven by the gradient of the stress (the Ekman components u_E and v_E):

$$u_g = -\frac{1}{\rho f} \frac{\partial p}{R \partial \phi},$$

$$v_g = \frac{1}{\rho f} \frac{\partial p}{R \cos \phi \partial \lambda}$$

and

$$u_E = \frac{1}{\rho f} \frac{\partial}{\partial z}(-\tau_{zy}), \qquad (3.23a)$$

$$v_E = -\frac{1}{\rho f} \frac{\partial}{\partial z}(-\tau_{zx}). \qquad (3.23b)$$

The Ekman components are mainly important in the boundary layers.

The vertical integration of Eqs. (3.23a) and (3.23b) from a depth $z = -\delta$, where δ is deep enough (on the order of 10 to 20 m) that $\tau_{zx}(-\delta)$ and $\tau_{zy}(-\delta)$ are practically zero, to the ocean surface at $z = 0$ leads to a mass transport:

$$M_{Ex} = \int_{-\delta}^{0} \rho u_E \, dz = \frac{1}{f}(-\tau_{0y}), \qquad (3.24a)$$

$$M_{Ey} = \int_{-\delta}^{0} \rho v_E \, dz = -\frac{1}{f}(-\tau_{0x}). \qquad (3.24b)$$

Here τ_{0x} and τ_{0y} are the components of the surface wind stress. These equations show that the mass transport vector $M_E = (M_{Ex}, M_{Ey})$ is at right angles to the surface wind stress vector τ_0 defined by

$$\tau_0 = \mathbf{i}\tau_{0x} + \mathbf{j}\tau_{0y}.$$

In vector form, Eq. (3.24) becomes

$$\mathbf{M}_E = -\frac{1}{f} \tau_0 \times \mathbf{k}. \qquad (3.25)$$

If we differentiate the two components of Eq. (3.24) with respect to x and y, respectively, and add we find

$$\frac{\partial}{\partial x} M_{Ex} + \frac{\partial}{\partial y} M_{Ey} = \text{curl}_z(-\tau_0/f).$$

Using conservation of mass within the Ekman layer the left-hand side of this equation reduces to $\rho w_E(-\delta)$, the vertical mass flux at the bottom of the Ekman layer:

$$\rho w_E(-\delta) = \text{curl}_z(-\tau_0/f). \tag{3.26}$$

Here $w_E(-\delta)$ is usually called the "Ekman pumping" vertical velocity.

Using the observed fact that the flow is generally geostrophic below the surface Ekman layer and away from the boundaries, we can now derive an expression for the net meridional flow in the interior of the ocean. We will use the (x,y,z,t) system for simplicity. First we eliminate the pressure by cross differentiation of the steady-state equations (3.22):

$$-fv = -\frac{1}{\rho}\frac{\partial p}{\partial x} + \frac{1}{\rho}\frac{\partial(-\tau_{zx})}{\partial z},$$

$$fu = -\frac{1}{\rho}\frac{\partial p}{\partial y} + \frac{1}{\rho}\frac{\partial(-\tau_{zy})}{\partial z},$$

which yields (assuming $\rho = $ constant)

$$\frac{\partial fv}{\partial y} + \frac{\partial fu}{\partial x} = \frac{1}{\rho}\text{curl}\left[\frac{\partial}{\partial z}(-\tau)\right]$$

or

$$\beta v + f\left(\frac{\partial u}{\partial y} + \frac{\partial v}{\partial y}\right) = \frac{1}{\rho}\frac{\partial}{\partial z}\text{curl}(-\tau),$$

where $\beta = df/dy = 2\Omega\cos\phi/R$. At 45° latitude $\beta = 1.62\times 10^{-11}$ m^{-1} s^{-1}. Integration with respect to z from the ocean bottom $(-H)$, where the flow is assumed to be zero, up to the height of sea level η leads to

$$\beta\int_{-H}^{\eta} v\, dz + f\int_{-H}^{\eta}\left(\frac{\partial u}{\partial x} + \frac{\partial v}{\partial y}\right)dz = \frac{1}{\rho}\text{curl}(-\tau_0).$$

Because of conservation of mass the second term on the left-hand side of the equation vanishes so that we can write for the northward mass flow M_y:

$$M_y = \int_{-H}^{\eta}\rho v\, dz = \frac{\text{curl}(-\tau_0)}{\beta}. \tag{3.27}$$

This is the expression for the net meridional flow of mass in the interior of the ocean gyres away from the lateral boundaries, the so-called Sverdrup transport. Thus one would expect to find a general equatorward flow in the subtropical gyres where $\text{curl}(-\tau_0) < 0$. The poleward return flow is found to occur in narrow western boundary currents such as the Gulf Stream and Kuroshio (see also discussions in Chap. 8).

3.3 VORTICITY EQUATION

3.3.1 Some definitions of vorticity

The vorticity is defined as the curl (rotational part) of the velocity field:

$$\zeta = \text{curl } \mathbf{c}. \tag{3.28}$$

Thus, in a Cartesian coordinate system (x,y,z) the vorticity components are related to the velocity components (u,v,w) as follows:

$$\zeta_x = \frac{\partial w}{\partial y} - \frac{\partial v}{\partial z}, \qquad (3.29a)$$

$$\zeta_y = \frac{\partial u}{\partial z} - \frac{\partial w}{\partial x}, \qquad (3.29b)$$

$$\zeta_z = \frac{\partial v}{\partial x} - \frac{\partial u}{\partial y}. \qquad (3.29c)$$

For a fluid in uniform rotation (rotating as a solid body) with constant angular velocity Ω_0, the linear velocity is given by

$$\mathbf{c} = \Omega_0 \times \mathbf{r},$$

so that

$$\text{curl } \mathbf{c} = \text{curl}(\Omega_0 \times \mathbf{r}) = 2\Omega_0,$$

i.e., the vorticity is twice the angular velocity. Using the Stokes and mean-value theorems, we will obtain a generalization of this same result, i.e., the average vorticity of each fluid element is twice the average angular velocity of the fluid. In fact, by Stokes theorem,

$$\iint_A \zeta \cdot \mathbf{n} \, dA = \oint \mathbf{c} \cdot d\mathbf{r} \equiv \Gamma_a,$$

the normal component of vorticity passing through an open surface A is equal to the circulation Γ_a of the velocity field around the boundary contour. For a surface element δA, application of the mean-value theorem leads immediately to

$$\bar{\zeta}_n = \bar{c}_t l/\delta A,$$

where $\bar{c}_t$ is the mean tangential velocity around the contour l. If we take for the elementary area a circle of radius r_0 then

$$\bar{\zeta}_n = \bar{c}_t \frac{2\pi r_0}{\pi r_0^2} = 2\frac{\bar{c}_t}{r_0},$$

so that the average vorticity $\bar{\zeta}_n$ is twice the angular velocity $\bar{c}_t/r_0$.

In the atmosphere, the large-scale flow is quasi-horizontal with the horizontal velocity several orders of magnitude larger than the vertical velocity. Because of this asymmetry, the vertical component of vorticity turns out to be the most relevant one, and we only have to consider the horizontal wind velocity $\mathbf{v}$. The vertical component of vorticity is $\zeta_z = \mathbf{k} \cdot \text{curl } \mathbf{v}$. In spherical coordinates, it takes the form

$$\zeta_z = \frac{1}{R \cos \phi} \frac{\partial v}{\partial \lambda} - \frac{1}{R} \frac{\partial u}{\partial \phi} + \frac{u \tan \phi}{R}. \qquad (3.30)$$

For an inertial nonrotating frame, the vorticity of a fluid is called absolute vorticity ζ_a and is the curl of the absolute velocity:

$$\zeta_a = \text{curl}(\mathbf{c} + \Omega \times \mathbf{r}_r) = \zeta + 2\Omega, \qquad (3.31)$$

where ζ is the relative vorticity. Thus, the absolute vorticity of a fluid element is the sum of the relative vorticity and the planetary vorticity 2Ω.

For the quasihorizontal motions of the atmosphere, the component of the planetary vorticity normal to the earth's surface, $2\mathbf{k}\cdot\mathbf{\Omega}$, is the Coriolis parameter $f = 2\Omega \sin \phi$. The vertical component of the absolute vorticity $\eta = \mathbf{k}\cdot\boldsymbol{\zeta}_a$ is then given by

$$\eta = f + \zeta_z. \tag{3.32}$$

If we consider the geostrophic wind [Eq. (3.15)] and assume that derivatives of f are negligible, the vorticity takes the form

$$\eta_g = \frac{1}{\rho f} \nabla^2 p + f \tag{3.33}$$

because

$$\text{curl}(\mathbf{k} \times \text{grad } p) = \mathbf{k} \text{ div grad } p \equiv \mathbf{k}\nabla^2 p.$$

In the (x,y,p) coordinate system the expression would be

$$\eta_g = \frac{g}{f} \nabla^2 z + f, \tag{3.34}$$

where z is the altitude.

3.3.2 General vorticity equation

To derive the equation of vorticity which gives the time rate of change of ζ we will apply the curl operation to the equation of motion (3.8):

$$\frac{d\mathbf{c}}{dt} = -2\mathbf{\Omega} \times \mathbf{c} - \frac{1}{\rho} \text{grad } p - \text{grad } \phi + \mathbf{F}.$$

However, the operation will be facilitated if we write this equation under the so-called Webber formulation. For this, we note that the material derivative

$$\frac{d\mathbf{c}}{dt} = \frac{\partial \mathbf{c}}{\partial t} + (\mathbf{c}\cdot\text{grad})\mathbf{c}$$

can be written in the equivalent form

$$\frac{d\mathbf{c}}{dt} = \frac{\partial \mathbf{c}}{\partial t} + \text{grad } \frac{1}{2}\mathbf{c}^2 + \boldsymbol{\zeta} \times \mathbf{c}$$

using the vector identity

$$(\mathbf{c}\cdot\text{grad})\mathbf{c} = (\text{curl } \mathbf{c}) \times \mathbf{c} + \text{grad } \tfrac{1}{2}\mathbf{c}^2$$

so that finally the equation of motion (3.8) can be written as

$$\frac{\partial \mathbf{c}}{\partial t} = -\text{grad}\left(\frac{\mathbf{c}^2}{2} + \phi\right) - [(\boldsymbol{\zeta} + 2\mathbf{\Omega}) \times \mathbf{c}]$$
$$- \frac{1}{\rho} \text{grad } p + \mathbf{F}. \tag{3.35}$$

Before taking the curl of this equation, we may note several facts that will simplify the process. First, the curl operation is commutative with the partial derivative with respect to time. Second, it may be recalled that for any scalar function B:

$$\text{curl grad } B = 0,$$

and third, that

$$\text{curl } (\alpha \text{ grad } p) = \text{grad } \alpha \times \text{grad } p,$$

since curl grad $p = 0$.

Now we can write the resulting vorticity equation

$$\frac{\partial \zeta}{\partial t} = -\mathrm{curl}(\zeta_a \times \mathbf{c}) - \mathrm{grad}\,\alpha \times \mathrm{grad}\,p + \mathrm{curl}\,\mathbf{F}. \qquad (3.36)$$

To put this equation in a more useful form, we begin by expanding the first term on the right-hand side. For two vectors **A** and **B**:

$$\mathrm{curl}(\mathbf{A} \times \mathbf{B}) = \mathbf{A}\,\mathrm{div}\,\mathbf{B} + (\mathbf{B} \cdot \mathrm{grad})\mathbf{A} - \mathbf{B}\,\mathrm{div}\,\mathbf{A} - (\mathbf{A} \cdot \mathrm{grad})\mathbf{B},$$

so that

$$\mathrm{curl}(\zeta_a \times \mathbf{c}) = \zeta_a\,\mathrm{div}\,\mathbf{c} + (\mathbf{c} \cdot \mathrm{grad})\zeta_a - (\zeta_a \cdot \mathrm{grad})\mathbf{c},$$

since ζ_a has zero divergence. Additionally, because Ω is constant in time, we can substitute $\partial \zeta_a / \partial t$ for $\partial \zeta / \partial t$.

Accordingly, the general vorticity equation can be written in the form

$$\frac{d\zeta_a}{dt} = (\zeta_a \cdot \mathrm{grad})\mathbf{c} - \zeta_a\,\mathrm{div}\,\mathbf{c} - \mathrm{grad}\,\alpha \times \mathrm{grad}\,p + \mathrm{curl}\,\mathbf{F}. \qquad (3.37)$$

This is a more useful form of the general vorticity equation than Eq. (3.36). It shows that the rate of change of the absolute vorticity following a fluid particle is equal to the sum of four terms. The first term is due to the variations of the velocity along a vortex line, and is called the tipping or stretching term. In fact, let us consider a vortex line associated with ζ_a. If the velocity varies along the vortex line, it will change its length and direction. The change in direction comes from the component of $\delta\mathbf{c}$ normal to ζ_a, the tipping or tilting term, and the change in magnitude from the component of $\delta\mathbf{c}$ parallel to ζ_a, the stretching term.

The second term, the divergence term, accounts for changes in the vorticity due to changes in the density of the fluid. If $\mathrm{div}\,\mathbf{c} > 0$ the fluid element expands, resulting in a decrease of its absolute vorticity (due to conservation of angular momentum). Similarly, if $\mathrm{div}\,\mathbf{c} < 0$, the fluid element contracts leading to an increase of vorticity.

The third term, the solenoidal term, exists only in baroclinic conditions, i.e., when the fluid density is not a sole function of pressure. For a baroclinic fluid the isosteric (constant volume) surfaces intersect the isobaric surfaces and the solenoidal term is a measure of their relative slope. The variation of the density along an isobaric surface will lead to a horizontal wind shear and will tend to produce an adjustment with vertical motions, thus generating rotation or vorticity. For a barotropic fluid, where α depends only on the pressure, the isobaric and the isosteric surfaces are parallel and the solenoidal term is zero.

Finally, the last term gives the contribution of the diffusive effects of friction to the time rate of change of vorticity.

3.3.3 Vorticity equation of the horizontal motion

As mentioned before, the large-scale motions in the atmosphere and oceans are predominantly horizontal, with the vertical velocity several orders of magnitude smaller than the horizontal velocity. Thus, the vertical component of vorticity becomes of principal importance.

We obtain the corresponding vorticity equation by taking the inner product of **k** with the general equation (3.37). Recalling that the Coriolis parameter $f = 2\Omega \sin \phi$ is a measure of the local vertical planetary vorticity 2Ω and that $f' = 2\Omega \cos \phi$ is the corresponding meridional component, we can write the vorticity equation for the vertical component as follows:

$$\frac{\partial \zeta_z}{\partial t} + u\frac{\partial}{\partial x}(\zeta_z + f) + v\frac{\partial}{\partial y}(\zeta_z + f) + w\frac{\partial}{\partial z}(\zeta_z + f)$$

$$= \zeta_x \frac{\partial w}{\partial x} + (\zeta_y + f')\frac{\partial w}{\partial y} + (\zeta_z + f)\frac{\partial w}{\partial z}$$

$$- (\zeta_z + f)\,\text{div } \mathbf{c} - \mathbf{k}\cdot(\text{grad } \alpha \times \text{grad } p) + \mathbf{k}\cdot\text{curl } \mathbf{F}. \tag{3.38}$$

After some simple mathematical manipulations, disregarding the very small term $f'\partial w/\partial y$, dropping the subscript z in ζ_z and taking into consideration expressions (3.29) of ζ in terms of the velocity components, we may write the equation in its usual and more compact form:

$$\frac{\partial}{\partial t}(\zeta + f) + (\mathbf{v}\cdot\text{grad})(\zeta + f) + w\frac{\partial(\zeta + f)}{\partial z}$$

$$= -(\zeta + f)\,\text{div }\mathbf{v} - \mathbf{k}\cdot\left(\text{grad } w \times \frac{\partial \mathbf{v}}{\partial z}\right)$$

$$- \mathbf{k}\cdot(\text{grad } \alpha \times \text{grad } p) + \mathbf{k}\cdot\text{curl } \mathbf{F}, \tag{3.39}$$

where the left-hand side is equal to $d(\zeta + f)/dt$ and the twisting term is expressed explicitly in terms of the horizontal gradient of w and the vertical wind shear $\partial \mathbf{v}/\partial z$.

Most applications of the vorticity equation involve further simplifications which can be justified by the judicious use of scale analysis. The final result of the filtering would be the same as when we assume that the fluid is barotropic ($\partial \mathbf{v}/\partial z = 0$) and frictionless, namely

$$\frac{d}{dt}(\zeta + f) = -(\zeta + f)\,\text{div }\mathbf{v}, \tag{3.40}$$

which is one of the most frequently used versions of the vorticity equation in planetary fluid dynamics.

If we assume as we did above that the fluid is incompressible, so that div $\mathbf{v} = -\partial w/\partial z$, the previous equation can be reduced to the form

$$\frac{d}{dt}(\zeta + f) - (\zeta + f)\frac{\partial w}{\partial z} = 0. \tag{3.41}$$

3.4 THERMODYNAMIC ENERGY EQUATION AND SOME APPLICATIONS

3.4.1 First law of thermodynamics

The equation of conservation of energy can be expressed by the first law of thermodynamics. Assuming that the air behaves like an ideal gas ($p\alpha = R_d T$, where R_d is the gas constant for dry air; see Sec. 3.5.1.1), the first law can be written

$$c_p \frac{dT}{dt} = Q + \alpha \frac{dp}{dt}, \tag{3.42}$$

where c_p is the specific heat of the air at constant pressure, Q equals the net heating rate per unit mass, and c_p is assumed to be constant, independent of the temperature. The heating term Q includes various diabatic effects, namely radiative heating (solar and infrared), latent heating, frictional heating, and turbulent and conductive heating near the earth's surface.

Again, using the ideal gas assumption, Eq. (3.42) takes the form

$$c_p \frac{d \ln T}{dt} - R_d \frac{d \ln p}{dt} = Q/T. \tag{3.43}$$

Since the rate of change of entropy per unit mass is given by

$$\frac{ds}{dt} = Q/T, \tag{3.44}$$

we see that

$$\frac{ds}{dt} = c_p \frac{d \ln T}{dt} - R_d \frac{d \ln p}{dt}. \tag{3.45}$$

An adiabatic process is defined as a process in which there is no exchange of heat between the system and its environment, so that Eq. (3.43) reduces to

$$c_p \frac{d \ln T}{dt} = R_d \frac{d \ln p}{dt}.$$

The integration of this equation between an initial state (p_0, T_0) and a final state (p, T) leads to Poisson's equation

$$(T/T_0) = (p/p_0)^\kappa, \tag{3.46}$$

where $\kappa = R_d/c_p$.

At this point, it is important to introduce the concept of potential temperature θ. This is the temperature that a parcel of air would attain in a reversible, adiabatic process if the parcel were displaced to a reference level $p_{00}(\simeq 1000 \text{ mb})$:

$$\theta = T (p_{00}/p)^\kappa. \tag{3.47}$$

With the concept of a parcel we mean an infinitesimal material element small enough to be regarded mathematically infinitely small and large enough to contain an ample number of molecules of air and dissolved materials to be physically representative of the environment.

By logarithmic differentiation of Eq. (3.47) with respect to time and using Eq. (3.45), we can express the entropy in terms of the potential temperature θ:

$$s = c_p \ln \theta + \text{const}, \tag{3.48}$$

where s is again the specific entropy.

Using the logarithmic differentiation of Eq. (3.47) with time, the first law of thermodynamics for the atmosphere (3.43) can also be written in terms of potential temperature in the form

$$c_p \frac{T}{\theta} \frac{d\theta}{dt} = Q. \tag{3.49}$$

An adiabatic ($Q = 0$) reversible process is isentropic ($ds/dt = 0$), and the potential temperature is also constant ($d\theta/dt = 0$).

The potential temperature implicitly takes into account the effects of the compressibility of the air. It is used in order to remove the cooling (warming) effects associated with the adiabatic expansion (compression), allowing the comparison of the temperature of air parcels at various levels in the atmosphere. This enables us to easily determine the state of hydrostatic equilibrium observed in the atmosphere.

3.4.2 Static stability

The static stability for dry air or moist unsaturated air (without phase transitions) will be discussed using the parcel method. In this method, a parcel is regarded as an individual system that does not mix with the surrounding air when it moves. The surrounding air is supposed to be in hydrostatic equilibrium.

The motion of the parcel can be regarded as adiabatic because it is very rapid and the heat conduction in air is relatively slow. Thus, the first law for the parcel can be written as

$$c_p dT = \alpha dp,$$

which under the hydrostatic assumption leads to a dry adiabatic lapse rate:

$$\gamma_d = -\left(\frac{dT}{dz}\right)_{ad} = \frac{g}{c_p},$$

i.e., approximately 1 °C/100 m.

If, after a small displacement in the vertical, the parcel tends to return to its initial position, the equilibrium is said to be *stable*. On the other hand, if it accelerates away from the initial position, the atmosphere is in an *unstable* equilibrium. Finally, the equilibrium is *neutral* when the net restoring force is zero.

The parcel in its motion is subject to gravity and pressure gradient forces, i.e.,

$$\frac{dw}{dt} = -g - \frac{1}{\rho}\frac{\partial p}{\partial z},$$

whereas the surrounding air satisfies the hydrostatic equilibrium condition

$$0 = -g - \frac{1}{\rho_A}\frac{\partial p}{\partial z},$$

where ρ and ρ_A are the density of the parcel and the environmental air, respectively. It is assumed that at each instant the pressure acting on the parcel is the same as the pressure of the environment at the same level ($p = p_A$). Thus, combining the two equations, we obtain

$$\frac{dw}{dt} = g(\rho_A - \rho)/\rho,$$

where $g(\rho_A - \rho)$ is the buoyancy force.

In terms of temperature, assuming again that the air is an ideal gas, the equation can be written

$$\frac{dw}{dt} = g(T - T_A)/T_A.$$

If the parcel starts from a level z_0, where its temperature is equal to the environmental temperature $T_A(z_0)$, the final temperature after moving a small distance z becomes

$$T(z) = T_A(z_0) - \gamma_d z,$$

where γ_d is again the dry adiabatic lapse rate. On the other hand, the temperature of the environmental air is

$$T_A(z) = T_A(z_0) - \gamma z,$$

where $\gamma = -\partial T_A/\partial z$ is the environmental lapse rate.

Thus, the above equation becomes

$$\frac{dw}{dt} = gz(\gamma - \gamma_d)/T_A. \qquad (3.50)$$

From this equation, we conclude that for upward ($z > 0$) or downward ($z < 0$) displacements the equilibrium is stable when $\gamma < \gamma_d$, neutral when $\gamma = \gamma_d$, and unstable when $\gamma > \gamma_d$. This criterion can also be expressed in terms of the vertical gradient of potential temperature. In fact, differentiating Eq. (3.47) logarithmically with respect to z and using $T = T_A$, Eq. (3.47) becomes

$$\frac{1}{\theta}\frac{\partial \theta}{\partial z} = \frac{1}{T_A}\frac{\partial T_A}{\partial z} - \frac{\kappa}{p}\frac{\partial p}{\partial z}$$

or using the definitions for γ and γ_d:

$$\frac{T_A}{\theta}\frac{\partial \theta}{\partial z} = -\gamma + \gamma_d. \qquad (3.51)$$

Thus, when θ increases with height ($\partial \theta/\partial z > 0$), the atmosphere is in a statically stable equilibrium (warm air over cold air). When $\partial \theta/\partial z < 0$, the atmosphere is unstable (cold air on top of warm air), and when $\partial \theta/\partial z = 0$ the atmosphere is in neutral equilibrium. It is also convenient to define the static stability as the quantity $(1/\theta)\partial \theta/\partial z$ because in a stable atmosphere the restoring force acting on the parcel moving vertically is given simply by

$$gz(\gamma - \gamma_d)/T_A = -gz\left(\frac{\partial \theta}{\partial z}\right)\Big/\theta.$$

Using the hydrostatic approximation and the (x,y,p,t) coordinate system, expression (3.51) for the static stability converts into

$$\frac{T_A}{\theta}\frac{\partial \theta}{\partial p} = \frac{\partial T_A}{\partial p} - \kappa\frac{T_A}{p}$$

$$= \frac{\alpha}{g}(\gamma - \gamma_d). \qquad (3.52)$$

When in a stably stratified atmosphere a parcel is displaced from its equilibrium level, an oscillatory motion about this level will result. In fact, the restoring force is proportional to the vertical displacement so that

$$\frac{d^2z}{dt^2} + \left(\frac{g}{\theta}\frac{\partial \theta}{\partial z}\right)z = 0.$$

The solution of this equation leads to a sinusoidal variation in z with a frequency (in units of rad s^{-1}) given by

$$N = \left(\frac{g}{\theta}\frac{\partial \theta}{\partial z}\right)^{1/2}, \qquad (3.53)$$

called the Brunt–Väisälä frequency.

3.4.3 Potential vorticity

Let us assume that the motion is adiabatic. Then a parcel of air remains on the same potential temperature surface and any isentropic surface is a material surface. The air contained between two isentropic surfaces remains confined to the same isentropic layer, which nevertheless can expand or contract. The thickness of this layer may be evaluated in terms of the pressure difference between the boundary isentropic surfaces. On the other hand, since on isentropic surfaces the pressure p and the specific volume α are related through the expression $\alpha p^{1-\kappa} = \text{const}$ (grad $\alpha \times$ grad $p = 0$), the isentropic surfaces do not intersect any isosteric–isobaric solenoids. Therefore, by Bjerknes theorem the motions along isentropic surfaces do not change the circulation Γ_a on these surfaces. A solenoid tube is a tube limited by two isobaric and two isosteric surfaces (see Sec. 3.5.3).

Now, if we combine the vorticity equation (3.40) with the continuity equation in the p system written in the form

$$\text{div } \mathbf{v} = -\frac{1}{\delta p}\frac{d}{dt}\delta p,$$

we obtain

$$\frac{1}{(\zeta+f)}\frac{d}{dt}(\zeta+f) - \frac{1}{\delta p}\frac{d}{dt}\delta p = 0,$$

or

$$\frac{d}{dt}\left(\frac{\zeta+f}{\delta p}\right) = 0. \tag{3.54}$$

The quantity in parentheses is called *potential vorticity*. This equation shows that for adiabatic and frictionless motion the potential vorticity of a fluid is conserved, i.e.,

$$\frac{\zeta+f}{\delta p} = \text{const.} \tag{3.55}$$

The "thickness" of the column δp limited by the surfaces θ and $\theta + \delta\theta$ can be expressed in terms of $\delta\theta$, since

$$\delta p = \frac{\partial p}{\partial \theta}\delta\theta$$

so that the potential vorticity as $\delta\theta$ remains invariant becomes

$$(\zeta+f)\frac{\partial \theta}{\partial p} = \text{const.} \tag{3.56}$$

This expression stresses the importance of the static stability $(\partial\theta/\partial p)$ in the dynamics of the atmosphere. According to this equation, the absolute vorticity as measured on a θ surface must increase if the stability decreases, and vice versa.

For a homogeneous fluid the potential vorticity reduces to

$$\frac{(\zeta+f)}{\delta z} = \text{const.}$$

As we can see the potential vorticity assumes different forms but none of them has even the dimension of vorticity. In essence, it constitutes a measure of the ratio of absolute vorticity to the effective depth of the vortex.

The last equation can illustrate the effect of vortex stretching or shrinking on the absolute vorticity. When a rotating atmospheric column remains at the same latitude (i.e., f remains the same), only changes in the height δz of the column can affect the relative vorticity. In the Northern Hemisphere, shrinking of the column (subsidence) is then associated with divergent flow at low levels, which the Coriolis force transforms into a clockwise rotation, i.e., negative relative vorticity. Vertical stretching of the column causes inward motion at low levels, which the Coriolis force transforms into counterclockwise circulation, i.e., a positive relative vorticity. For example, vortex stretching is the mechanism which explains the formation of lee-side depressions when a westerly air flow passes over a mountain range.

In general, the variation of f must also be taken into account. Since for a poleward displacement f increases, negative vorticity is developed to keep the potential vorticity invariant. Similarly an equatorward displacement of a column of constant depth will cause a positive vorticity to develop.

3.4.4 The thermodynamic energy equation and the local rate of change of temperature

Let us again consider the thermodynamic energy equation in the form (3.42). If we note that $dp/dt \equiv \omega$ this equation can also be written in the form

$$\frac{dT}{dt} = \kappa \frac{T}{p} \omega + \frac{Q}{c_p}. \tag{3.57}$$

This equation shows that the time rate of change of temperature of an individual parcel of air as it moves through space is due to adiabatic expansion or compression and to diabatic heating or cooling. This form of the equation is the appropriate one to use in problems where it is necessary to follow parcels that have some special properties, but for most problems in the atmosphere it is not necessary to keep track of individual air parcels. It is sufficient to know the distribution of certain quantities as continuous functions of the spatial coordinates and of time. It is therefore convenient to have a new version of the previous equation expanding the total derivative of temperature in terms of the local time rate of change and the advection like in expansion (3.5). The resulting equation is then

$$\frac{\partial T}{\partial t} = -\mathbf{v}\cdot\text{grad } T + \omega\left(\kappa \frac{T}{p} - \frac{\partial T}{\partial p}\right) + Q/c_p, \tag{3.58}$$

where $\mathbf{v}$ is the two-dimensional horizontal velocity vector (u,v). This equation shows that the local time rate of change of temperature depends on three main factors: the diabatic heating or cooling (Q/c_p), the horizontal advection ($-\mathbf{v}\cdot\text{grad } T$), and a vertical motion term containing the static stability. The term in parentheses is the static stability [see Eq. (3.52)], which strongly affects the vertical motion. Thus, when $\gamma < \gamma_d$ (the stable case), ascending motion results in local cooling and descending motion in local warming. The more stable the equilibrium, the larger the local rise of temperature resulting from a given rate of downward motion ω. When $\gamma > \gamma_d$ (unstable conditions), ascending motion will lead to local heating and descending motion to local cooling. Finally, when $\gamma = \gamma_d$ (neutral conditions), the vertical motions will not influence the temperature.

3.5 EQUATION OF STATE

In this section, we will give a general description of the equations of state for the atmosphere and oceans. We will regard them as two-component (binary) systems formed by dry air and water substance for the atmosphere, and liquid water and dissolved salts for the oceans. The atmosphere is considered mainly in the gaseous phase and the oceans as a diluted solution.

3.5.1 Atmosphere

3.5.1.1 Moist atmosphere

If we consider the air as a homogeneous, single-phase system (no clouds) we require, according to Gibb's phase rule, three independent variables to specify its thermodynamic state. We will assume that the various components behave as ideal gases and obey Dalton's law (i.e., the total pressure equals the sum of the partial pressures of the constituent gases, as if each gas occupied the total volume at the same temperature).

If the pressure and specific volume are given by p_d and α_d, the ideal gas equation for the dry air is

$$p_d \alpha_d = R_d T,$$

where R_d (about 287 J kg^{-1} K^{-1}) was defined earlier as the gas constant for dry air. The corresponding equation for water vapor is

$$e \alpha_v = R_v T,$$

where e is the pressure and α_v is the specific volume ($= 1/\rho_v$). The gas constant for water vapor R_v is related to R_d by

$$\frac{R_d}{R_v} = \frac{m_w}{m_d} = \frac{18.0}{28.9} = 0.622,$$

where m_d is the "apparent molecular weight" of dry air (i.e., the weighted mean of the molecular weights of the atmospheric gases, mainly nitrogen and oxygen; $m_d = 28.9$), and m_w is the molecular weight of water vapor ($m_w = 18.0$).

According to Dalton's law, the pressure p of the moist air is then given by

$$p = p_d + e.$$

If we combine the three previous equations we obtain for the air density ρ:

$$\rho = \rho_d + \rho_v$$

$$= \frac{p-e}{R_d T} + 0.622 \frac{e}{R_d T} = \frac{p}{R_d T}\left(1 - 0.378 \frac{e}{p}\right).$$

This equation for the moist air can be rewritten in the form of an ideal gas law:

$$p = \rho R_d T_V, \tag{3.59}$$

where

$$T_V = T/(1 - 0.378 e/p) \tag{3.60}$$

is called the *virtual temperature*. This fictitious temperature represents the temperature to which the dry air has to be raised in order to have the same density as the density of the moist air at the same pressure. Moist air is less dense than dry air so that the virtual temperature is always greater than the actual temperature. The thermodynamic state of the atmosphere is defined by any three of the four variables p, ρ, T, and e, since they are linked by one equation.

The amount of water vapor for a given volume of air may be expressed in different ways. Thus, we will define the specific humidity q as the ratio of the mass M_v of water vapor to the mass of the moist air M for the same volume of moist air:

$$q = M_v/M = \rho_v/\rho = 0.622 e/(p - 0.378 e) \approx 0.622 e/p. \tag{3.61}$$

Another commonly used humidity parameter is the relative humidity U, defined as the ratio of the actual specific humidity at a given temperature and pressure to the saturation specific humidity at the same temperature and pressure:

$$U = q/q_s = e/e_s, \qquad (3.62)$$

where e_s is the saturated vapor pressure. Usually the relative humidity is given as a percentage, and varies between 0% (dry air) and 100% (saturated air).

Expression (3.60) can now be rewritten using a series expansion in the usual form

$$T_V \simeq T(1 + 0.378e/p) = T(1 + 0.61q), \qquad (3.63)$$

where q is expressed in g of water vapor per kg of moist air. Thus, the equation of state for moist air (3.59) takes the usual form

$$p = \rho R_d T(1 + 0.61q). \qquad (3.64)$$

3.5.1.2 Clausius–Clapeyron equation

When air at a certain temperature is saturated ($U = 100\%$), the water pressure attains its maximum possible value e_s, and we can define the saturation specific humidity as $q_s = 0.622\, e_s/p$.

The saturation pressure varies with temperature according to the Clausius–Clapeyron equation (see, e.g., Wallace and Hobbs, 1977)

$$\frac{de_s}{dT} = \frac{L}{T(\alpha_2 - \alpha_1)}, \qquad (3.65a)$$

where L is the latent heat of the phase transition and α_1 and α_2 are the specific volumes of the two phases.

For the evaporation and sublimation, in which the specific volume of the vapor is much larger than the specific volume of the condensed phase, we can write ($\alpha_2 \gg \alpha_1$)

$$\frac{de_s}{dT} \approx \frac{L}{T\alpha_v} = 0.622 \frac{L e_s}{R_d T^2}. \qquad (3.65b)$$

This is the expression commonly used in meteorological applications, such as numerical modeling. The integration of Eq. (3.65b) leads to

$$e_s \propto \exp\left(-0.622 \frac{L}{R_d T}\right), \qquad (3.66)$$

i.e., the saturation vapor pressure increases exponentially with increasing temperature. This is the reason why warm air generally contains much more water vapor than cold air.

3.5.1.3 Adiabatic processes in saturated air

The adiabatic lifting of saturated air will lead to condensation of the water vapor with the release of latent heat $-L dq_s$, where $-dq_s$ is the amount of water vapor condensed. The change of phase is associated with a variation of entropy $ds = -L dq_s/T$. From Eq. (3.48) we see that

$$ds = c_p \frac{d\theta}{\theta}.$$

Thus

$$-d\left(\frac{Lq_s}{c_p T}\right) = \frac{d\theta}{\theta}$$

or

$$\theta = \theta_e \exp\left(-\frac{Lq_s}{c_p T}\right),$$

where θ_e is a constant of integration that can be obtained for the limit $q_s/T \to 0$. The quantity θ_e is called the equivalent potential temperature which is the potential temperature of an air parcel when all moisture is condensed and the latent heat released is used to warm the parcel:

$$\theta_e = \theta \exp\left(\frac{Lq_s}{c_p T}\right). \tag{3.67}$$

For saturated, air the static stability can be expressed in terms of $\partial \theta_e/\partial z$ as we did in Sec. 3.4.2 for dry air. Thus, when $\partial \theta_e/\partial z > 0$, the saturated air is stable, when $\partial \theta_e/\partial z = 0$, it is neutral, and when $\partial \theta_e/\partial z < 0$, it is unstable.

When the lapse rate γ is between the dry adiabatic and saturated lapse rates, $\gamma_s < \gamma < \gamma_d$, the equilibrium is conditionally unstable.

Let us consider the stability of a layer when it is lifted to saturation. If the layer is stable after lifting to saturation it is said to be potentially stable, and if it becomes unstable the layer is potentially unstable. In the first case, the equivalent temperature in the layer increases with height ($\partial \theta_e/\partial z > 0$) and in the second case, it decreases ($\partial \theta_e/\partial z < 0$), as it does for saturated air. The equivalent potential temperature is conserved during adiabatic processes involving saturated or dry air.

3.5.2 Oceans

If we consider the oceans as a binary system consisting of a homogeneous mixture of water and dissolved salts and assume that it is a monophasic, liquid system, we find, according to Gibb's phase rule, that the oceans have three degrees of freedom. Thus the thermodynamic state is specified by three independent intensive variables, e.g., temperature, salinity (S), and pressure, and the equation of state for the oceans is of the form:

$$\rho = \rho(T, S, p). \tag{3.68}$$

Because the observed values of density range between about 1000 and 1040 kg m^{-3} it is sometimes convenient to express the density in terms of a quantity

$$\sigma(T, S, p) \equiv \rho(T, S, p) - \rho_0,$$

where $\rho_0 = 1000$ kg m^{-3}.

However, more often a quantity σ_t is used as defined by

$$\sigma_t(T, S) \equiv \rho(T, S, p_0) - \rho_0, \tag{3.69}$$

which is evaluated at the surface pressure p_0, and therefore does not include the effects of pressure.

Unlike for the atmosphere, an analytic form of the equation of state has not been found for the oceans because of the complex chemical composition of sea water and because no general expression for the equation of state of a liquid exists. In case of the oceans, the equation of state has to be established empirically (see, e.g., Fofonoff, 1962; Levitus, 1982) using polynomial expressions of T, S, and p with the coefficients of thermal expansion, saline contraction, and isothermal compressibility. Thus, the density cannot be expressed completely in terms of temperature, salinity, and pressure.

As in the atmosphere, it is sometimes convenient to use the potential temperature θ instead of the *in situ* temperature T. Because sea water is compressible, a parcel coming to the surface will expand and thus will tend to cool. We can define the potential temperature as the temperature a fluid element would attain if it were brought to the surface adiabatically and without exchange of salt with the environment. This temperature is important in accounting for the effects of compressibility when water masses at different depths are compared or when vertical displacements are considered:

$$\theta(z) = T(z) - \int_{p_0}^{p(z)} \left(\frac{\partial T}{\partial p}\right)_{ad} dp, \tag{3.70}$$

where $(\ )_{ad}$ indicates an adiabatic process without exchange of heat and salt with the environment.

The potential density is defined in an analogous way by the expression

$$\rho_\theta(z) = \rho(z) - \int_{p_0}^{p(z)} \left(\frac{\partial \rho}{\partial p}\right)_{ad} dp = \rho(z) - \int_{p_0}^{p(z)} c_s^{-2} dp \tag{3.71}$$

and can be interpreted as the density of a fluid element if it were brought up to the reference level p_0, i.e., the ocean surface, adiabatically and without changing its salinity. The quantity c_s denotes the speed of sound and is given by the well-known relation

$$c_s^2 = \left(\frac{\partial p}{\partial \rho}\right)_{ad}. \tag{3.72}$$

Because of the very large variations in pressure in the vertical and the nonlinear dependence of the density on T and S, we will make use of a local reference level instead of the ocean surface in order to compute a representative measure of the static stability in the deep ocean. Thus, the local potential density $\rho_{\theta L}(z)$ is defined as the density a parcel of water would attain if it were brought up adiabatically and without change in its salinity from level z to level $z + \delta z$:

$$\rho_{\theta L}(z) = \rho(z) - \int_{p(z+\delta z)}^{p(z)} c_s^{-2} dp. \tag{3.73}$$

In order to analyze the equilibrium conditions and the static stability in the oceans, we will make use of the parcel method, following a procedure similar to the one used before in the atmosphere. Let us consider a vertical displacement δz of a "parcel" of water from a level z to the level $z + \delta z$ when no heat and salt are exchanged with the environment. The resulting buoyancy force at the level $z + \delta z$ will be $F = g\rho(z + \delta z) - g\rho_{\theta L}(z)$ where $\rho(z + \delta z)$ is the density of the environmental water and $\rho_{\theta L}(z)$ is the final density of the parcel after displacement from level z to level $z + \delta z$. The acceleration at the level $z + \delta z$ will then be

$$\frac{d^2\delta z}{dt^2} = \frac{g}{\rho(z)}\{\rho(z + \delta z) - \rho_{\theta L}(z)\} \equiv g\frac{\delta \rho}{\rho(z)}, \tag{3.74}$$

where $\delta \rho$ indicates not a geometric difference in density but a difference defined by expression (3.74); it is used as a measure of the local stability of a parcel (Neumann and Pierson, 1966, p. 139). Expressing ρ and $\rho_{\theta L}$ in terms of the initial density $\rho(z)$ and using Eq. (3.73), we find

$$\frac{d^2\delta z}{dt^2} = g\left(\frac{1}{\rho}\frac{\partial \rho}{\partial z} + \frac{g}{c_s^2}\right)\delta z \equiv \frac{g}{\rho}\frac{\delta \rho}{\partial z}\delta z \tag{3.75}$$

$$= -gs^*\delta z.$$

The term g/c_s^2 gives the correction due to the compressibility of sea water. The expression in parentheses equals the negative of the static stability

$$s^* = -\frac{1}{\rho}\frac{\partial \rho}{\partial z} - \frac{g}{c_s^2} = -\frac{1}{\rho}\frac{\delta \rho}{\partial z}. \tag{3.76}$$

As for the atmospheric case, Eq. (3.75) is of the form

$$\frac{d^2\delta z}{dt^2} + N^2\delta z = 0, \tag{3.77}$$

where N is the buoyancy frequency or the Brunt–Väisälä oscillation frequency (in units of rad s^{-1}) defined by the expression

$$N^2 = -g\left(\frac{1}{\rho}\frac{\partial \rho}{\partial z} + \frac{g}{c_s^2}\right) = -\frac{g}{\rho}\frac{\delta \rho}{\partial z} = gs^*. \tag{3.78}$$

In a stably stratified column the term $-(1/\rho)(\partial \rho/\partial z)$ is greater than the compression term g/c_s^2 so that N^2 becomes positive and N real. In this case, the solution of Eq. (3.77) is oscillatory and it represents buoyancy oscillations. On the other hand, when $-(1/\rho)(\partial \rho/\partial z)$ is smaller than g/c_s^2, the value of N^2 becomes negative and N imaginary. This leads to unstable conditions where the water parcel continues its rising or sinking motion. Finally, when $-(1/\rho)(\partial \rho/\partial z) = g/c_s^2$, N^2 vanishes and the equilibrium is neutral. In this case, the parcel motion is neither accelerated nor decelerated. Thus, we find that the quantity N^2 represents a useful measure of the stability of the water column with respect to small vertical displacements.

3.5.3 Barotropy and baroclinicity

The vertical distribution of mass in the atmosphere can be represented by a collection of constant specific volume surfaces separated by layers of one specific volume unit thickness. These surfaces ($\alpha = 1/\rho =$ constant) are referred to as *isosteric* surfaces. Analogously, the vertical pressure distribution can be represented by a collection of constant pressure (isobaric) surfaces separated by layers of one pressure unit thickness. Since the slope of the isobaric surfaces is very small, they may be considered to be almost horizontal. In general, the two sets of surfaces intersect in space defining a continuous family of unit isosteric-isobaric tubes, the so-called isosteric-isobaric *solenoids*. A cross section is presented in Fig. 3.2a, in which each tube is

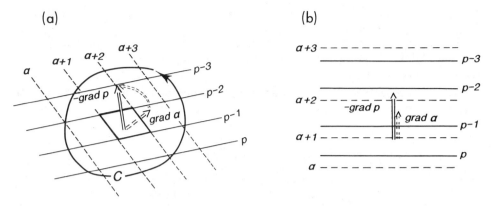

FIGURE 3.2 Schematic diagram showing isosteric-isobaric solenoids (a) enclosed by curve C in a baroclinic atmosphere. In (b) the atmosphere is barotropic and the isosteric and isobaric surfaces are parallel with no solenoids.

represented by a quadrilateral. The number of tubes per unit area expresses the intensity of the solenoidal field. From the geometrical meaning of the cross product we may say that this number is given by the magnitude of the baroclinicity vector **N**. This vector is defined by

$$\mathbf{N} = \text{grad } \alpha \times (-\text{grad } p)$$

which as we have seen in (3.36) equals

$$\mathbf{N} = -\text{curl}(\alpha \text{ grad } p)$$

since curl grad $p = 0$.

In synoptic applications, the isosteric-isobaric solenoids are more easily seen in terms of temperature and pressure so that

$$\mathbf{N} = -\text{grad } T \times \frac{R}{p} \text{ grad } p$$

or, in terms of potential temperature and temperature,

$$\mathbf{N} = -c_p \text{ grad}(\ln \theta) \times \text{grad } T = -\text{grad } s \times \text{grad } T,$$

where s is the entropy (see Eq. 3.48). In these cases, the α, ρ, T, or θ surfaces are "inclined" with respect to the p surfaces and the stratification of the atmosphere is called *baroclinic*. The number of solenoids per unit area measures the degree of baroclinicity. On the other hand, when $\mathbf{N} = 0$ (no solenoids), both sets of surfaces are parallel (see Fig. 3.2b). This state of stratification is called *barotropic*.

In summary, we may say that atmosphere is barotropic when surfaces of constant specific volume (density, temperature or potential temperature) coincide with surfaces of constant pressure. Barotropy is then a state of zero baroclinicity. Mathematically speaking, the vectors grad α and grad p are then parallel and proportional to each other so that the specific volume (or density, temperature, potential temperature) is only a function of pressure as is the case in, e.g., an isothermal or adiabatic atmosphere (see Eq. 3.47). When the atmosphere remains barotropic in time the atmosphere is called *autobarotropic*. However, as Eqs. (3.64) and (3.68) show, the atmosphere and oceans are, in general, baroclinic since the density depends on temperature as well as on pressure and specific humidity or salinity.

We make use of these basic concepts throughout the book. They are important, e.g., in the vorticity (Secs. 3.3.3 and 3.4.3), the vertical wind shear (Sec. 7.4.3) and the energetics (Secs. 13.3.3.5 and 14.1). As we will see, the vertical wind shear (thermal wind) is proportional to the horizontal component of the baroclinicity vector **N**. Furthermore, from the Bjerknes theorem, we know that the acceleration of the absolute circulation along a curve is equal to the number of solenoids enclosed by the curve (see Sec. 7.4.3).

In the oceans, it is more common to use the potential density with relation to pressure as an indicator of baroclinicity, but the conclusions are similar to those for the atmosphere.

3.6 EQUATION OF WATER VAPOR

The balance equation for water vapor can be written in the form

$$\frac{dq}{dt} = s(q) + D, \tag{3.79}$$

where q is the specific humidity, $s(q)$ represents the sources and sinks of water vapor for the parcel of air considered, and $D = -\alpha \operatorname{div} \mathbf{J}_q^D$ the molecular and turbulent eddy diffusion of water vapor into the volume. Diffusion can usually be neglected. The term $s(q)$ can be represented by $e - c$, the difference between the rate of evaporation plus sublimation and the rate of condensation per unit mass. Usually the rate of condensation c equals the precipitation rate out of the volume plus the rate at which water or snow and ice is stored in the clouds. A frequent simplification made in modeling is that the water vapor falls out in the form of precipitation as soon as it condenses. In this pseudoadiabatic process, the equivalent potential temperature θ_e is conserved. The evaporation at the earth's surface will be included through the term $\mathbf{J}_q^D$.

3.7 SUMMARY OF THE BASIC EQUATIONS IN LAGRANGIAN AND EULERIAN FORM

The seven basic hydrodynamic and thermodynamic equations describing the behavior of the atmosphere are summarized below in Lagrangian form:

$$\frac{d\rho}{dt} = -\rho \operatorname{div} \mathbf{c};$$

$$\frac{du}{dt} = \frac{\tan \phi}{R} uv - \frac{uw}{R} + fv - f'w - \frac{1}{\rho} \frac{\partial p}{R \cos \phi \partial \lambda} + F_\lambda;$$

$$\frac{dv}{dt} = -\frac{\tan \phi}{R} u^2 - \frac{vw}{R} - fu - \frac{1}{\rho} \frac{\partial p}{R \partial \phi} + F_\phi;$$

$$\frac{dw}{dt} = \frac{u^2}{R} + \frac{v^2}{R} + f'u - \frac{1}{\rho} \frac{\partial p}{\partial z} - g + F_z;$$

$$c_p \frac{dT}{dt} = Q + \alpha \frac{dp}{dt};$$

$$\frac{dq}{dt} = s(q) + D;$$

$$p = \rho RT(1 + 0.61q).$$

These equations are of the general form

$$\frac{dA}{dt} = F(A, B, C, \ldots, x, y, z, t),$$

where $A, B, C, \ldots$ denote the climatic variables, and the function F the source and sink terms.

Due to the difficulties involved, we usually do not study the properties of a fluid following the motions of the individual particles according to a Lagrangian scheme. Instead, it is more convenient to use an Eulerian approach and to study the behavior of the fluid at a fixed point in space and time. This can be accomplished by combining the expansion

$$\frac{dA}{dt} = \frac{\partial A}{\partial t} + (\mathbf{c}\cdot\text{grad})A$$

with the equation of continuity (3.1). This leads to the general balance equation

$$\frac{\partial \rho A}{\partial t} = \rho \frac{dA}{dt} - \text{div}\,\rho A\mathbf{c}. \tag{3.80}$$

From Eq. (3.80) we see that the local rate of change of A per unit volume results from the generation or destruction of A, and the net inflow across the boundaries into the volume.

Using Eq. (3.80) the seven basic equations can be immediately written in a local, Eulerian framework, stressing their nonlinearity, as follows:

$$\frac{\partial \rho}{\partial t} = -\,\text{div}\,\rho\mathbf{c}; \tag{3.81}$$

$$\frac{1}{\rho}\frac{\partial \rho u}{\partial t} = -\frac{1}{\rho}\text{div}\,\rho u\mathbf{c} + \frac{\tan\phi}{R}uv - \frac{uw}{R} + fv - f'w - \frac{1}{\rho}\frac{\partial p}{R\cos\phi\partial\lambda} + F_\lambda; \tag{3.82a}$$

$$\frac{1}{\rho}\frac{\partial \rho v}{\partial t} = -\frac{1}{\rho}\text{div}\,\rho v\mathbf{c} - \frac{\tan\phi}{R}u^2 - \frac{vw}{R} - fu - \frac{1}{\rho}\frac{\partial p}{R\partial\phi} + F_\phi; \tag{3.82b}$$

$$\frac{1}{\rho}\frac{\partial \rho w}{\partial t} = -\frac{1}{\rho}\text{div}\,\rho w\mathbf{c} + \frac{u^2}{R} + \frac{v^2}{R} + f'u - \frac{1}{\rho}\frac{\partial p}{\partial z} - g + F_z; \tag{3.82c}$$

$$\frac{1}{\rho}\frac{\partial \rho T}{\partial t} = -\frac{1}{\rho}\text{div}\,\rho T\mathbf{c} + Q/c_p + \kappa\frac{T}{p}\frac{dp}{dt}; \tag{3.83}$$

$$\frac{1}{\rho}\frac{\partial \rho q}{\partial t} = -\frac{1}{\rho}\text{div}\,\rho q\mathbf{c} + (e - c) + D; \tag{3.84}$$

$$p = \rho RT(1 + 0.61q).$$

The first six of the seven basic equations [(3.1), (3.9a), (3.9b), (3.9c), (3.42), and (3.79)] in the Lagrangian framework and Eqs. (3.81)–(3.84) in the Eulerian framework are evolution or prognostic equations because they express the time derivatives of dependent variables in terms of their present values. On the other hand, the last equation, the equation of state, is a diagnostic equation because it does not contain time derivatives.

These equations form a closed system, provided that Q and $\mathbf{F}$ are known functions or depend on the other variables. With appropriate boundary conditions, one can then describe the evolution of the system starting from a given initial state. As mentioned in Chap. 2, knowledge of certain basic parameters, such as the rotation rate of the earth, the orbital characteristics, the chemical composition, and gravity, is of course also required. The boundary conditions will specify the mechanical and thermal interactions between the atmosphere and the other components of the climatic system, such as the oceans, the continents, and the incoming and outgoing radiation.

The application of the general conservation principles of physics to the climatic system in later chapters will allow us to derive certain requirements that govern the behavior of the climatic system. It will also enable us to describe the main features of the general circulations and to explain some of the processes that are responsible for the observed phenomena. The principles that we will use are based on the conservation of mass, angular momentum, energy, and water substance.

CHAPTER 4
Various Decompositions of the Circulation

The atmospheric and oceanic circulations exhibit some regularities when regarded on the large scale. However, on a regional basis they also show anomalies that require analysis and quantification. The anomalies occur both in time and space. In this chapter we will develop some techniques that permit the study of these anomalies and of their contributions to the general circulation.

4.1 TRANSIENT AND STATIONARY EDDIES

4.1.1 Time and horizontal resolutions of the circulation

The fields that characterize the state of the atmosphere and oceans are highly variable in time and space. However, the climate is defined, to a large extent, by the average conditions which suggests the use of averages over certain time intervals, such as monthly, seasonal, and yearly averages.

The understanding of the general circulation of the atmosphere requires the study of the statistics of various orders for the important meteorological variables. Following Starr and White (1954), these statistics can be defined both in space and time, and in a mixed space–time domain. We represent the time average of a quantity A for a specific interval τ by an overbar:

$$\overline{A} \equiv \frac{1}{\tau} \int_0^\tau A \, dt \tag{4.1}$$

and denote the deviation from the average by A'. The instantaneous value of A is given by

$$A = \overline{A} + A'. \tag{4.2}$$

Of course, $\overline{A'} = 0$.

The average product of two quantities A and B is given by

$$\overline{AB} = \overline{(\overline{A}+A')(\overline{B}+B')}$$
$$= \overline{A}\,\overline{B} + \overline{A'B'}. \tag{4.3}$$

In this expression, $\overline{A'B'}$ is the covariance of A and B in time, and we define

$$\overline{A'B'} = r(A,B)\,\sigma(A)\,\sigma(B),$$

where r is the temporal correlation coefficient and σ the temporal standard deviation. The mean of the product of A and B is equal to the product of the means only when the quantities A and B vary independently of each other in time [i.e., when $r(A,B)$ vanishes].

Fields such as wind velocity or temperature are also not uniform in space, varying both as a function of latitude and longitude. However, meteorological conditions are generally more uniform along a latitude circle than in the north–south direction. To a first approximation, we may assume a "zonally" symmetric distribution with respect to the axis of rotation, as would be expected from the distribution of the daily averaged incoming solar radiation. Thus, it is convenient to define zonal-mean values at each latitude circle in order to assess the north–south variability. This can be achieved by introducing a zonal-average operator defined by

$$[A] \equiv \int_0^{2\pi} A\,d\lambda/2\pi \tag{4.4}$$

and the departure from this average A^*, so that

$$A = [A] + A^* \tag{4.5}$$

and, of course, $[A^*] = 0$.

Combining the space and time expansions we obtain

$$A = [\overline{A}] + \overline{A}^* + [A]' + A'^*.$$

The term $[\overline{A}]$ represents the zonally symmetric part of the steady time-average quantity, e.g., easterly trade winds at low latitudes and the westerly winds at midlatitudes. The second term, $\overline{A}^* = \overline{A} - [\overline{A}]$ gives the asymmetric part of the time-average quantities, such as the monsoon circulations and the longitudinal land–sea temperature contrast. The term $[A]' = [A] - [\overline{A}]$ represents the instantaneous fluctuations of the symmetric part, such as the fluctuations of the zonal-mean circulation (e.g., the index cycle). The last term A'^* indicates the instantaneous, zonally asymmetric part, such as the traveling low- and high-pressure systems shown on weather maps.

For the horizontal vector velocity field $\mathbf{v} = u\mathbf{i} + v\mathbf{j}$, we can write

$$\mathbf{v} = \overline{\mathbf{v}} + \mathbf{v}' = [\overline{\mathbf{v}}] + \overline{\mathbf{v}}^* + [\mathbf{v}]' + \mathbf{v}'^*,$$

where $\overline{\mathbf{v}}$ represents the stationary and v' the transient motions. The two components of $[\overline{\mathbf{v}}]$, i.e., $[\overline{u}]$ and $[\overline{v}]$, are known as the mean zonal and meridional circulations, $\overline{\mathbf{v}}^*$ the stationary (sometimes called standing) eddy circulations, $[\overline{\mathbf{v}}]'$ the transient zonal-mean circulations, and $\mathbf{v}'^*$ the transient asymmetric circulations.

The zonal average product of A and B is given by

$$[AB] = [A][B] + [A^*B^*]. \tag{4.6}$$

In this expression, $[A^*B^*]$ is the zonal or spatial covariance of A and B, and we again define

$$[A^*B^*] = r^*(A,B)\,\sigma^*(A)\,\sigma^*(B),$$

where r^* is now the spatial correlation coefficient and σ^* the spatial standard deviation.

In many cases, we need to consider time- and zonal-mean quantities of the form
$$[\overline{AB}].$$
Since the two operators are permutable,
$$[\overline{AB}] = \overline{[AB]}.$$
Let us first consider the zonal averages:
$$AB = [A][B] + A^*[B] + [A]B^* + A^*B^*,$$
so that
$$[AB] = [A][B] + [A^*B^*].$$
But
$$[A] = \overline{[A]} + [A]' \text{ and } [B] = \overline{[B]} + [B]',$$
so that
$$[AB] = \overline{[A]}\,\overline{[B]} + \overline{[A]}[B]' + [A]'\overline{[B]} + [A]'[B]' + [A^*B^*]$$
or, taking the time average,
$$\overline{[AB]} = \overline{[A]}\,\overline{[B]} + \overline{[A]'[B]'} + \overline{[A^*B^*]}. \tag{4.7}$$
Let us give the northward flux of sensible heat as an example to illustrate the meaning of the various terms in Eq. (4.7). Using $A = v$ and $B = c_p T$ in Eq. (4.7), we find
$$c_p\,\overline{[vT]} = c_p\,\overline{[v]}\,\overline{[T]} + c_p\,\overline{[v]'[T]'} + c_p\,\overline{[v^*T^*]}. \tag{4.8}$$
The total northward transport of heat on the left-hand side of the equation is decomposed into three terms: the transports by the (steady) mean meridional circulation, the transient mean meridional circulation, and the spatial eddy circulations, respectively. This expansion is very convenient in diagnosing long-time integrations of numerical general circulation models because it provides an efficient way of accumulating the statistics during the course of the integration.

Let us now analyze $[\overline{AB}]$. We have
$$\overline{AB} = \overline{A}\,\overline{B} + \overline{A'B'}$$
$$= [\overline{A}][\overline{B}] + [\overline{A}]\overline{B}^*$$
$$+ \overline{A}^*[\overline{B}] + \overline{A}^*\overline{B}^* + \overline{A'B'}$$
so that
$$[\overline{AB}] = [\overline{A}][\overline{B}] + [\overline{A}^*\overline{B}^*] + [\overline{A'B'}]. \tag{4.9}$$
Again taking the northward flux of sensible heat as an example, we may substitute $A = v$ and $B = c_p T$ in Eq. (4.9) leading to the more common and useful expansion in observational studies of the general circulation:
$$c_p[\overline{vT}] = c_p[\overline{v}][\overline{T}] + c_p[\overline{v}^*\overline{T}^*] + c_p[\overline{v'T'}]. \tag{4.10}$$
As before, the total heat flux is decomposed into three but somewhat different transport terms: terms associated with the mean meridional circulation, the stationary eddies, and the transient eddies, respectively.

Let us compare expansions (4.7) and (4.9). The left-hand sides are equal, as are the first terms on the right-hand side, so that the sums of the remaining two terms must be equal. However, individually they are very different, both computationally and physically.

By taking the sums of the last two terms and further expanding

$$\overline{[A]'[B]'} + \overline{[A^*B^*]} = \overline{[A]'[B]'} + [\overline{A}^*\overline{B}^*] + \overline{[A'^*B'^*]},$$

$$[\overline{A}^*\overline{B}^*] + [\overline{A'B'}] = [\overline{A}^*\overline{B}^*] + \overline{[A]'[B]'} + [\overline{A'^*B'^*}],$$

we see that the expressions are indeed the same. In our example of the northward flux of sensible heat, the two last terms in Eqs. (4.8) and (4.10) can be written as the sum of three terms:

$$c_p \overline{[v]'[T]'} + c_p [\bar{v}^*\bar{T}^*] + c_p [\overline{v'^*T'^*}].$$

These terms represent the transports associated with the transient meridional circulations, the stationary eddies, and the transient asymmetric eddies, respectively. We should note that the transient as well as the stationary fluxes depend on the chosen averaging period.

Expansions (4.7) and (4.9) can be simplified when $A = B$:

$$\overline{[A^2]} = \overline{[A]}^2 + \overline{[A]'^2} + \overline{[A^{*2}]} \tag{4.11}$$

and

$$[\overline{A^2}] = [\overline{A}]^2 + [\overline{A}^{*2}] + [\overline{A'^2}]. \tag{4.12}$$

If we substitute $A = \mathbf{v}$ in Eq. (4.11) or (4.12) we find how much the various spatial and temporal components contribute to the total wind variance or, in other words, to the kinetic energy:

$$[\overline{\mathbf{v}\cdot\mathbf{v}}] = [\bar{\mathbf{v}}]\cdot[\bar{\mathbf{v}}] + [\bar{\mathbf{v}}^*\cdot\bar{\mathbf{v}}^*] + \overline{[\mathbf{v}]'\cdot[\mathbf{v}]'} + [\overline{\mathbf{v}'^*\cdot\mathbf{v}'^*}]. \tag{4.13}$$

Besides the computational and statistical importance of these expressions for studies of the general circulation, they also have a deeper physical meaning when interpreted in terms of the circulations. Indeed, as we will show later these expansions permit us to identify the mechanisms responsible for the various modes of transport.

4.1.2 Vertical resolution of the circulation

The various meteorological quantities also vary with height or pressure. We may define an operator for the mass-weighted vertical average of A:

$$\langle A \rangle \equiv \int_0^\infty \rho A \, dz \Big/ \int_0^\infty \rho \, dz,$$

or in pressure coordinates

$$\langle A \rangle = \int_0^{p_0} A \, dp \Big/ p_0. \tag{4.14}$$

We denote the departure from the vertical mean by A'', so that

$$A = \langle A \rangle + A'' \tag{4.15}$$

and $\langle A'' \rangle = 0$.

For the wind field, we then have

$$\mathbf{v} = \langle \mathbf{v} \rangle + \mathbf{v}'',$$

where $\langle \mathbf{v} \rangle$ is the "barotropic" component of the motion and $\mathbf{v}''$ the "baroclinic" component. In a barotropic atmosphere, $\mathbf{v}$ is constant with height [see thermal wind equation (7.6)] so that $\mathbf{v} = \langle \mathbf{v} \rangle$ and $\mathbf{v}'' = 0$.

The average product of two quantities A and B is given by

$$\langle AB \rangle = \langle (\langle A \rangle + A'')(\langle B \rangle + B'') \rangle$$
$$= \langle A \rangle \langle B \rangle + \langle A''B'' \rangle. \qquad (4.16)$$

In this expression $\langle A''B'' \rangle$ is the covariance of A and B in the vertical.

Simultaneously applying the time and vertical mean operators to the product of A and B, we obtain

$$\overline{\langle AB \rangle} = \overline{\langle A \rangle}\, \overline{\langle B \rangle} + \overline{\langle A''\overline{B}'' \rangle} + \overline{\langle A \rangle'\langle B \rangle'} + \overline{\langle (A'')'(B'')' \rangle}. \qquad (4.17)$$

The first two terms on the right-hand side represent the time-average barotropic and baroclinic contributions, respectively, whereas the last two terms are due to transient barotropic and baroclinic perturbations, respectively.

For the oceans, similar expansions in time and space can be used, with the exception of the zonal average operation. Since the integration along a latitude circle is now interrupted by the presence of continents, the zonal average has to be replaced by a sum over the contributions by the individual oceans:

$$[A] \equiv \sum_i \frac{1}{L_i} \int_{\lambda_1}^{\lambda_2} A\, R\, \cos \phi\, d\lambda, \qquad (4.18)$$

where L_i represents the length of the ith ocean sector along the latitude ϕ.

4.2 SPECTRAL ANALYSIS OF METEOROLOGICAL FIELDS

4.2.1 Spectral analysis in space and time

Spectral analysis has been extensively used in many domains of meteorology. Among these we may refer to the study of turbulence in the surface boundary layer, time series analysis of local meteorological quantities, numerical weather prediction, and the study of the general circulation of the atmosphere.

In Appendix A, we will show in more detail why the use of the spectral analysis is not only justifiable but constitutes an adequate and natural method for studying the general circulation of the atmosphere (e.g., Saltzman, 1957; Hayashi, 1982). The Fourier transform in time leads to the spectral components in the frequency domain, whereas the Fourier transform in space leads to the spectral components in the wave number domain. In Appendix A, we confine ourselves mainly to the spectral analysis in the space domain. Formally, the mathematical treatment is similar in both domains so that it is only necessary to substitute time t for longitude λ, frequency ω for wave number k, and period T for wavelength L.

The configurations of most of the meteorological fields on isobaric levels are almost symmetric with respect to the earth's axis of rotation. This explains the common practice in general circulation studies of using a decomposition of the meteorological field into a zonal-mean field around latitude circles and an irregularly varying eddy field. As we have seen in the previous section, this approach has led to the use of generalized versions of the classic Reynolds expansion for turbulent flows in the study of the various processes which occur in the atmosphere. However, the Reynolds formulation does not give much information as to the nature and behavior of the individual eddies in the atmosphere. These eddies result from the superposition of disturbances with many different scales and frequencies. Thus, to understand the phenomena involved we need to obtain a further resolution by separating the different scales and by identifying each one with a physical process whenever possible.

The undulatory nature of the distribution of the meteorological variables (e.g., the temperature and the moisture content of the atmosphere) and the wave-like character of the flow suggest immediately the resolution into a spectrum of spatial scales or modes by one-dimensional Fourier analysis along latitude circles.

This method of resolution permits the separation of the eddies into simple wave disturbances. Furthermore, from a mathematical point of view, this type of resolution appears to be particularly convenient due to the cyclic continuity along a latitude circle. Other formulations are, of course, possible, such as a spatial expansion in spherical harmonics (Kubota, 1954; Bourke, 1988). It seems, however, that the zonal harmonic analysis, being much simpler to apply, will provide enough insight into the nature of the physical features that describe the processes occurring in the atmosphere.

In this new form of representation each perturbation is a simple sinusoidal curve, whose zonal wave number determines its scale; the amplitude gives its intensity and the phase angle defines its longitudinal location. Instead of the standard usage in physics where the wave number k is given by $k = 2\pi/L$ and L is the wavelength, we will follow here the more common meteorological nomenclature, where the wave number k indicates the number of waves along the latitude circle, i.e., $k = 2\pi R \cos \phi/L$.

The zonally averaged value corresponds to the component of wave number zero. Since the wave number represents the number of complete sine waves along a latitude circle, it follows that the scale of an eddy component is inversely proportional to the wave number. According to standard usage, the eddy wave components may be grouped into three main categories based on the wave number k; long waves for k between 1 and 5 (ultralong if $k < 3$); synoptic waves for k between 6 and 10; and short waves for $k \geqslant 11$. This classification is, of course, somewhat arbitrary.

Spectral analysis has become a very important tool for the study of the dynamical processes involved in the maintenance of the general circulation of the atmosphere and for the study of its energetics. It enables us to determine the scale and intensity of the dominant fluctuations at various isobaric levels and their seasonal variations through studying the amplitude and phase spectra. From the variance spectra, we can infer the behavior of the eddy wave components in the balance of such important quantities as kinetic energy and available potential energy (Saltzman, 1957). The cospectral analysis allows us to study the role of the various disturbances in achieving the net meridional transports of angular momentum, enthalpy, water vapor, and potential energy.

Frequently meteorological disturbances do not travel along latitude circles (see, e.g., Hoskins and Karoly, 1981; Hoskins *et al.*, 1983) and also lose their identity not too far away from their place of origin. In those cases, instead of a spectral analysis along a latitude circle, a more localized type of analysis is required, such as spatial correlation methods. For example, Wallace and Gutzler (1981) have used such techniques to reveal some of the dominant oscillation or so-called teleconnection patterns in the Northern Hemisphere.

4.2.2 Limitations of sampling

A meteorological function $f(x,t)$ is generally known only at discrete locations in space and time. Usually this function is uniformly sampled so that its values are given at equal intervals. When the time-dependent part, $f(x_0,t)$, is sampled only at regular intervals Δt, the continuous variable t has to be replaced by a discrete variable $t = j\Delta t$. Similarly, the space-dependent part, $f(x,t_0)$, can be discretized with an equal interval Δx, in which case x has to be replaced by the discrete variable $x = n\Delta x$. In each case, a certain amount of information is lost, since intermediate values are no longer represented. Intuitively, however, one has the feeling that the amount of information lost will become negligible if the sampling interval is sufficiently small. The question to be answered is then: How can we determine that interval so that the lost information can be disregarded in evaluating the spectral components relevant for our problem?

Furthermore, we are dealing with a sample of finite length that is known for a finite time period. Outside of these intervals the values are not known but are usually assumed either to be constant or periodic with a period equal to the sampling length.

We are faced, then, with two separate difficulties: the sampling problem and the truncation problem. Still another difficulty may arise from the roundoff of the data, which can be regarded to be equivalent to the addition of low-power white noise.

The answer to the sampling problem is given by the *Nyquist theorem*: no information is lost if the sampling interval is smaller than $1/(2f_{max})$, where f_{max} is the maximum frequency present in the series. It is called the Nyquist frequency: for a sampling interval Δt it is given by $f_N = 1/(2\Delta t)$. If this condition is not met the spectrum will be distorted, due to the misrepresentation of high frequencies, leading to what is called aliasing of the spectrum.

The truncation problem is more difficult to circumvent. Fortunately, it will not arise in a spatial analysis over the whole globe, since there we have a natural condition of periodic continuity in case mountains do not interfere. However, under other conditions, as in time-series analyses, truncation may introduce discontinuities at the limits of the sampling interval leading to the so-called *Gibbs phenomenon* (Sommerfeld, 1953; Bracewell 1978). The effects of truncation show up through the building up of an overshoot of the initial function at the limits of the interval when it is synthesized from the spectral components. The Gibbs phenomenon can be reduced by using a weighting function known as the "data window," but at the cost of some distortions of the spectral components (Guillemin, 1948). Recently, a new technique has been introduced to solve the truncation problems based on the maximum entropy method (Ulrych and Bishop, 1975), where the function outside of the sampling interval is assumed to be random instead of constant or periodic. The spectra obtained by this method do not show the Gibbs phenomenon and are superior to the classical spectra in terms of spectral resolution and accuracy.

4.3 EMPIRICAL ORTHOGONAL FUNCTION ANALYSIS

The empirical orthogonal function (EOF) analysis, sometimes referred to as eigenvector or principal components analysis, provides a convenient method for studying the spatial and temporal variability of long time series of data over large areas. This method splits the temporal variance of the data into orthogonal spatial patterns called empirical eigenvectors.

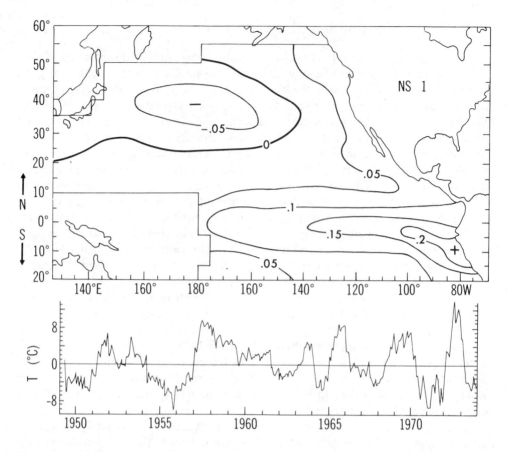

FIGURE 4.1. Example of an empirical orthogonal function (EOF) analysis for the sea surface temperature anomalies in the North Pacific Ocean. The first EOF field shown in the top portion of the figure explains 23.1% of the total nonseasonal variance in the sea surface temperature. The time series of the multiplication coefficient at the bottom of the figure shows the temporal behavior of the first EOF between January 1949 and December 1973 (from Weare *et al.*, 1976).

Since its introduction in the atmospheric sciences by Lorenz (1956) eigenvector analysis has been widely used to describe geophysical fields in hydrology, oceanography, and solid earth geophysics, besides meteorology and climatology. This approach enables one to identify a set of orthogonal spatial modes, such that, when ordered, each successive eigenvector explains the maximum amount possible of the remaining variance in the data. The eigenvectors are arranged in decreasing order according to the percentage of variance explained by them. Each eigenvector pattern is associated with a series of time coefficients that describes the time evolution of the particular spatial mode. The eigenvector patterns that account for a large fraction of the variance are, in general, considered to be physically meaningful and connected with important centers of action. Recently the use of EOF's has become widespread in identifying patterns of large-scale climate fluctuations. Since the modes are orthogonal in nature, any two modes are spatially uncorrelated and their time variations are also uncorrelated. Thus, in this sense, no one mode carries within itself any relationship to any other mode (except that two modes could be uncorrelated and in quadrature with each other).

The principal advantages of the eigenfunctions are that they provide the most efficient way of compressing geophysical data in both space and time and they may be regarded as uncorrelated (independent) modes of variability of the fields.

From a mathematical point of view, the EOF analysis results from the possibility of reducing a large number of correlated quantities in time and space into a small number of orthogonal functions that are linear combinations of the original observations and account for a large percentage of the total variance. Eigenvectors may be used in the same way as orthogonal functions. However, unlike most other orthogonal representations the EOF's are derived directly from the data themselves rather than being a predetermined orthogonal set of analytical functions (Fourier analysis, Tchebycheff polynomials, etc.). Furthermore, the first few eigenvectors usually explain a much higher fraction of the total variance than the same number of other orthogonal functions. Thus, fewer empirical functions are usually required to explain the same amount of variance in the data.

The EOF's are the eigenvectors of the data covariance matrix whose elements are formed from the difference of the observations from their long-term means. Thus, the EOF's depend on the coherence of the departures from their normal values. The contribution of any component to the total variance in the field is given by the associated eigenvalue which provides a measure of its relative importance. In general, a large portion of the total variance can usually be represented by a small number of modes, with the remainder representing minor features, smaller-scale fluctuations, and noise. A comparison of the results from different samples is needed to decide which modes best represent the dominant behavior of a given field. Also, a number of tests have been designed to determine where a break occurs between those modes which are statistically significant and those which are not (e.g., Overland and Preisendorfer, 1982). The significant modes have a good chance of being physically meaningful as well. The higher order EOF components are most likely affected by sampling fluctuations and usually account only for a small part of the total variance.

Figure 4.1 shows, as an example of an EOF analysis, a map of the first mode of the variability in the monthly mean sea surface temperature over the Pacific Ocean between 20 °S and 55 °N during the period 1949–1973 [as published by Weare et al. (1976)]. This first mode explains 23.1% of the total nonseasonal variance, and displays the mode of variation characteristic of the important El Niño–Southern Oscillation (ENSO) phenomenon. An interesting feature is the fact that the temperatures near western North America vary in phase whereas those in the central Pacific vary out of phase with the temperature anomalies in the eastern Equatorial Pacific. The time series of the coefficient of the first eigenmode is presented at the bottom of Fig. 4.1. It shows the typical time variation of the ENSO phenomenon. These points will be discussed further in Chap. 16.

By introducing a complex EOF representation (Hardy and Walton, 1978; Salstein et al., 1983) one can apply the EOF analysis to two-dimensional vector fields such as the horizontal wind, the water vapor transport, etc. Also, in order to improve the physical interpretation of such fields, new techniques have been developed to alter the EOF's by forming new eigenvectors from linear combinations of the old basis vectors, essentially performing a rotation in the vector space of EOF's. These techniques can also reduce the dependence of the EOF representation on the structure of the chosen domain. It should be pointed out, however, that such a new set no longer sequentially maximizes the variance of the data. These methods and their interpretation are discussed by Richman (1986).

A detailed analysis of the mathematics of the EOF approach to both scalar and vector fields is presented in Appendix B.

CHAPTER 5

The Data

To gain a deeper understanding of so complex a system as the climate we must rely on observations, theory, and experiments (see Fig. 1.1).

Using the observations together with the balance equations, it is possible to investigate the mechanisms by which the various processes, such as the transports of angular momentum, water vapor, and energy, occur in the atmosphere and oceans. Through these diagnostic techniques, we will be able to describe not only the processes that maintain the climatic state, but also to infer the magnitude of some of the external constraints or forcings. Furthermore, the insertion of the atmospheric observations in the climatic equations may lead to results for other lesser-known parts of the climatic system. This approach may be called the "observational budget" or "observational balance" method. It constitutes an important tool for diagnostic studies of the behavior of the atmosphere and oceans. We have been following this approach in various studies over the last three decades. The results of our studies based on surface and upper air data from 1963–73 constitute the principal source of information for the atmospheric statistics in the present book.

5.1 OBSERVATIONAL NETWORKS

Most of our present knowledge of the physical and dynamical structure of the atmosphere and oceans is based on the *in situ* observations described below. However, as we have mentioned before (see Fig. 1.1) different technologies, e.g., satellites, are becoming available to improve and supplement the conventional system of observing the earth. Therefore, it is appropriate to describe the various types of observations needed to define the state of the climatic system.

5.1.1 Atmospheric data

The main sets of observations of the atmosphere can be grouped into *in situ* surface and upper air data, and remote satellite data. The surface data over land include the pressure, temperature, specific humidity, cloud cover, and rate of precipitation as measured in the world meteorological network shown in Fig. 5.1. The number of land-based surface stations is at least one order of magnitude greater than the number of upper air stations. The network grew from about 60 stations mainly located in western Europe and the United States in 1875 [see Fig. 5.1(a)] to a truly

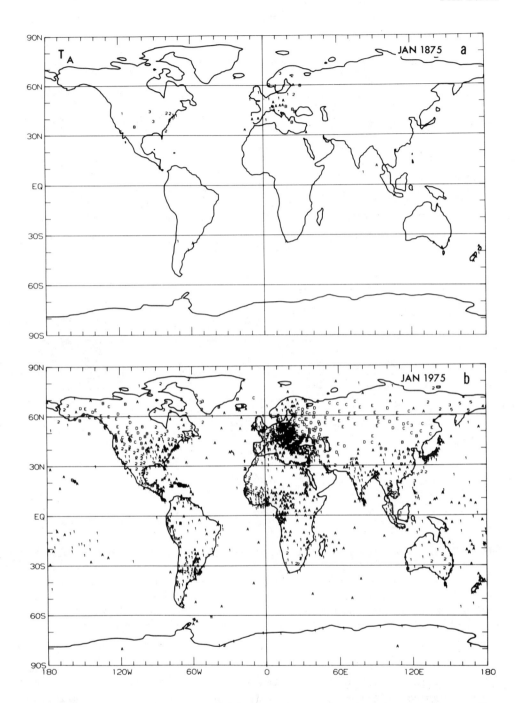

FIGURE 5.1. Network of land-based surface meteorological stations that exchange monthly-mean data internationally on a regular basis. The network has grown from about 100 stations mainly located in Western Europe and the United States in 1875 (a) to a truly global network of more than 1700 stations in 1975 (b). The numbers on each map indicate positive temperature departures (°C) from the 1950–1979 normal, and the letters negative departures. (We may note that practically all statements of regional to global heating and/or cooling trends in the climate are based on the surface air temperature reports of these stations.)

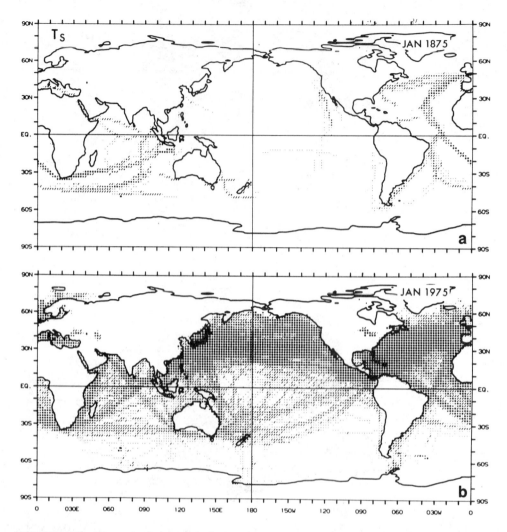

FIGURE 5.2. Network of meteorological surface ship reports complementing the land-based network shown in Fig. 5.1. Shown are maps of the distribution of $2° \times 2°$ ocean squares with checked ship reports of sea surface temperature during the month of January for the years 1875 (a) and 1975 (b). The number of observations is shown by a dot ($N = 1,2$), a slash ($N = 3-5$), a plus ($N = 6-10$), or an asterisk ($N \geqslant 11$). Somewhat surprisingly, the ocean surface records for the late 19th century provide a better global picture of the climate at that time than the records based on the land stations [Fig. 5.1(a)].

global network of about 7000 stations that currently report on a daily basis in the world meteorological communications network. In Fig. 5.1(b) a subset of about 1700 stations is shown for which monthly averages are readily available for climatological research.

Over the oceans, the observations include the sea surface temperature, salinity, atmospheric temperature, pressure, humidity, wind direction, and wind speed. Most of the observations over the oceans have been taken by commercial ships. The actual data distribution over the oceans is shown in Figs. 5.2(a) and 5.2(b). These figures illustrate the evolution of the global distributions and show that the early coverage over the oceans [Fig. 5.2(a)] might have been somewhat better than over land [Fig. 5.1(a)].

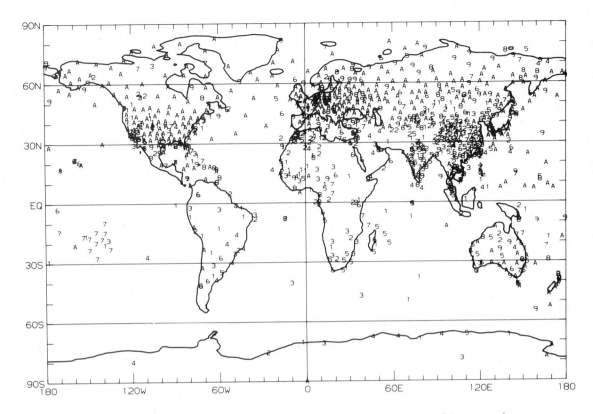

FIGURE 5.3. Network of upper-air (rawinsonde) stations used to compute most of the upper-air fields and statistics shown in this book. The plotted values indicate the number of years of wind and temperature observations available at 500 mb for January and range between 1 and 10 ($= A$) years, covering the 10-year period May 1963 through April 1973. The lowest numbers indicate relatively poor reporting stations, whereas the symbol A indicates a complete record. Of the 1093 stations only about 600 station soundings came in on the average each day.

After World War II an international effort was begun to launch rawinsondes twice daily at the major airports of the world for general aviation and weather-forecasting purposes. A rawinsonde or radiosonde is a meteorological device attached to a balloon that measures the important meteorological quantities at various altitudes. During each rawinsonde ascent the temperature, pressure, and humidity are sampled from the ground to about 20- to 30-km altitude. In addition, by tracking the displacements of the balloon the horizontal wind components are obtained. The number of reporting stations grew from a few hundred in the early 1950s to about 800 stations in the 1980s; these report twice daily to the world meteorological communications network. The data coverage over the inhabited regions of the continents is now sufficient for the purpose of large-scale climate research. However, extensive regions without data are found over the oceans, except where there are island stations and in the North Atlantic where there are a number of weather ships.

The basic atmospheric input data used in this book are contained in two sets. Set 1 contains all available daily reports from the global rawinsonde network for the 10-year period May 1963 through April 1973. This sample contains the bulk of the information for our studies. The station coverage during 1963–73 is indicated in Fig. 5.3 on a latitude–longitude map and for the polar regions in Fig. 5.4 on two polar stereographic maps. The levels used in our analyses are given in Table 5.1 in terms of

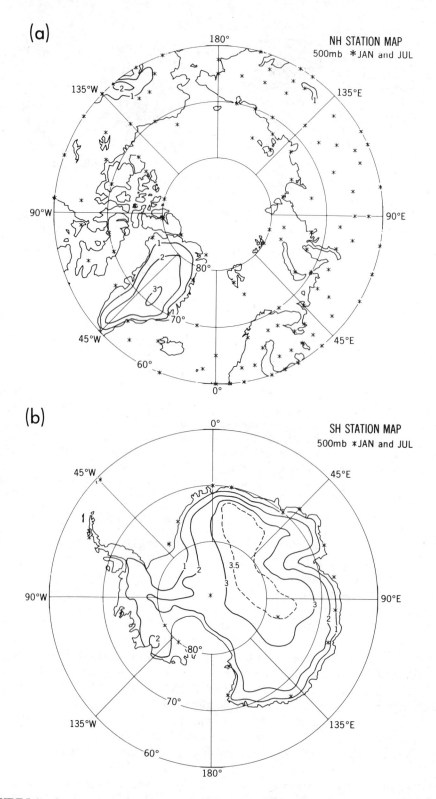

FIGURE 5.4. Surface configuration and distribution of rawinsonde stations for the north (a) and south (b) polar regions. The stations are plotted at the nearest grid point of a 2.5° latitude by 5° longitude grid. Some height contours of the surface topography are shown in km's.

TABLE • 5.1. Values of the geopotential height and temperature in the NMC standard atmosphere at the different pressure levels used in our data analyses. The NMC standard atmosphere represents typical conditions near 45 °N.

		Z_{SA} (gpm)	T_{SA} (°C)
	Surface	...	...
1	1000 mb	113	14.2
2	950	540	...
3	900	988	...
4	850	1 457	5.5
5	700	3 011	−4.6
6	500	5 572	−21.3
7	400	7 181	...
8	300	9 159	−44.6
9	200	11 784	−56.6
10	100	16 206	−56.6
11	50	20 632	−56.0

pressure and the corresponding height in the NMC standard atmosphere. A typical example of the distribution of the input data and the number of good years available for the rawinsonde stations at 500 mb is presented in Fig. 5.3. The lack of upper air data over the southern oceans is clear. Thus, some of the results between 30 °S and 70 °S latitude must be regarded as tentative. In the tropics of the Northern Hemisphere, there are also some large data gaps over the eastern North Pacific and the Atlantic Oceans. Set 2 supplements the first set over the oceans at the lowest level of analysis (1000 mb/surface). It contains all available daily surface ship reports for the same 10-year period.

5.1.2 Oceanic data

The subsurface ocean data used were collected largely from research vessels on a nonoperational basis. The spatial coverage, as shown in Fig. 5.5, is more uniform than the coverage in the atmosphere. However, there are usually only a few soundings available at the same location making it difficult to study temporal variability below the ocean surface. For example, it appears practically impossible to study the oceanic heat storage for individual years, except in certain regions of the North Atlantic and North Pacific Oceans. We must emphasize that our present inability to monitor the oceanic heat storage is one of the greatest obstacles in understanding long-term climatic change and low-frequency variability.

The standard oceanographic "station" data obtained by research vessels are temperature, salinity, oxygen content, and concentrations of various nutrients, measured at many levels in the vertical between the surface and the ocean bottom [see Figs. 5.6(a) and 5.6(b)]. Fortunately, most variability is found in the upper 100 m of the oceans, where one can sample easily and frequently, using bathythermograph data to measure the vertical temperature profiles. The recent introduction of expendable bathythermographs (XBT's) on commercial ships has greatly increased the data base for the upper 400 m of the oceans as shown in Fig. 5.6. The decrease in the number of

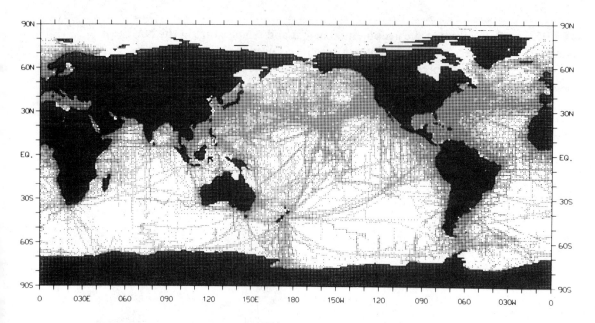

FIGURE 5.5. Network of subsurface ocean data. Shown are those 1° latitude × 1° longitude squares that contain any historical temperature data taken by oceanographic research vessels during the three-month period February to April, irrespective of year. A small dot indicates a square containing one to four observations, and a large dot a square containing five or more observations (from Levitus, 1982).

the data since the early 1970s is due to the considerable lag between the time of taking the observations and the time of assembling and archiving them on a world-wide basis. The apparent annual fluctuation in the number of data in Fig. 5.6 is due to a seasonal bias in taking the observations, namely fewer observations are taken during the Northern Hemisphere winter than during its summer. Because the year-to-year variability is smaller in the oceans than in the atmosphere, especially below the surface layer, the historical data are probably sufficient to make first estimates of such quantities as the seasonal cycle of the oceanic heat storage, one of the crucial factors affecting the climate.

Current velocities are difficult to measure from a ship since they are usually very small (on the order of a few cm s^{-1}) except in the major ocean currents. This fact severely limits our present knowledge of the dynamical structure of the oceans.

5.1.3 Satellite data

The role of satellites in climate research has expanded greatly since the early 1960s when cloud pictures were essentially the only useful product, as can be seen in Table 5.2. In fact, measurements of net incoming and outgoing radiation fluxes at the "top" of the atmosphere during the last decade provide the first reliable measurements of the basic driving force of the climatic system.

Further, satellites are providing useful information from other spectral bands besides the ultraviolet, visible, and infrared bands. For example, the microwave band is used to study the precipitable water and liquid water content of the atmosphere over the oceans, wind stresses at the ocean surface, the distribution and extent of sea

THE DATA   77

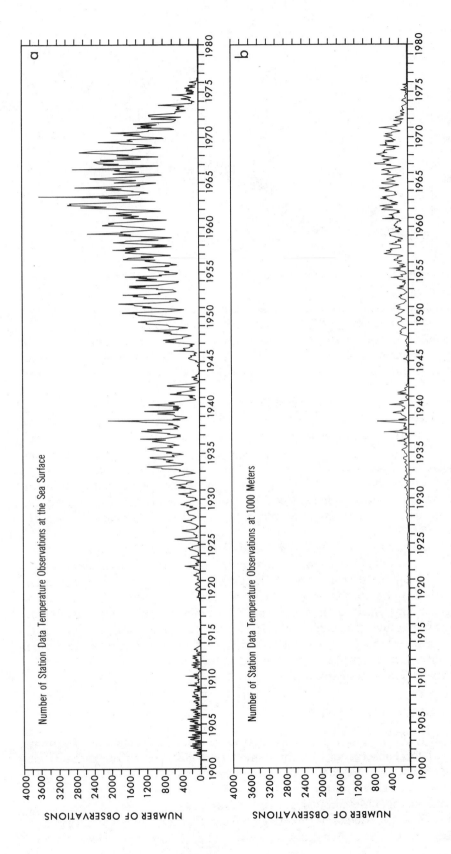

FIGURE 5.6. Time series of the number of station data temperature observations at the sea surface (a) and at 1000-m depth (b) for the period 1901 through 1976, and time series of the number of MBT and XBT casts (c) since 1940. The station data often cover the entire depth of the ocean, whereas the MBT records generally give reliable temperature data only down to about 250-m depth, and the XBT records down to, at times, 1000-m depth (from Levitus, 1982).

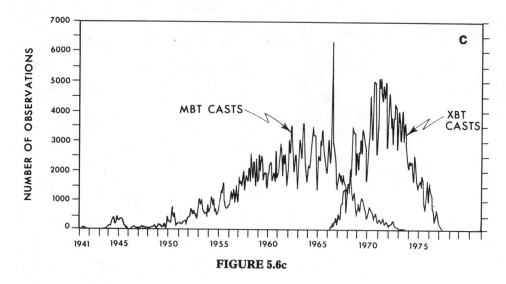

FIGURE 5.6c

ice, the height of sea level, the distribution of pollutants, etc. The height of sea level is of major importance for getting a handle on the velocity fields in the oceans. Another example is the determination of the nature and concentration of aerosols, e.g., sulfuric acid from volcanic eruptions (such as Mount St. Helens in March 1980 and El Chichón in April 1982) and dust in the atmosphere using backscattered ultraviolet radiation.

In addition, through inversion techniques in the various absorption bands of CO_2 and H_2O, useful temperature and humidity profiles in the atmosphere are now obtained to supplement the rawinsonde network observations over the oceans (Kaplan, 1959; Houghton, 1985). So far only mean temperatures for relatively thick layers and winds at two cloud levels have been inferred from satellite information. Therefore, the vertical resolution is limited, and the data are, in many cases, not compatible with the observations obtained from rawinsondes at the different levels. In our analyses, we have only used the net radiation data from satellites from the period 1964 through 1978 (see Table 5.2).

The amount of work involved in our studies of the climatic system is quite substantial. For example, for an average number of 1000 stations taking observations twice a day during the 10-year period for five parameters (u, v, T, z, and q) at 11 levels in the vertical, required handling a minimum of $1000 \times 2 \times 10 \times 365 \times 5 \times 11 = 4 \times 10^8$ pieces of information. In the oceans, about one million soundings of three parameters were processed at about 30 levels leading to a total of 10^8 pieces of information. It is clearly a formidable task to handle such a mass of data and to ensure quality control, since they are collected by countless individuals in many countries using somewhat different instruments.

5.1.4 International field projects

Naturally, climate is a topic that affects all people and all nations. Its study requires both local observations and international and global organization so that the individual observations are compatible with each other and can be used to obtain a uniform global description of the climate.

The International Meteorological Organization (IMO), founded in 1873, and its successor, the World Meteorological Organization (WMO), founded in 1950 as a technical agency of the United Nations, with its headquarters in Geneva, Switzer-

TABLE ● 5.2. Chronological list of the main earth-orbiting satellites from which radiation measurements are used for global budget studies (after Stephens et al., 1981, updated).[a,b]

Month	1964	1965	1966	1968	1969	1970	1975	1976	1977	1978	1979–1983	1984	1985	1986	1987	1988
Jan		Ex(1030)			E7	N3		N6	N6	N6		N7	N7,ERBE	N7,ERBE	N7,ERBE	ERBE
Feb		Ex(1035)			E7			N6	N6	N6		N7	N7,ERBE	N7,ERBE	N7,ERBE	ERBE
Mar		Ex(1040)			E7			N6	N6	N6		N7	N7,ERBE	N7,ERBE	N7,ERBE	
Apr					N3(1130)[c]			N6	N6	N6		N7	N7,ERBE	N7,ERBE	N7,ERBE	
May					N3			N6	N6			N7	N7,ERBE	N7,ERBE	N7,ERBE	
Jun			N2(1130)[c]		N3			N6	N6			N7	N7,ERBE	N7,ERBE	N7,ERBE	
Jul	Ex(0830)				N3		N6(1145)[c]	N6	N6			N7	N7,ERBE	N7,ERBE	ERBE	
Aug	Ex(0855)				N3		N6	N6	N6			N7	N7,ERBE	N7,ERBE	ERBE	
Sep	Ex(0915)				N3		N6	N6	N6			N7	N7,ERBE	N7,ERBE	ERBE	
Oct	Ex(0940)			E7(1430)			N6	N6	N6	N7		N7	N7,ERBE	N7,ERBE	ERBE	
Nov	Ex(1005)			E7			N6	N6	N6	N7		N7,ERBE	N7,ERBE	N7,ERBE	ERBE	
Dec	Ex(1030)		E3(1440)	E7				N6	N6	N7		N7,ERBE	N7,ERBE	N7,ERBE	ERBE	

[a] The approximate local time at which each satellite crossed the equator during daylight hours appears in parentheses. Ex, experimental; N2, Nimbus 2; N3, Nimbus 3; N6, Nimbus 6; N7, Nimbus 7; E3, Essa 3; E7, Essa 7; ERBE, Earth Radiation Budget Experiment.
[b] Resolution≃half-power diameter: Experimental, 1280 km, 11.5°; ESSA 3, Nimbus 2, averaged to 10° grid; ESSA 7, 2200 km, 20°; Nimbus 3 averaged to 10° grid; Nimbus 6, 1100 km, 10° (analyzed from 16°).
[c] Albedo corrected for diurnal variation of reflection with directional reflectance model.

land, have been providing the standards for taking, collecting, and distributing the observations and facilitating their exchange among the different nations. On a routine, daily basis a large fraction of the meteorological and oceanographic observations taken is now distributed to the various forecasting centers using the Global Telecommunication System (GTS).

We will now review a few of the relevant international projects that have led to major advances in the observational network and in our knowledge of the atmosphere (see, e.g., WMO, 1973). Going back in time, the first large-scale projects were the two International Polar Years in 1882–83 and 1932–33, intended mainly to increase the knowledge of the arctic regions. Then during the International Geophysical Year in 1957–58, principal emphasis was given not only to the arctic region but also to the antarctic region where many new meteorological stations (including one at the South Pole itself) were established, as well as to the tropical belt.

Following the more recent development of meteorological satellites, numerical weather prediction (NWP) models, and global communication networks, for the first time truly global projects, such as the World Weather Watch, became feasible. Thus, in 1968, the Global Atmospheric Research Program (GARP) planned and coordinated jointly by the intergovernmental WMO and the more scientific International Council of Scientific Unions (ICSU) came into being. The First GARP Global Experiment (FGGE, also called Global Weather Experiment) took place in 1979. Its major objectives were:

(1) to obtain a better understanding of atmospheric motion for the development of more realistic models for weather prediction;

(2) to assess the ultimate limit of predictability of weather systems.

One of the other goals of FGGE has been to design the optimum composite meteorological observing system for routine numerical weather prediction of the larger-scale features of the general circulation. The observing system consisted of the routine land-based and ocean weather stations, ships, aircraft, geostationary and polar orbiting satellites, drifting buoys south of 20 °S, and constant-level balloons in the tropics. It also included some special intense observing periods and regional experiments such as in the Asian monsoon and polar regions (see, e.g., Fleming *et al.*, 1979a and 1979b).

With the growth of our abilities to monitor and understand the global atmosphere, the realization of the interconnectedness of the atmosphere with the other parts of the climate system has also grown. In turn, this has led to a renewed interest in climatic fluctuations beyond the one-to-two-week time scale of weather phenomena. Thus, as a successor to GARP the World Climate Program was established in 1979 with a research component that has as its major objectives to determine

(1) to what extent climate can be predicted and

(2) the extent of man's influence on climate.

As one of its activities the Tropical Ocean Global Atmosphere (TOGA) experiment was launched in 1985 partly in response to the extraordinary El Niño–Southern oscillation event in 1983 (see Chap. 16). Another activity of the World Climate Program is the International Satellite Cloud Climatology Project (ISCCP).

Finally, for the near future in the 1990s a World Ocean Circulation Experiment (WOCE) and a Global Energy and Water Cycle Experiment (GEWEX) among several other experiments are being planned under the general umbrella of the International Geosphere–Biosphere Program (IGBP). It is clear that great progress both in the data bases and in our understanding of natural and manmade climatic changes can be expected in the next few decades.

5.2 DATA PROCESSING TECHNIQUES

In this section and in Sec. 5.3, we will discuss the specific procedures that were used by us and our colleagues to obtain the statistics presented in this book. Some other procedures that are also frequently used will be discussed in Sec. 5.4.

5.2.1 Atmospheric data

The accuracy of the basic data is subject to some uncertainties not only due to instrumental limitations but also due to the processing and transmitting of the observations. It seems reasonable to assume that these errors are random in character and uncorrelated. This implies that, in the case of linear quantities, the statistics evaluated for a given station would become progressively more reliable as longer averaging periods are used. For further discussion see Oort (1978).

The surface and upper air data sets were carefully checked for erroneous reports at several phases in the data processing scheme. The most important checks were an initial, gross test for unreasonable meteorological values and, at a later stage, the application of a cutoff criterion for those values of x which fell outside the range $[\bar{x} - 4\sigma(x), \bar{x} + 4\sigma(x)]$. The seasonal averages $\bar{x}$ and the seasonal standard deviations $\sigma(x)$ were computed from the 10-year time series at each station and for each isobaric level and season. In the case of the surface ship data, all reports within a grid square of 2° latitude × 2° longitude were considered to represent values for a hypothetical station located in the center of the square. From the checked time series, monthly mean statistics were computed at each station for each of the 120 months under study, following the same mathematical techniques used previously in several of our projects (e.g., Starr, 1968; Rosen et al., 1979; and Oort, 1983).

To give an idea of the quantities computed we present in Tables 5.3(a) and 5.3(b) the basic station statistics in matrix form at the various pressure levels. From the matrix of the accumulated monthly sums the covariance matrix was obtained. Since both types of matrices are symmetric, the diagonal terms in Table 5.3(b) represent the variances. As discussed previously, the variances and covariances are directly related to the role of the transient eddies in climate.

There are unavoidable missing data in the records creating additional difficulties that we have to resolve. It seems that ten days of upper air observations per month are sufficient to represent the prevailing atmospheric conditions during that month. In the case of the ocean surface data, a lower criterion was used with a minimum of three days of observations since the surface network is much more dense over most of the globe (see Fig. 5.2) than the upper air network (see Fig. 5.3). The reason to choose the lower cutoff criterion stemmed from the additional information contributed by data from neighboring 2° squares (Oort, 1983).

For the study of the long-term mean conditions used in the present book average station values were calculated for each calendar month, which enabled us to study the seasonal cycle in great detail. This procedure implies that the year-to-year variations

TABLE • 5.3(a). Monthly sums evaluated at each station and each level.

N	Σu	Σv	ΣT	Σz	Σq
...	Σu^2	Σuv	ΣuT	Σuz	Σuq
...	...	Σv^2	ΣvT	Σvz	Σvq
...	...	...	ΣT^2	...	...
...	...	...	...	Σz^2	...
...	...	...	...	...	Σq^2

TABLE • 5.3(b). Monthly means, variances and covariances evaluated at each station and at each level.[a]

$\bar{u}$	$\bar{v}$	$\bar{T}$	$\bar{z}$	$\bar{q}$
$\overline{u'u'}$	$\overline{u'v'}$	$\overline{u'T'}$	$\overline{u'z'}$	$\overline{u'q'}$
...	$\overline{v'v'}$	$\overline{v'T'}$	$\overline{v'z'}$	$\overline{v'q'}$
...	...	$\overline{T'T'}$	...	...
...	...	...	$\overline{z'z'}$	...
...	...	...	...	$\overline{q'q'}$

[a] Note that the generic matrix term $\overline{A'B'}$ is computed as follows: $\overline{A'B'} = \overline{AB} - \bar{A}\bar{B} = \Sigma(AB)/N - (\Sigma A/N)(\Sigma B/N)$.

are not included in the transient eddy statistics. However, the eddy statistics will contain the most important fluctuations with time scales of one month or less.

The 10-year average station statistics were then used as input for the global analyses of the different quantities at the standard levels (see Table 5.1).

The various maps, zonal-mean cross sections, and profiles presented in this book are based on the point values of a 2.5° latitude × 5° longitude grid. Most of the horizontal maps were somewhat smoothed before drafting by twice applying a two-dimensional five-point median smoother followed by a weak Laplacian smoother (see Sec. 5.3). The required seasonal and annual-mean grid point values were calculated from the 12 monthly analyses. As discussed in Sec. 4.1, we should note that all fluctuations within a year are considered as transient eddies, so that the annual-mean statistics include the contributions from both synoptic eddies and the "eddies" associated with the normal seasonal variation. If one is interested in annual-mean statistics where only the synoptic systems are included as transient eddies and the effects of the annual cycle are removed, a good measure of the annual mean can be obtained by taking a straight average of the DJF and JJA transient eddy statistics.

The topography of the earth's surface was taken into account in all computations so that only grid points above the mountains contribute to the zonal- and vertical-mean estimates. To accomplish this, the topography was expressed in pressure coordinates, assuming that it is invariant throughout the year and equal to the mean annual value. To show the relative importance of the topography at various levels, we present in Fig. 5.7 a meridional cross section of the fraction of points above the earth's surface.

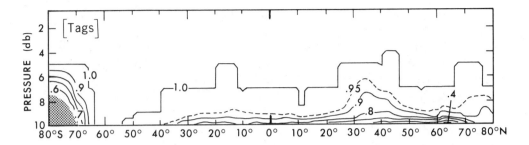

FIGURE 5.7. Latitude-pressure diagram of the east–west fraction of grid points above the earth's surface. A value of 1.0 indicates that all points along the latitude circle lie above the earth's surface.

5.2.2 Oceanic data

As mentioned before, the biggest problem in studying the climate of the oceans is the general lack of subsurface data in certain areas of the globe. In addition, we have to consider the quality of the data, which requires various checks. For example, unreasonable and spurious values have to be eliminated, and some vertical and horizontal consistency checks such as static stability checks and some standard deviation cutoff criteria, have to be applied in order to obtain representative statistics. Generally speaking, very similar filtering criteria were used as those described earlier in the case of the atmospheric data.

After completion of the error checking, all available historical oceanographic soundings were used to generate 1° latitude by 1° longitude square "station" statistics. These were assigned to the center of each square for monthly, seasonal, and annual-mean values. Of course, the number of observations tends to decrease with depth. However, in general, they appear to be sufficiently dense to give representative long-term mean analyses of the various fields (Levitus, 1982).

Since the basic ship observations are not regularly distributed in space and time in data-sparse regions, some uncertainty is bound to occur in the quality of the statistics with regard to certain scales of oceanographic phenomena. It may happen that the few available observations only sample a very uncommon situation, such as a detached vortex in the mid-ocean that came from a region with completely different oceanographic properties. Examples are cutoff eddies coming from the Gulf Stream or Kuroshio, or lenses of water in the middle of the oceans originating from coastal waters or Mediterranean Seas.

5.2.3 Satellite data

The basic data set used here consists of reflected solar and infrared radiation measurements from polar orbiting, sun-synchronous satellites taken during the years 1964 through 1979. The albedo measurements of the atmosphere–earth system give the information needed to calculate the incoming solar radiation, assuming a solar constant of 1360 W m^{-2}. Combined with simultaneous measurements of outgoing infrared radiation, the net radiation balance at the top of the atmosphere F_{TA} was evaluated on a monthly basis. For most satellite sensors, the ground resolution was low, of the order of 1000×1000 km^2. More detailed information on the satellite data that have been included in the present study as well as the local time of observation is presented in Table 5.2. For completeness the table was extended to the late 1980s although only the 1964–1979 data were used here. All satellites were in sun-synchronous orbits, thus sampling at just one local time during daylight hours. However, the set includes late morning, near noon, and early afternoon orbits so that diurnal sampling bias is minimized. For further information concerning the satellite data and error estimates see Stephens *et al.* (1981).

5.3 OBJECTIVE ANALYSIS METHODS

The purpose of the analyses to be presented in this book is to give a three-dimensional description of the climate in the atmosphere and oceans, by converting the information contained in the station data onto a regular horizontal grid. To accomplish this in the atmosphere, we used an objective analysis scheme developed mainly by Welsh (Harris *et al.*, 1966; Rosen *et al.*, 1979a), and in the oceans a simple iterative scheme (Cressman, 1959).

5.3.1 Atmospheric analyses

Our 10-year ensemble of station statistics from the period May 1963 through April 1973 was used as input for the analyses. The analysis requires an initial guess value at each grid point. The zonal average of the station data in a 10°-wide latitudinal belt is used as first guess for all grid points in the latitude belt. If there are stations close to a grid point, a simple interpolation scheme is used to correct the first guess. This correction is the departure from the initial guess averaged over the neighboring station values. The same correction is applied to all grid points which lie within a specified distance of the corrected grid point (taken as an internal boundary point) leading, after smoothing, to the second-stage analysis.

For the grid points that have no nearby reporting stations the field of station departures from the second-stage analysis is extended by solving a Poisson's equation. This can be accomplished for each grid point by relaxation methods (e.g., Richtmeyer and Morton, 1967) with constraints imposed by the internal boundary points (the observations) and by the boundary conditions at the perimeter of the grid

$$\nabla^2 T(i,j) = F(i,j), \tag{5.1}$$

where T is the interpolated value at grid point (i,j), F is the forcing function, and ∇^2 is the Laplacian operator. This operator can be defined in finite difference form as

$$\nabla^2 T(i,j) = \{T(i,j+1) + T(i,j-1) + T(i+1,j) + T(i-1,j) - 4T(i,j)\}/4.$$

From Eq. (5.1) we note that the forcing function defines the shape of the field, and is chosen to be the Laplacian of the second-stage analysis of the departures.

Of course, when $F = 0$ the Poisson's equation reduces to a Laplace equation. Its solution may be physically represented by a membrane stretched between the already corrected internal grid point values and the boundary values. This would correspond to a linear interpolation scheme, whereas with the Poisson's equation a nonlinear interpolation is possible due to the forcing field. The final step in the analysis cycle is to apply a smoothing operator (filter) at all grid points in order to minimize small-scale noise which may be artificially introduced by the data themselves or the analysis scheme. This sequence of steps is repeated a few times in order to get a stable analysis.

The analyses were performed on a modified National Meteorological Center (NMC) grid leading to arrays with an average grid point distance of about 430 km. The hemispheric analyses were done on polar stereographic maps. Finally, global latitude–longitude maps, with a 2.5°-latitude and 5°-longitude mesh, were constructed by interpolation from the two hemispheric maps.

5.3.2 Reliability of the atmospheric analyses

To illustrate how the analysis scheme works, the following experiment was performed. Four fictitious data points were introduced in the Northern Hemisphere at 70 °N, 0°; 70 °N, 90 °E; 70 °N, 180 °E; and 70 °N, 90 °W with values of 10, 10, − 10, and − 10, respectively, and a similar set of 4 points for the Southern Hemisphere. Next, an analysis was done using the scheme described above. The resulting analysis is presented in Fig. 5.8. It shows that the central values are somewhat reduced in absolute magnitude from the original value of 10.0 to a value of 9.2, and that the radius of influence for isolated data points tends to be quite large.

To further assess the influence of the number and distribution of the stations used several experiments were conducted. In these experiments, 10% of the station values randomly chosen were withheld, and the remaining 90% were objectively analyzed. This experiment was repeated several times using different, nonoverlapping random samples. The root-mean-square (rms) departures of the analyzed values from the withheld station values in each case can then be regarded as a measure of the uncertainty involved in our analyses or as a measure of the errors in our analyses associated with the nonuniform distribution of the stations over the globe.

The resulting rms errors are presented in Table 5.4 in the columns "A" for two levels and the Northern and Southern Hemispheres for various quantities. For com-

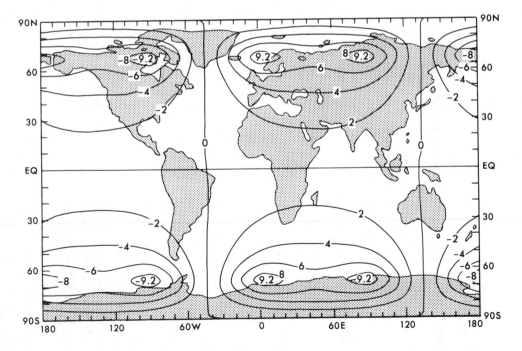

FIGURE 5.8. Test result of the analysis procedure for four selected data points in each hemisphere. The original values were + 10 and − 10 (from Oort, 1983).

TABLE • 5.4. Estimated values of the typical rms errors due to spatial data gaps (A),[a] the typical rms variations in the analyzed climatological mean fields (B),[b] and their ratios (C) in the Northern and Southern Hemispheres. Ratios larger than 1 are printed in bold face indicating that spatial variations on the analyzed maps cannot be trusted for the particular quantity.

	Northern Hemisphere								Southern Hemisphere						Units
	850 mb			200 mb					850 mb			200 mb			
	A	B	C	A	B	C			A	B	C	A	B	C	
u	1.6	3.9	0.41	3.6	14.4	0.25			2.6	4.6	0.57	5.4	13.4	0.40	m s^{-1}
v	1.4	1.6	0.85	2.4	3.8	0.62			1.8	1.6	**1.12**	3.1	2.6	**1.17**	m s^{-1}
T	1.2	11.4	0.11	0.9	3.9	0.23			1.5	11.6	0.13	1.2	4.9	0.24	°C
z	10	68	0.15	27	418	0.06			15	117	0.13	45	510	0.09	gpm
q	0.7	3.4	0.21	...	...	...			1.4	3.5	0.40	...	...	...	g kg^{-1}
u'^2	11	17	0.63	40	54	0.75			18	16	**1.09**	50	63	0.79	m^2 s^{-2}
v'^2	10	18	0.54	40	65	0.61			14	18	0.79	63	65	0.97	m^2 s^{-2}
T'^2	5.4	11.3	0.48	3.7	10.4	0.36			4.2	7.9	0.53	4.8	9.9	0.48	°C^2
z'^2	600	2506	0.23	3150	7835	0.40			940	2288	0.41	4940	7718	0.64	gpm^2
q'^2	1.2	1.8	0.65	...	...	...			1.9	1.8	**1.06**	...	...	...	g^2 kg^{-2}
$v'u'$	7	3.6	**1.94**	23	22	**1.04**			10	4.6	**2.17**	30	26	**1.17**	m^2 s^{-2}
$v'T'$	4.6	7.5	0.61	7.6	5.6	**1.35**			5.2	7.9	0.66	8.9	6.4	**1.39**	m s^{-1} °C
$v'q'$	1.6	2.3	0.70	...	...	...			2.5	2.0	**1.22**	...	...	...	m s^{-1} g kg^{-1}

[a] Estimates were made by computing the rms error for 9 samples where 10% of the station values were withheld from the analyses. The rms error was computed from the differences between the actual (withheld) station values and the analyzed values at the location of the withheld station. The values of the rms error are averages for July 1973 and January 1974.
[b] The entries in the columns B represent 10-year mean climatological rms departure values (see the text) averaged for January and July.

parison, estimates of the typical spatial variability in the time-mean climatological fields are given in Table 5.4 in the columns "B." The measure of variability used here is the rms departure from the global-mean value

$$\text{rms}(\bar{x}) = \left\{ \sum_{i,j} (\bar{x} - \tilde{\bar{x}})^2 \cos \phi_i \bigg/ \sum_{i,j} \cos \phi_i \right\}^{1/2},$$

where $\tilde{\bar{x}}$ is the area-weighted global-mean value of the time-mean field of x, and the summations are performed over all grid points (i,j) in either the Northern or the Southern Hemisphere.

Table 5.4 also shows the ratios of the rms error and the rms of the analyzed climatological fields in columns "C" as a measure of the reliability of the analyses. We will assume, as a first indicator of the skill of our analyses, that a ratio of 1 separates the reliable from the unreliable analyses. In fact, ratios larger than 1 (in bold face) indicate that the typical errors due to spatial data gaps tend to be larger than the actual variations shown on the analyzed maps, and thus that there is practically no skill in the detailed spatial features on these maps.

As a general conclusion, we can say that the rms errors are greater in the Southern than in the Northern Hemisphere, as one would expect. Further, the spatial variations of the linear quantities $\bar{u}, \bar{T}, \bar{z}$, and $\bar{q}$, but not $\bar{v}$, are fairly well measured through the rawinsonde network. Especially the temperature ($\bar{T}$) and height ($\bar{z}$) fields are very reliable with ratios less than 0.25.

As far as the transient eddy fields are concerned we notice that, in general, the large-scale features in the Northern Hemisphere can be trusted, except for the transient eddy momentum flux ($\overline{v'u'}$) at all levels, and the transient eddy heat flux ($\overline{v'T'}$) at the upper levels (200 mb). In the Southern Hemisphere, we find overall larger ratios and, except for $\overline{v'T'}$ at 850 mb, all transient eddy flux fields seem to be unreliable in that hemisphere.

In summary, the values of the ratios in columns C of Table 5.4 indicate how accurately we can define the spatial patterns on the horizontal maps of the various quantities based on the rawinsonde network as shown in Figs. 5.3 and 5.4. This knowledge will be especially useful when we discuss the observed mean state of the atmosphere in Chap. 7.

5.3.3 Oceanic analyses

In the oceans, a somewhat different scheme of analysis, as described in Levitus (1982), was applied. This approach was followed in view of the higher resolution required to resolve the important features of the oceans, such as the strong gradients near the coasts and near the equator. The following simple iterative scheme developed by Cressman (1959) was used to generate the analyses on a 1°-latitude × 1°-longitude grid. A first guess, $F(i,j)$, was formed with the zonal mean values for each ocean basin. An influence radius R was then specified. Where there was an observed mean value, the difference Q_s between this value and the first guess field was computed. A correction $C(i,j)$ was evaluated at all grid points (i,j) as the distance-weighted average of the differences in the points that lie within a circle of radius R around the grid point. Thus

$$C(i,j) = \sum_{s=1}^{n} W_s Q_s \bigg/ \sum_{s=1}^{n} W_s, \qquad (5.2)$$

where the weighting function

$$W_s = e^{-4r^2/R^2} \quad \text{for } r \leqslant R,$$
$$W_s = 0 \quad \text{for } r > R, \qquad (5.3)$$

and r = distance between the observation and the grid point.

At each grid point the analyzed value $G(i,j)$ is then computed as the sum of the first guess and the correction

$$G(i,j) = F(i,j) + C(i,j).$$

The new field $G(i,j)$ can now be used as a first guess and the procedures described above can be repeated using a smaller influence radius R. Only a few iterations are required to obtain a stable analysis. For further details, see Levitus (1982).

5.4 OTHER ATMOSPHERIC DATA SETS

So far we have discussed those sets of climatological analyses that are based on direct observations and that are usually obtained through one of the mathematical interpolation or extrapolation techniques described above. However, there are other methods of interpolation based on numerical weather prediction (NWP) models. These methods use the physical constraints imposed by the basic thermo-hydrodynamic equations as derived in Chap. 3 (see especially Secs. 3.2.2 and 3.7). They thereby provide a sound physical basis for filling in the empty spaces in between the observations with model data. Of course, one of the caveats is the capacity of the mathematical model to capture the real evolution of the weather all over the globe. Nevertheless, through continual model improvements the daily NWP analyses have become very valuable as research tools. They are also easy to use since, in contrast with the real observations, the grid point data cover the globe uniformly and weather systems are not lost in the large data-void regions of the world (see Fig. 5.3).

We may note that there are no fundamental differences between the models used in day-to-day weather prediction and those used in climate research, the general circulation models (GCM's, see Chap. 17). Nevertheless, the NWP models require a higher resolution and somewhat less emphasis on parameterization of the physical processes than the climate models. In the case of numerical weather prediction, models are being integrated over relatively short periods of up to ten days at present, and up to a month on an experimental basis. Here, the initial atmospheric conditions remain important throughout the integration. On the other hand, in the second case the models are integrated over years or decades, and the surface and top boundary conditions instead of the initial conditions become the factors that determine the climate statistics.

The flow of data in both our observational approach (method A) and in the NWP approach (method B) are shown on the left- and right-hand sides of Fig. 5.9, respectively. As mentioned before, in method A twice daily station reports are averaged for each station over the month, and the various circulation statistics are computed at each station separately. To obtain analyzed fields on a regular grid covering the globe, the values at grid points are obtained by mathematically fitting a surface through the time-averaged station values, as discussed before. This method tends to work reasonably well north of 30 °S, especially over the Northern Hemisphere land areas where abundant rawinsonde data are available. The reason that it works is that the time-averaging process removes much of the synoptic-scale variability connected

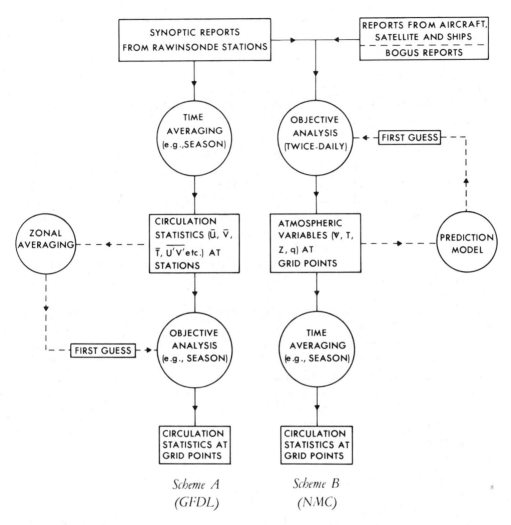

FIGURE 5.9. Flow diagram illustrating two alternative processing schemes A and B for compiling general circulation statistics (from Lau and Oort, 1981).

with the "weather" and, therefore, tends to yield fairly smooth large-scale fields. The important influence of the weather systems is then included in the transient eddy fields, which exhibit coherent large-scale patterns.

In method B usually six-hour forecasts are used to interpolate between data points on a daily basis. In the model, it is possible to include other nonstandard data besides the rawinsonde reports, such as aircraft reports, traveling ship reports, and satellite data in a continuous four-dimensional data assimilation (see, e.g., Bengtsson et al., 1982). Obviously, if assimilated correctly, these additional data should significantly improve the generated daily weather maps, and indeed much progress continues to be made to improve the basic analyses. The generated daily fields of the basic variables (the two wind components, temperature, geopotential height, and humidity) are then averaged, and similar circulation statistics are derived as in method A.

Sometimes "bogus" reports and/or empirical corrections are included in scheme B to compensate for known deficiencies in the models. In principle, this is probably an undesirable procedure. However, in practice, it may be necessary to correct for model biases (at least temporarily), as in the case of mountain-generated gravity waves that are not adequately resolved in the present NWP models (e.g., Wallace et al., 1983; Palmer et al., 1986).

Valuable hemispheric and (recently) global upper air analyses have been generated using method B at many weather centers in the world, e.g., at NMC in Washington, D.C., the European Centre for Medium Range Weather Forecasts (ECMWF) in Reading, England, the British Meteorological Office in Bracknell, England, and the Bureau of Meteorology in Melbourne, Australia.

In view of the limits in predictability (see Fig. 17.4 and, e.g., Lorenz, 1969; Miyakoda et al., 1972), it is interesting to note that even a six-hour forecast may show systematic model biases, connected, e.g., with mountains and intense thermal convection (Heckley, 1985; Arpe et al., 1985; Molteni and Tibaldi, 1985). Another practical difficulty in the use of the analyses produced by method B is that there have been many adjustments in the NWP models, designed to improve their forecasting performance. Therefore, it is difficult to assess whether temporal variations in the results of method B are due to real changes in climate or due to changes in the analysis scheme. On the other hand, method A has produced more consistent analyses, which constitutes an obvious advantage.

For a comparison of the general circulation statistics obtained using methods A and B we can refer to Lau and Oort (1981,1982), Rosen and Salstein (1980), Rosen et al. (1985), and Karoly and Oort (1987). The present consensus appears to be that the two methods have both their strengths and weaknesses, and that neither one is clearly preferable above the other. In the future, though, one would expect that method B should become superior in view of the sizeable improvements that are presently being realized in the NWP models and because, in principle, an interpolation based on the dynamics and physics of the system should be more appropriate than one based purely on mathematical techniques.

CHAPTER 6

Radiation Balance

6.1 INTRODUCTION

6.1.1 Nature of solar and terrestrial radiation

The major energy sources and sinks for the earth as a whole are the solar radiation and the terrestrial radiation, respectively. The radiant energy travels in the wave form at the speed of light c (2.9973×10^8 m s^{-1} in a vacuum). It can be characterized by the wavelength of propagation λ or by the frequency ν (cycles s^{-1}), which are related by the expression $\lambda \nu = c$.

The solar radiation covers the entire electromagnetic spectrum from gamma and x rays, through ultraviolet, visible, and infrared radiation, to microwaves and radiowaves. However, the most significant portion of the spectrum associated with radiative energy transfer in the climate system ranges from the ultraviolet to the near infrared as shown in Fig. 6.1.

Sometimes the wave number n is used instead of the frequency ν to describe the radiation, so that $n = \nu/c = 1/\lambda$. Thus, when $\lambda = 1$ μm, the wave number $n = 10\,000$ cm^{-1} or 10^6 m^{-1}.

Essentially all energy that enters the earth's atmosphere comes from the sun since the upward conduction of heat from the earth's interior (due to radioactive decay) is negligible. The incoming solar radiation is partly absorbed, partly scattered, and partly reflected by the various gases of the atmosphere, aerosols, and clouds. The remainder that reaches the earth's surface is largely absorbed by the oceans, lithosphere, cryosphere, and biosphere, and only a small part is reflected. According to the first law of thermodynamics the absorbed energy can be transformed into internal energy (heat) or it can be used to do work against the environment, appearing as potential or kinetic energy.

In order to maintain the earth in its long-term observed state of quasi-equilibrium, the amount of absorbed energy must be balanced by an equal amount of energy moving into outer space. The outgoing energy is also in the form of radiant energy emitted by the earth's surface and atmosphere. In fact, we know (based on Prévost's work in the late 18th century) that all bodies having a temperature above absolute zero K emit radiant energy over a large range of wavelengths. As we will see, the

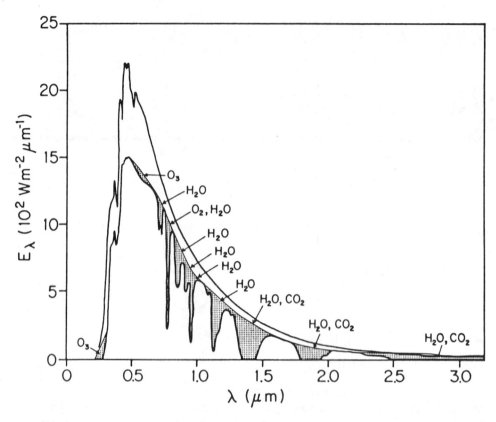

FIGURE 6.1. Spectral distribution of solar irradiation at the top of the atmosphere and at sea level for average atmospheric conditions for the sun at zenith. The shaded areas represent absorption by various atmospheric gases. The unshaded area between the two curves represents the portion of the solar energy backscattered by the air, water vapor, dust, and aerosols and reflected by clouds. For the curve at the top of the atmosphere the integral $\int_0^\infty E_\lambda \, d\lambda \simeq 1360 \text{ W m}^{-2}$ represents the solar constant (adapted from Gast, 1965).

higher the temperature of the emitting body, the larger the amount of emitted energy and the shorter the wavelength of its peak. Thus, due to the large difference in solar and terrestrial temperature, the incoming solar radiation has its maximum emission in the visible range ($\sim 0.5 \, \mu$m), whereas the outgoing terrestrial radiation has its peak in the infrared portion of the spectrum ($\sim 10 \, \mu$m). Most of the solar energy of interest for the energetics of the climate system ranges from 0.1 to 2.0 μm in the ultraviolet, visible, and near infrared regions, while most of the outgoing terrestrial radiation to space is in the region between 4.0 and 60 μm well in the infrared range of the electromagnetic spectrum [see Fig. 6.2 (a)]. This gives the rationale for the breakdown of the radiant energy that heats and cools the climate system into two types: the short-wave or solar radiation with $\lambda < 4 \, \mu$m and the long-wave or terrestrial radiation with $\lambda \geqslant 4 \, \mu$m. We may add that the mathematical and physical treatments of the two types of radiation are also quite different.

Since the solar radiation comes from a very distant point-like source, the sun, it can be treated as parallel unidirectional radiation. On the other hand, the terrestrial radiation comes from all directions since each molecule acts as an individual "minuscule sun" for thermal diffuse radiation. Furthermore, the terrestrial emission is neg-

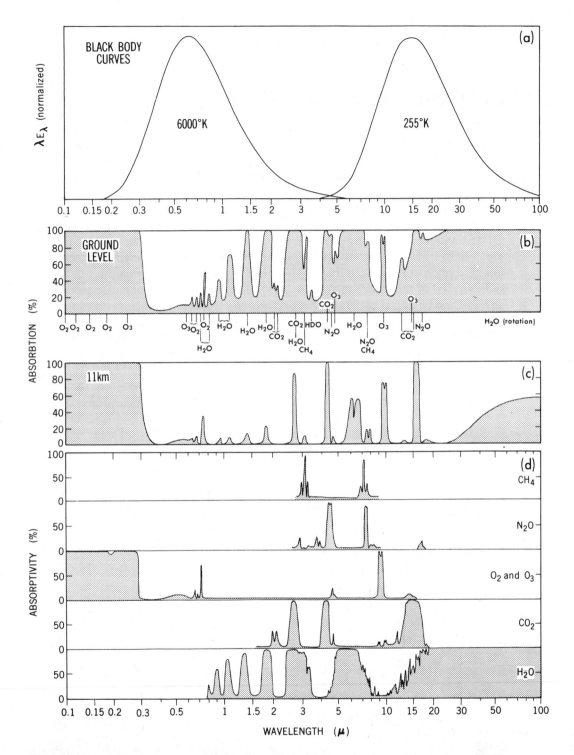

FIGURE 6.2. Black body curves for the solar radiation (assumed to have a temperature of 6000 K) and the terrestrial radiation (assumed to have a temperature of 255 K) (a); absorption spectra for the entire vertical extent of the atmosphere (b) and for the portion of the atmosphere above 11 km (c) after Goody (1964); and absorption spectra for the various atmospheric gases between the top of the atmosphere and the earth's surface (d) after Howard *et al.* (1955) [updated with data from Fels and Schwarzkopf (1988, personal communication) between 10 and 100 μm].

ligible at the wavelengths of solar emission so that only absorption has to be considered. However, emission and absorption are equally important at the frequencies of terrestrial radiation and need to be considered simultaneously.

6.1.2 Global radiation balance

In order to get a global view of how, in the long run, the climatic system maintains its state of quasi-equilibrium, we present in Fig. 6.3 a preliminary diagram of the global radiation balance for the earth. The figure summarizes the annual-mean radiation balance for the climate system as a whole, separating it into the solar and terrestrial radiation budgets on the left- and right-hand sides of Fig. 6.3, respectively. The numbers are scaled so that the incident solar radiation flux is set at 100 units.

Of the 100 units of incoming solar radiation shown on the left-hand side of Fig. 6.3, 16 units are absorbed by stratospheric ozone, tropospheric water vapor and aerosols, 4 units by clouds, and 50 units by the earth's surface. The remaining 30 units of solar radiation are backscattered by the air (6 units), reflected by clouds (20 units), and reflected by the earth's surface (4 units). These 30 units do not participate in the physical and chemical processes that occur in the climate system.

Of the 50 units of solar radiation absorbed at the earth's surface, the right-hand side of Fig. 6.3 shows that 20 units are emitted as long-wave radiation into the atmosphere and 30 units are transferred upward into the atmosphere by turbulent and convective processes in the form of sensible heat (6 units) and latent heat (24 units). Of the 20 units of emitted long-wave radiation, 14 units are absorbed in the atmosphere mainly by water vapor and carbon dioxide and 6 units are emitted directly into space.

Considering now the atmosphere by itself, it absorbs 20 units of solar radiation plus 44 units of energy coming from the earth's surface. This total absorbed amount

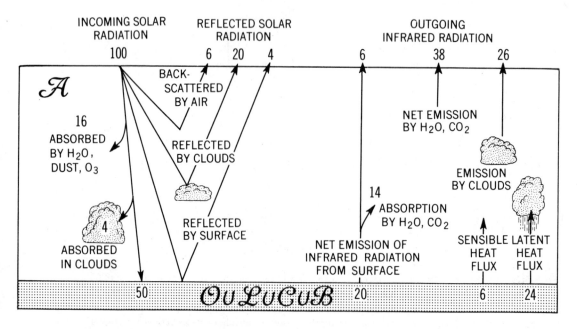

FIGURE 6.3. Schematic diagram of the global radiation budget in the climatic system. A value of 100 units is assigned to the incoming flux of solar energy.

of 64 units is then balanced by the emission into space of infrared radiation by water vapor and carbon dioxide (38 units) and by clouds (26 units). Adding the 6 units of radiation passing directly through the atmosphere from the earth's surface, we find a total loss of 70 units in the form of long-wave radiation at the top of the atmosphere required to balance the 70 units of absorbed short-wave solar radiation.

We will present a more detailed regional analysis of the radiation budget in the following sections and will return to the global energy budget in discussing the ocean–atmosphere heat engine in Chap. 14.

6.2 PHYSICAL RADIATION LAWS

6.2.1 Planck's law

By definition, a black body is a perfect absorber. It also emits the maximum possible amount of energy at a given temperature. The amount and quality of the energy emitted by a black body are uniquely determined by its temperature, as given by Planck's law. This law states that the monochromatic radiance (intensity) of radiation $B_\lambda(T)$ (energy per unit time per unit area per unit solid angle per unit wavelength) emitted by a black body at temperature T is expressed by

$$B_\lambda(T)\,d\lambda = \frac{2hc^2}{\lambda^5[\exp(c\,h/k\lambda T) - 1]}\,d\lambda, \tag{6.1}$$

where $h = 6.63 \times 10^{-34}$ J s is the Planck constant and $k = 1.38 \times 10^{-23}$ J K^{-1} is the Boltzmann constant. This law can also be written in terms of frequency using $\lambda = c/\nu$:

$$B_\nu(T)\,d\nu = \frac{2h\nu^3}{c^2[\exp(h\nu/kT) - 1]}\,d\nu. \tag{6.2}$$

We note that black body radiation is isotropic, i.e., the intensity is independent of the direction.

6.2.2 Stefan–Boltzmann law

The total radiance (or intensity) of a black body can be obtained by integrating the Planck law over the entire wavelength domain from 0 to ∞:

$$B(T) = \int_0^\infty B_\lambda(T)\,d\lambda \sim T^4. \tag{6.3}$$

This emission can be integrated over all angles of a hemisphere covering a horizontal surface, leading to the total flux (energy per unit time) coming from all angles at that surface:

$$\int B(T)\cos\theta\,d\omega\,da = \sigma T^4\,da, \tag{6.4}$$

where σ is the Stefan–Boltzmann constant ($\sigma = 5.670 \times 10^{-8}$ W m^{-2} K^{-4}), θ is the angle between the incoming beam of solar radiation and the vertical, $d\omega = \sin\theta\,d\theta\,d\lambda$ is a solid angle element, and da is an element of surface area. Noting that $B(T)$ is independent of the direction, the integration of the left-hand side of Eq. (6.4) over the entire hemisphere leads to

$$2\pi B(T) \int_0^{\pi/2} \cos\theta \sin\theta \, d\theta = \pi B(T)$$

so that

$$\pi B(T) = \sigma T^4. \tag{6.5}$$

Thus, the flux density (energy per unit area per unit time) emitted by a black body is proportional to the fourth power of the absolute temperature.

6.2.3 Wien displacement law

By setting the derivative of $B_\lambda(T)$ (with respect to λ) in Eq. (6.1) equal to zero, the wavelength of maximum emission λ_{max} can be obtained leading to the Wien displacement law

$$\lambda_{max} T = A = \text{const.} \tag{6.6}$$

When λ is in μm and T in K we find that $A = 2898\,\mu$m K. The Wien displacement law states that for black body radiation the wavelength of maximum emission is inversely proportional to the absolute temperature. Using this relation the temperature of a black body can be determined from the wavelength of maximum monochromatic radiation. The law shows that the radiation emitted from the surface ($T \approx 293$ K) peaks at about 9.9 μm, well in the infrared region of the spectrum. Of course, the radiation emitted from the upper troposphere ($T \approx 255$ K) peaks at a somewhat higher wavelength. Taking the observed value of $\lambda_{max} = 0.474\,\mu$m for the solar radiation we find the sun's surface temperature to be on the order of 6110 K [see Fig. 6.2(a)].

We may note that both the Wien and Stefan–Boltzmann laws were obtained prior to Planck's law by other means. However, now they may be regarded as corollaries of Planck's law.

6.2.4 Kirchhoff's law

The previous laws deal essentially with the intensity emitted by a black body. However, in general, a medium will not only absorb but also reflect part of the incident radiation and transmit the remainder. Thus, in terms of the ratios of the absorbed, reflected, and transmitted radiation with respect to the monochromatic intensity of the radiation I_λ incident upon a layer, we may write

$$a_\lambda + r_\lambda + \tau_\lambda = 1, \tag{6.7}$$

where $a_\lambda = I_{\lambda a}/I_\lambda$ is the absorptivity, $r_\lambda = I_{\lambda r}/I_\lambda$ the reflectivity (or albedo at wavelength λ), and $\tau_\lambda = I_{\lambda \tau}/I_\lambda$ the transmissivity of the layer.

Kirchhoff's law states that in thermodynamic equilibrium and at a given wavelength the ratio of the intensity of emission I_λ to the absorptivity a_λ of any substance does not depend on the nature of the substance. It depends only on the temperature and the wavelength:

$$I_\lambda / a_\lambda = f(\lambda, T). \tag{6.8a}$$

In case of a black body $a_\lambda = 1$ for all values of λ. Therefore, the ratio $f(\lambda, T)$ is equal to $B_\lambda(T)$, the black body intensity for a given temperature and wavelength. For any real body a_λ is less than 1, so that $I_\lambda < B_\lambda(T)$. When we assume that a_λ is the same at all wavelengths we define what is known as a gray body.

In order for I_λ to be different from zero it is necessary that both $B_\lambda(T)$ and a_λ be different from zero. Thus, for a body to be able to emit energy at a given wavelength and at a given temperature it is necessary that a black body also emit energy at that temperature and that the body be able to absorb it. Since the emissivity ϵ_λ is defined as the ratio of the emitted intensity to the Planck function, Kirchhoff's law can also be expressed by

$$\epsilon_\lambda = a_\lambda \tag{6.8b}$$

so that any selective absorber of radiation at wavelength λ is also a selective emitter of radiation at the same wavelength.

6.2.5 Beer–Bouger–Lambert law

This law expresses the change in radiation intensity I_λ due to the absorption of the radiation. Let us consider a parallel beam of radiation with intensity I_λ passing through an absorbing medium. The intensity of radiation after traversing a layer of thickness ds in the direction of propagation is $I_\lambda + dI_\lambda$, and

$$dI_\lambda = -k_{\lambda a} I_\lambda \rho\, ds, \tag{6.9}$$

where ρ is the density of the medium and $k_{\lambda a}$ is the absorption coefficient (absorption cross section in units of area per unit mass) for radiation of wavelength λ. Integration of this equation between $s = 0$ and $s = s_1$ yields the emergent intensity $I_\lambda(s_1)$ so that

$$I_\lambda(s_1) = I_\lambda(0)\exp\left(-\int_0^{s_1} k_{\lambda a}\rho\, ds\right), \tag{6.10}$$

where $I_\lambda(0)$ is the intensity at $s = 0$. When the medium is homogeneous, $k_{\lambda a}$ is independent of s, and Eq. (6.10) then expresses the Beer–Bouger–Lambert absorption law. A similar law is valid for the scattering of a parallel beam passing through the atmosphere. In this case, we must use a scattering coefficient $k_{\lambda s}$. When absorption and scattering occur simultaneously we may write $k_\lambda = k_{\lambda a} + k_{\lambda s}$, where k_λ is called the extinction coefficient. The transmissivity τ_λ of the atmosphere at a given wavelength is then given by

$$\tau_\lambda = I_\lambda / I_{\lambda 0} = \exp\left(-\int_0^\infty k_\lambda \rho\, ds\right), \tag{6.11a}$$

where k_λ is the extinction coefficient along a slant path ds in the direction of propagation through the atmosphere. If the zenith angle of the sun is Z, $ds = dz \sec Z$ and Eq. (6.11a) can be written as

$$\tau_\lambda = I_\lambda / I_{\lambda 0} = \exp\left(-\sec Z \int_0^\infty \rho k_\lambda\, dz\right)$$
$$= \exp(-\sec Z\, u), \tag{6.11b}$$

where the quantity

$$u = \int_0^\infty k_\lambda \rho\, dz \tag{6.12}$$

is called the optical depth or optical thickness.

For normal incidence ($Z = 0$) and when $u = 1$ we find $\tau = e^{-1} = 0.37$ which implies that the initial intensity $I_{\lambda 0}$ is decreased by 63%. If $u = 2$ we find $\tau = e^{-2} = 0.14$ or a decrease in intensity of 86%. During normal atmospheric conditions the optical depth is much less than 1, but for very thick, dark clouds it can be much greater than 1.

6.3 SOLAR RADIATION

6.3.1 Solar spectrum and solar constant

Most solar radiation that affects the climate system is in the ultraviolet, visible, and near infrared regions of the spectrum (see Fig. 6.1). Indeed, 99% of the solar energy reaching the earth has a wavelength between 0.15 and 4.0 μm with 9% in the ultraviolet ($\lambda < 0.4\,\mu$m), 49% in the visible ($0.4 < \lambda < 0.8\,\mu$m), and 42% in the infrared ($\lambda > 0.8\,\mu$m) (Houghton, 1985).

Observations over many years show that the intensity of solar radiation has not changed substantially. For this reason we may introduce a quantity known as the solar constant S. The solar constant is defined as the amount of solar radiation incident per unit area and per unit time on a surface normal to the direction of propagation and situated at the earth's mean distance from the sun. The value of the solar constant is about $1360\,\mathrm{W\,m^{-2}}$. Figure 6.1 gives the spectral distribution of the solar radiation both at the top of the atmosphere and at ground level. The spectrum at the top of the atmosphere closely resembles the spectrum given by Planck's law for a black body with a temperature of about 6000 K. The shaded area in Fig. 6.1 indicates the absorption of the radiation due to various gases when it travels vertically through the atmosphere under clear conditions. The remaining portion between the two curves represents the reduction of the solar radiation due to scattering.

The radiation coming directly from the sun received at the earth's surface on a unit area normal to the solar beam is called direct solar radiation. The amount of scattered radiation coming from all angles, except for the solid angle subtended by the solar disk, is called diffuse solar radiation. The sum of both components as received on a horizontal surface is called global solar radiation.

In further discussions, we will start with the solar radiation at the top of the atmosphere. We will then study the interactions of the solar radiation with the various atmospheric constituents and clouds. Finally, we will evaluate the amount of solar radiation that reaches the earth's surface and its transformations.

6.3.2 Distribution of solar radiation at the top of the atmosphere

The distribution of solar radiation at the top of the atmosphere depends on the geometry of the globe, its rotation, and its elliptical orbit around the sun. Thus, it is a function of the tilt of the axis to the plane of the ecliptic, the eccentricity of the orbit, and the longitude of the perihelion. The mean distance between the sun and the earth is 1.496×10^{11} m and is known as the Astronomical Unit (AU). The maximum earth–sun distance (aphelion) is 1.521×10^{11} m ($= 1.017$ AU), and the minimum distance (perihelion) is 1.471×10^{11} m ($= 0.983$ AU). The eccentricity of the earth's orbit is defined by the ratio of the maximum deviation from a circular orbit and the mean radius itself; the eccentricity has a present value of 0.0167. The present value of the tilt or obliquity of the ecliptic is 23° 27′.

As the solar radiation enters the atmosphere it is depleted by absorption and scattering. The absorbed radiation is directly added to the heat budget, whereas the scattered radiation is partly returned to space and partly continues its path through the atmosphere where it is subject to further scattering and absorption. Maximum depletion of the solar beam tends to occur at high latitudes were the path length through the atmosphere is longest, and minimum depletion in the intertropical regions where the path length is shortest.

The irradiance F_{SW} on a horizontal surface depends on the sun's zenith angle Z:

$$F_{SW} = F_{SW}^0 \cos Z,$$

where F_{SW}^0 is the irradiance normal to the solar beam. Applying spherical trigonometry to the so-called ZPS (zenith-pole-sun) spherical triangle, we find that

$$\cos Z = \sin \phi \sin \delta + \cos \phi \cos \delta \cos h, \qquad (6.13)$$

where ϕ is the latitude, δ the solar declination, and h the hour angle from the local meridian (where $h = 0$). At sunset and sunrise (hour angle $h = H$) $Z = \pi/2$ so that

$$\sin \phi \sin \delta + \cos \phi \cos \delta \cos H = 0 \qquad (6.14a)$$

or

$$\cos H = - \tan \phi \tan \delta, \qquad (6.14b)$$

and the length of day (in radians)

$$2H = 2 \cos^{-1}(- \tan \phi \tan \delta). \qquad (6.15)$$

The length of day ($= 24 H/\pi$ hours) is uniquely defined by the latitude and the time of year through the solar declination (see, e.g., Iqbal, 1983).

The distribution of solar radiation incident on a horizontal plane at the top of the atmosphere depends, of course, not only on $\cos Z$ but also on the earth–sun distance through the inverse square law

$$f(d) = (d_m/d)^2,$$

where d is the actual distance and d_m the mean distance between the sun and the earth. As we have seen before, the term $\cos Z$ depends on the day of year, time of day, and latitude so that for a given instant

$$F_{SW} = S (d_m/d)^2 \cos Z, \qquad (6.16)$$

where S is the solar constant.

Under present astronomical conditions the function $f(d)$ reaches extreme values of 1.0344 in early January and 0.9646 in early July. Secular variations in the orbital parameters of the earth may constitute a major cause for climate changes as those experienced during the Pleistocene ice ages (Milankovitch, 1941; Berger, 1978; Imbrie and Imbrie, 1979).

The total daily insolation at the top of the atmosphere can be determined by integration of Eq. (6.16) over a day:

$$Q_0 = S (d_m/d)^2 \int_{\text{time of sunrise}}^{\text{time of sunset}} \cos Z \, dt. \qquad (6.17)$$

Using Eq. (6.14a) and $dt = (12/\pi) \, dh$, Eq. (6.17) becomes

$$Q_0 = \frac{24}{\pi} S (d_m/d)^2 \left(\int_0^H \sin \phi \sin \delta \, dh \right.$$
$$\left. + \int_0^H \cos \phi \cos \delta \cos h \, dh \right)$$
$$= \frac{24}{\pi} S (d_m/d)^2 (H \sin \phi \sin \delta + \cos \phi \cos \delta \sin H).$$

Again using Eq. (6.14a) we find

$$Q_0 = \frac{24}{\pi} S (d_m/d)^2 \sin\phi \sin\delta (H - \tan H), \quad (6.18)$$

where Q_0, S, and H are given in units of W m^{-2} day, W m^{-2}, and radians, respectively (for more details see Iqbal, 1983). The values of Q_0 as a function of time of year and latitude are presented in Fig. 6.4.

Next, we will consider in more detail the interactions between solar radiation and the atmospheric constituents.

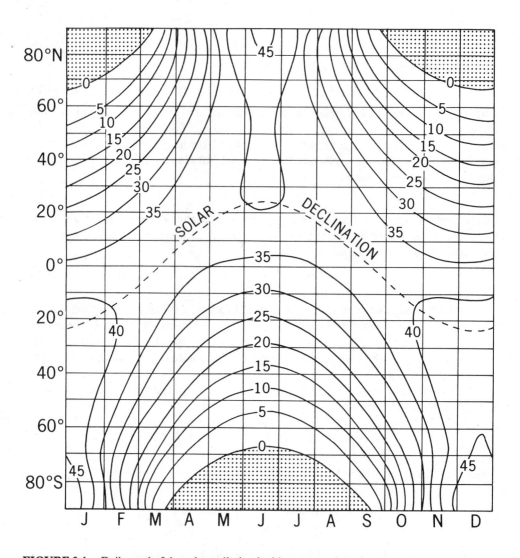

FIGURE 6.4. Daily total of the solar radiation incident on a unit horizontal surface at the top of the atmosphere as a function of latitude and date in 10^6 J m^{-2} (1×10^6 J m^{-2} for one day ≈ 11.6 W m^{-2}). Shaded areas represent the areas that are not illuminated by the sun (from Wallace and Hobbs, 1977; after List, 1951).

6.3.3 Aerosols

When solar radiation penetrates into the atmosphere, it is absorbed and scattered not only by atmospheric gases but also by aerosols and clouds. Aerosols are defined as suspensions of liquid or solid particles in the air, excluding cloud droplets and precipitation. The mean radius of the aerosol particles ranges from about 10^{-4} to $10\,\mu m$. The aerosol particles in the atmosphere are due mainly to two processes: (1) direct injection of dust, soot, sea salt particles, etc., from volcanoes, forest fires, human activities, and other natural processes, and (2) chemical reactions of gaseous materials within the atmosphere, such as the transformation of SO_2 into HSO_4 or sulfates, NO_x into nitric acids, and so on.

The aerosols can be classified according to their origin, size, and atmospheric distribution (Houghton, 1985). The very small particles with mean radius (0.001 to $0.1\,\mu m$) are called Aitken particles, the particles between 0.1 and 1 μm large particles, and those between 1 and 10 μm giant particles. The large and giant particles are responsible for the turbidity (haziness) of the atmosphere. The concentration of aerosols is usually expressed as the number of particles per cm^3. The concentrations are generally greater over the continents than over the oceans. The concentration of Aitken nuclei ranges from values of $150\,000\;cm^{-3}$ in cities to $400\;cm^{-3}$ over the oceans. The concentrations of large and giant particles are much smaller than those of the Aitken particles, and strongly depend on the type of air mass, being greatest in moist, tropical air masses.

Aerosols can influence climate in two main ways. First, aerosol particles can alter the solar energy by absorbing or scattering the radiation both in cloud-free and cloudy conditions, thereby affecting the energy balance of the earth. Second, the particles can serve as condensation and freezing nuclei playing an important role in the process of cloud formation and precipitation.

6.3.4 Absorption of solar radiation

The solar spectrum shows a large number of absorption lines and bands, some resulting from absorption in the sun's atmosphere and others from absorption by the gases of the earth's atmosphere. The main atmospheric gases absorbing solar energy are water vapor (H_2O), carbon dioxide (CO_2), ozone (O_3), oxygen (O_2), nitrogen (N_2) and their oxides (N_2O, NO_2), and methane (CH_4) [see Figs. 6.1 and 6.2(b)–6.2(d)].

Radiation with wavelengths shorter than $0.3\,\mu m$ (ultraviolet, x radiation) is mainly absorbed in the high atmosphere above 20 km by O_3, O_2, O, and N_2. Ionization due to extreme ultraviolet and x radiation occurs at very high levels forming and maintaining the ionosphere. Below 40 km in the stratosphere the absorption of solar radiation is due to O_2 and O_3, mainly in the near ultraviolet. The combination of all these absorptions explains the abrupt cutoff at the wavelength of about $0.3\,\mu m$ of the solar radiation entering the troposphere [as shown in Fig. 6.2(c)].

Ozone is one of the most important stratospheric constituents absorbing solar ultraviolet radiation. It is formed in the stratosphere and mesosphere by photochemical processes and has a maximum concentration between about 20- and 25-km height. However, the concentration and vertical distribution of ozone vary with latitude and season, and on shorter time and space scales as well. Depleting it would result in an increase of the ultraviolet radiation at the earth's surface reducing agricultural production, disrupting the oceanic food chain, increasing skin cancer, etc. Thus, the ozone layer can be regarded as a global sunscreen protecting the biosphere from possible damage by excess ultraviolet radiation.

In the troposphere, the absorption of solar radiation is rather weak; it occurs in the visible and near infrared (0.55 μm $< \lambda <$ 4.0 μm) regions of the spectrum due primarily to H_2O, CO_2, and clouds. Besides absorption, we must also consider the depletion of solar radiation by atmospheric scattering.

In the atmosphere, there are solid and liquid particles (dust, smoke, ions, etc.) that range in size from clusters of a few molecules to clusters with a radius of tens of μm's forming the aerosols. In the lower atmosphere, aerosols heat the troposphere by the absorption of solar energy and decrease the amount of radiation that reaches the surface. They also increase the planetary albedo mainly due to the backscattering of solar radiation. The net effect of aerosols is to increase the temperature of the stratosphere and to decrease the temperature of the troposphere. It has been suggested that changes in volcanism can involve changes in the aerosol content of the atmosphere which, in turn, may alter the energy balance of the climate system. Therefore, aerosols may be an important factor in explaining climate variations in past ages and may also affect the future climate.

6.3.5 Scattering of solar radiation

The light in the atmosphere is diverted or scattered from its direction of propagation when it encounters particles or inhomogeneities, such as air molecules, aerosols, and clouds. The scattering occurs because the index of refraction of the particles differs from that of the homogeneous medium in which they are imbedded (Houghton, 1985). Although the frequency of the scattered radiation does not change, its phase and polarization may change substantially from those of the incident radiation.

Scattering of solar radiation does not lead to a conversion of radiant energy into heat as does absorption. The radiant energy is merely dispersed in all directions as if the particles act as a new source of radiation. Because some of the solar energy is scattered backwards and sideways, the amount of energy that reaches the surface is reduced. Thus, both absorption and scattering lead to a depletion of solar radiation. The radiation that finally reaches the surface is partly reflected and partly absorbed by ocean waters, soil, vegetation, snow, and ice. A large portion of the latter energy is used to evaporate water into the atmosphere, whereas the remainder is transferred down into the ocean by conduction and turbulent heat exchange and up into the atmosphere by similar processes and by the emission of long-wave radiation. The fraction of the total incident solar radiation that is reflected and backscattered is called the albedo, as we have seen before. The albedo of the earth, as measured at the top of the atmosphere, increases with latitude, and changes seasonally (as we will discuss later).

Scattering occurs for particles of all sizes, including molecules. Rayleigh developed the scattering theory for particles with diameters that are small compared with the wavelength of the incident radiation. He showed that the amount of scattering is inversely proportional to the fourth power of the wavelength ($\sim \lambda^{-4}$), so that the shorter the wavelength the more radiation will be scattered. Thus we may explain that the blue color of the sky is due to the more intense scattering by the atmospheric molecules in the blue than in the red part of the spectrum. In fact, the sky is made visible through the scattering process. On the other hand, sunsets and sunrises appear reddish because the shorter (blue) wavelengths in the direct light are removed by scattering during the long path through the atmosphere, leaving the remaining reddish colors of the spectrum.

The shorter radiation is also scattered by particles (dust, smoke, and ions) and impurities that form the aerosol. When the dimensions of the particles increase, the λ^{-4} law ceases to be valid, dispersion is less selective with respect to λ, and Mie scattering theory should be used (van de Hulst, 1957). Mie theory is more generally valid and contains Rayleigh scattering and geometrical optics as limiting cases. When the particles are sufficiently large the dispersion of radiation approaches a $1/\lambda$ dependence, leading to diffuse reflection. This explains why cloud drops and ice crystals reflect or refract radiation in all directions. In general, for large particles the change in direction of the incident radiation may be explained by geometrical optics, such as diffraction, reflection, refraction, or a combination of these effects, producing coronae, haloes, rainbows, etc.

The diffuse radiation of "white" sunlight is also white due to multiple reflections and refractions, explaining the whitish color of clouds and fog.

6.3.6 Effects of clouds on solar radiation

It is well known that clouds play a very important role in the radiation balance and in climate. They affect the albedo, absorptivity, and transmissivity of the incident radiation, as we will show later when we discuss the distributions of cloudiness and albedo. However, the detailed radiative properties of clouds are not well known. They change substantially with cloud type and cloud form. For example, the albedo of a thin stratus is about 30%, whereas the albedo of a thick stratus may vary between 60% and 70%. Nimbostratus clouds have albedo values around 70% and cirrus clouds have much lower albedo values on the order of 20% (Houghton, 1985). The thickness of a cloud as well as the zenith angle of the sun seem to be important factors. Inside the clouds multiple scattering dominates the depletion of solar radiation. The absorption inside clouds occurs in drops, ice crystals, and, to a lesser extent, in water vapor for the near infrared region of the spectrum. Thus the clouds are strong absorbers of terrestrial radiation.

6.3.7 Solar radiation at the earth's surface

We now consider the solar energy that reaches the surface of the earth after passing through the atmosphere. An important fraction of this energy is reflected back into the atmosphere. The surface albedo A_{sfc}, defined as the ratio of the reflected over the incoming radiation, depends strongly on the nature of the surface, vegetation cover, snow cover, etc. Table 6.1 gives albedo values for the averaged solar spectrum in the case of some representative land surfaces. For a water surface the value of the albedo is generally much lower than for a land surface. Depending on the solar zenith angle Z, it changes from 2% to 6% when Z varies between 0° and 60°. The mean value of the surface albedo is on the order of 0.15, whereas the planetary albedo at the top of the atmosphere was found to be about 0.30, mainly due to the high albedo of clouds and the backscattering of the atmosphere.

The surface albedo also depends on the wavelength of the incident radiation, mainly in the case of green vegetation, which strongly reflects in the near infrared. The spectral dependence of the albedo with the type of vegetation is important when we consider photosynthesis, since the growth of plants is a function not only of the amount of radiation but also of its spectral composition. Usually there is strong absorption in the ultraviolet and visible regions of the spectrum while in the near infrared the absorption is small and, in fact, most of the energy is reflected. This has a

TABLE • 6.1. Albedo (%) for various surfaces for radiation in the visible part of the spectrum from Houghton (1985).

Sand	18–28
Grassland	16–20
Green crops	15–25
Forests	14–20
Dense forests	5–10
Fresh snow	75–95
Old snow	40–60
Cities	14–18

beneficial effect on the plants since it prevents overheating in a region of the spectrum where the solar radiation is not required.

The rate of direct conversion of solar radiation by photosynthesis, important as it may be for the biosphere, is relatively small when compared with the total flux of radiation reaching the earth's surface. Moreover, most of the energy stored by photosynthesis will be returned to the oceans or atmosphere by oxidation of organic matter and only a minute fraction will be left in sediments as potential fossil fuels. The fraction of absorbed energy that penetrates into the subsurface layers depends on the thermal conductivity of the materials of the layers. As different substances conduct heat away from the surface at different rates, the resulting vertical temperature profiles are also different. Penetration rates are faster and the heat reaches greater depths in water (~ 100 m) than in land (a few m's) because the heat transfer in soils can only be through molecular conduction, whereas in water (and air) it may occur through turbulent eddy exchange and convection (see Chap. 10).

6.4 TERRESTRIAL RADIATION

6.4.1 Introduction

The absorption of short-wave solar radiation by the atmosphere and the earth's surface leads to heating of the climate system. Thus, according to the Stefan–Boltzmann law [Eq. (6.5)] and the Wien law [Eq. (6.6)], all components of the climate system emit radiant energy in the long-wave region of the spectrum (infrared, microwave, etc.). This is why the terrestrial radiation is sometimes called thermal radiation. This has been stated already in the late 18th century by Prévost when he said that "all bodies have to lose energy." Today we would present the Prévost principle as follows: All bodies with a temperature above 0 K emit energy whose quality (in terms of ν or λ) depends on their temperature.

In the mean, the short-wave solar radiation absorbed by the planet earth must be returned to space as long-wave terrestrial radiation. Thus practically all exchange of energy between the earth and outer space is through radiative transfer.

As we have seen before, all real bodies emit and absorb less radiant energy than a black body at the same temperature and wavelength. Their emissivities (or absorptivities) are less than 1 and vary with the wavelength (Kirchhoff law). In solids and liquids, the variation of emissivity with wavelength is small, but it is very large for gases. Gases in atomic form absorb (and emit) radiant energy only in very narrow distinct wavelength intervals that result from quantized changes in electronic states. They are called spectral absorption lines. However, molecular gases show spectral absorption bands, each one formed by a large number of very close lines. The location of the bands in the spectrum and their strength depend on the molecular structure of the gas. The absorptivities of the gases vary greatly with wavelength, and the absorption spectra are highly irregular and discontinuous.

The net effect of the absorption and emission of long-wave terrestrial radiation in the troposphere becomes clear when we compare Figs. 6.2(b) and 6.2(c) for $\lambda \gtrsim 4\,\mu$m, which give the atmospheric absorption spectra for the entire vertical column and for the layer above 11-km altitude, respectively. The absorption spectrum for the entire vertical column results from the superposition of the absorption spectra of the individual constituents of the atmosphere shown in Fig. 6.2(d).

In solids and liquids, the atoms and molecules are so closely packed that they do not show the discontinuous spectra typical of gases. Thermal vibrations and rotations of the molecules produce continuous emission (absorption) spectra with an emissivity slightly less than 1. It is therefore convenient to treat the radiant energy emission separately for the earth's surface and the atmosphere.

Let us start by considering the long-wave (thermal) emissivity at the surface. Various types of surfaces have slightly different emissivities: on the order of 0.92 for land, 0.98 for vegetation, and 0.96 for water. For most applications the emissivity can be assumed to be uniform both for land and water. In some cases it can even be assumed that the earth's surface acts like a black body for long-wave radiation.

6.4.2 Absorption and emission spectra of atmospheric gases

The absorption and emission of radiation in gases occurs at specific wavelengths according to their atomic and molecular structure. Isolated atoms and molecules, as we may consider the gases of the atmosphere to be, can only absorb and emit energy at certain discrete energy states and can only undergo discrete changes between these states. As predicted by quantum mechanics the energies involved in the transitions are quantized. Thus, the interaction of an atom or molecule with electromagnetic radiation, such as light, can only take place at well-defined frequencies that are characteristic of that molecule and of the corresponding pair of energy values between which the transition is taking place. This can be expressed in quantum-mechanical terms by

$$W_i - W_k = h\nu, \qquad (6.19)$$

where W_i and W_k are the energies of an energetically higher and lower state, respectively, ν is the frequency of radiation, and h the Planck constant. In other words, the radiant energy has to be in resonance with the energy gap in order to make the molecule "jump." If a molecule changes from a state of lower energy to one of higher energy it needs to absorb the necessary energy quanta (photons) $h\nu$ out of the radiation. If the molecule changes its energy from a higher to a lower state, the energy difference is liberated, also in the form of photons. The frequency of absorption and emission of photons has the same value, given by Eq. (6.19).

Each chemical element or combination of elements has a characteristic absorption and emission spectrum showing the frequencies at which they absorb or emit radiation. The spectra are essentially discontinuous and consist of lines. The spectral distributions are explained by the fundamental principles of atomic physics and quantum mechanics, involving quantum-mechanical selection rules whose treatment is beyond the scope of this book. When an atom absorbs radiant energy, the electrons in the atom can go to a higher discrete orbit with some restrictions based on the selection rules and the conservation of angular momentum. When the electrons return to their original base state (ground state), they emit energy at a frequency determined by the energy difference in the two orbits. In summary, we can say that an atom has certain preferred (natural) frequencies of emission based on its structure, just as a pendulum has a natural period depending on its length.

The atmospheric absorption spectra consist of many lines that correspond to electronic energy transitions characteristic of each particular atomic species. In the case of an atomic emission spectrum it is, of course, necessary that the higher energy states of the atom be populated, so that the atoms can emit energy when an electron moves to an orbit closer to the nucleus. Thus, the emission lines can result from either transitions to the ground state or from transitions between excited states. If radiation penetrating a gas cannot excite the atoms or molecules in the gas its energy will not be absorbed or emitted (Kirchhoff's law).

The absorption of a molecular gas occurs in bands which consist of a large number of closely spaced spectral lines. The molecular spectral bands in the ultraviolet and visible ranges are due to electronic energy transitions. However, at the relatively low temperatures of the atmosphere and the earth's surface electronic bands are not important because the higher energy states are not sufficiently populated.

6.4.3 Rotational and vibrational bands

In this section, we give a simplified account of the infrared absorption (and emission) spectra of the principal absorbing gases of the atmosphere.

The emission spectra of molecules are usually more complex than those of individual atoms because they have more degrees of freedom. For example, the atoms of a diatomic molecule can rotate about a common axis leading to rotational energy. Another mode can occur when two or more atoms vibrate towards or away from each other. According to quantum mechanics there are certain preferred modes of behavior for each gas that give the specific emission characteristics for that gas. Although the energy amounts involved in the quantized vibrational and rotational energy transitions are much smaller than those in the electronic transitions, they are of paramount importance in the infrared and microwave regions of the spectrum that dominate the emission of terrestrial radiation.

Diatomic molecules can only have rotational or vibrational spectra if the rotation or vibration results in an oscillating electric dipole moment. Because the most abundant molecules in the atmosphere, N_2 and O_2, have no electric dipoles due to their symmetric charge distribution, they show no vibrational or simple rotational spectra. Their absorption and emission spectra are caused by electronic transitions, and are therefore in the ultraviolet and visible regions of the electromagnetic spectrum.

The principal atmospheric gases that are active in the long-wave range of the spectrum are H_2O, CO_2, and O_3. They are all triatomic. Their structure and their principal vibrational modes are shown schematically in Fig. 6.5.

The atoms of water vapor have a triangular configuration with an oxygen atom at the apex. There are three orthogonal axes of rotation passing through the center of mass of the molecule with different moments of inertia about each axis. The combination of the rotational and vibrational states leads to a very complex and irregular absorption spectrum for water vapor. Vibrational bands occur at higher energies

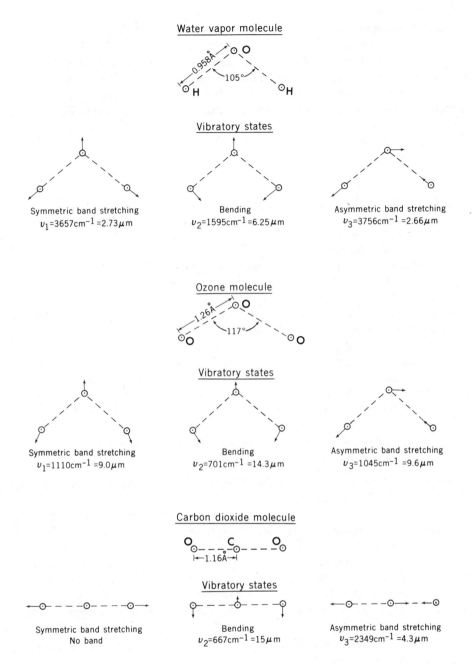

FIGURE 6.5. The structures of the water vapor, ozone, and carbon dioxide molecules and their vibratory states (from Houghton, 1985).

(higher frequencies) than the rotational bands, and the energies of vibrational quanta are two orders of magnitude larger than those of the rotational quanta. Water vapor has a pure rotational band extending upward from 14 μm and centered at 65 μm, and has several vibrational–rotational bands in the 1–8-μm region [see Figs. 6.2(d) and 6.5].

Carbon dioxide is another important constituent of the atmosphere causing strong absorption in the far infrared. Because CO_2 is a linear symmetric molecule, its rotation does not produce an oscillating dipole moment and it has no rotational bands. Its absorption spectrum shows maxima at 2, 3, and 4 μm and in the 13–17-μm region.

Ozone, which is abundant in the stratosphere, contributes to the absorption of the terrestrial radiation mainly in one band centered at 9.6 μm, which is of a vibrational–rotational nature [see Figs. 6.2(d) and 6.5]. In the troposphere, ozone is only a minor tracer and its influence in the radiative exchanges can be disregarded. Therefore, the principal absorbers in the troposphere are water vapor and carbon dioxide, whereas in the stratosphere all three are important.

6.4.4 Spectral lines–Lorentz formula

The spectral lines in the emission (absorption) spectrum of an isolated molecule have a very small but finite width on the frequency scale as a consequence of Heisenberg's uncertainty principle (which states the impossibility of having exact simultaneous measurements of both the position and velocity of a particle).

As we know from the kinetic theory of gases, the molecules of a real gas are in rapid and random motion. The speeds obey Maxwell's distribution law when collisions are very frequent. A spectral line results from the emission by a large number of molecules undergoing the same transition that is characteristic of the observed line. Monochromatic emission is almost never observed because in real situations the energy levels change slightly due to external perturbations such as collisions and due to random motions of the molecules. The collisions between emitting molecules and between emitting and nonemitting molecules lead to a broadening of the emission line. The line broadening depends on the rate at which collisions take place, the concentration of the molecules (density), and the mean molecular speed ($\sim \sqrt{T}$). In terms of thermodynamic quantities, the collision broadening is proportional to the product $\rho \sqrt{T}$, or to $p/\sqrt{T}$. As the range of atmospheric pressure is much greater than that of $\sqrt{T}$, this effect is called pressure broadening.

Let us now consider another important broadening process arising from random motions of the molecules so that some move towards the observer and others move away from the observer. Due to the Doppler effect, the molecules moving toward or away from the observer at a speed v appear to emit at a higher or lower frequency $\nu = \nu_0(1 \pm v/c)$. The resulting emission line occupies a much wider spectral interval than the basic line; this is called Doppler broadening. Since this process depends on the mean velocity v of the molecules, the Doppler broadening is also proportional to $\sqrt{T}$.

The most commonly used formula for the shape of a line is the Lorentz line shape

$$k_\nu = \frac{S}{\pi} \frac{\alpha}{(\nu - \nu_0)^2 + \alpha^2}, \tag{6.20}$$

where k_ν is the absorption coefficient, S is the line intensity defined by $S = \int k_\nu \, d\nu$, ν_0 is the central frequency of the line, and α is the half-width (defined as one-half of the distance between the two frequencies where k_ν equals half of the value at the center of the line). The half-width of the line is directly proportional to the pressure and inversely proportional to $\sqrt{T}$:

$$\alpha = \alpha_0 \frac{p}{p_0} \left(\frac{T_0}{T}\right)^{1/2}, \tag{6.21}$$

where α_0 is the half-width at a standard pressure p_0 and standard temperature T_0.

At levels below 30 km the width of the emission line is largely determined by pressure broadening, and the Lorentz line shape is acceptable. However, in the upper stratosphere and mesosphere the Doppler effect becomes important so that both broadening processes must be taken into consideration, as is the case with the Voigt line shape (Liou, 1980).

6.4.5 Transmissivity functions

As mentioned before the spectral bands of the various atmospheric emitters (absorbers) are very irregularly distributed. Moreover, each band contains thousands of emission lines with different intensities. Due to the rapid variation of the absorption coefficient k_ν with frequency in the vibrational and rotational portions of the infrared spectrum, it is almost impossible to take into account the fine structure of the lines and the details of the path in the absorbing medium. Thus, it is not feasible to obtain the transmissivity by direct line-by-line integration over the spectrum.

To circumvent this difficulty, we consider the radiation integrated over a finite spectral interval $\Delta\nu$ for which an effective transmissivity function can be obtained by theory or experiment. The transmissivity function is usually evaluated by averaging over a relatively large spectral interval $\Delta\nu$ that contains several absorption lines:

$$\tau_{\Delta\nu} = \frac{1}{\Delta\nu} \int_{\Delta\nu} \exp\left(-\int_0^s k_\nu \rho \, ds\right) d\nu. \tag{6.22}$$

In order to evaluate τ, we must know how the absorption coefficient varies with frequency within the spectral interval $\Delta\nu$. Accepting that in the lower atmosphere the lines are well represented by the Lorentz shape, we find from Eqs. (6.20) and (6.21) that k_ν depends on the pressure and temperature along the path. If the atmosphere is assumed to be homogeneous, the absorption coefficient is independent of the path and the integrations in ν and s can be done independently.

6.4.6 Band models

The basic quantitative information on absorption spectra comes from laboratory experiments using conditions that, to some degree, resemble those observed in the atmosphere. However, in the laboratory the paths are homogeneous and it is not possible to duplicate the continuous changes in pressure and temperature observed in the atmosphere.

Another approach has been to use theoretical band models that are idealized representations in the spectral intervals containing one or several lines with a given shape, usually the Lorentz shape. The choice of the particular band model to be used depends on the type of absorber and its spectral characteristics.

The most common models in use are the single line band model, the regularly spaced (Elsasser) band model, and the statistical (Goody) band model. The first model assumes a frequency interval $\Delta\nu$ that is very large compared with the half-width of the single spectral lines. This model can be used in the case of various well-separated lines within the interval, as is the case for the atmosphere above 20–30 km. The Elsasser band model consists of an array of regularly shaped overlapping lines of uniform intensity. This model is suggested by the carbon dioxide absorption spectrum. The Goody stochastic model assumes a line distribution that is random in frequency with a normal distribution of the intensity of the lines. This model was suggested by the apparent random positions of the water vapor lines in the infrared absorption spectrum (Goody, 1964).

6.4.7 Nonhomogeneous paths

The use of band models allows the integration of τ_ν over frequency when the absorption coefficient k_ν is independent of the path, i.e., for a homogeneous atmosphere (constant temperature and pressure). However, in the real atmosphere the Lorentz shape varies strongly with pressure and, to a lesser degree, with temperature so that the nonhomogeneity of the real atmosphere must be taken into account. This necessitates certain physical adjustments in the formulation.

A simplified procedure is given by the Curtis–Godson approximation which permits the application of the homogeneous path formulation to the case of a nonhomogeneous path. In the Curtis–Godson approximation, a mean value of the half-width $\bar{\alpha}$ is used in the Lorentz line shape instead of the actual half-width α when integrating along an inhomogeneous path:

$$\bar{\alpha} = \int_0^s S(T)\alpha(p,T)\rho\, ds \bigg/ \int_0^s S(T)\rho\, ds. \qquad (6.23)$$

The Curtis–Godson approximation becomes even simpler when the temperature dependence is disregarded, so that

$$\bar{\alpha} = \int_0^s \alpha(p)\rho\, ds \bigg/ \int_0^s \rho\, ds \qquad (6.24a)$$

or

$$\bar{p} = \int_0^s p\, \rho\, ds \bigg/ \int_0^s \rho\, ds, \qquad (6.24b)$$

since according to Eq. (6.21) $\alpha \propto p$. In this case one may use the homogeneous path data at a mean pressure value, which is the averaged pressure along the path weighted by $\rho\, ds / \int \rho\, ds$.

6.5 RADIATIVE TRANSFER

6.5.1 Schwarzchild equation

We will now analyze the transfer of terrestrial radiation through an absorbing and emitting atmosphere. At these infrared wavelengths absorption and emission are equally important and they must be taken into account simultaneously. From Kirchhoff's law [Eq. (6.8)] we know that an absorbing slab in the atmosphere also emits radiation proportionally to the absorption at the same frequency. Let us consider a beam of monochromatic radiation traveling through an absorbing medium. The change in intensity due to absorption after traversing a distance ds in the direction of propagation is $-k_\lambda I_\lambda \rho\, ds$; see Eq. (6.9). By Kirchhoff's law, the intensity emitted in the direction of the beam is $k_\lambda B_\lambda(T)\rho\, ds$. Therefore, the net balance of the intensity for the radiative transfer through the slab ds including absorption and emission for a given wavelength will be

$$dI_\lambda = -k_\lambda I_\lambda \rho\, ds + k_\lambda B_\lambda(T)\rho\, ds$$

or

$$\frac{dI_\lambda}{dm} = -k_\lambda[I_\lambda - B_\lambda(T)], \qquad (6.25)$$

where $dm = \rho\, ds$ is the element of mass crossed by the radiation. This is Schwarzchild's equation. It shows that it is theoretically possible to determine the intensity of radiation at any point of the atmosphere provided that the distribution of absorbing mass and the absorption coefficients are known.

Although Eq. (6.25) is the basic equation of radiative transfer, for infrared radiation it is often more convenient for practical applications to derive a different form of the equation that explicitly incorporates the transmissivity function (or the emissivity). In that form an integration over relatively broad spectral intervals is easier to perform, and more efficient computational procedures can be used.

6.5.2 Radiative transfer equation

Let us consider an atmospheric layer bounded by horizontal planes at z_0 and z_1, and let us assume that the absorbing constituents are uniformly stratified in the horizontal direction, as shown in Fig. 6.6. We then want to evaluate the intensity of the radiation that emerges from the layer at the reference level z_1. By convention, the optical depth or optical path length $u = \int k\rho\, ds$ for the upcoming radiation is considered positive downward starting at the reference level z_1 where $u = 0$. We will assume that the intensity of the vertically upward radiation at the bottom level z_0 is I_0.

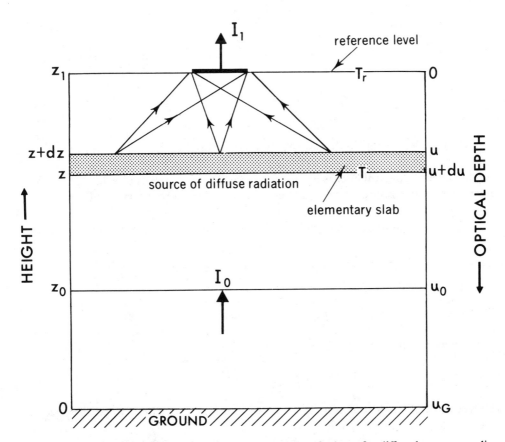

FIGURE 6.6. Model used to calculate the transmissivity of a layer for diffuse long-wave radiation. I_0 and I_1 are the directional intensities of radiation coming from below at levels z_0 and z_1, respectively. Integration of I_1 over all angles of the hemisphere leads to the upward radiation flux $F^{\uparrow}(z_1)$ at level z_1.

Let us consider an elementary slab dz with optical depth $du = k\rho\,ds$ at the level z. Denoting the temperature at that level by T, the intensity of radiation dI emitted upwards is given by $B(z)\,k\rho\,dz$. However, the radiation emitted by the slab will be partly absorbed before it reaches the reference level z_1 so that the transmitted radiation is

$$dI = \tau(z, z_1)\, B(z)\, k\rho\, dz. \tag{6.26}$$

The quantity $\tau(z, z_1)$ is the transmissivity between levels z and z_1, defined before as

$$\tau(z, z_1) = \exp\left(-\int_z^{z_1} k\rho\, dz\right).$$

Using this last expression, Eq. (6.26) can be written in the form

$$\frac{dI}{dz} = B(z)\frac{d\tau(z, z_1)}{dz}. \tag{6.27}$$

Equation (6.27) may also be written in the traditional form as a function of the optical path length u rather than of z:

$$\frac{dI}{du} = B(T, u)\frac{d\tau(u)}{du}. \tag{6.28}$$

Integration of Eq. (6.27) from z_0 to z_1 yields

$$I_1 = I_0 \tau(z_0, z_1) + \int_{\tau(z_0, z_1)}^{1} B(z)\, d\tau. \tag{6.29}$$

If we include the integration over frequency this equation can be generalized to the form

$$I_1 = \int_0^\infty I_{0\nu}\tau_\nu(z_0, z_1)\, d\nu + \int_0^\infty \int_{\tau_\nu(z_0, z_1)}^{1} B_\nu(z)\, d\tau_\nu\, d\nu. \tag{6.30}$$

We mentioned before that in theory the net flux for all frequencies can be obtained through line-by-line integration over the spectrum, but that in practice a summation over all spectral lines is impossible. Thus, to circumvent this difficulty, it was necessary to use appropriate band models and frequency intervals of sufficiently large width for the computations.

For a given frequency interval $\Delta\nu_i$, Eq. (6.30) is then transformed into

$$I_1 = I_{0i}\tau_i(z_0, z_1) + \int_{\tau_i(z_0, z_1)}^{1} B_i(z)\, d\tau_i. \tag{6.31}$$

The subscript i on the different quantities indicates that the corresponding values are spectrally averaged over the interval $\Delta\nu_i$. Thus, $B_i(z)$ is the spectrally averaged value of the Planck function in the interval $\Delta\nu_i$, defined by

$$B_i(z) = \frac{1}{\Delta\nu_i}\int_{\Delta\nu_i} B_\nu(z)\, d\nu.$$

However, the Planck function can often be considered constant when the spectral interval $\Delta\nu_i$ is relatively small.

As mentioned before, the thermal radiation emitted by the earth's surface and by the atmosphere is not in a parallel beam but diffuse. Thus, the radiation reaching the reference level z_1 comes from all directions, and we must integrate Eqs. (6.30) and (6.31) over all solid angles of a hemisphere centered at the specified area element on the plane $z = z_1$. In other words, we must transform the directional intensity I_1 into an upward flux $F_i(z_1)$:

$$F_i^\uparrow(z_1) = \pi I_{0i} \tau_i^*(z_0, z_1) + \int_{\tau_i^*(z_0,z_1)}^1 \pi B_i(z) d\tau_i^*. \tag{6.32}$$

The quantity π comes from the hemispheric integration over all angles below the plane $z = z_1$.

The transmissivity τ in Eq. (6.31) has been replaced by τ^*, the hemispheric spectrally averaged transmissivity or the diffuse transmissivity. It is often approximated by scaling the parallel beam optical length by a factor 1.66, so that

$$\tau^*(u) = \tau(1.66u). \tag{6.33}$$

Choosing now the earth's surface as level z_0 in Eq. (6.32), the upward flux at level z_1, for the frequency interval $\Delta\nu_i$ is given by

$$F_i^\uparrow(z_1) = \pi B_i(0) \tau_i^*(0, z_1) + \int_{\tau_i^*(0,z_1)}^1 \pi B_i(z) d\tau_i^*, \tag{6.34a}$$

where the ground level ($z = 0$) is assumed to act as a black body, i.e., $I_{0i} = B_i(0)$.

The corresponding downward flux of radiation $F_i^\downarrow(z_1)$ at the reference level z_1 is given by a similar expression as Eq. (6.34a) but with different limits of integration, namely $\tau = 1$ and $\tau^*(z_t, z_1)$ where z_t designates a very high level in the atmosphere where the concentration of the absorbers becomes negligible:

$$F_i^\downarrow(z_1) = \int_{\tau^*(z_t,z_1)}^1 \pi B_i(z) d\tau_i^*. \tag{6.34b}$$

The total fluxes integrated over all frequency intervals are then given by

$$F(z) = \sum_i F_i(z) \Delta\nu_i. \tag{6.35}$$

We should note that the use of the previous equations may be severely limited by assuming that the Planck function B is constant in the spectral interval $\Delta\nu_i$. One way to avoid this limitation is to integrate Eq. (6.32) by parts (Houghton, 1985) so that

$$F_i^\uparrow(z_1) = \pi [I_{0i} - B_i(z_0)] \tau_i^*(z_0, z_1)$$
$$+ \pi B_i(z_1) - \pi \int_{z_0}^{z_1} \tau_i^*(z, z_1) \frac{dB_i(z)}{dz} dz. \tag{6.36}$$

This procedure is helpful because $dB_i(z)/dz = [\partial B_i(T)/\partial T] dT/dz$ is an analytic function whereas τ_i is an empirical quantity.

An alternative approach for calculating the fluxes is to use the so-called emissivity method, in which the integration of the emission over spectral intervals may become more efficient. If we define absorptivity as $a_i^* = 1 - \tau_i^*$ and assume the earth's surface to radiate as a black body, the previous equation (6.36) can be written in the form

$$F_i^\uparrow(z_1) = \pi B_i(z_0) + \pi \int_0^{z_1} a_i^*(z, z_1) \frac{dB_i(z)}{dz} dz. \tag{6.37}$$

Summing over all frequency intervals i, the total upward infrared radiation becomes

$$F^\uparrow(z_1) = \pi B(z_0) + \pi \int_0^{z_1} \bar{\epsilon}(z, z_1) dB(z),$$

where the weighted emissivity $\tilde{\epsilon}$ is given by

$$\tilde{\epsilon}(z,z_1) = \sum_i a_i^*(z,z_1) \frac{dB_i(z)}{dB(z)}$$

and $B(z)$ is integrated over all wave numbers $[\pi B(z) = \sigma T^4(z)]$. Of course, $\tilde{\epsilon}$ depends on the nature and concentration of the main emitters and on the pressure and temperature.

6.6 RADIATION BALANCE OF THE ATMOSPHERE

The heating or cooling of an atmospheric layer due to the change in net solar and terrestrial radiation with height can be calculated using the principle of conservation of energy. Let us consider a layer of the atmosphere between levels z and $z + \Delta z$ where the net vertical fluxes of radiation are $F(z)$ and $F(z + \Delta z)$, respectively. Then we find that

$$\rho c_p \Delta z \left(\frac{\partial T}{\partial t}\right)_{rad} = \frac{\partial F_{net}}{\partial z} \Delta z$$

or

$$\left(\frac{\partial T}{\partial t}\right)_{rad} = \frac{1}{\rho c_p} \frac{\partial F_{net}}{\partial z}, \qquad (6.38)$$

where $F_{net} = F^\downarrow - F^\uparrow$.

If we express Eq. (6.38) in °C/day, we obtain

$$\left(\frac{\partial T}{\partial t}\right)_{rad} = \frac{8.64 \times 10^4}{\rho c_p} \frac{\partial F_{net}}{\partial z},$$

where the divergence is given in W m^{-3}, c_p in J kg^{-1} K^{-1}, and ρ in kg m^{-3}.

The application of this expression to the long-wave radiation component shows that this component generates a net cooling on the order of 2.5 °C per day. On the average, the heating of the atmosphere by absorption of solar radiation is only on the order of 0.5 °C/day, and therefore does not compensate for the cooling by long-wave radiation. The maintenance of a steady state in the atmosphere is only possible through the transfer of sensible heat (enthalpy) and latent heat (evaporation–condensation) from the earth's surface to the atmosphere.

In the daytime and with clear skies, the net radiation balance is dominated by the short-wave solar radiation but at night the balance is, of course, entirely due to the long-wave radiation. Because the concentrations of some of the absorbers (e.g., H$_2$O and O$_3$) vary strongly with height the solar and terrestrial radiation fluxes are also strongly height dependent. In general, in clear and calm nights the long-wave radiative flux increases with height, leading to a flux divergence and a cooling of the atmosphere.

Clouds have a strong effect on the transfer of long-wave radiation because they modify the emissivity of the atmosphere at certain wavelengths. Clouds are almost completely opaque for infrared radiation. They act as if they close the atmospheric window, preventing the escape of long-wave radiation into space. This effect is large enough to strongly influence the surface temperature. For example, under cloudy conditions the nocturnal temperature drop will be much reduced as compared with clear night conditions. Even the presence of a thin layer of cirrus clouds can be enough to cause an increase in surface air temperature because of the additional long-wave radiation emitted by the clouds.

Manabe and Strickler (1964) used a simple one-dimensional climate model to study the contributions by H_2O, CO_2, and O_3 to the atmospheric heating and cooling rates (see Chap. 17). Their model was based on the radiative transfer equations given earlier in this chapter taking into consideration the main absorption bands of the absorbers. Their results for the case that the cloudiness is equal to the global mean value are shown in Fig. 6.7.

The troposphere shows a net radiative cooling mainly due to water vapor, which is compensated by the latent and sensible heating associated with moist convection. In the stratosphere, there is a strong heating due to the absorption of ultraviolet solar radiation by ozone and to a much smaller extent, due to the absorption of long-wave terrestrial radiation in the 9.6-μm band [see Fig. 6.2(b)]. We may note that the long-wave heating is only possible because of the low concentrations of O_3 in the troposphere. The cooling in the stratosphere is due to long-wave emission mainly by CO_2 and to a lesser extent by water vapor and ozone. In the one-dimensional model of Manabe and Strickler, the heating and cooling in the stratosphere compensate because the stratosphere is assumed to be in radiative equilibrium.

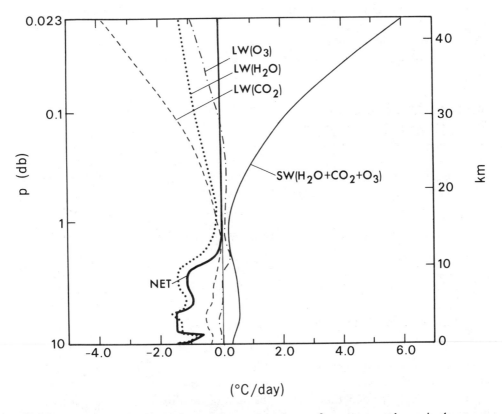

FIGURE 6.7. Vertical distributions of the computed rate of temperature change in the atmosphere for thermal equilibrium due to various absorbers. LW(H_2O), LW(CO_2), and LW(O_3) show the rate of temperature change due to long-wave radiation emission and absorption by water vapor, CO_2, and O_3, respectively. SW($H_2O + CO_2 + O_3$) shows the rate of temperature change due to the absorption of solar radiation by these three gases. "Net" means the net rate of temperature change due to all components (adapted from Manabe and Strickler, 1964).

In the 8–12-μm region the atmosphere is almost transparent for long-wave terrestrial radiation, with the exception of the absorption at 9.6 μm by ozone [see Figs. 6.2(b) and 6.2(c)]. This region of the spectrum is therefore known as the atmospheric spectral window. It is also the range where the long-wave radiation for the atmosphere is a maximum [see Eq. (6.6)]. It is interesting to note that, through absorption in the spectral window, small increases of CO_2 or of other trace gases may have a large impact on the climate. In fact, Ramanathan *et al.* (1985) have brought attention to the fact that some minor trace gases, such as chlorofluorocarbons, methane (CH_4), and nitrous oxides (N_2O), also have absorption lines in the spectral window. The recent increase in the concentrations of these trace gases can lead to additional trapping of the radiation emitted by the earth's surface and thereby to heating of the troposphere and cooling of the stratosphere.

Several other studies have been made using observed distributions of the absorbers in the atmosphere as, e.g., the study by Dopplick (1979). His results show a net radiative cooling throughout the year at almost all latitudes in the troposphere. However, in the stratosphere he finds heating at low latitudes and cooling at mid and high latitudes. All the studies mentioned so far are limited by an incomplete knowledge of the concentrations of the various absorbers in the upper troposphere and stratosphere and by the difficulties of incorporating the effects of clouds.

Most of the atmospheric gases have bands in the microwave region of the spectrum. These bands are not important in the transfer of infrared radiation in the atmosphere, because in the microwave region the thermal radiative fluxes are very small. However, they are being used to infer the vertical profiles of temperature, moisture, and liquid water in clouds through remote sensing from satellites.

6.7 RADIATION BALANCE AT THE EARTH'S SURFACE

The net flux of radiation at the earth's surface results from a balance between the solar and terrestrial radiation fluxes:

$$F_{\text{rad}}^{\text{sfc}} = F_{\text{SW}} + F_{\text{LW}}.$$

The short-wave and long-wave radiation balance can be expressed by

$$F_{\text{SW}} = F_{\text{SW}}^{\downarrow} - F_{\text{SW}}^{\uparrow}$$

and

$$F_{\text{LW}} = F_{\text{LW}}^{\downarrow} - F_{\text{LW}}^{\uparrow}.$$

Therefore, the overall radiation balance becomes

$$F_{\text{rad}}^{\text{sfc}} = F_{\text{SW}}^{\downarrow} - F_{\text{SW}}^{\uparrow} + F_{\text{LW}}^{\downarrow} - F_{\text{LW}}^{\uparrow},$$

where the downward and upward arrows denote the incoming and outgoing radiation components, respectively.

The incident solar radiation $F_{\text{SW}}^{\downarrow}$ is the sum of the direct and diffuse solar radiation. It has a pronounced diurnal and seasonal variation, and is also strongly affected by clouds. The outgoing short-wave solar radiation is the part reflected by the surface $F_{\text{SW}}^{\uparrow} = A_{\text{sfc}} F_{\text{SW}}^{\downarrow}$, where A_{sfc} is the surface albedo so that the net short-wave radiation is

$$F_{\text{SW}} = (1 - A_{\text{sfc}}) F_{\text{SW}}^{\downarrow}.$$

The incoming long-wave radiation $F^\downarrow_{LW}$ comes from the atmosphere and depends on the vertical temperature profile, the clouds, and the vertical distribution of the absorbers. It does not show a significant diurnal variation. The outgoing long-wave radiation $F^\uparrow_{LW}$ is given by the Stefan–Boltzmann law, assuming a given emissivity ϵ for the earth's surface. As expected, it follows the diurnal cycle in surface temperature with a maximum value in the early afternoon and a minimum value in the early morning. The incoming and outgoing long-wave radiation components have the same order of magnitude so that the net long-wave radiation flux is small compared to the net solar flux. However, it becomes important at night when $F_{SW} = 0$. Usually $|F^\downarrow_{LW}| < |F^\uparrow_{LW}|$ implying a long-wave radiative cooling of the surface.

The net radiation flux at the surface is then given by

$$F^{sfc}_{rad} = F^\downarrow_{SW}(1 - A_{sfc}) - \epsilon\sigma T^4_{sfc} + F^\downarrow_{LW}. \tag{6.39}$$

This net radiation heats the surface.

The infrared radiation emitted by the surface $\epsilon\sigma T^4_{sfc}$ is strongly absorbed by water vapor and carbon dioxide in the atmosphere. In turn, the atmosphere will re-emit the absorbed energy both upward and downward. The downward component will be absorbed by the earth's surface and heat it. Thus, the temperature of the earth will be higher than it would be if the atmosphere were transparent for the long-wave radiation. This so-called greenhouse effect was mentioned earlier in Chap. 2.

The observed world-wide increase in the concentrations of CO_2 and other trace gases in the atmosphere must enhance the greenhouse effect and may lead to a net increase in surface temperature over the earth.

The main part of the energy absorbed at the surface is used to evaporate water, another part is lost to the atmosphere as sensible heat, and a smaller part is lost to the underlying layers or used to melt snow and ice. Thus, there are essentially four types of energy fluxes at the earth's surface. They are the net radiation flux F_{rad}, the (direct) sensible heat flux $F^\uparrow_{SH}$, the (indirect) latent heat flux $F^\uparrow_{LH}$, and the heat flux into the subsurface layers $F^\downarrow_G$. Under steady conditions the balance equation for the energy is given by

$$F^{sfc}_{rad} - F^\uparrow_{SH} - F^\uparrow_{LH} - F^\downarrow_G - F_M = 0, \tag{6.40}$$

where F_M is the energy involved in melting snow and ice or in freezing water. This energy budget equation will be analyzed in more detail in Sec. 10.2.

6.8 OBSERVED RADIATION BALANCE

As is well known, the radiational energy of the sun constitutes the basic driving force for the atmospheric and oceanic general circulations. Thus, the most important external factor for the earth's climate is the total incoming solar radiation. As mentioned before in Sec. 6.3, the incoming solar radiation is characterized by the solar constant, the obliquity, the eccentricity, and the longitude of perihelion for the earth's orbit around the sun. Variations in these astronomical factors may be closely linked to observed variations in the earth's climate at time scales of thousands of years and longer (see Figs. 2.5 and 2.7). However, they do not seem important at the decadal time scale which we are concerned with in this book.

6.8.1 Radiation balance of the earth

One of the most important scientific contributions of meteorological satellites has been the measurement of the radiation budget at the top of the atmosphere. This involves measuring both the solar short-wave input and the terrestrial long-wave output components. Measurements of the planetary albedo combined with the known impinging solar radiation at the top of the atmosphere supply the necessary information on the first component of the radiation budget, the net solar input. Uncertainties in the present albedo values are related to inadequate diurnal sampling and incomplete knowledge of how to extrapolate from measurements at one zenith angle to full half-sphere values. The second component of the radiation budget involves measuring the long-wave, terrestrial radiation. Less uncertainty seems to be involved in this determination, except for effects associated with the diurnal cycle in cloudiness.

By and large, the earth as a whole is in radiative equilibrium averaged over a period of several years. In other words, as much energy must be leaving the system in the form of long-wave radiation as is entering in the form of short-wave radiation:

$$F_{TA} = \int_{top} (1-A) F^{\downarrow}_{SW} \, ds - \int_{top} F^{\uparrow}_{LW} \, ds \approx 0, \tag{6.41}$$

where F_{TA} = net flux of radiation at the top of the atmosphere, $F^{\downarrow}_{SW}$ = incoming solar flux, and $F^{\uparrow}_{LW}$ = long-wave flux to space.

Since the average albedo of the earth is on the order of 30%, an amount of solar radiation given by

$$(\pi R^2 / 4\pi R^2)(1-A) S = 238 \text{ W m}^{-2} \tag{6.42}$$

is absorbed in the atmosphere and oceans, and later re-emitted as long-wave terrestrial radiation, where the solar constant $S \simeq 1360 \text{ W m}^{-2}$. The value of 238 W m^{-2} is a useful reference number for our later studies of the radiational energy available for the atmospheric and oceanic energetics.

Assuming that there is a balance between the amount of solar energy received and the amount of energy emitted by the earth as a whole and that the earth radiates as a black body, we can compute the so-called radiative equilibrium temperature T_e of the earth from the Stefan–Boltzmann law (6.5) so that

$$\sigma T_e^4 = 238 \text{ W m}^{-2}$$

or

$$T_e = 255 \text{ K or } -18 \text{ °C}.$$

However, Eq. (6.39) shows that due to the existence of the atmosphere with gases that absorb and emit the long-wave radiation, the surface temperature of the earth T_{sfc} is greater than the effective emission temperature T_e. Thus,

$$T_{sfc} = T_e + \Delta T, \tag{6.43}$$

where ΔT represents the atmospheric greenhouse effect on the surface temperature. Since the mean surface temperature of the earth is 288 K, the greenhouse effect due to the existence of the atmosphere is $\Delta T = 33$ K.

If $\overline{F}_{TA} \neq 0$, the earth would be subject to cooling or heating. However, as far as we can tell, in view of the present uncertainties in the data (~ 10 W m^{-2}), this does not seem to be the case. A possible exception is the global surface heating associated with the observed increases in CO_2 and other trace gases that is generally expected to become noticeable in the near future. For example, an annual excess of 1% of absorbed over emitted flux ($\overline{F}_{TA} = 2.38$ W m^{-2}) would be equivalent to a heating of about 7 K of the entire atmosphere or a heating of 1 °C of the top 25 m of the world oceans if the excess were maintained during the period of one year.* Such an imbalance in the climatic system should be detectable from the global atmospheric and oceanic temperature records. We may note that the heating due to geothermal processes can be neglected because it is estimated to be on the order of 0.06 W m^{-2}, or less than 0.03% of the absorbed solar radiation.

The next question is whether the earth is also in radiative equilibrium averaged over shorter periods, i.e., over the period of a month or season. First of all one has to consider the annual variation in the earth–sun distance from 0.983 AU in January to 1.017 AU in July. Assuming no annual variation in the global albedo, one would expect a difference in the absorbed radiation inversely proportional to the earth–sun distance squared [see Eq. (6.16)]. Thus, one would expect a difference in solar input of $238/(0.983)^2 - 238/(1.0167)^2 = 16.1$ W m^{-2} between January and July equivalent with a value of about 7% of the net available radiational energy.

Both the computed and the actually observed annual variations are shown in Fig. 6.8. Differences between the two curves are caused by variations in the global albedo. The passage of the sun twice a year over the relatively dark tropics during the time of the equinoxes is connected with a greater absorption of radiation by the earth. On the other hand, during the solstices more reflection of sunlight and less absorption occurs since the sun then illuminates the ice-covered polar caps. The combined effects lead to a semiannual variation in the net radiation curve. Furthermore, there is an asymmetry in the response of the two hemispheres caused by the large seasonal variation in the area covered by snow in the Northern Hemisphere compared with the relatively minor variations inthe Southern Hemisphere (see Fig. 9.4). Besides snow and ice, changes in cloudiness (see Fig. 7.29) and vegetation will also affect the planetary albedo. Nevertheless, the variation of the incident solar radiation at the top of the atmosphere is mainly due to the variation in the earth–sun distance.

Offhand, one might expect a compensation through stronger long-wave cooling of the earth in January than July to ensure radiative equilibrium at the monthly time scale. However, satellite data show that this is not the case. In fact, infrared cooling is observed to be stronger in July than in January, as illustrated in Fig. 6.9. This process may be classified as a positive feedback effect of the earth. The reason lies in the asymmetry of the two hemispheres. The Northern Hemisphere atmosphere is subject to much greater seasonal variations in its temperature than the Southern Hemisphere atmosphere because the major continents are located in the Northern Hemisphere. Therefore, the seasonal variation in global temperature tends to follow the northern seasons leading to a greater global heat loss in July than in January. Cloudiness will also play a role but it is certainly less dominant than temperature.

Combining the estimates of the gain in solar radiation and the loss in terrestrial radiation in Fig. 6.9, we find that the net radiation flux at the top of the atmosphere undergoes an annual cycle with January–July differences of about 25 W m^{-2}. We may speculate that if the major continents were moved from their present position in

---

*The global excess in radiation after one year would be 5.12×10^{14} m$^2 \times 2.38$ W m^{-2} $\times 3.15 \times 10^7$ s $= 3.84 \times 10^{22}$ J. Since the area of the world ocean is 3.61×10^{14} m^2, the specific heat of water is 4200 J kg^{-1} °C^{-1} and the density of sea water is about 10^3 kg m^{-3}, an equivalent average heating of 1 °C would take place for an ocean layer of 3.84×10^{22} J$/(3.61 \times 10^{14}$ m$^2 \times 4200$ J kg^{-1} °C$^{-1} \times 10^3$ kg m$^{-3} \times 1$°C$) = 3.84 \times 10^{22}/$ (15.1×10^{20}) m ≈ 25 m.

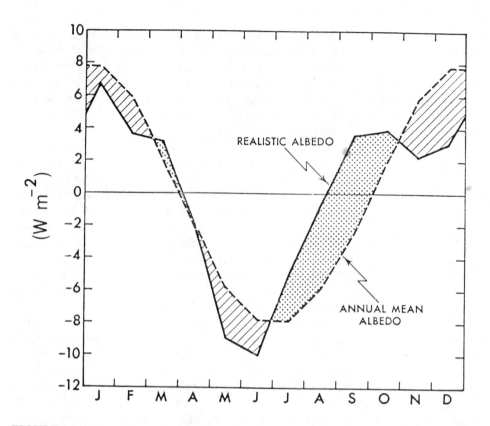

FIGURE 6.8. Seasonal variation of the net incoming solar radiation as observed by satellites at the top of the atmosphere (solid curve) and of the computed solar radiation assuming an idealized annual-mean value of the albedo throughout the year (dashed curve). The annual-mean value of the incoming solar radiation (241 W m^{-2}) has been subtracted out. The differences between the two curves must be due to seasonal differences in cloudiness and snow and ice cover, affecting the albedo of the earth–atmosphere system.

the Northern to the Southern Hemisphere, the annual cycle in global radiation would be drastically changed and radiative equilibrium might prevail over the course of a year.*

With improved satellite observations of the solar constant, the global albedo and the global outgoing long-wave flux new opportunities will arise in the near future to study the presently unknown, but possibly significant, global imbalances that might occur at time scales from months to years. Conceivably such imbalances would be important in interannual climatic fluctuations and longer-term climatic changes.

Finally, we should add that the observed excess or deficit in radiative heating of the earth during the course of a normal year (as shown in Fig. 6.9) has to show up as an increased or decreased energy level of the earth itself (Ellis *et al.*, 1978). We will come back to this issue in Chap. 13 (see Fig. 13.22).

---

*We should note that the more recent Nimbus and ERBE data (e.g., Weare and Soong, 1990) confirm the short-wave radiation values shown in Fig. 6.9, but give a reduced annual range on the order of 8 W m^{-2} for the global long-wave flux values.

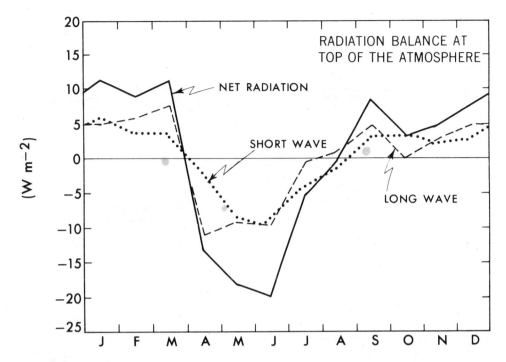

FIGURE 6.9. Radiation balance at the top of the atmosphere for the solar (short-wave), terrestrial (long-wave), and net radiation components as a function of month of the year. Shown are the departures from the annual-mean values.

6.8.2 Global distribution of the radiation balance

We will now discuss a series of radiation maps (Figs. 6.10 and 6.11) that are based on satellite observations reduced by Campbell and Vonder Haar (1980).

The incoming solar radiation that becomes available for driving the climatic system is strongly modulated by the albedo. Figure 6.10(a) shows the distribution of the albedo over the globe with values increasing monotonically from the equatorial regions where it has the lowest values (on the order of 20% and less) to the polar regions where the values are the largest (ranging between 60% and 95%). The maxima observed in the polar regions are mainly associated with the snow cover and the high angle of incidence of the solar radiation. The minimum values in the intertropical zone are located over the oceans with somewhat higher values over the subtropical continents, in general agreement with the corresponding cloud distributions (see Fig. 7.28). Worth mentioning is the high albedo over the Sahara desert where the cloudiness attains a minimum. During December–February (not shown) the subtropical minima in the albedo are mainly located in the Southern Hemisphere and during June–August they shift northward into the Northern Hemisphere.

The available solar radiation, given by $(1 - A)F^{\downarrow}_{SW}$, is presented in Fig. 6.10(b) for annual-mean conditions. As expected, the figure shows that the available solar radiation is negatively correlated with the albedo. The highest values of about 350 W m^{-2} tend to occur over the intertropical oceans, decreasing almost monotonically to values less than 100 W m^{-2} at the poles.

The global distribution of the infrared radiation emitted by the earth is shown in Fig. 6.10(c) for the annual-mean case. The highest values (on the order of 270–280

122 PHYSICS OF CLIMATE

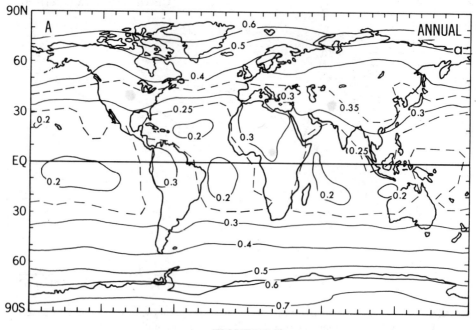

FIGURE 6.10a

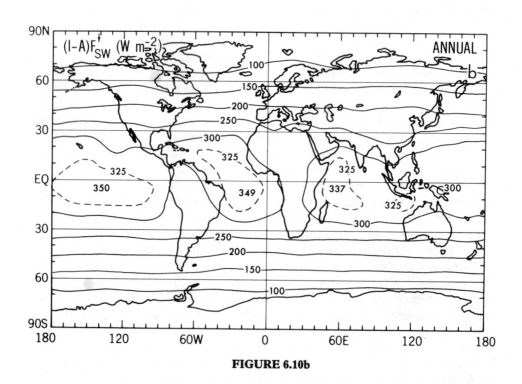

FIGURE 6.10b

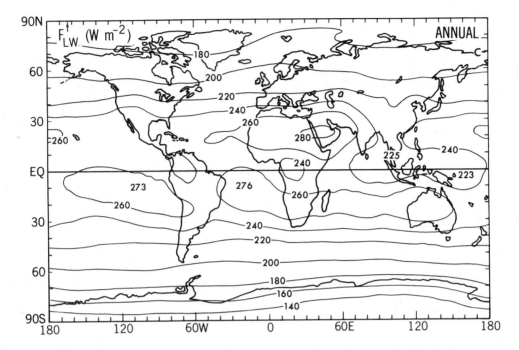

FIGURE 6.10. Global distributions of the albedo A (a), absorbed solar radiation $(1-A)F^{\downarrow}_{SW}$ (b), and outgoing terrestrial radiation $F^{\uparrow}_{LW}$ (c) for annual-mean conditions (based on data from Campbell and Vonder Haar, 1980).

W m^{-2}) are observed over the subtropical latitudes decreasing gradually towards the poles where they reach values of 160 W m^{-2} and less. The seasonal distributions (not presented here) show similar patterns with two double maxima straddling the equator. The tropical values tend to be negatively correlated with the cloud distribution. Thus, when high clouds are present the emission temperatures are lower leading to lower values of the emitted radiation (Stefan–Boltzmann law). The global distribution of emitted radiation is very different from that of the absorbed solar radiation which shows much stronger latitudinal gradients.

Finally, Fig. 6.11 shows the net radiation at the top of the atmosphere for the year and the northern winter and summer seasons. The annual picture basically shows a zonal pattern with energy input ranging from about 60 to 70 W m^{-2} near the equator and energy losses of about 100 W m^{-2} near the south pole and 120 W m^{-2} near the north pole. For the year as a whole, Fig. 6.11(a) shows that the ocean regions generally gain more energy than the land regions, pointing to the need for a net annual transport of energy by the atmospheric circulation from the oceans into the land. Of particular interest is the strong negative anomaly over the North African desert, requiring an appreciable influx of atmospheric energy or, in other words, requiring adiabatic heating by compression of the air to compensate for the radiational cooling. This is mainly a summer phenomenon.

Predominant zonality of the net incoming radiation is clear in the winter hemispheres in Figs. 6.11(b) and 6.11(c). The summer hemisphere patterns are broken up with minima of net radiation over the continents and maxima over the oceans. The reasons for these continent–ocean differences are that the values of the surface albedo are larger over land than over water leading to more reflection of solar radiation over land, and that the infrared radiation losses during summer tend to be greater over

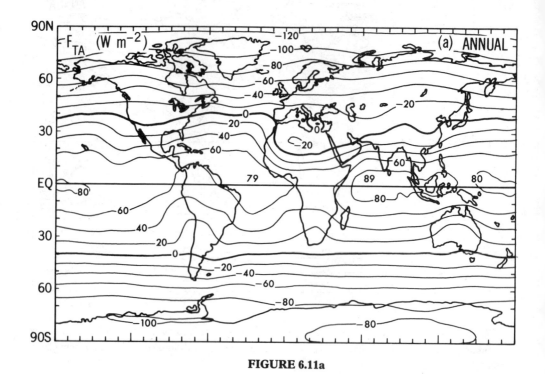

FIGURE 6.11a

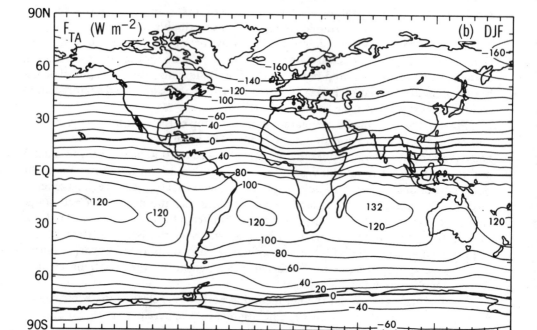

FIGURE 6.11b

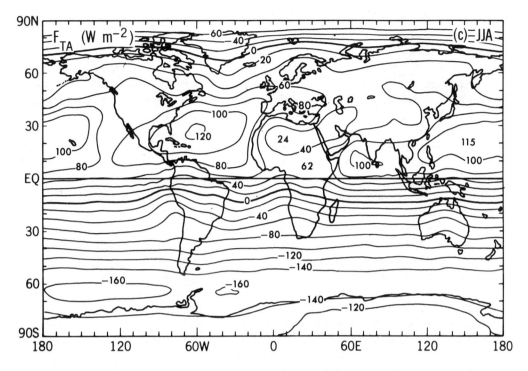

FIGURE 6.11. Global distributions of the net incoming radiation F_{TA} at the top of the atmosphere for annual (a), northern winter (b), and northern summer (c) mean conditions in W m^{-2} (based on data from Campbell and Vonder Haar, 1980).

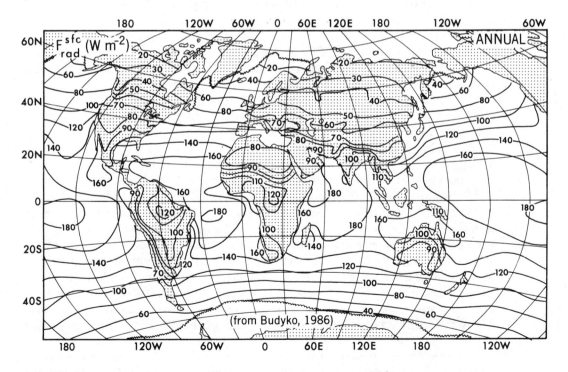

FIGURE 6.12. Global distribution of the net downward radiation flux F_{rad}^{sfc} at the earth's surface for annual-mean conditions in W m^{-2}. The annual-mean ice boundary is indicated by a wiggly line. Above the ice in winter F_{rad}^{sfc} can be negative (i.e., upward) (after Budyko, 1986).

the relatively warm continents than over the cool oceans. Much of the excess of summer radiation over the oceans is absorbed in the oceans themselves.

By subtracting estimates of the seasonal heat storage in the oceans and in the overlying atmosphere from the net radiation values, Campbell and Vonder Haar (1983) were able to determine the need for a substantial atmospheric transport of energy from the land to the ocean regions during summer, and from the oceans to the land during winter.

The strongest north–south energy fluxes in the atmosphere should occur in the winter hemisphere where the meridional gradients in radiation are greatest. It is also of interest to note that the largest heat losses during winter are found not in the vicinity of the poles but near 65° latitude, which is possibly connected with the occurrence of ice-free water at those latitudes throughout the year so that heat losses can go on continuously.

The annual-mean net radiation flux at the earth's surface is presented in Fig. 6.12 as calculated by Budyko (1986) from Eq. (6.39) using some direct surface observations over land and oceans (see also Fig. 6.3). The radiation flux decreases with latitude from values of 160 to 180 W m^{-2} near the equator to values on the order of 20–40 W m^{-2} poleward of 60° latitude. Over most of the globe the net surface radiation is downward. However, over the polar regions in winter we may find a net radiation loss at the surface, when the solar influx tends to be very small or zero. In general, the values are higher over the oceans than over the continents at the same latitude. The highest values on the map are on the order of 180 W m^{-2}. They occur in the intertropical regions over the oceans in agreement with the distribution of the total solar radiation absorbed in the atmosphere plus oceans shown in Fig. 6.10(b). Secondary equatorial maxima are found over the continents. The lowest values in the tropics occur over the deserts, which is to be expected from the high values of the surface albedo, the low values of cloudiness and humidity, and the high surface temperatures.

Zonal-mean radiation profiles at the top of the atmosphere based on the data from Campbell and Vonder Haar (1980) are shown in Figs. 6.13 and 6.14. The solar radiation available [Fig. 6.13(a)] exhibits a strong gradient in the winter hemisphere decreasing from a value of about 475 W m^{-2} in the subtropics of the summer hemisphere to a zero value at the winter pole, and only a weak gradient toward the summer pole. A considerable part of this radiation is reflected back to space as shown in Fig. 6.13(b). This is especially the case at high latitudes, where the albedo [see Fig. 6.14(a)] is very large ($> 70\%$) because of the greater angle of incidence of the incoming radiation and, to some extent, because of snow and ice coverage.

Finally, the curves for the absorbed solar radiation available for driving the earth's general circulation are given in Fig. 6.14(b). The main qualitative difference with the original solar energy curves in Fig. 6.13(a) is the decrease of available radiation over the summer pole, leading to an appreciable north–south gradient of absorbed solar radiation even in the summer hemisphere. The annual curves in Figs. 6.13(a) and 6.14(b) have practically the same shape with a shift of 100 W m^{-2} that is almost uniform over latitude. In analyzing the implications of these curves, it is good to recall that, for the globe as a whole, the annual average absorbed solar radiation is 238 W m^{-2} with an expected seasonal variation of about 8 W m^{-2} amplitude (see Fig. 6.8) due to the eccentricity of the earth's orbit around the sun.

The profiles of outgoing terrestrial radiation at the top of the atmosphere in Fig. 6.14(c) show a high plateau between about 30 °N and 30 °S with a slight dip over the ITCZ, mainly due to extensive cloudiness, and low values at high latitudes. The atmosphere over the Antarctic seems to lose less infrared radiation than the Arctic

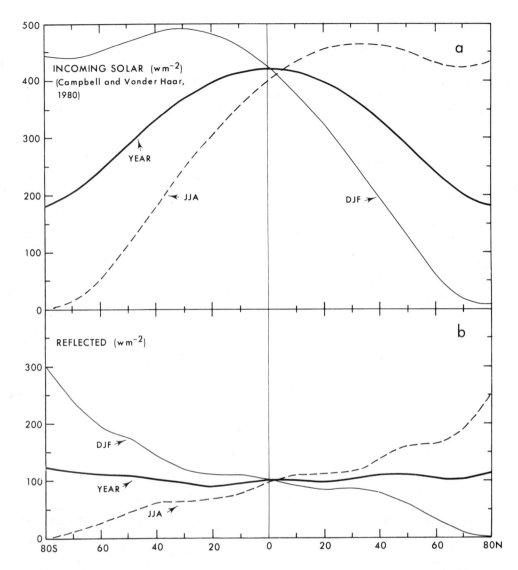

FIGURE 6.13. Meridional profiles of the zonal-mean incoming (a) and reflected solar radiation (b) at the top of the atmosphere in W m^{-2} for annual, DJF, and JJA mean conditions (based on data from Campbell and Vonder Haar, 1980). No corrections were made to insure global radiation balance.

atmosphere, presumably because of the high ice cap in the Antarctic (Bowman, 1985; Nakamura and Oort, 1988). However, the north–south gradients in outgoing long-wave radiation are always weak. The net meridional heating profiles given in Fig. 6.14(d) are obtained by subtracting the emitted terrestrial from the absorbed solar radiation profiles. The shape of the resulting profiles closely resembles the shape of the profiles of absorbed radiation although variations in cloudiness and surface albedo over deserts, forests, oceans, etc., also play a role in shaping these profiles. However, it is beyond the scope of the present book to further analyze these factors.

A synthesis of the hemispheric and global radiation components of the energy budget is presented in Table 6.2. The upper portion of the table gives the results without correction for global annual balance. It shows a net annual excess radiation

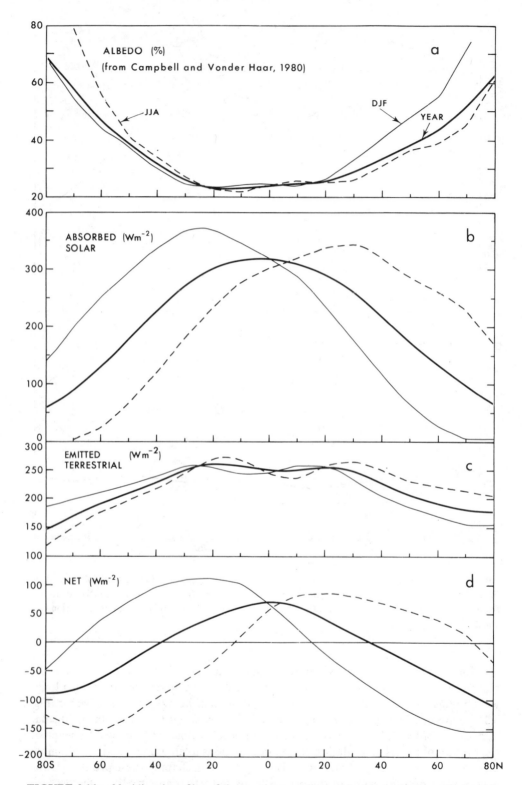

FIGURE 6.14. Meridional profiles of the zonal-mean albedo (a), absorbed solar radiation (b), emitted terrestrial radiation (c), and net radiation (d) at the top of the atmosphere for annual, DJF, and JJA mean conditions (based on data from Campbell and Vonder Haar, 1980). No corrections were made for global radiation balance.

TABLE • 6.2. Hemispheric and global radiation values at the top of the atmosphere in units of W m^{-2} after Oort and Peixoto (1983).

	Annual			DJF			JJA		
	NH	SH	Globe	NH	SH	Globe	NH	SH	Globe
A. No correction									
Incoming solar	345	344	345	243	467	355	445	225	335
Reflected solar	105	104	103	74	146	110	137	70	98
Absorbed solar	240	243	241	169	321	245	308	165	237
Outgoing terrestrial	232	233	232	222	238	230	242	230	236
Net	8.2	10.1	9.2	−53	83	15.2	66	−64	0.9
Albedo (%)	30.5	29.5	30.0	30.3	31.2	30.9	30.8	26.6	29.4
B. With correction[a]									
Reflected solar	108	104	106	76	149	112	140	61	101
Absorbed solar	237	240	239	167	317	242	304	164	234
Outgoing terrestrial	237	239	238	227	244	236	248	235	242
Net	−0.2	1.7	0.8	−60	74	6.7	56.5	−71.5	−7.5
Albedo (%)	31.2	30.2	30.7	31.1	32.0	31.7	31.6	27.2	30.1

[a] Corrected for global balance by mutiplying reflected solar and outgoing terrestrial radiation by 1.025.

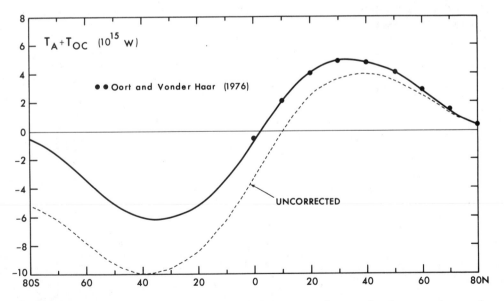

FIGURE 6.15. Meridional profiles of the annual transport of energy by the atmosphere and oceans in 10^{15} W calculated from radiation requirements. The dashed curve is obtained from uncorrected data starting the integration at the North Pole, and the solid curve from the same data after correction for global balance (see Table 6.2) (after Oort and Peixoto, 1983).

input of 9.2 W m^{-2} for the globe. This annual imbalance could be correct for periods of several years if the excess energy were stored in the oceans or cryosphere (Saltzman, 1983). However, the imbalance must be spurious for a long-term mean. In the present context we will assume the observed global imbalance of 9.2 W m^{-2} to be a measure of errors in the data. Some correction is then required to preserve the climatic balance as shown, e.g., in the bottom part of Table 6.2. The correction used here (Campbell and Vonder Haar, 1980) reduces the global imbalance to almost zero. The difference between the corresponding estimates in the two sections of Table 6.2 gives some idea of the uncertainties involved in the evaluation of the radiation terms.

The net radiation profile for annual-mean conditions shown in Fig. 6.14(d) enables us to estimate the total poleward energy transport in the ocean–atmosphere system, $T_A + T_{OC}$, needed to maintain the observed temperature structure. The actual calculations proceed by integration of the net radiation values with respect to latitude starting with a zero transport at one of the poles:

$$T_A + T_{OC} = -\int_{\phi'=\phi}^{\pi/2} F_{TA} 2\pi R^2 \cos \phi' \, d\phi'. \tag{6.44}$$

The results for the raw data uncorrected for global balance (top portion of Table 6.2) are shown in Fig. 6.15 as a dashed curve, and for the corrected data (bottom portion of Table 6.2, with an additional small correction to ensure exact global balance) as a solid curve. It is clear that a correction for global annual-mean radiation balance is essential in order to arrive at reasonable values of the required energy transport. For example, a global excess of incoming over outgoing radiation of 10 W m^{-2} would lead to a fictitious southward energy flux of -2.6×10^{15} W at the equator and -5.1×10^{15} W at the south pole. This example shows that great care must be taken to obtain exact global balance.

The corrected curve in Fig. 6.15 shows almost no asymmetry with respect to the equator. The largest poleward transports calculated are about 5.0×10^{15} W near 30 °N and about -6.0×10^{15} W near 35 °S. A small cross-equatorial transport of energy in the atmosphere and oceans from the Northern to the Southern Hemisphere seems to be necessary.

CHAPTER 7

Observed Mean State of the Atmosphere

In this chapter, we present a general description of the mean state of the atmosphere as obtained from observations. Since the atmosphere is a thermo-hydrodynamical system it can be characterized by its composition, its thermodynamical state as specified by three thermodynamical variables (e.g., pressure, temperature, and specific humidity), and its three-dimensional motion field. A complete specification of the state of the atmosphere should also include the global distributions of other variables, such as cloudiness, aerosols, and so on, because they influence the large-scale behavior of the atmosphere. Although precipitation, evaporation, and runoff are fluxes and not state variables, they are intimately connected with the state of the atmosphere and are, therefore, included in this chapter. Of course, they are also basic elements in the hydrological cycle that will be discussed later. Furthermore, not only the averages of the atmospheric variables but also higher moments of the probability distributions, such as variances and covariances, should be taken into account. However, for the sake of simplicity we will first deal with the simple averaged fields leaving the discussion of the higher modes of variation to later chapters.

It seems appropriate to start with the analysis of the total mass of the atmosphere because the mass is one of the fundamental quantities that characterize the system and, also, because the mass distribution is closely related to the pressure field.

7.1 ATMOSPHERIC MASS AND PRESSURE

7.1.1 Mass balance

We assume that the total mass of the atmosphere taken over a year is practically constant. The addition of gases by volcanic eruptions, the escape of light gases to outer space, and the surface exchange of gases, such as oxygen and carbon dioxide, are very small, and can be disregarded in this context. However, we must realize that the mass of water vapor is an important component of the total mass of the atmosphere, and that it is subject to short-period fluctuations. We will discuss this later in more detail.

Let us consider the continuity equation of mass (3.3). Integration of this equation for a polar cap defined by a volume V and bounded by a conceptual latitudinal wall, the earth's surface, and the top of the atmosphere leads, by the use of Gauss' divergence theorem, to

$$\int_V \frac{\partial \rho}{\partial t} dV = -\int_V \text{div } \rho \mathbf{c} \, dV = -\int_S \rho \mathbf{c} \cdot \mathbf{n} \, ds$$
$$= \int_{\text{wall}} \rho v \, ds, \qquad (7.1)$$

where $\mathbf{n}$ is the unit vector directed outward across the boundary surface S. Thus,, the time rate of change of total mass in a polar cap equals the net inflow of mass across the wall.

In the long-term mean, there should be no accumulation of mass so that

$$\int_{\text{wall}} \overline{\rho v} \, ds = 0.$$

In other words, there is no net flow of mass across a latitudinal wall. Of course, this same result holds not only for a polar cap but also for any other fixed volume. On the other hand, on a monthly or seasonal time scale there can be small, but measurable changes of mass in the volume.

We will show that the mass distribution in the atmosphere can be obtained most readily by measuring the pressure distribution. Assuming that the equilibrium in the atmosphere is hydrostatic ($\partial p/\partial z = -\rho g$), we can obtain the pressure at a given level z by integration:

$$p(z) = g \int_z^\infty \rho \, dz. \qquad (7.2a)$$

Thus, under hydrostatic conditions the surface pressure p_0 will give a good measure of the total atmospheric mass in a unit vertical column. Substituting the equation of state in the hydrostatic equation gives $dp/p = -g \, dz/(R_d T_V)$. Integration of this expression between the surface z_0 and level z then leads to

$$p(z) = p_0 \exp\left\{-\int_{z_0}^z \frac{g}{R_d T_V} dz\right\}, \qquad (7.2b)$$

where T_V is the virtual temperature defined in Eq. (3.63). This expression shows that the pressure decreases rapidly (exponentially) with height so that the air is most dense near the ground and approaches zero at the "top" of the atmosphere. For practical purposes 30 km (or the 10-mb level) is sometimes used as the top of the atmosphere for the analyses; this corresponds to including 99% of the atmospheric mass. Equation (7.1) can be rewritten in the form

$$\frac{\partial}{\partial t} \frac{1}{g} \int_{\text{sfc}} p_0 \, ds = \int_{\text{wall}} \rho v \, ds. \qquad (7.3)$$

This shows that pressure changes can only occur through a net inflow or outflow of mass through the boundaries.

The rapid dropoff of density with height distinguishes the atmosphere from the oceans, where the density has relatively small variations in the vertical since air is much more compressible than water.

7.1.2 Distribution of mass in terms of pressure

Figure 7.1(a) shows an annual-mean map of the geopotential height anomalies of the 1000-mb pressure level $z_{1000} - z_{1000}^{SA}$ which is practically equivalent with a mean sea-level pressure map more commonly used in synoptic weather practice. Where the 1000-mb height field intersects or dips below the mountains, extrapolated values of z_{1000} have been used. The conversion to sea-level pressure p_{SL} can be done by using the hydrostatic balance condition $dp = -\rho g \, dz$ in terms of

$$p_{SL} = (\rho_{sfc} g \, z_{1000}/100) + 1000 \approx 0.121 z_{1000} + 1000,$$

where p_{SL} is in mb (1 mb = 10^2 Pa), z_{1000} is in gpm, the density $\rho_{sfc} \approx 1.23$ kg m^{-3}, and $g = 9.8$ m s^{-2}. Although the values in mountainous terrain are extrapolated to sea level, leading to fictitious pressures, this type of analysis has traditionally been used by meteorologists in order to follow the migrating weather systems. Over the oceans and low-level terrain the map accurately represents the mass distribution but, of course, over mountainous terrain one has to go back to the originally measured surface pressure before reduction to sea level.

The subtropics near 30° latitude are dominated in both hemispheres by semipermanent high-pressure cells, the so-called subtropical highs, or anticyclones (clockwise rotation). These cells are bordered by a more or less continuous low-pressure zone near the equator (the intertropical convergence zone, ITCZ) and poleward by low-pressure belts, in which the Icelandic and Aleutian lows are imbedded. On the other hand, the polar regions show predominance of high-pressure cells.

The Northern and Southern Hemispheres are very different with respect to their physiography. In fact, using all available surface data, the annual-mean surface pressures over the Northern and Southern Hemispheres are found to be rather different, namely 983.6 and 988.0 mb. The globally averaged pressure at the earth's surface is about 985.8 mb (Oort, 1983). In terms of total atmospheric mass, assuming the area of

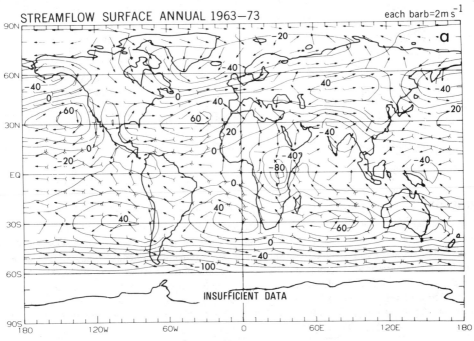

FIGURE 7.1a

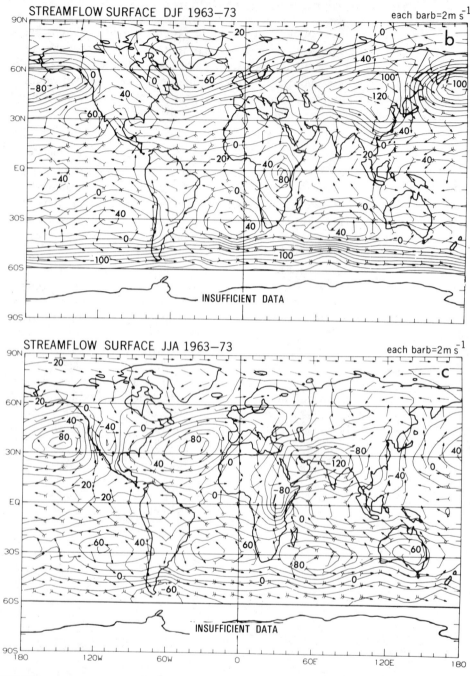

FIGURE 7.1. Global distributions of the height anomalies of the 1000-mb pressure field, $\bar{z}_{1000} - z_{1000}^{SA}$, in gpm for annual-mean (a), northern winter (b), and northern summer (c) mean conditions. The quantity z_{1000}^{SA} ($= 113$ gpm) represents the height of the 1000-mb level according to the NMC standard atmosphere. Also shown are vector plots of the surface winds. For geostrophic flow the arrows should parallel the isolines. Each barb on the tail of an arrow represents a wind speed of 2 m s^{-1}. The isoheight lines can also be interpreted as isobars for the surface pressure reduced to sea level. Since 1 gpm is equivalent to about 0.121 mb, we can, e.g., relabel an isoheight line of $+40$ gpm by $(40 + 113) \times 0.121 + 1000 = 1018.4$ mb, and an isoheight line of -40 gpm by $(-40 + 113) \times 0.121 + 1000 = 1008.8$ mb. Note that 1 gpm $\approx$ 1 m.

each hemisphere to be 2.56×10^{14} m^2 and $g = 9.80$ m s^{-2}, the annual-mean values for the Northern Hemisphere, Southern Hemisphere, and globe are 2.57, 2.58, and 5.15×10^{18} kg, respectively. Our values agree well with the values estimated by Trenberth *et al.* (1987).

Figures 7.1(b) and 7.1(c) present the mean surface-pressure conditions for December–February and June–August. The large subtropical anticyclones in each hemisphere show a slight tendency to move toward the pole during summer. These highs are especially well developed during northern summer over the North Atlantic and North Pacific Oceans. In the northern high latitudes the low-pressure systems intensify during winter, whereas in the Southern Hemisphere they undergo only a small seasonal variation. These Southern Hemisphere low-pressure systems form an almost continuous zonal belt around Antarctica with very low values of the surface pressure. The largest seasonal pressure variations are found over the Asian continent where a strong anticyclone develops over Siberia during winter and a low-pressure system forms during summer north of the Indian subcontinent. This change is associated with the monsoon cycle over Southeast Asia and the movement of the ITCZ. A similar situation takes place over the North American continent, but it is less intense: the annual variations of the surface pressure over Siberia exceed 25 mb (or $\Delta z_{1000} \gtrsim 200$ gpm) whereas over the American continent the values do not exceed 10 mb (or $\Delta z_{1000} \lesssim 80$ gpm).

In order to get a better idea of the magnitude of the seasonal pressure variations, we present in Fig. 7.2 zonal-mean profiles of the sea-level pressure for the year and the extreme seasons. Extremely low-pressure values, on the order of 985 mb, are found in the belt around Antarctica. The seasonal shifts in the high-pressure belts near 30 °N and 30 °S and their intensification during the winter season are clearly shown. The mean global values of the surface pressure reduced to sea level in Fig. 7.2 are higher than the corresponding values estimated using the actual surface pressure given before (because of the fictitious layer of air that is now added in mountainous areas between the elevated earth's surface and sea level).

The seasonal variations in the hemispheric and global-mean surface pressures are shown in Fig. 7.3(a). The annual range is estimated to be 2.4 mb in the Northern Hemisphere, 2.7 mb in the Southern Hemisphere, and 0.15 mb for the globe. The hemispheric values imply the existence of important shifts of mass across the equator.

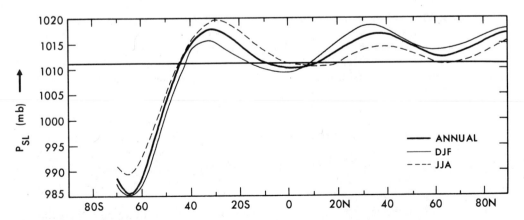

FIGURE 7.2. Meridional profiles of the zonal-mean sea level pressure for annual-mean conditions after Trenberth (1981) and for the DJF and JJA seasons after Oort (1983) in mb.

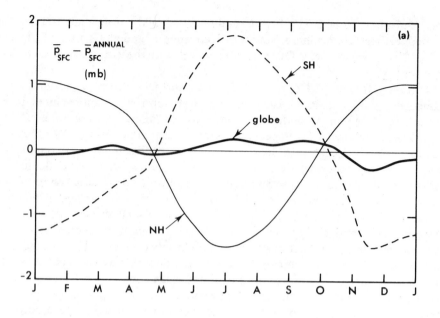

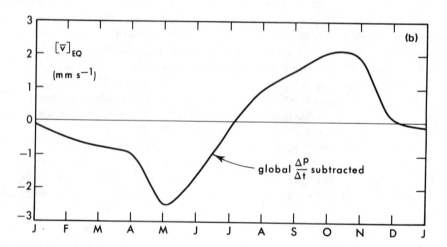

FIGURE 7.3. (a) Annual variation in the average surface pressure (i.e., actual, not reduced to sea level) over the two hemispheres and the globe in mb from Oort (1983). Plotted are departures from the annual-mean values. (b) Annual variation in the vertical-mean meridional velocity (in mm s^{-1}) across the equator as inferred from the hemispheric pressure changes shown in (a). The actual cross-equatorial transports of mass in units of 10^9 kg s^{-1} can be obtained by multiplying the plotted velocity values by 0.4.

Using Eq. (7.3) we can infer the required net meridional velocity at various latitudes and throughout the year:

$$\langle [v] \rangle = \int_{z_0}^{\infty} [\rho v] dz \bigg/ \int_{z_0}^{\infty} \rho \, dz.$$

For example, if we assume an average pressure increase of 1 mb to take place during one month over the entire Northern Hemisphere, the mean velocity across the equator would be

$$\langle [v] \rangle = 2\pi R^2 \frac{1 \text{ mb}}{1 \text{ month}} \frac{1}{1012 \text{ mb} \times 2\pi R}$$
$$= 0.002 \text{ m s}^{-1},$$

where p_0 is taken as 1012 mb. The equatorial values of $\langle [v] \rangle$ shown in Fig. 7.3(b) were derived using similar computations based on the actually observed pressure changes.

The seasonal variation of the global surface pressure in Fig. 7.3(a) shows a net positive residual value during the northern summer semester. This appears to disagree with our basic assumption of a constant atmospheric mass. The reason is simple. We have to take into consideration the contributions to the total mass in the atmosphere by the added water vapor associated with variations in the net evaporation minus precipitation at the earth's surface. Observations of the precipitable water in the atmosphere show an excess of water vapor on the order of 0.2 g kg^{-1} during Northern Hemisphere summer due to the higher values of evaporation and higher surface temperatures over the continents. This excess in water vapor mass is compatible with the observed global pressure difference of 0.2 mb.

7.2 MEAN TEMPERATURE STRUCTURE OF THE ATMOSPHERE

7.2.1 Global distribution of the temperature

The temperature distribution in the atmosphere is of fundamental importance for defining the thermodynamic state and, ultimately, the wind structure in the atmosphere. The mean surface temperature distributions are shown in Figs. 7.4(a) and 7.4(b) for January and July that represent the seasonal extremes.

As expected from the distribution of absorbed solar radiation in Fig. 6.10(b), the highest temperatures are found in the intertropical regions where the largest amounts of solar radiation are received during the course of the year. In the equatorial regions, the meridional temperature gradients are very small because of the small gradients in insolation. The lowest temperatures occur over the polar regions of minimum annual solar insolation. The isotherms are much more zonally uniform in the Southern Hemisphere associated with the predominance of oceans over land.

We find a strong equator-to-pole temperature gradient reaching the highest values over the Northern Hemisphere continents during winter. The effects of warm and cold ocean currents are also evident on the maps. We note that the steepest horizontal temperature gradients are found in middle latitudes. The influence of the land–sea distribution, the nature of the land surface, and the surface topography are all evident in the configuration of the mean isotherms as well as in the strong gradients near the coasts and mountainous regions, such as the Rocky Mountains, the Tibetan Plateau, Andes, and Antarctica. The coldest regions are found over the northern parts of the continents in the Northern Hemisphere during winter (northeastern Siberia and Canada) and over Antarctica.

The annual range of temperature is an important aspect of climate because it highlights the land–sea contrasts. The January minus July temperature difference map shown in Fig. 7.4(c) is a good measure of the annual range. In the equatorial zone the values are very small because the solar irradiation does not change substantially

138 PHYSICS OF CLIMATE

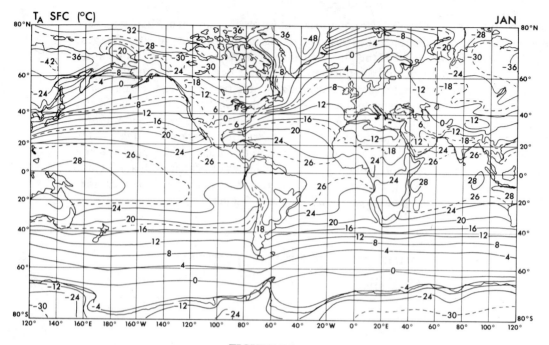

FIGURE 7.4a

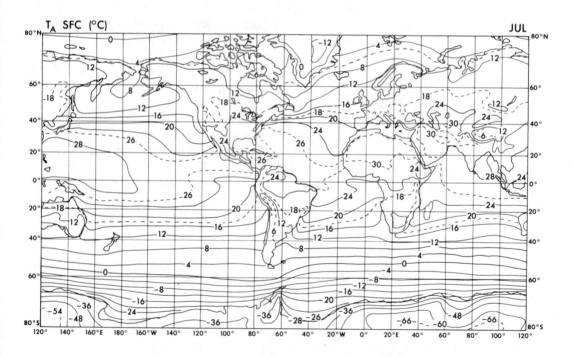

FIGURE 7.4b

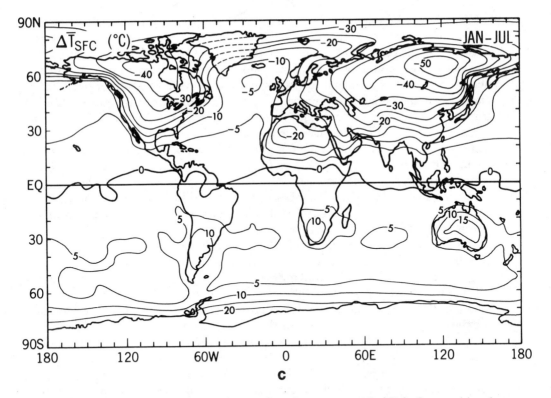

FIGURE 7.4. Horizontal distributions of the surface air temperature (in °C) for January (a) and July (b) after National Climatic Data Center (1987), and for the January–July difference (c) based on the 1963–73 analyses in Oort (1983).

throughout the year. The fluctuations are also relatively small for the large waterbodies of the oceans due to the high value of the specific heat and the strong mixing in the surface layer of the ocean leading to a large thermal inertia of the oceans. Extreme values (negative in the Northern and positive in the Southern Hemisphere) occur over all continents which become alternately hot in summer and cold in winter. The lowest values are found in central Siberia (-50 °C) and Northern Canada (-40 °C). Another local extremum ($\Delta T = -20$ °C) is found over the Sahara. In the Southern Hemisphere, the extreme values are much smaller (10–15 °C), and are observed over the southern parts of the continents. The polar regions are subject to large fluctuations on the order of 30 °C.

7.2.2 Vertical structure of the temperature

The influence of the continents and oceans diminishes with height so that at midtropospheric levels (500 mb) the isotherms are more uniform along a latitude circle. A characteristic feature of the lower and middle troposphere temperature fields is the pattern of large-scale standing waves mainly with wave numbers one and two. The strongest pole-to-equator temperature gradients are found in middle latitudes and are most intense during winter. In the stratosphere above the tropopause the horizontal gradients of temperature are reversed since the lowest temperatures are observed over the equator.

To discuss the vertical temperature structure of the atmosphere the zonal-mean distribution of temperature $[\overline{T}]$ is given in Fig. 7.5 for the annual-mean and extreme seasons. On the right-hand side of the figure vertical profiles depict the hemispheric and global-mean conditions. The figure shows a rapid decrease of temperature with height in the troposphere and reverse conditions in the lower stratosphere. For example, near the equator the temperature difference between the ground and the tropopause (at about 17-km height) is on the order of 105 K. In the lower stratosphere the temperature increases slightly with altitude in middle latitudes and becomes almost isothermal in the polar regions. The tropopause varies in altitude from about 10 km (250 mb) at the poles to 17 km (~100 mb) at the equator. The vertical temperature gradients are strongest in the tropical upper troposphere due to subsidence around the deep cumulonimbus clouds in the ITCZ and the evaporation and radiational cooling at the tops of these clouds. The very low winter temperatures that are observed at 50 mb in the Southern Hemisphere are associated with the polar night jet. Such low temperatures are also found in the Northern Hemisphere but above the 50-mb level so that they do not appear in the cross sections in Fig. 7.5.

Due to its importance for studying the dynamical processes in the atmosphere, we will discuss the distribution of the zonal-mean potential temperature $[\overline{\theta}]$ defined before in Eq. (3.47). The annual-mean cross section of $[\overline{\theta}]$ in Fig. 7.6(a) shows approximate symmetry with respect to the equator similar to the temperature cross section in Fig. 7.5. Minimum values of potential temperature are observed in the polar regions near the earth's surface. The values of $[\overline{\theta}]$ increase with height. The greatest vertical gradients $\partial \theta / \partial z$ occur in the stratosphere, as shown in Fig. 7.6(c). In the intertropical regions, the horizontal gradients in θ are fairly small as compared with those at mid and high latitudes.

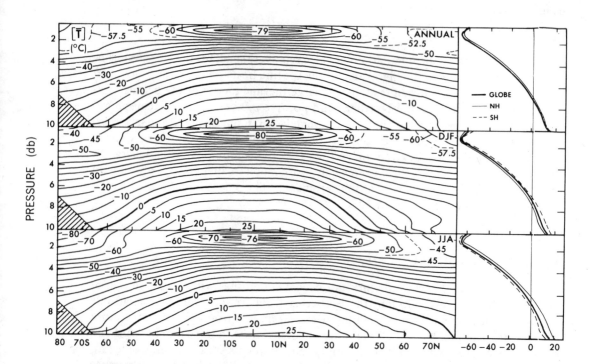

FIGURE 7.5. Zonal-mean cross sections of the temperature for annual-mean, DJF, and JJA conditions in °C. Vertical profiles of the hemispheric and global mean temperatures are shown on the right.

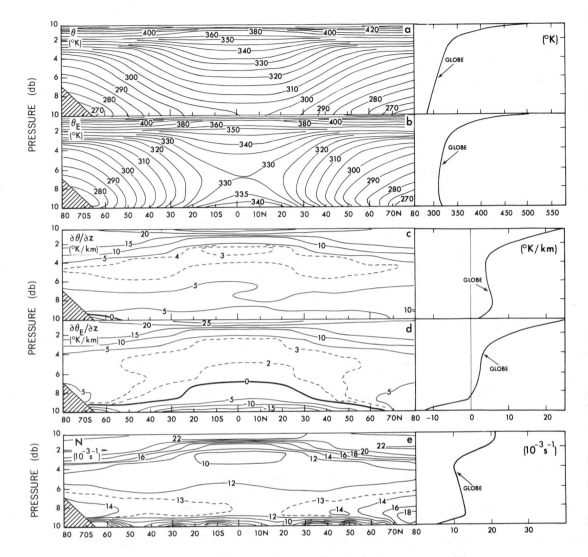

FIGURE 7.6. Zonal-mean cross sections of the potential temperature (a) in K, equivalent potential temperature (b) in K, vertical gradient of potential temperature (c) in K/km, vertical gradient of equivalent potential temperature (d) in K/km, and Brunt–Väisälä frequency in 10^{-3} rad s^{-1} for annual-mean conditions. Vertical profiles of the global mean values are shown on the right.

From the vertical distributions of T and θ, we can infer the distribution with height of the static stability. As we have seen in Chap. 3, the static stability is given by Eq. (3.51):

$$\frac{T}{\theta}\frac{\partial \theta}{\partial z} = \gamma_d - \gamma.$$

Since $\gamma_d = 1\,°C/100$ m and $\gamma = 0.6\,°C/100$ m, as can be deduced from the mean profiles in Fig. 7.5, the atmosphere is found to be in stable equilibrium. The same conclusion is reached when we consider the values of $\partial[\bar{\theta}]/\partial z$ given in Fig. 7.6(c), since they are positive almost everywhere. However, locally in certain special situations the atmosphere may become unstable ($\gamma > \gamma_d$) and dry convection may occur.

Figure 7.6(e) shows a zonal-mean cross section of the Brunt–Väisälä frequency N [see Eq. (3.53)] for small-scale vertical oscillations that may occur superimposed on the large-scale atmospheric flow patterns. The values of N are on the order of 10^{-2} rad s^{-1} which corresponds with a period of oscillation ($=2\pi/N$) of about 10 min. In the troposphere N varies from values less than 10×10^{-3} rad s^{-1} near the surface to values of 14×10^{-3} rad s^{-1} or more at high latitudes near the 850-mb level. In the stratosphere N increases to values larger than 20×10^{-3} rad s^{-1}. Since N is also a measure of the stability, its cross section resembles that of $\partial\theta/\partial z$ in Fig. 7.6(c).

A zonal-mean cross section of the equivalent potential temperature θ_e [see Eq. (3.67)] is given in Fig. 7.6(b). It shows that the highest values of θ_e occur in the intertropical regions with a bimodal profile in the vertical and that the meridional distribution of θ_e is almost symmetric with respect to the equator. Since the equivalent potential temperature and the potential temperature are related through Eq. (3.67): $\theta_e = \theta \exp(Lq_s/c_p T)$, the two cross sections show some resemblance. The main differences are found in the lower and middle troposphere of the tropics, where the saturation specific humidity q_s is greatest. In the upper troposphere and stratosphere, they are about the same because of the strong decrease in water vapor with height and $q_s \to 0$. The mean profiles of θ and θ_e on the right-hand side of Figs. 7.6(a) and 7.6(b) are different in the troposphere, where θ_e has substantially higher values. The almost constant value of θ_e in the troposphere confirms the concept that θ_e is a conservative property both with and without precipitation provided that the transformations are pseudoadiabatic, i.e., if all condensed water is precipitated out.

The cross section of $\partial\theta_e/\partial z$ in Fig. 7.6(d) is a measure of the conditional (or moist convective) instability (see Sec. 3.5.1.3). It differs substantially from the cross section of $\partial\theta/\partial z$ for dry air. The values of $\partial\theta_e/\partial z$ are negative in the lower troposphere, and the zero line reaches the 700-mb level in the tropics, indicating strong conditional instability. This shows how important deep convection is in the tropics. We should note that we are dealing here with zonally averaged quantities and that local convection may reach up to very high levels near the tropopause, especially in the ITCZ.

The horizontal temperature gradients are displayed in Fig. 7.7. They are weaker than the vertical temperature gradients but are nevertheless very important because they are a measure of the available potential energy which is the main energy source for the general circulation of the atmosphere. Furthermore, they are also a measure of the baroclinicity that plays an important role in the development of weather systems, as we shall see later. The seasonal distributions of the horizontal temperature gradients show that the winter–summer differences are much smaller in the Southern than in the Northern Hemisphere. The midlatitude patterns are displaced equatorward from their annual-mean positions in winter and poleward in summer, in phase with the changes in solar inclination.

7.2.3 Variability of the temperature

We will now analyze the variability in space and time of the temperature. Thus, Figs. 7.8(a) and 7.8(b) show zonal-mean cross sections of the variability in temperature in terms of zonal-mean standard deviations for the year and extreme seasons. The variability in time in Fig. 7.8(a) is associated with the alternation of different air masses at a given location. As expected the variability is large (on the order of 5 °C or greater) in mid to high latitudes and small (less than 3 °C) in the tropics. The variability is also more intense in the Northern than in the Southern Hemisphere mainly in

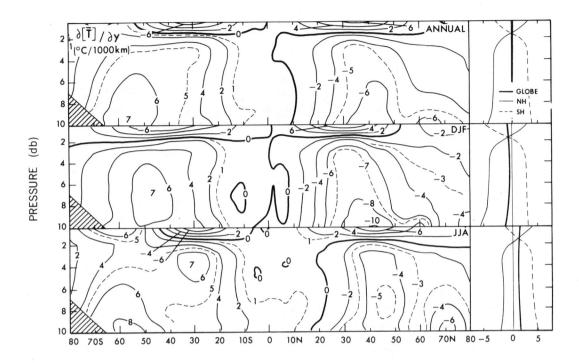

FIGURE 7.7. Zonal-mean cross sections of the meridional temperature gradient in °C/1000 km for annual, DJF, and JJA mean conditions. Vertical profiles of the hemispheric and global mean values are shown on the right.

the lower troposphere, as the vertical profiles on the right-hand side of Fig. 7.8(a) confirm. The variability in the upper atmosphere is due to variations in the height of the tropopause. If we assume that the annual-mean values are well represented by the average of the summer and winter values the yearly variance is given by

$$\overline{T'^2_{\text{yr}}} \approx \tfrac{1}{2}\left\{ \overline{T'^2_{\text{DJF}}} + \overline{T'^2_{\text{JJA}}} + \tfrac{1}{2}(\overline{T}_{\text{DJF}} - \overline{T}_{\text{JJA}})^2 \right\}. \quad (7.4)$$

This expression includes the variability associated with the annual cycle, and is, therefore, larger than the average of the seasonal variances, especially near the earth's surface in the Northern Hemisphere. The spatial (stationary eddy) variability in the annual-mean temperature field in Fig. 7.8(b) shows low values on the order of 1 °C or less south of about 20 °N. High values (greater than 2 °C) are found mainly at mid and high latitudes of the Northern Hemisphere associated with the pronounced land–sea contrast at those latitudes.

The meridional profiles of the vertical- and zonal-mean quantities shown in Fig. 7.9 synthesize the main findings described earlier. It is interesting to point out that for the temporal variability of the temperature $\sqrt{\overline{T'^2}}$, the annual profile far exceeds the December–February and June–August profiles as expected from the large amplitude of the annual cycle in the mean temperature in Fig. 7.9(a) [see Eq. (7.4)]. The stationary eddy profiles in Fig. 7.9(c) show the most significant contributions during northern winter with a peak value of almost 4 °C near 60 °N.

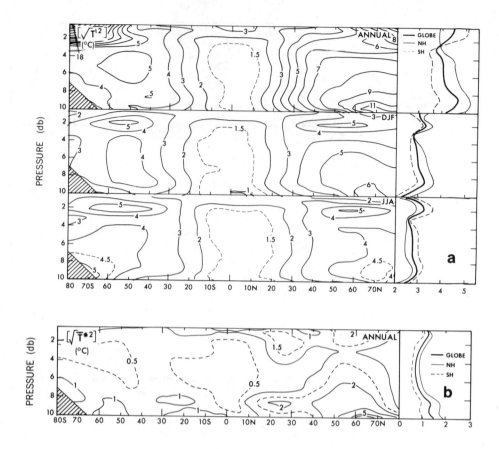

FIGURE 7.8. (a) Zonal-mean cross sections of the day-to-day standard deviation of temperature in °C for the year, DJF, and JJA. Vertical profiles of the hemispheric and global mean values are shown on the right. (b) Zonal-mean cross section of the east–west standard deviation of the annual-mean temperature field in °C.

7.3 MEAN GEOPOTENTIAL HEIGHT STRUCTURE OF THE ATMOSPHERE

7.3.1 Vertical structure of the geopotential

Because the various balance equations in meteorology are usually first derived in the (x, y, z, t) coordinate system and then, for practical use, transformed in the (x, y, p, t) coordinate system, it is important to clearly show how the two systems are related. Thus, Figs. 7.10(a) and 7.10(b) illustrate the relationship between constant height and constant pressure surfaces in the atmosphere. The meridional slope in each figure indicates the strength of the geostrophic component of the zonal flow since $[u_g] = -(g/f)\partial[z]/R\partial\phi$ [see Eq. (3.16b)]. We note that, in the mean, the pressure surfaces slope upward from the pole to the equator [Fig. 7.10(b)], indicating that westerly winds must prevail throughout most of the atmosphere. According to the figures, these westerly winds are greatest in middle latitudes and increase with altitude within the troposphere reaching a maximum near 200 mb just below the tropopause.

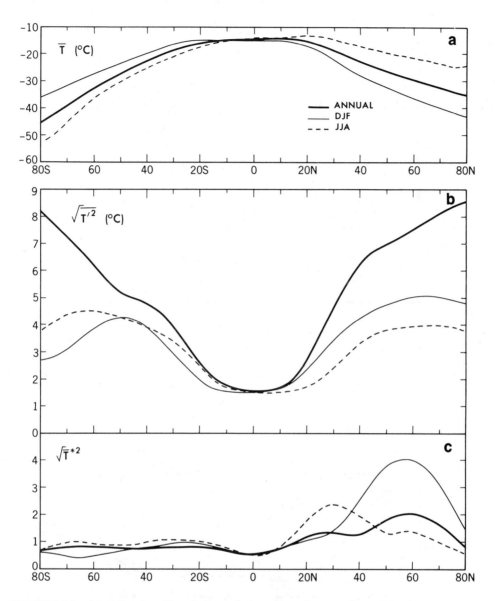

FIGURE 7.9. Meridional profiles of the vertical- and zonal-mean values of the time-mean temperature (a), the day-to-day standard deviation of the temperature (b), and the east–west standard deviation of the time-mean temperature (c) in °C for annual, DJF, and JJA mean conditions.

Cross sections of the zonal-mean geopotential height as a function of pressure and latitude are shown in Fig. 7.11. This type of presentation is important since it shows the distribution of the geopotential (divided by g) as a function of pressure. The variations of $[\bar{z}]$ with latitude are apparent because they are presented here as variations with respect to a reference atmosphere, namely the NMC standard atmosphere:

$$\Delta[z] = [z] - z_{\mathrm{SA}}.$$

146 PHYSICS OF CLIMATE

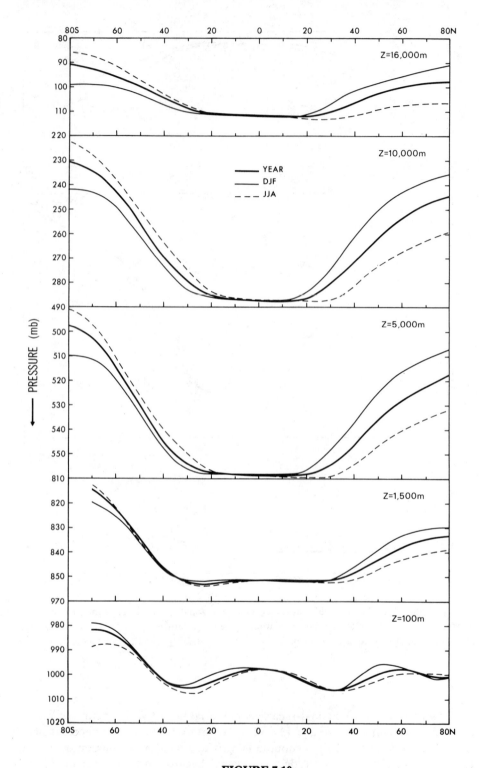

FIGURE 7.10a

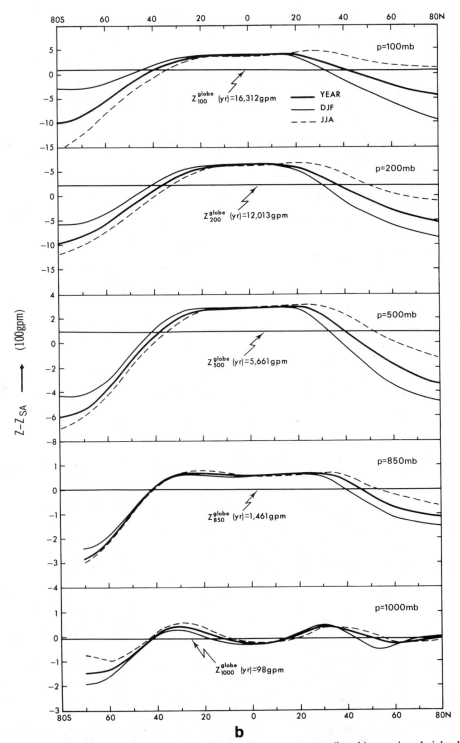

FIGURE 7.10. (a) Meridional profiles of the zonal-mean pressure (in mb) at various heights between 0.1 and 16 km. (b) Meridional profiles of the zonal mean height (in units of 100 gpm ≈ 100 m) for various pressure levels between 1000 and 100 mb.

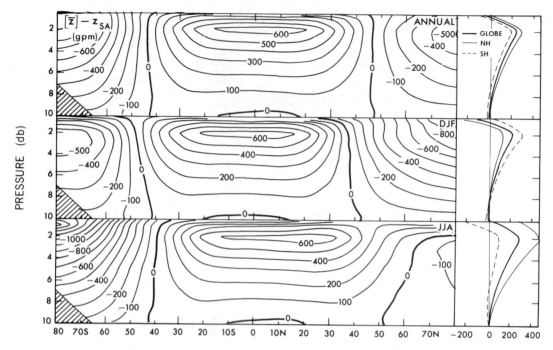

FIGURE 7.11. Zonal-mean cross sections of the geopotential height in gpm (1 gpm ≈ 1 m) for annual, DJF, and JJA mean conditions. Shown are the departures from the NMC standard atmosphere (see Table 5.1). Vertical profiles of the hemispheric and global-mean values are shown on the right.

The values of the NMC standard atmosphere were shown before in Table 5.1; they are only a function of pressure (i.e., barotropic) and represent typical conditions near 45 °N so that $\Delta[z] \approx 0$ near that latitude.

In a hydrostatic atmosphere, the geopotential height is given by

$$z(p) = z(p_0) + g^{-1} \int_p^{p_0} R_d T_V \, d \ln p, \qquad (7.5)$$

where T_V is the virtual temperature of the air as defined before in Eq. (3.63). Therefore,

$$z(p) - z_{SA} = z(p_0) - z_{SA}(p_0) + g^{-1} \int_p^{p_0} R_d (T_V - T_{SA}) \, d \ln p,$$

where the vertical distribution of T_{SA} is prescribed by the NMC standard atmosphere (see Table 5.1). Thus, the variations in Fig. 7.11 are related to the temperature structure of the atmosphere and, to a lesser extent, to the differences in surface pressure. Where the underlying atmosphere is colder than the standard atmosphere, the $[\bar{z}] - z_{SA}$ values are negative and, where the underlying atmosphere is warmer, they are positive. We notice that the baroclinic regions of the atmosphere, where there is a strong meridional gradient in $[\bar{z}] - z_{SA}$, are located in mid and high latitudes.

The curves on the right-hand side of Fig. 7.11 give the vertical profiles for the globe and each hemisphere, showing a maximum near the tropopause. The largest interhemispheric differences are found in the June–August season.

7.3.2 Variability of the geopotential height

The variability of the geopotential height can be studied not only through the seasonal variation of the mean geopotential height, but also through the temporal standard deviation $\sqrt{\overline{z'^2}}$ and the spatial stationary eddy deviation $\sqrt{\bar{z}^{*2}}$. Meridional profiles of their zonal- and vertical-mean values are shown in Fig. 7.12.

The profiles of $\langle[\bar{z}]\rangle$ in Fig. 7.12(a) are related to the mean atmospheric temperature. They show a maximum in the equatorial region and a monotonic decrease toward the poles. The profiles shift seasonally with the sun and the minimum values are obtained in the winter season. On the other hand, the transient eddy profiles in Fig. 7.12(b) show a minimum slightly to the north of the equator and maxima in the polar regions.

Most profiles show strong meridional gradients in extratropical latitudes, that are associated with the fluctuations in the zonal wind through the geostrophic relation. As expected, the yearly values of the transient component are much larger than the seasonal ones because they include the annual cycle, which is so large that it tends to mask the synoptic variations. The east–west variance of the mean height in Fig. 7.12(c) shows a large asymmetry between the two hemispheres, as the temperature does [see Fig. 7.9(c)] with much more zonal symmetry in the Southern than in the Northern Hemisphere, because of the reduced land–sea temperature contrast in the Southern Hemisphere. Overall, the stationary eddy contributions are much smaller than the transient ones.

7.4 MEAN ATMOSPHERIC CIRCULATION

7.4.1 Introduction

As we have seen before, the large-scale motions in the atmosphere are, to a high degree of approximation, horizontal. In the vertical direction, the pressure gradient almost balances the gravity force [see Eq. (3.13)] so that the vertical accelerations are negligible, and the vertical component of velocity is small everywhere. In the horizontal direction, the principal forces in the free atmosphere are the pressure gradient and the Coriolis forces leading to a quasigeostrophic equilibrium. Thus, the motions are practically parallel to the pressure contours (isobars) and the wind speeds are inversely proportional to the spacing between the contours [see Eq. (3.15)]. The isobars are thus approximate streamlines for the flow outside the equatorial region. Indeed, the surface winds, shown before in Fig. 7.1, tend to parallel the height contour lines with high pressure on the right-hand side in the Northern Hemisphere, and on the left-hand side in the Southern Hemisphere. There is a tendency for convergence into the equatorial zone and into the low-pressure belts near 60° latitude, and a tendency for divergence from the subtropical high-pressure cells and from the polar regions. Ageostrophic effects in the form of flow into low-pressure areas and flow out of high-pressure areas are due to friction and small-scale turbulent effects in the surface boundary layer [see Eq. (3.18)]. The angle of deflection is a manifestation of the quasibalance among the pressure force, the friction force, and the Coriolis force.

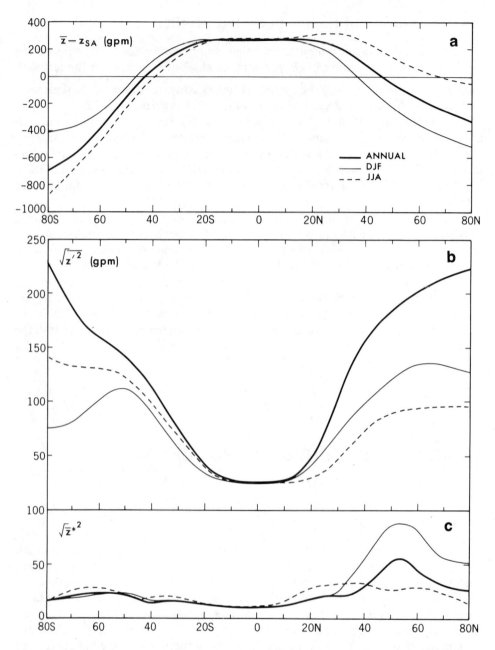

FIGURE 7.12. Meridional profiles of the vertical- and zonal-mean geopotential height departures (a), day-to-day standard deviations of the geopotential height (b), and east–west standard deviations of the time-mean geopotential height (c) in gpm (1 gpm $\approx$ 1 m) for annual, DJF, and JJA mean conditions.

7.4.2 Global distribution of the circulation

The atmospheric winds vary considerably with location on the globe and with the seasons. The surface winds were shown before in Fig. 7.1. The observed winds in the free atmosphere at 200 mb, where the maximum velocities in the tropospheric jet streams are usually found, are shown in Fig. 7.13 (indicated by arrows) together with

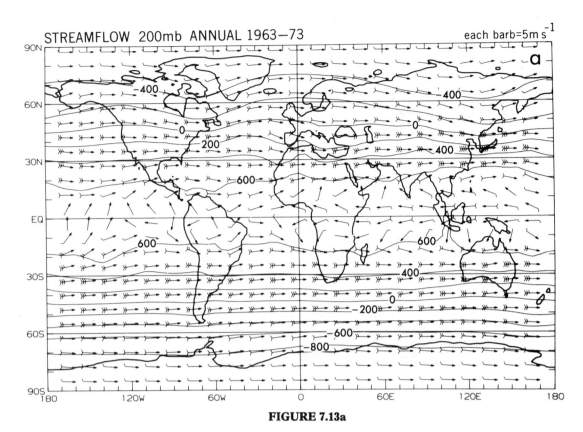

FIGURE 7.13a

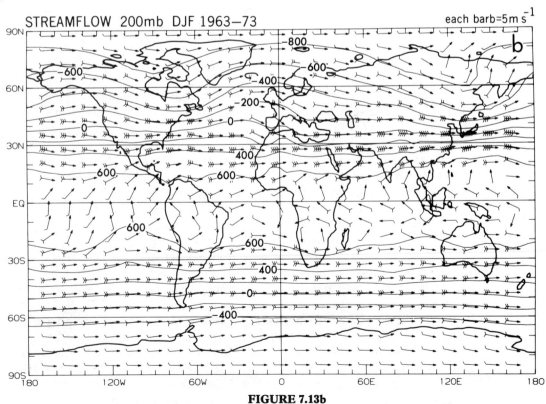

FIGURE 7.13b

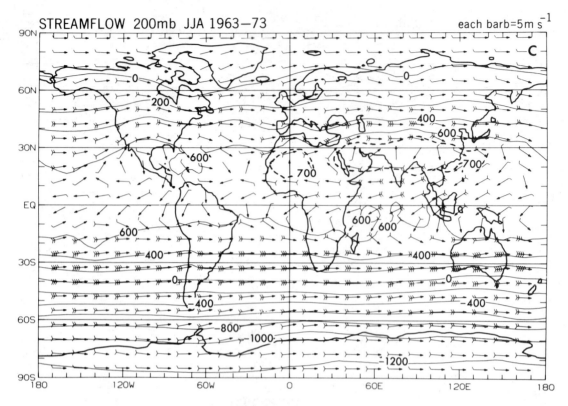

FIGURE 7.13. Global distributions of the height of the 200-mb pressure field, $\bar{z}_{200} - 11\,784$ gpm, in gpm for annual-mean (a), northern winter (b), and northern summer (c) mean conditions. Also shown are vector plots of the 200-mb winds. For geostrophic flow the arrows should parallel the isolines. Each barb on the tail of an arrow represents a wind speed of 5 m s^{-1}.

the geopotential height field. The vectors are generally parallel to the isoheight lines confirming the near-geostrophic balance of the flow. We find in each hemisphere a broad zonal current with superposed large-amplitude disturbances, the long planetary standing waves. The circulation is predominantly from west to east, and is stronger and more zonal in the Southern than in the Northern Hemisphere. In the tropics the circulation is much weaker than in mid and high latitudes.

The seasonal maps for summer and winter are shown in Figs. 7.13(b) and 7.13(c) for 200 mb. They show an intensification of the jet streams in the winter season. In the summer of the Northern Hemisphere a closed circulation is prevalent over the southern part of Asia accompanied by a northward shift of the jet stream. As expected, the standing waves at 200 mb are more pronounced in the Northern Hemisphere, and strongest during winter. Over the northern continents they show a preferred wave-number-two configuration with two troughs (just to the east of the American and Asian continents) and two ridges (just to the west of Europe and North America). It appears that the phase and amplitude of these planetary waves strongly depend on the forcing at the surface. Thus, the wave pattern is somewhat dependent on the season as shown by the latitudinal shifts of the troughs and ridges with the seasons. Comparing the 200-mb heights maps in Fig. 7.13 with the 1000-mb height maps in Fig. 7.1, we notice that there is a westward tilt with elevation of the high-latitude troughs and ridges in the troposphere. The tilt is associated with the advection of

OBSERVED MEAN STATE OF THE ATMOSPHERE 153

cold air upstream of the surface lows and advection of warm air upstream of the surface highs.

A map of the annual-mean zonal wind component at 200 mb, $\bar{u}_{200}$, is shown in Fig. 7.14(a). In midlatitudes, between about 20° and 50° latitude, one finds strong maxima in the westerly winds associated with the jet streams. These westerlies lead to a superrotation of the atmosphere as a whole on the order of 6 m s^{-1} relative to the

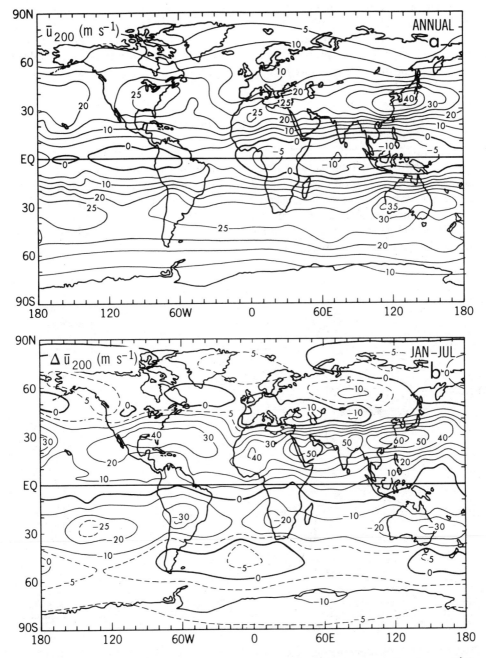

FIGURE 7.14. Global distribution of the 200-mb zonal wind component $\bar{u}_{200}$ in m s^{-1} for annual-mean conditions (a) and for the January–July difference (b).

solid earth. Again the patterns in the Southern Hemisphere are more zonally uniform than in the Northern Hemisphere, as would be expected from the greater homogeneity of the earth's surface in that hemisphere. We should mention that the westerly winds between about 50 and 70 °S are probably underestimated due to poor data coverage over the southern oceans (see Fig. 5.3). The data gaps may also be a contributing factor in making the patterns in the Southern Hemisphere more uniform. The major contributions to the zonal momentum in the Northern Hemisphere stem from the midlatitude jets over eastern North America, Asia, and the adjacent oceans.

There are seasonal shifts of about 10° latitude in the belt of westerlies toward the summer pole. The westerlies are strongest in the winter hemisphere related to the increase in the pole-equator temperature gradient. These seasonal shifts can be inferred from the difference map in Fig. 7.14(b) and are also seen in the mean profiles of Fig. 7.20 (to be discussed in more detail later). The largest seasonal differences in the $\bar{u}_{200}$ field are located near 30° latitude in both hemispheres.

7.4.3 Vertical structure of the circulation

The vertical and meridional distributions of the mean zonal flow are presented in Fig. 7.15 with, on the right-hand side, vertical profiles of hemispheric and global mean conditions. The similarities between the two hemispheres for the annual mean are quite striking. For example, the zonal circulation in both hemispheres is dominated by a westerly jet maximum of about 25 m s^{-1} near 200 mb. However, there are also differences since in the Southern Hemisphere the winds at all levels between 35° and 60° latitude are consistently stronger, by about 5 m s^{-1}, than in the Northern

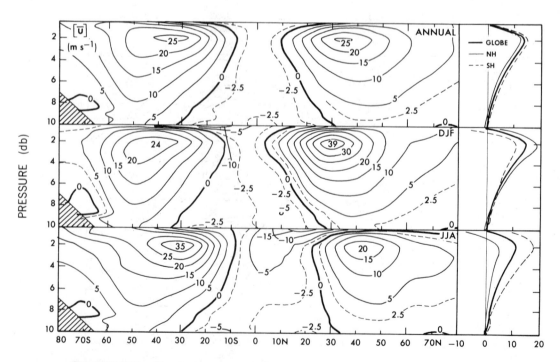

FIGURE 7.15. Zonal-mean cross sections of the zonal wind component in m s^{-1} for annual, DJF, and JJA mean conditions. Vertical profiles of the hemispheric and global mean values are shown on the right.

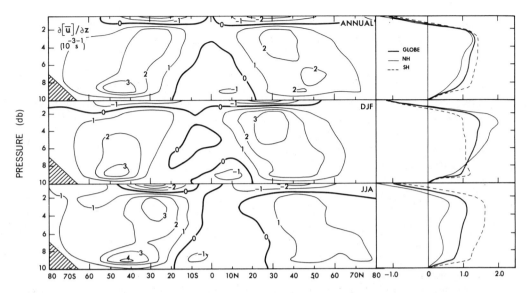

FIGURE 7.16. Zonal-mean cross sections of the vertical gradient of the zonal wind component (in 10^{-3} s^{-1}) for annual, DJF, and JJA mean conditions. Vertical profiles of the hemispheric and global mean values are shown on the right.

Hemisphere, reflecting the greater strength of the zonal winds in the Southern Hemisphere. Yet the winds in the Southern Hemisphere high latitudes should still be somewhat stronger than our analyses indicate because of the occurrence of large data gaps over the oceans and the bias due to a possible, selective loss of balloons during strong winds at some stations in the Southern Hemisphere [see the geostrophic wind estimates by Van Loon *et al.* (1971), Swanson and Trenberth (1981), and Oort (1983)]. Figure 7.15 also shows that easterly winds dominate in the tropical regions. The relatively weak surface easterlies in the equatorial zone are associated with the doldrums.

We further find that the surface easterlies and westerlies occupy almost equal areas and that, therefore, the global surface integrals are close to zero. We shall come back to this in the angular momentum discussion in Chap. 11. The broad maxima in $[\bar{u}]$ near 200 mb are located just above the regions where the north–south temperature gradient (i.e., the baroclinicity of the atmosphere) is a maximum.

A cross section of the vertical wind shear, $\partial[\bar{u}]/\partial z$, is shown in Fig. 7.16. Since, in general, the winds vary with height we can define a fictitious vector $\mathbf{v}_T$, the so-called thermal wind, so that $\mathbf{v}_T \equiv \mathbf{v}(z + \Delta z) - \mathbf{v}(z) \approx (\partial \mathbf{v}/\partial z)\Delta z$. As we will show next, the vertical shear is related to the temperature gradient through the thermal wind equation. This equation can be derived assuming that the horizontal flow is geostrophic (see Eq. 3.15). The shear is then given by

$$\frac{\partial}{\partial z}(\rho \mathbf{v}_g) = \frac{1}{f} \mathbf{k} \times \mathrm{grad}\, \frac{\partial p}{\partial z}$$

or with the hydrostatic approximation (3.13)

$$\frac{\partial \mathbf{v}_g}{\partial z} = -\frac{g}{\rho f} \mathbf{k} \times \mathrm{grad}\, \rho - \frac{1}{\rho}\frac{\partial \rho}{\partial z}\mathbf{v}_g.$$

Using the equation of state after logarithmic differentiation and again Eq. (3.15), we obtain finally for the wind shear vector

$$\frac{\partial \mathbf{v}_g}{\partial z} = \frac{g}{f}\mathbf{k} \times \frac{\mathrm{grad}\, T}{T} + \frac{1}{T}\frac{\partial T}{\partial z}\mathbf{v}_g. \qquad (7.6a)$$

Scale analysis of the last equation, the thermal wind equation, for large-scale flow shows that the first term on the right hand side dominates so that

$$\frac{\partial \mathbf{v}_g}{\partial z} \approx \frac{g}{fT} \mathbf{k} \times \text{grad } T. \tag{7.6b}$$

Thus the vertical shear at a given level is proportional to the gradient of temperature and parallel to the isotherms. The geostrophic wind will increase with height ($\partial \mathbf{v}_g/\partial z > 0$) with the warm air on the right of the shear vector in the Northern Hemisphere and on the left in the Southern Hemisphere.

The thermal wind equation (7.6b) can be also written in the p-system in terms of the potential temperature θ using Eq. (3.47):

$$\frac{\partial \mathbf{v}_g}{\partial p} \approx -\frac{1}{\rho\theta} \mathbf{k} \times \text{grad } \theta. \tag{7.6c}$$

According to the thermal wind equation, the zonal wind increases with height $[\partial u_g/\partial z \approx -(g/fT)(\partial T/R\partial \phi) > 0]$ in midlatitudes where the meridional temperature gradient is most pronounced. Of course, in regions where the atmosphere is barotropic the winds do not change with height ($\mathbf{v}_T = 0$) because grad $T = 0$ (see Sec. 3.5.3). However, Fig. 7.16 shows that the atmosphere is in general baroclinic.

Let us show how the thermal wind is related to the baroclinicity vector $\mathbf{N}$ (see Sec. 3.5.3). If we take the cross product of $\mathbf{N}$ with the vertical unit vector $\mathbf{k}$ we find

$$\mathbf{k} \times \mathbf{N} = -\frac{1}{\rho T}(\mathbf{k} \cdot \text{grad } p)\text{grad } T + \frac{1}{\rho T}(\mathbf{k} \cdot \text{grad } T)\text{grad } p.$$

Using the hydrostatic approximation this equation becomes

$$\mathbf{k} \times \mathbf{N} = \frac{g}{T} \text{grad } T + \frac{1}{\rho T}\frac{\partial T}{\partial z} \text{grad } p.$$

If we again cross multiply with $\mathbf{k}$ we find

$$\mathbf{k} \times (\mathbf{k} \times \mathbf{N}) \equiv -\mathbf{N} + N_z \mathbf{k} = \frac{g}{T} \mathbf{k} \times \text{grad } T + \frac{1}{\rho T}\frac{\partial T}{\partial z} \mathbf{k} \times \text{grad } p.$$

Using the geostrophic wind equation (3.15) and the thermal wind equation (7.6a), noting that $\mathbf{N}_h = \mathbf{N} - N_z \mathbf{k}$ is the horizontal vector component of $\mathbf{N}$, we find

$$-\mathbf{N}_h = \frac{g}{T} \mathbf{k} \times \text{grad } T + \frac{f}{T}\frac{\partial T}{\partial z} \mathbf{v}_g = f\frac{\partial \mathbf{v}_g}{\partial z}.$$

Of course, $\mathbf{N}_h$ represents the number of solenoids in a vertical plane across the solenoidal tubes and measures the baroclinicity in that plane. Thus, as we have seen before, in a barotropic atmosphere ($\mathbf{N}_h = 0$) the winds do not change with height.

The meridional component of the zonal-mean circulation is shown in Fig. 7.17. Outside the tropics the evaluation of the mean meridional circulation from the observed $\bar{v}$ field becomes unreliable and even unusable in the Southern Hemisphere. In view of this difficulty, we have used indirectly computed values of $[\bar{v}]$ poleward of 15° latitude, in both hemispheres (see Oort and Peixoto, 1983). For the region between 15 °S and 15 °N, the directly measured $[\bar{v}]$ values were used because there the time-mean meridional winds are fairly uniform and the observational network seems ade-

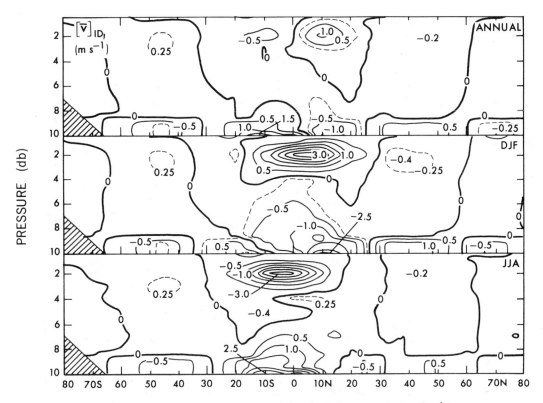

FIGURE 7.17. Zonal-mean cross sections of the meridional wind component in m s^{-1} for annual, DJF, and JJA mean conditions. The $[\bar{v}]$ values between 10° S and 10° N were computed directly from the rawinsonde analyses, and poleward of 20° latitude indirectly from momentum balance.

quate to monitor the dominant zonally symmetric circulation. Between 10° and 20° latitude the values derived from the two methods were averaged with weights varying between 0 and 1.

Although small, the meridional component of the wind plays an important role in maintaining the zonal winds [fv term in the zonal equation of motion (3.18a)]. The lower branches of the tropical Hadley circulations are directly associated with the northeast and southeast trade winds, whereas the lower branches of the Ferrel cells are associated with the prevailing westerlies in midlatitudes, and those of the polar cells with the polar easterlies.

Cross sections of the vertical velocity $[\bar{\omega}]$ can be computed from the $[\bar{v}]$ values in Fig. 7.17 using the continuity equation of mass in zonally averaged form (7.7) and the streamfunction approach (7.8) to be discussed next. The resulting vertical velocity patterns in Fig. 7.18 show for the annual-mean strong rising motions of 2×10^{-4} mb s^{-1} (or about 3 mm s^{-1}) centered near 5 °N, associated with the mean position of the ITCZ. This equatorial belt is flanked in each hemisphere by sinking motions between about 10° and 40° latitude, followed by rising motions between 50° and 70° latitude, and weak sinking motions poleward of 70° latitude. In northern winter we find the strongest rising motions between 10 °N and 20 °S and the strongest sinking motions between 10 and 35 °N, whereas in northern summer the strongest rising motions are found between 10 °S and 20 °N and the strongest sinking motions between 10 and 40 °S. The vertical profiles on the right-hand side of Fig. 7.18 are basically symmetric for both hemispheres with a sign reversal between winter and summer. Extreme values occur between about 400 and 500 mb.

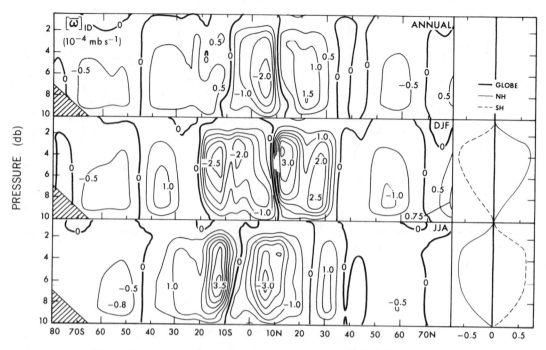

FIGURE 7.18. Zonal-mean cross sections of the vertical velocity ω in 10^{-4} mb s^{-1} for annual, DJF, and JJA mean conditions as computed from the meridional velocities in Fig. 7.17 using conservation of mass. Vertical profiles of the hemispheric and global mean values are shown on the right.

The $[\bar{v}]$ data can also be used to construct streamlines indicating the mean overturnings of mass in a north–south cross section. A stream function can be computed using the continuity equation of mass (3.4) in zonally averaged form:

$$\frac{\partial [\bar{v}]\cos\phi}{R\cos\phi\partial\phi} + \frac{\partial [\bar{\omega}]}{\partial p} = 0. \tag{7.7}$$

Thus, we may introduce the so-called Stokes stream function ψ given by the equations

$$[\bar{v}] = g\frac{\partial \psi}{2\pi R \cos\phi\partial p}, \tag{7.8a}$$

$$[\bar{\omega}] = -g\frac{\partial \psi}{2\pi R^2 \cos\phi\partial\phi}. \tag{7.8b}$$

We can now calculate the ψ field from the observed $[\bar{v}]$ distribution by vertical integration of Eq. (7.8a) starting at the top of the atmosphere where we assume that $\psi = 0$. The meridional gradient of ψ then gives the vertical component $[\bar{\omega}]$ of the zonally symmetric overturnings.

The computed annual-mean and seasonal streamlines are shown in Fig. 7.19. The circulation cells in these cross sections are often called "mean meridional circulations." It is clear that the annual-mean cells in the tropics represent an average of two very different winter and summer patterns, and that they are perhaps only representative for conditions during the transition seasons in spring and fall. The annual-mean picture thus shows a somewhat idealized situation of three cells in each hemisphere, one in the tropical regions (the so-called Hadley cell), another in midlatitudes (the Ferrel cell), and finally a third one in the polar regions (the polar cell).

OBSERVED MEAN STATE OF THE ATMOSPHERE

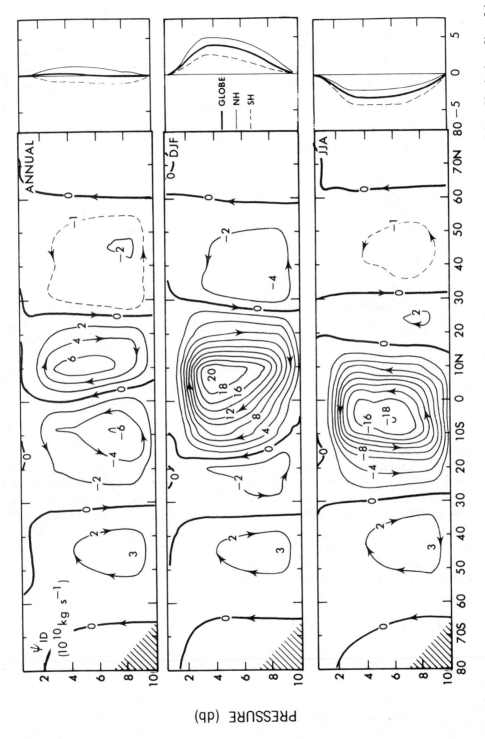

FIGURE 7.19. Zonal-mean cross sections of the mass stream function in 10^{10} kg s^{-1} for annual, DJF, and JJA mean conditions. Vertical profiles of the hemispheric and global mean values are shown on the right.

The seasonal cross sections presented in Fig. 7.19 are very interesting because they show the strong intensification and predominance of the winter Hadley cell and the almost disappearance of the summer Hadley cell in each hemisphere. The intensification leads to a shift of the low-level circulation into the summer hemisphere. The Ferrel cells also appear to be more intense in the winter season, the differences being more pronounced in the Northern Hemisphere as one would expect. The vertical profiles on the right-hand side of Fig. 7.19 reflect the seasonal behavior of the Hadley cells, and show the antisymmetry of the profiles between the two seasons.

The direct Hadley cells in the tropics in the three cross sections of Fig. 7.19 are much stronger than the indirect Ferrel cells in middle latitudes. At these latitudes the circulation is dominated by almost horizontal, wave-like flows, and the Ferrel cells are only small statistical residues which result after zonal averaging of large, almost compensating, northward and southward flows in the quasistationary atmospheric waves. The direct polar cells are quite weak. In the annual mean, the southern Hadley cell penetrates across the equator reinforcing the upward motions characteristic of the mean ITCZ.

In the Hadley cells, there is a rising of warm (light) and moist air in the equatorial region and a descent of colder (heavier) air in the subtropics leading to a thermally driven direct circulation. However, in the Ferrel cells there is a rising of relatively cold air in high latitudes and a sinking of relatively warm air in the lower midlatitudes leading to a thermally indirect circulation in which cold air is forced to rise. In a direct circulation with the lowering of the center of mass there is a production of kinetic energy, whereas in an indirect cell with the net rising of the center of mass there is a consumption of kinetic energy. Further discussions on the mean meridional circulations and their maintenance will be given in Sec. 14.5.

The vertical motion patterns associated with the mean meridional circulations also indicate the principal climate zones, namely the equatorial rainy zone, the subtropical arid and desert-like zones, the moist mid to high-latitude zones, and the dry polar zones.

As a final remark, we want to emphasize that the observed three-cell regime in the annual-mean atmosphere results, of course, from the imposed pole-to-equator gradient in net radiation combined with the rotation of the earth. For different rotation rates, such as is the case for the other planets, we can expect a different number and configuration of cells. For example, in the low-rotation case, we may have one direct cell in each hemisphere as Hadley proposed in 1735 to explain the trade winds, while in the high-rotation case we may have many more cells such as may occur for Jupiter (see, e.g., Williams and Holloway, 1982).

7.4.4 Variability of the circulation

To show the temporal and spatial variability of the wind components, profiles of the vertical- and zonal-mean standard deviations for u and v are presented in Fig. 7.20.

It is of interest to mention the close symmetry between the two hemispheres. Further, for both u and v the temporal standard deviations are of the same magnitude as $[\bar{u}]$ or even larger at certain latitudes. This feature, as well as the nearly identical patterns for the temporal standard deviations of u and v, clearly point out the turbulent character of the atmospheric general circulation. The larger winter–summer contrast in the $\sqrt{\overline{u'^2}}$ curves in the Northern Hemisphere reflects the larger variability of the northern jet streams. In terms of the transient eddy kinetic energy,

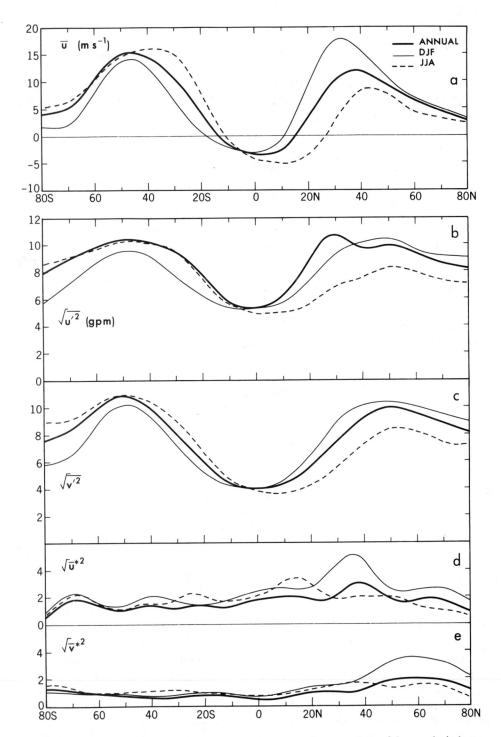

FIGURE 7.20. Meridional profiles of the vertical- and zonal-mean values of the zonal wind component (a), the day-to-day standard deviations of the zonal (b) and meridional (c) wind components, and the east–west standard deviations of the zonal (d) and meridional (e) wind components, all in m s^{-1} for annual, DJF, and JJA mean conditions.

$(\overline{u'^2} + \overline{v'^2})/2$, we find an approximate equal partitioning of energy between the u and v components, i.e., $\overline{u'^2} \approx \overline{v'^2}$. The winter standard deviations are somewhat larger than the summer ones, as one would expect.

The yearly curves in Fig. 7.20 sometimes lie above both seasonal profiles due to the extra variance resulting from the winter–summer differences included in the yearly standard deviation [see Eq. (7.4)].

7.5 MEAN KINETIC ENERGY IN THE ATMOSPHERE

7.5.1 Global distribution of the kinetic energy

An important quantity to characterize the atmospheric circulations and to clarify their variability in time and space is the kinetic energy. In fact, using

$$u = [\bar{u}] + \bar{u}^* + u',$$
$$v = [\bar{v}] + \bar{v}^* + v',$$

and expansion (4.13) we may write

$$[\overline{u^2}] = [\bar{u}]^2 + [\bar{u}^{*2}] + [\overline{u'^2}],$$
$$[\overline{v^2}] = [\bar{v}]^2 + [\bar{v}^{*2}] + [\overline{v'^2}].$$

Thus, the mean total kinetic energy per unit mass can be written as the sum of the transient eddy, stationary eddy, and zonal-mean components:

$$K = K_{TE} + K_{SE} + K_M, \tag{7.9a}$$

where

$$K = \tfrac{1}{2}[\,\overline{u^2} + \overline{v^2}\,],$$
$$K_{TE} = \tfrac{1}{2}[\,\overline{u'^2} + \overline{v'^2}\,], \tag{7.9b}$$
$$K_{SE} = \tfrac{1}{2}[\bar{u}^{*2} + \bar{v}^{*2}], \tag{7.9c}$$

and

$$K_M = \tfrac{1}{2}([\bar{u}^2] + [\bar{v}^2]). \tag{7.9d}$$

Vertically averaged, annual-mean maps of the transient eddy kinetic energy $\tfrac{1}{2}(\overline{u'^2} + \overline{v'^2})$ and the total kinetic energy $\tfrac{1}{2}(\overline{u^2} + \overline{v^2})$ are shown in Figs. 7.21(a) and 7.21(b), respectively. The first map mainly reflects the dominant storm tracks in midlatitudes, whereas in the second map the midlatitude maxima are reinforced due to the steady components of the subtropical and polar jet streams. There are two distinct peaks in the Northern Hemisphere kinetic energy just east of Japan and east of the North American continent, whereas a more continuous belt of high kinetic energy is found between 30 and 60 °S in the Southern Hemisphere.

7.5.2 Vertical structure of the kinetic energy

Zonal-mean cross sections of the transient eddy, stationary eddy, mean and total kinetic energy per unit mass, as well as the vertical profiles of their hemispheric- and global-mean values are displayed in Fig. 7.22. The annual-mean total kinetic energy

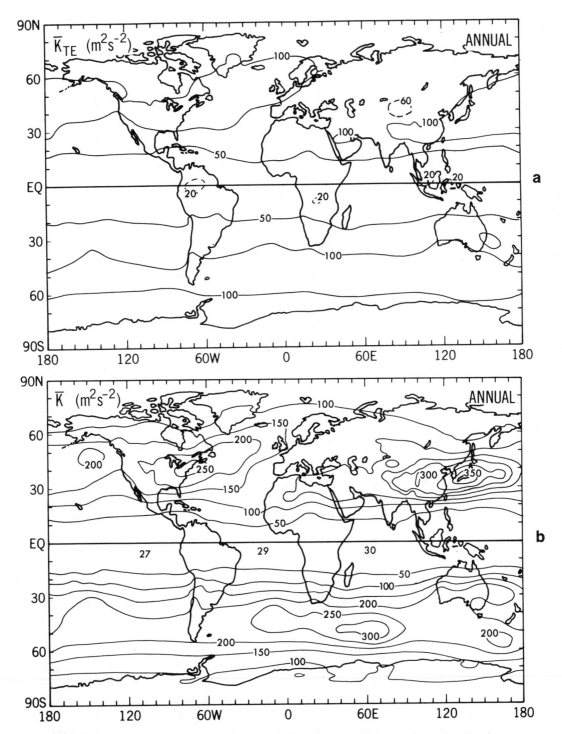

FIGURE 7.21. (a) Global distribution of the vertical-mean value of the transient eddy kinetic energy in $m^2\ s^{-2}$ for annual-mean conditions. (b) Global distribution of the vertical-mean total kinetic energy in $m^2\ s^{-2}$ for annual-mean conditions.

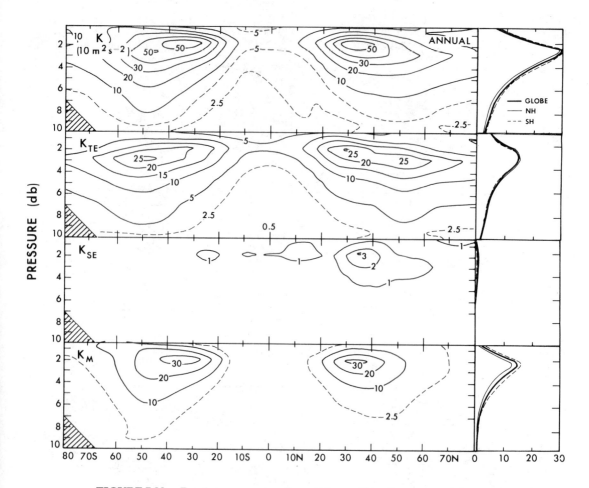

FIGURE 7.22. Zonal-mean cross sections of the total kinetic energy and the transient eddy, stationary eddy, and mean meridional components of the kinetic energy, all in $m^2 \, s^{-2}$, for annual-mean conditions. Vertical profiles of the hemispheric and global mean values are shown on the right.

shows a similar pattern in both hemispheres. The 200-mb maxima in the total kinetic energy at 35° latitude are mainly due to the subtropical jets. It is clear that K_{TE} and K_M are the main contributors to the total kinetic energy. The stationary eddies only contribute significantly in the Northern Hemisphere. The broad midlatitude maxima in the K_{TE} cross sections are due to both the daily meandering of the polar and subtropical jet streams and the seasonal shifts in the latitude of the subtropical jet.

The mean vertically integrated values are shown in the meridional profiles of Fig. 7.23. They clearly show the differences between K_{SE}, K_{TE}, and K_M. The largest seasonal variations occur in K_M. They are much more pronounced in the Northern than in the Southern Hemisphere, with a summer–winter difference in the Northern Hemisphere peak value by a factor of 3 and with high values in the Southern Hemisphere throughout the year. However, the variations in K_{TE} are also not negligible, in contrast with those in K_{SE}.

As a final comment, we should mention that the midlatitude values of kinetic energy in the Southern Hemisphere as computed from the rawinsonde network probably represent considerable underestimates of the actual values.

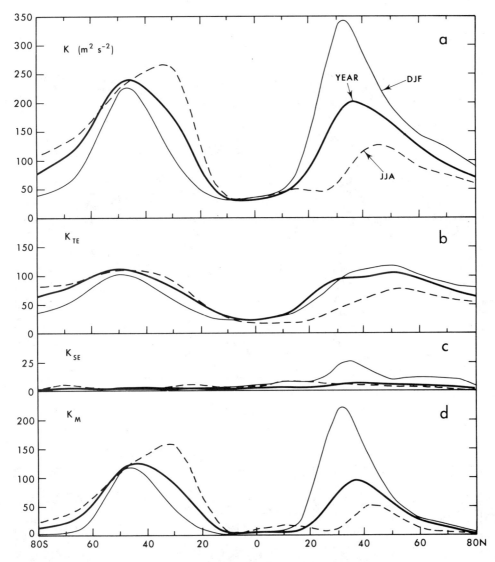

FIGURE 7.23. Meridional profiles of the zonal- and vertical-mean values of the total kinetic energy (a) and the transient eddy (b), stationary eddy (c), and mean meridional (d) components of the kinetic energy in $m^2\ s^{-2}$ for annual, DJF, and JJA mean conditions (from Oort and Peixoto, 1983).

7.6 PRECIPITATION, EVAPORATION, RUNOFF, AND CLOUDINESS

7.6.1 Precipitation

Precipitation is one of the principal climatic elements. It is highly variable in space and time. Nevertheless, its average values are fairly stable and can be represented well in map form. So we will start with a discussion of Fig. 7.24 which shows

166   PHYSICS OF CLIMATE

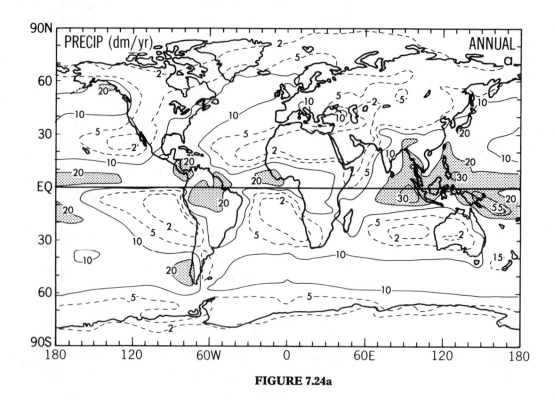

FIGURE 7.24a

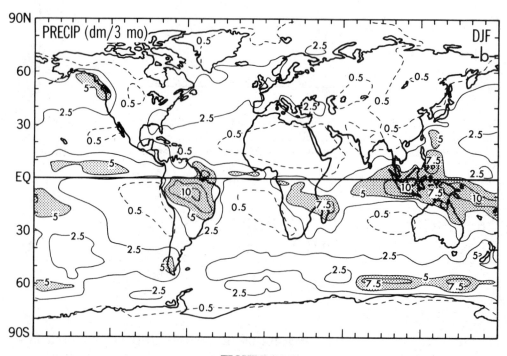

FIGURE 7.24b

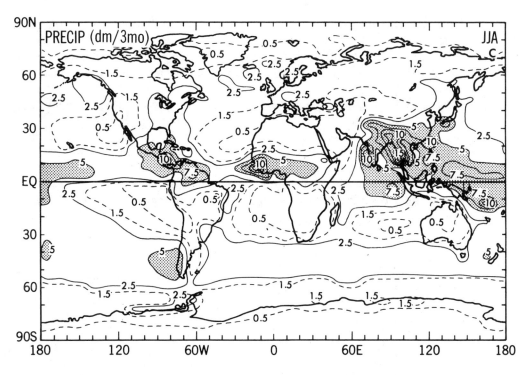

FIGURE 7.24. Global distributions of the precipitation rate for annual-mean conditions (a) in dm yr^{-1}, and for DJF (b) and JJA (c) conditions in dm (3 months)$^{-1}$, based on data from Jaeger (1976). Note that a precipitation rate of 1 m yr^{-1} corresponds to a release of latent heat in the atmosphere of about 79 W m^{-2}.

the distribution of precipitation according to Jaeger (1976) based on the surface station network [see Fig. 5.1(b)]. Note that the units on the maps are dm yr^{-1} for the annual mean and dm (3 months)$^{-1}$ for the seasons. The annual and seasonal distributions clearly reveal the influence of the oceans and continents. We should also mention that the precipitation data are very uncertain over the oceans because of the difficulties in directly measuring precipitation from ships.

The most significant features of the distribution are the high rainfall in equatorial latitudes associated with the strong convection in the ITCZ. Especially noteworthy are the very high values of precipitation over the equatorial regions in South America, Africa, and Indonesia and in the equatorial Pacific Ocean where precipitation may exceed 3 m yr^{-1}. During the course of the year the ITCZ migrates north and south in phase with the solar insolation which explains the shift of the maxima in Figs. 7.24(b) and 7.24(c). There is a striking contrast in the seasonal maps over southeast Asia, which is mainly due to the Indian southwest monsoon dominating the summer circulation over the Horn of Africa, India, and southeast Asia.

Subsidence and low precipitation rates often less than 0.2 m yr^{-1} dominate in many of the subtropical regions which are under the influence of the large semipermanent anticyclones. Large parts of the subtropical continents, such as in Africa and Australia, are covered by deserts, where the precipitation is very low. During the annual cycle the high-pressure centers migrate north and south causing summer dryness or semi–arid conditions in each hemisphere at the poleward side and winter dryness at the equatorial side of their annual-mean positions.

There is a secondary maximum in precipitation over midlatitudes where the polar fronts and the associated disturbances predominate. Here, precipitation is abundant during all seasons, except on their equatorial border where dryness prevails during the summer season when the high-pressure anticyclones move poleward, such as in the Mediterranean region.

Over the polar regions the moisture content of the atmosphere is very low, and the amounts of precipitation are less than 0.2 m yr^{-1} during all seasons.

The land–sea contrasts as well as seasonal differences become even clearer when we compare the zonal-mean profiles in Fig. 7.25. The seasonal shifts of the ITCZ are found to be more pronounced over land than over the oceans. The seasonal migration of the ITCZ is the determining factor in the existence of the marginal climates bordering the arid subtropical regions, such as are found in the Sahel belt in Africa.

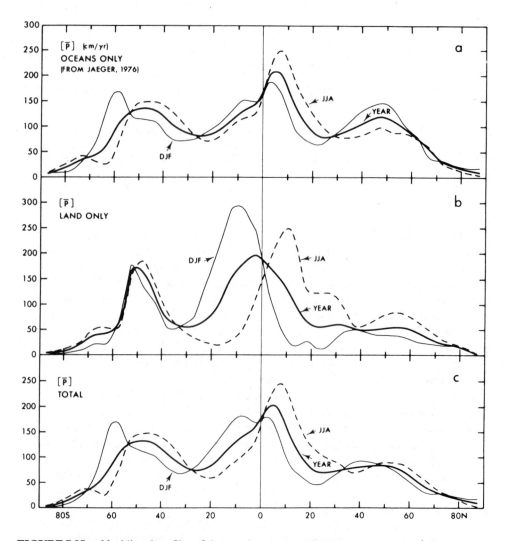

FIGURE 7.25. Meridional profiles of the zonal-mean precipitation rate in cm yr^{-1} for the ocean area (a), the land area (b), and the total land plus ocean area (c) for annual, DJF, and JJA mean conditions based on data from Jaeger (1976).

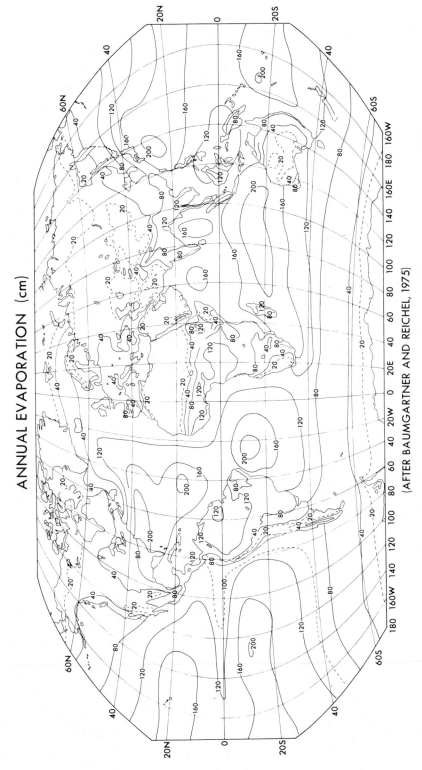

FIGURE 7.26. Global distribution of the annual-mean evaporation rate in cm yr^{-1} after Baumgartner and Reichel (1975).

7.6.2 Evaporation

The evaporation rate depends on many factors. The most important ones are the incoming radiation, temperature, wind speed, humidity, stability of the air, and the availability of water. Evaporation is often measured with a shallow circular pan. However, such measurements are much influenced by local conditions and local exposure. Thus, useful as they may be for local purposes, such as assessing evaporation from water reservoirs, small lakes, and irrigated areas, they are of little use in computing the water budget over larger regions of the earth.

Using surface ship data the evaporation over the oceans can also be evaluated from an approximate, empirically derived expression:

$$\overline{E} = -\rho C_w \overline{|\mathbf{v}|(q_a - q_s)},$$

where ρ is the density of air, C_w the coefficient of eddy diffusion (≈ 0.0013), q_s the saturated specific humidity at the sea surface temperature, and q_a the specific humidity at a standard height of about 10 m above the surface. The foundations of this method will be discussed in Chap. 10.

Figure 7.26 shows the annual-mean evaporation as estimated by Baumgartner and Reichel (1975) using a variety of methods. The map shows that the highest values of evaporation occur over the subtropical oceans, where the oceanic "deserts" are found. The effects of warm and cold ocean currents and land–sea differences are very important, as illustrated by the midlatitude (mainly winter) maxima in evaporation, on the order of 2 m yr^{-1}, over the relatively warm Gulf Stream and Kuroshio currents to the east of the two major continents. Over the equatorial oceans, where precipitation is abundant, evaporation is less intense due to weaker winds and relatively low sea surface temperatures in the oceanic upwelling regions. Over the continents maximum evaporation occurs in the equatorial belt, mainly due to the higher precipitation and higher temperatures observed there.

Annual and seasonal profiles of the zonal-mean oceanic evaporation based on the bulk aerodynamic method are shown in Fig. 7.27 together with Baumgartner and Reichel's (1975) annual-mean estimates. The differences reflect the considerable uncertainties involved in estimating the evaporation fields. The zonal-mean profiles summarize the main aspects of the behavior of evaporation over the oceans. They reveal that the hemispheric evaporation rates tend to be higher during winter than during summer mainly due to the stronger surface winds in winter.

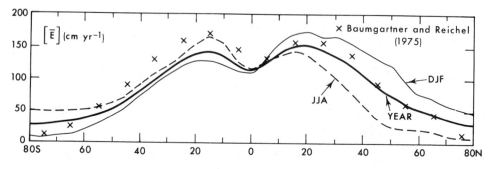

FIGURE 7.27. Meridional profiles of the zonal-mean evaporation rate (in cm yr^{-1}) over the oceans computed using Eq. (10.38) and our 1963–73 surface data. Baumgartner and Reichel's (1975) ocean values from Fig. 7.26 have been added for comparison.

7.6.3 Surface runoff

After considering the precipitation and evaporation fields separately it is useful to compare the behavior of these two quantities since they are closely related elements of climate and hydrology. Thus, we present in Table 7.1 the zonal-mean annual averages of precipitation and evaporation for 10° latitude belts as well as the hemispheric and global averages according to Baumgartner and Reichel (1975). Similar estimates of the same quantities for individual continents and oceans are presented in Table 7.2. The tables include also the $P - E$ differences, the so-called discharge or runoff, and the values of the evaporation ratio E/P and the runoff ratio $(P - E)/P$. These last two quantities are of interest since they are sometimes used as climatic indices or in hydrology studies. Table 7.2 further includes the estimated river discharge R_0 from the continents as measured by the peripheral runoff that reaches the oceans.

The $P - E$ values in Table 7.1 show an excess of precipitation over evaporation at mid and high latitudes as well as in the equatorial zone between 10°S and 10°N,

TABLE • 7.1. Estimated mean annual values of the precipitation rate P, evaporation rate E, runoff rate $(P - E)$, evaporation ratio E/P (an aridity index), and runoff ratio $(P - E)/P$ for 10° latitude belts, the hemispheres, and the globe from Baumgartner and Reichel (1975). For comparison, Sellers' (1965) estimates for P and E, and Peixoto and Oort's (1983) independent estimates of $P - E$ as computed from Table 12.1 are shown in parentheses.

	Surface area	P	E	$P - E$	E/P	$(P - E)/P$
80–90 °N	3.9	46 (120)	36 (42)	10 (93)	0.78	0.22
70–80 °N	11.6	200 (185)	126 (145)	74 (124)	0.63	0.37
60–70 °N	18.9	507 (415)	276 (333)	231 (224)	0.54	0.46
50–60 °N	25.6	843 (789)	447 (469)	396 (250)	0.53	0.47
40–50 °N	31.5	874 (907)	640 (641)	234 (156)	0.73	0.27
30–40 °N	36.4	761 (872)	971 (1002)	−210 (23)	1.28	−0.28
20–30 °N	40.2	675 (790)	1110 (1246)	−435 (−435)	1.64	−0.64
10–20 °N	42.8	1117 (1151)	1284 (1389)	−167 (−322)	1.15	−0.15
0–10 °N	44.1	1885 (1934)	1250 (1235)	635 (478)	0.66	0.34
0–10 °S	44.1	1435 (1445)	1371 (1304)	64 (144)	0.96	0.04
10–20 °S	42.8	1109 (1132)	1507 (1541)	−398 (−342)	1.36	−0.36
20–30 °S	40.2	777 (857)	1305 (1416)	−528 (−312)	1.68	−0.68
30–40 °S	36.4	875 (932)	1181 (1256)	−306 (−128)	1.35	−0.35
40–50 °S	31.5	1128 (1226)	862 (895)	266 (150)	0.76	0.24
50–60 °S	25.6	1003 (1046)	553 (520)	450 (278)	0.55	0.45
60–70 °S	18.9	549 (418)	229 (174)	320 (245)	0.42	0.58
70–80 °S	11.6	230 (82)	54 (45)	176 (98)	0.23	0.77
80–90 °S	3.9	73 (30)	12 (0)	61 (32)	0.16	0.84
0–90 °N	255.0	970 (1009)	897 (944)	73 (39)	0.92	0.07
0–90 °S	255.0	975 (1000)	1048 (1064)	−73 (−39)	1.07	−0.07
Globe	510.0	973 (1004)	973 (1004)	...	1.00	...
Units	10^6 km^2	mm yr^{-1}	mm yr^{-1}	mm yr^{-1}	...	...

TABLE • 7.2. Estimated mean annual values of the precipitation rate P, evaporation rate E, runoff rate $P-E$, river runoff rate R_0 from continents into the oceans, evaporation ratio E/P, and runoff ratio $(P-E)/P$ for the various continents and oceans from Baumgartner and Reichel (1975). For comparison, estimates of P, E, and $P-E$ from Sellers (1965) have been added in parentheses.

Region	Surface area ($10^6\,km^2$)	P ($mm\,yr^{-1}$)	E ($mm\,yr^{-1}$)	$P-E$ ($mm\,yr^{-1}$)	R_0 ($mm\,yr^{-1}$)	E/P	$(P-E)/P$
Europe	10.0	657 (600)	375 (360)	282 (240)	...	0.57	0.43
Asia	44.1	696 (610)	420 (390)	276 (220)	...	0.60	0.40
Africa	29.8	696 (670)	582 (510)	114 (160)	...	0.84	0.16
Australia	8.9	803	534	269	...	0.67	0.33
[without islands]	7.6]	[447 (470)]	[420 (410)]	[27 (60)]	...	[0.94]	[0.06]
North America	24.1	645 (670)	403 (400)	242 (270)	...	0.62	0.38
South America	17.9	1564 (1350)	946 (860)	618 (490)	...	0.60	0.40
Antarctica	14.1	169 (30)	28 (0)	141 (30)	...	0.17	0.83
All land areas	148.9	746 (720)	480 (410)	266 (310)		0.64	0.36
Arctic Ocean	8.5	97 (240)	53 (120)	44 (120)	307	0.55	0.45
Atlantic Ocean	98.0	761 (780)	1133 (1040)	−372 (−260)	197	1.49	−0.49
Indian Ocean	77.7	1043 (1010)	1294 (1380)	−251 (−370)	72	1.24	−0.24
Pacific Ocean	176.9	1292 (1210)	1202 (1140)	90 (70)	69	0.93	0.07
All oceans	361.1	1066 (1120)	1176 (1250)	−110 (−130)	110	1.10	−0.10
Globe	510.0	973 (1004)	973 (1004)	0 (0)	...	1.10	0

whereas a deficit of precipitation is found in the subtropics of each hemisphere between about 10° and 40° latitude. In the long-term mean, the excess or deficit in each belt has to be compensated by a net meridional divergence or convergence of water in the particular belt. The runoff ratio $(P-E)/P$ gives an idea of the fraction of the precipitation that is involved in the runoff. The values of the evaporation ratio E/P show clearly the high aridity of the subtropics with ratios larger than 1.

We should stress that there is not always a close agreement between the values of P and E published by different authors, as demonstrated by the comparisons in Tables 7.1 and 7.2 with Sellers (1965) values based on similar observations. Usually the individual values of P and E are adjusted subjectively in a rather arbitrary way by assuming certain global and zonal constraints to yield an overall balance between P and E. This procedure imposes serious limitations on the usefulness of the estimates. The differences are even larger when we compare the results from two different methods, such as in the case of $P-E$ where the independent, aerological estimates are given in parentheses (see column 5, Table 7.1).

Over the globe as a whole, evaporation must balance precipitation in the long-term mean. The precipitation in the two hemispheres is almost the same whereas a large difference is found for the evaporation (about 150 $mm\,yr^{-1}$). The higher values of evaporation in the Southern Hemisphere result because this hemisphere is largely covered by oceans. The Northern Hemisphere shows a positive water balance ($P-E = 73\,mm\,yr^{-1}$) whereas in the Southern Hemisphere a net negative value of $-73\,mm\,yr^{-1}$ is found. Thus, we are led to the conclusion that a flow of water in the liquid form must take place across the equator from the Northern into the South-

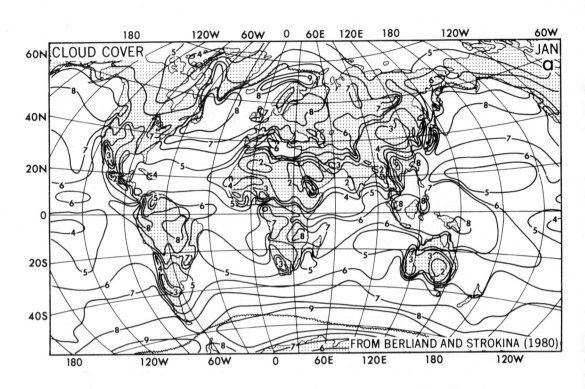
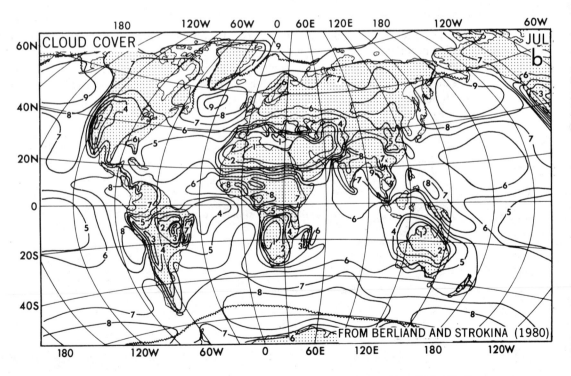

FIGURE 7.28. Global distributions of the cloud cover (in tenths) for January (a) and July (b) after Berlyand and Strokina (1980).

ern Hemisphere. As we will see later, an equal amount of water in the vapor form has to be exported in the opposite direction to maintain the balance of the water substance.

As seen from Table 7.2, the quantities P, E, and $P - E$ are not the same for the different oceans and continents. This is not only due to physiographic differences between them but also due to the differences in areal extent. For example, South America shows the highest P, E, and $P - E$ values in agreement with what we have seen before on the corresponding global maps. On the other hand, Australia, Africa, and Antarctica show very small runoff $(P - E)$ values. Overall, the precipitation and evaporation tend to be smaller over the continents than over the oceans, except for the extremely high values of P and E over South America and the extremely low values of P and E over the Arctic Ocean. The mean value of $P - E$ over all continents is estimated to be 266 mm yr^{-1}. This surplus of condensed water must be transported by rivers and glaciers from the continents into the oceans where a deficit of -110 mm yr^{-1} is found. When the surplus over land and the deficit over the oceans are multiplied by the appropriate areal factors they must balance. The table further shows that $P - E$ is positive for the Arctic and Pacific Oceans and strongly negative for the Atlantic and Indian Oceans (the "dry" oceans) leading to the net deficit for all oceans combined. The implications of these estimates of $P - E$ taken together with the values of the observed river discharge are that a net transfer of water must occur from the Pacific and Arctic Oceans into the Atlantic and Indian Oceans. For example, water transport by the rivers from the surrounding continents into the Atlantic is estimated to be on the order of $R_0 = 197$ mm yr^{-1} so that the equivalent of $E - P - R_0 = 175$ mm yr^{-1} must come from the Pacific and Arctic Oceans. Further, for the Indian Ocean which only receives 72 mm yr^{-1} from continental runoff an inflow of 179 mm yr^{-1} must take place from the Pacific Ocean. The net excess of precipitation over evaporation over the continents must be maintained by a net influx of water in vapor form from the large ocean sources.

The crude picture of the global water transfers in the climate system sketched so far will be further extended in Chap. 12.

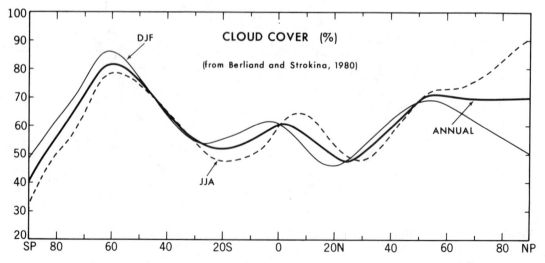

FIGURE 7.29. Meridional profiles of the zonal-mean cloud cover (in %) for annual, DJF, and JJA mean conditions based on data from Berlyand and Strokina (1980).

7.6.4 Cloudiness

Clouds influence the energetics of the atmosphere in at least two major ways. First, they release large amounts of latent heat during the condensation process while the condensed water is removed from the atmosphere through precipitation. Second, by scattering, absorption, reflection, and emission of radiation clouds strongly influence the atmospheric radiation budget (see Chap. 6), and thereby the energy budget of the earth.

In view of the obvious connections between evaporation, clouds, and precipitation we shall now consider the spatial distribution of the cloudiness. Figures 7.28(a) and 7.28(b) show the estimated mean cloudiness conditions for January and July. Of course, the cloud cover is high in the equatorial belt, which is clearly associated with the strong convection in the ITCZ. Other important features are the minima in cloudiness in subtropical latitudes suggesting that these are zones of downward or only weak upward motion. The cloud cover increases markedly with latitude poleward of the subtropics reaching maximum values near 50°–60° latitude associated with the polar fronts.

By and large, the cloudiness is higher over the oceans than over the continents. The minima in the subtropics shift poleward from winter to summer in each hemisphere. The characteristic variations with longitude must be related with the land–sea distribution, topography, and the semipermanent circulation features, such as the ITCZ, monsoons, polar fronts, and storm tracks. The lowest values of cloud cover are observed over the continental deserts as discussed before in connection with the albedo.

The meridional profiles of the zonal-mean cloudiness in Fig. 7.29 synthesize the main characteristics of the maps and the seasonal differences. Of interest are the very large seasonal differences in cloud cover over the Arctic Ocean where the highest values are observed during summer when warm and moist air masses invade the cold polar regions leading to low-level stratus clouds.

CHAPTER 8

Observed Mean State of the Oceans

As we have seen before the oceans are a very important component of the climatic system. Indeed, with their high heat capacity the oceans store large amounts of solar energy that can be released later in the form of sensible and latent heat into the atmosphere. The oceans are also an important vehicle to transport energy from low to high latitudes, thereby reducing the north–south gradient of temperature. Through these processes the oceans play a crucial role as moderators of the earth's climate. Furthermore, they are the main source of the water that falls as precipitation over the continents, supplying the runoff for the major rivers in the world. Through air–sea interaction the oceans are also important in the formation and modification of certain types of air masses.

The major surface currents in the world ocean are shown schematically in Fig. 8.1. They are thought to be instrumental in making the climate more amenable in certain regions of the globe which otherwise might be inhospitable. For example, the relatively mild climate of northwestern Europe compared with the climate of other regions in the same latitude belt may be associated with the advection of warmer water in the Gulf Stream system.

It is clear that the study of the climate of the oceans is not only important *per se* but also crucial for understanding the climate in the atmosphere.

8.1 MEAN TEMPERATURE STRUCTURE OF THE OCEANS

8.1.1 Global distribution of the temperature

The horizontal temperature distribution at the ocean surface is shown in Fig. 8.2(a). As expected, the highest values are found over the tropical regions with maxima over the western equatorial Pacific and the Indian Oceans, whereas the strongest north–south gradients are found in mid to high latitudes, being most pronounced in the Southern Hemisphere.

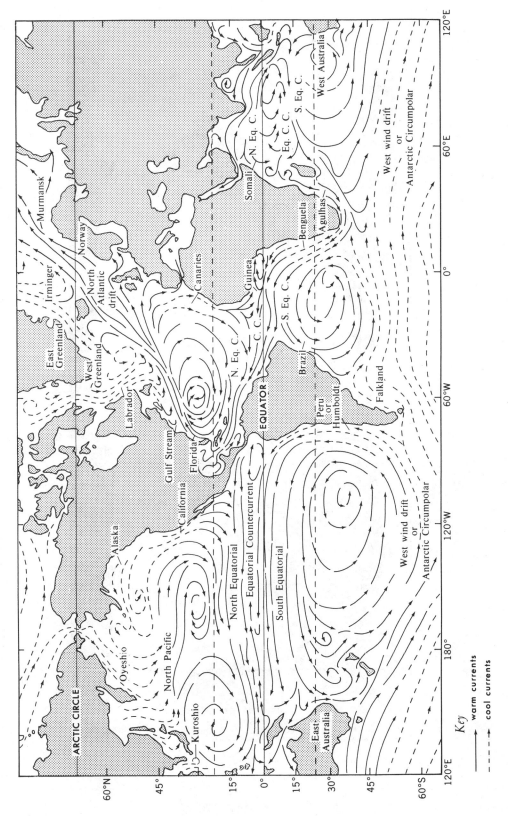

FIGURE 8.1. A map of the major surface currents in the world ocean during northern winter (from Tolmazin, 1985).

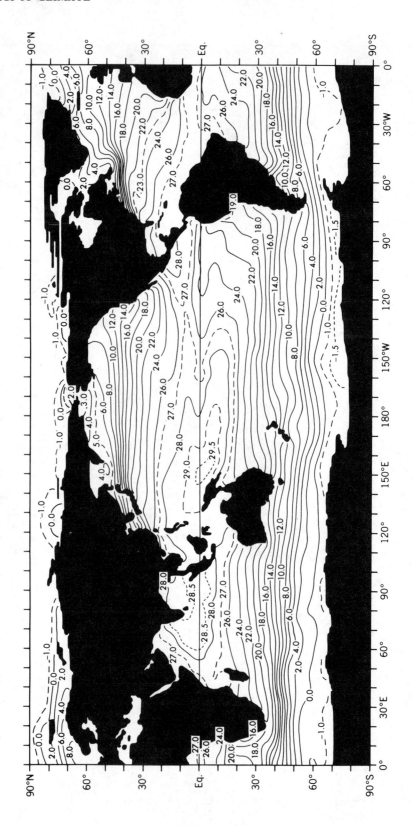

FIGURE 8.2a

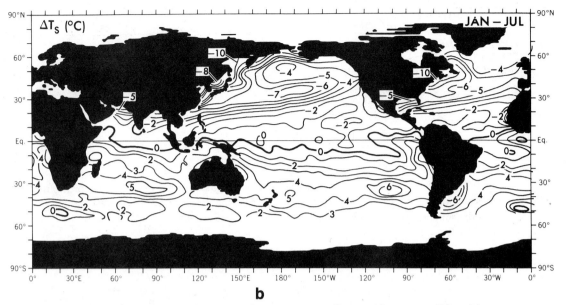

FIGURE 8.2. Global distributions of the sea surface temperature for annual-mean conditions (a) from Levitus (1982) and the January–July difference of the sea surface temperature (b) in °C.

To first order, the sea surface isotherms are zonal in character but with some distortion due to the influence of the continents. This influence manifests itself in the warm poleward currents (e.g., Gulf Stream, Brazil, and Kuroshio Currents) and the cold equatorward current systems (e.g., the Labrador, Canary Islands, Benguela, California, and Peru currents). Thus comparing both sides of the oceans, we find that in the subtropics the waters on the west sides tend to be warmer than those on the east sides, whereas in high latitudes the opposite tends to occur. This is evident in the North Atlantic Ocean where the Norwegian Sea is anomalously warm due to the poleward transport of heat in the Gulf Stream system.

The low temperatures in the cold equatorward currents on the eastern side of the subtropical gyres are reinforced by upwelling of even colder, nutrient-rich water from great depths. The main regions in which this coastal upwelling occurs are clearly depicted in Fig. 8.2 by the equatorward dipping of the isotherms, namely the Canary current (10–40 °N), the California current (25–40 °N), the Benguela current (10–30 °S), the Peru current (5–45 °S), and (during the southwest summer monsoon) the Somali current regions (0–15 °N). The upwelling in these regions is associated with local wind regimes and is most intense when there is a strong wind blowing equatorward (poleward in case of the Somali current) parallel to the coast, leading to a large Ekman transport away from the coast and directed to the right of the wind in the Northern and to the left in the Southern Hemisphere [see Fig. 8.3 and Eq. (3.24a)]. These wind regimes are connected with the position of the large subtropical atmospheric anticyclones which move poleward in summer and equatorward in winter.

To obtain an order of magnitude estimate of the vertical velocity at the bottom of the upwelling layer, we will assume that w_E is uniform in a strip of width L_x along the coast which is assumed to be oriented in the north–south direction (see Fig. 8.3). Then the amount of mass transported across an imaginary vertical "wall" of length L_y at a distance L_x from the coast [see Eq. (3.24a)] is such that

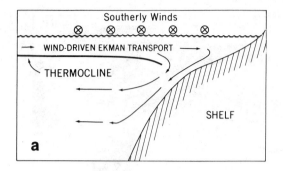
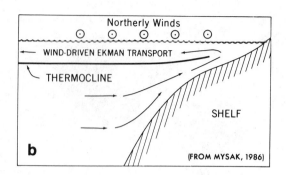

FIGURE 8.3. (a) Downwelling along a north–south coast line due to southerly (northward) longshore winds ($\tau_{0y} < 0$) after Mysak (1986). (b) Upwelling along a north–south coast line due to northerly (southward) longshore winds ($\tau_{0y} > 0$) after Mysak (1986).

$$M_{Ex}L_y = \frac{(-\tau_{0y})}{f} L_y. \qquad (8.1)$$

Due to the conservation of mass, this flux has to be compensated by vertical upwelling of the same magnitude:

$$M_{Ex} \approx -\rho w_E L_x$$

or

$$w_E \approx \frac{-M_{Ex}}{\rho L_x}. \qquad (8.2)$$

For example, for a southward wind of 10 m s^{-1} blowing over a distance $L_y = 1500$ km along the coast [see Fig. 8.3(b)] the bulk aerodynamic formula (10.30) gives a wind stress $\tau_{0y} = 0.16$ N m^{-2}, so that at a latitude $\phi = 25°$, where $f = 0.62 \times 10^{-4}$ s^{-1}, the transport (8.1) becomes

$$M_{Ex}L_y \approx -3.9 \times 10^9 \text{ kg s}^{-1}.$$

Thus the total water exported away from the coast would amount to about 4×10^9 kg s^{-1} or a volume transport of about 4 Sv (1 Sverdrup = 10^6 m^3 s^{-1} or 10^9 kg s^{-1}). Further, if we take $L_x = 40$ km and $\rho = 1.025 \times 10^3$ kg m^{-3} we find from Eq. (8.2) for the upwelling velocity

$$w_E = 6.3 \times 10^{-5} \text{ m s}^{-1} \approx 5.5 \text{ m day}^{-1}.$$

The relatively low equatorial temperatures are a consequence of the dominant easterly surface winds that generate an Ekman mass transport away from the equator. The reason for this Ekman divergence is that the Coriolis parameter f changes sign at the equator so that $M_{Ey} = \tau_{0x}/f > 0$ just north of the equator and $M_{Ey} < 0$ just south of it, leading to $(\partial M_{Ey}/\partial y) > 0$. Due to the poleward increase of $|f|$ the Ekman transport decreases rapidly with latitude restricting the meridional extent of the upwelling to a narrow zone around the equator.

We should mention that the observed negative temperatures (down to nearly -2 °C) near the Antarctic and Arctic in Fig. 8.2(a) can exist due to the lowering of the freezing point of sea water associated with the dissolved salts (Raoult's law).

To show how important the annual cycle in the sea surface temperature (SST) is we present in Fig. 8.2(b) a map of the difference between the mean January and July temperatures. The largest differences of about − 10 °C occur in the Northern Hemisphere near 45° latitude close to the east coasts of the continents. These features can be linked to the prevailing tracks of the migrating air masses formed over the Eurasian and North American continents. Because these air masses are warm in summer and very cold and dry in winter, a strong interaction takes place, mainly at the east coasts of the continents, between the air and the underlying ocean water with intense exchange of sensible and latent heat.

As expected, the seasonal variations in the tropical and equatorial latitudes are quite small, at most on the order of 2 °C. In the Southern Hemisphere, the range in temperature is much smaller than in the Northern Hemisphere with maximum values of 6 °C or less because of the limited land coverage. The temperature differences are also more zonally uniform, with the exception of the coastal regions of Africa and South America, presumably associated with variations in up- and down-welling.

Let us compare Fig. 8.2(b) with a similar map for the atmospheric surface temperature range presented before in Fig. 7.4(c). The comparison shows that the annual range in oceanic temperatures is, in general, much smaller than the range in atmospheric surface temperature, which can be as high as 50 °C, over the northern continents.

8.1.2 Vertical structure of the temperature

The zonal-mean values of the ocean temperatures are shown in Fig. 8.4 as a function of latitude and depth, while their hemispheric and global-mean values are given on the right-hand side of the figure. The cross section shows a prominent broad wedge of warm water in each hemisphere penetrating downward from about 15° latitude near the surface to about 35° latitude at 1000-m depth. As shown later by the meridional circulation in Fig. 8.20, this feature is associated with the Ekman convergence and downwelling in the surface layers between the trade winds and midlatitude westerlies. Similarly, the region of relatively cold water in tropical latitudes is caused by Ekman divergence and upwelling from deeper layers connected with the persistent trade winds.

From the inspection of the vertical gradients in the upper portion of Fig. 8.4, we notice that the main or permanent thermocline is generally deepest in subtropical latitudes. In the inner tropics, the thermocline is shallower with smallest values underneath the meteorological equator. In high and subpolar latitudes, the main thermocline is not well defined as shown by the very small vertical gradients of temperature. In fact, it appears as if the thermocline has reached the surface at these latitudes. In northern high latitudes, the cause of the weak vertical gradients is the strong cooling in the North Atlantic Ocean during winter leading to deep convection, often down to the bottom of the ocean. The associated sinking waters form an important component of the general thermohaline circulation of the world ocean (see Fig. 8.20) and are a source of the North Atlantic deep water. In the Arctic Ocean, convection is limited in spite of the low surface temperatures because of the low surface salinities (due to river inflow of fresh water) that tend to stabilize the vertical density profile. In the southern high latitudes, there is strong sinking of very cold, saline water around the periphery of Antarctica forming the Antarctic bottom water. Another belt of deep subsidence is found just north of the Antarctic circumpolar current and is associated with the strong Ekman convergence.

Further inspection of the cross section in Fig. 8.4 shows that the "deep" water below 2000 m is fairly homogeneous with very weak horizontal and vertical temperature gradients except at high southern latitudes. The peculiar kinks in the isotherms

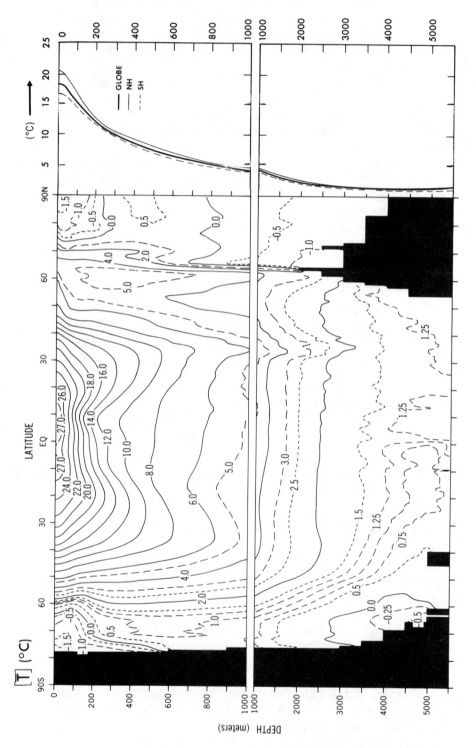

FIGURE 8.4. Zonal-mean cross section of temperature for the top ocean layer 0–1000 m and the layer below 1000 m in °C for annual-mean conditions. Vertical profiles of the hemispheric and global mean temperatures are shown on the right (after Levitus, 1982).

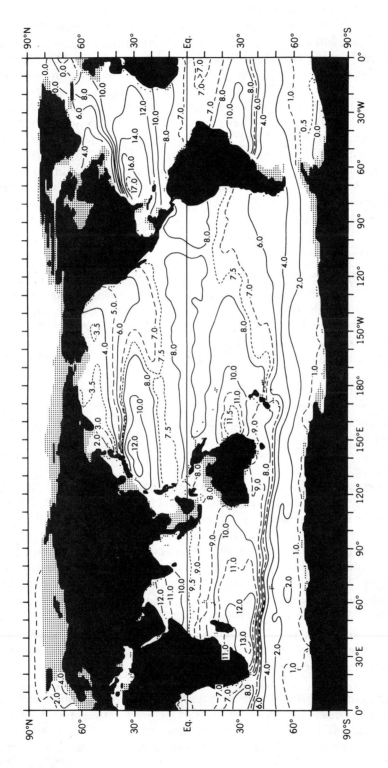

FIGURE 8.5. Global distribution of the annual-mean temperature in °C at 500-m depth (from Levitus, 1982).

below 1000-m depth between 30 and 40 °N latitude are related to the outflow of very saline Mediterranean Sea water that after sinking to about 1000-m depth spreads out laterally over great distances in the North Atlantic Ocean.

The temperature distribution at 500-m depth presented in Fig. 8.5 gives an idea of the behavior of the oceans below the main thermocline layer. It also shows the transition between the variable surface layer and the less stable, but more homogeneous deep ocean waters. Inspection of the map shows that the oceans at this level are still baroclinic as implied by the strong thermal gradients. In equatorial latitudes relatively low temperatures are observed but the most significant features are the areas of high temperature near 30° latitude in both hemispheres with highest values near the western portion of the oceans. These warm areas are associated with the surface Ekman convergence and downwelling [see Eq. (3.26)] in the subtropical gyres. In midlatitudes of the Southern Hemisphere, a strong zonality begins to show at this level. At greater depths (not shown here), the zonality becomes an even more dominant feature.

8.1.3 Variability of the temperature

To assess the temporal variability of the ocean temperature, maps of its standard deviation $\sigma(T)$ are shown for the surface, 150- and 500-m depth in Figs. 8.6(a)–8.6(c). These maps contain the effects of all transient phenomena with time scales less than one year so that seasonal, as well as, real transient eddy effects are included. Comparing the subsurface distributions in Figs. 8.6(b) and 8.6(c) with the surface distribution in Fig. 8.6(a), we find that the strong surface variability in midlatitudes connected with the major ocean currents is somewhat reduced in strength with depth but that it is still clearly present down to a depth of, at least, 500 m. However, it may be more surprising that in the tropics, equatorward of about 20° latitude, the variability actually tends to increase with depth in the first few 100 m's. This increase is probably largely associated with strong seasonal variations in the depth of the thermocline in the western tropical oceans, that occur well below the surface mixed layer at a depth of a few 100 m.

The nonseasonal contribution to the temperature variability at the surface is shown in Fig. 8.7 where the annual signal was removed to get a better representation of the eddies with time scales of one month or less as an indicator of the mesoscale eddy activity in the oceans. Comparing Figs. 8.7 and 8.6(a) we notice that a sizeable fraction of the total surface variability is due to nonseasonal variations, such as the meandering eddies in the Antarctic circumpolar current and in the Gulf Stream, Kuroshio, Agulhas, and Falkland boundary currents, where they leave the coast.

Besides showing the spatial distribution of the temporal variability, the maps in Figs. 8.6 and 8.7 are also important from the point of view of the ocean energetics since the quantity $\sigma^2(T) = \overline{T'^2}$ is the variance of temperature. This variance gives an approximate measure of the transient eddy available potential energy in the oceans, as we will discuss further in Chap. 14.

A comprehensive picture of the variability of the sea surface temperature in time and space can also be obtained from the inspection of Fig. 8.8. It contains zonal-mean profiles, for all oceans combined, of the surface temperature [Fig. 8.8(a)] and its standard deviations for the transient eddy [Fig. 8.8(b)] and stationary eddy [Fig. 8.8(c)] components during the annual mean and extreme DJF and JJA seasons. The variability is larger in the Northern than in the Southern Hemisphere as expected from the greater land–sea contrast in the Northern Hemisphere. In the equatorial

OBSERVED MEAN STATE OF THE OCEANS

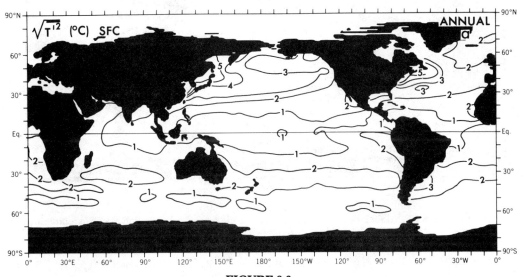

FIGURE 8.6a

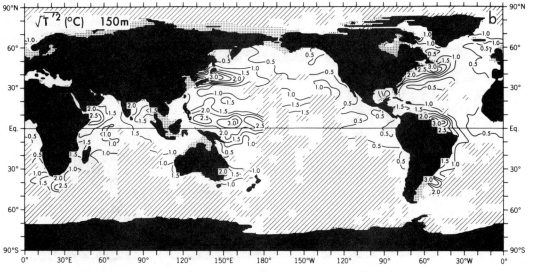

FIGURE 8.6b

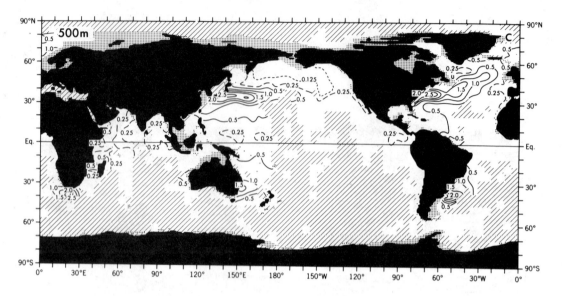

FIGURE 8.6. Global distributions of the day-to-day standard deviation of ocean temperature at the surface (a), 150- (b) and 500-m (c) depth in °C for annual-mean conditions. All fluctuations with periods from a few days to a year are included in the calculation of the standard deviation. Stippled areas in (b) and (c) indicate the added land areas at 150 and 500-m depth, respectively. Shaded areas indicate regions with insufficient data (from Levitus, 1982).

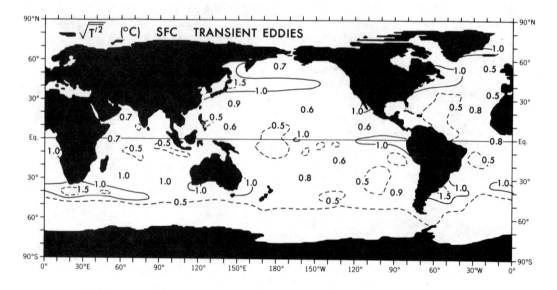

FIGURE 8.7. Global distribution of the day-to-day standard deviation of the sea surface temperature in which the annual cycle has been removed. Thus only fluctuations with periods from a few days to one month are included, the so-called synoptic-scale or mesoscale eddies.

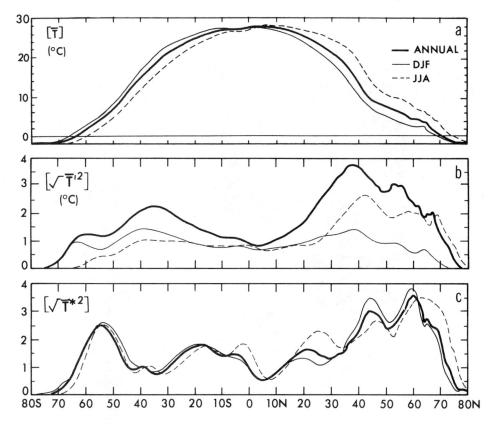

FIGURE 8.8. Meridional profiles of the zonal-mean values of the sea surface temperature (a), its day-to-day (transient eddy) standard deviation (b), and its east–west (stationary eddy) standard deviation (c) in °C for annual, DJF, and JJA mean conditions.

regions, the variability is much smaller than at middle and high latitudes. It is interesting to note that the spatial variability in the time-mean temperature field [Fig. 8.8(c)] is almost as large as the transient eddy variability [Fig. 8.8(b)].

8.2 MEAN SALINITY STRUCTURE OF THE OCEANS

8.2.1 Global distribution of the salinity

The global distribution of salinity (in ‰) at the ocean surface is shown in Fig. 8.9. It shows the highest salinity values between about 15° and 35° latitude in all subtropical oceans, where most of the evaporation occurs (see Figs. 7.26 and 7.27). Minima in salinity are found just north of the equator at the mean position of the ITCZ, where precipitation far exceeds evaporation (see Figs. 7.24 and 7.26), and at high latitudes where melting of sea ice and snow become important. More localized minima in salinity are found where the major rivers, such as the Amazon, Congo, and Ganges, flow into the oceans or seas.

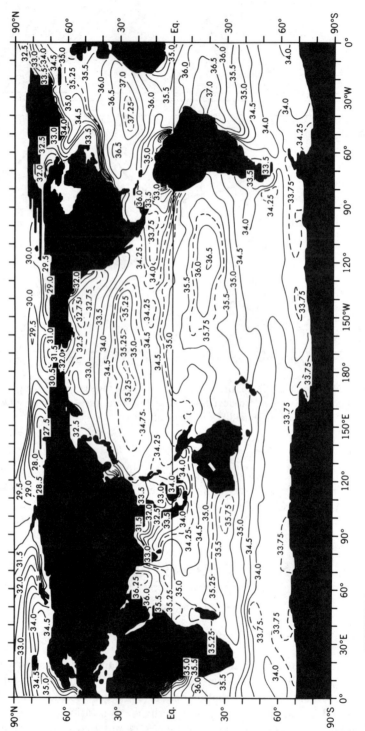

FIGURE 8.9. Global distribution of the annual-mean salinity in ‰ at the sea surface (from Levitus, 1982).

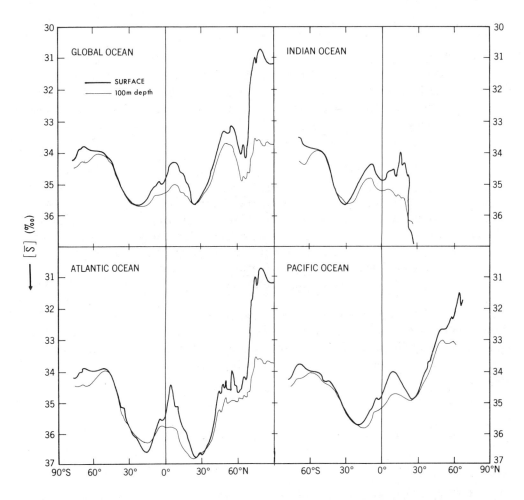

FIGURE 8.10. Meridional profiles of the zonal-mean values of the salinity at the surface and 100-m depth for the Atlantic, Pacific, Indian, and global oceans in ‰. Note that the salinity is plotted increasing downward.

By and large, the Atlantic Ocean is more saline than the Pacific Ocean as shown even more clearly in the profiles in Fig. 8.10. These differences are associated with the net deficit of precipitation over evaporation of -37 cm/yr over the Atlantic and the net excess of 9 cm/yr over the Pacific Ocean as given before in Table 7.2. The average salinity in the North and Central Atlantic Oceans away from the coasts is very high ranging between 35.5‰ and 37.3‰, while in the South Atlantic Ocean the salinity is slightly lower. The Central Indian Ocean waters are, in general, fresher than the Atlantic waters, and range between 34.0‰ and 35.7‰. In contrast with the situation in the South and North Atlantic Ocean, the South Pacific is more saline than the North Pacific. The low salinity values just north of the equator in all oceans are due to the large precipitation (see Fig. 7.24) in the ITCZ. The extremely low salinities in the Arctic Ocean compared to those analyzed near Antarctica are due to the large runoff by the various rivers that flow from the Eurasian and American continents into the Arctic basin.

8.2.2 Vertical structure of the salinity

At 1000-m depth, the distribution of salinity, as shown in Fig. 8.11, is very different from that at the surface, particularly in the North Atlantic and in the Antarctic Convergence Zone around 55 °S. The highest values of salinity at this level are found in the eastern part of the North Atlantic and are due to the intrusion of very saline Mediterranean water of originally more than 37‰ at the surface.

The vertical distribution of salinity for annual-mean conditions is presented in Fig. 8.12. The cross section shows that the broad wedges of warm water penetrating near 30° latitude as shown before in Fig. 8.4 are also saline. The salinity tends to become more uniform in the horizontal direction with increasing depth. Low salinity values are generally found between about 500- and 1500-m depth, whereas below 2000 m the salinity values are again relatively high and homogeneous ranging between 34.69‰ and 34.95‰, which is consistent with the configuration of the mean vertical profiles presented on the right-hand side of Fig. 8.12. At high latitudes in the Arctic Ocean we find a shallow layer of very fresh water with a strong halocline centered at about 50-m depth as was already evident from a comparison of the surface and 100-m depth curves in Fig. 8.10.

As before in the temperature section we can clearly recognize the influence of the warm, saline Mediterranean waters in the very high salinity values between 30 and 40 °N, and the influence of the relatively cold, fresh Antarctic waters in the Antarctic bottom water in the lower left-hand corner of Fig. 8.12.

8.3 MEAN DENSITY STRUCTURE OF THE OCEANS

8.3.1 Global distribution of the density

Another important parameter to define the climate of the oceans is the density ρ since its structure is closely connected with the stability as well as the three-dimensional velocity fields in the oceans. It is a function of temperature, salinity, and pressure $\rho = \rho(T,S,p)$. The density changes due to compressibility are far larger than the density changes due to thermal expansion and saline contraction for the range of temperatures and salinities observed in the world ocean. The horizontal distributions of the annual-mean density at the surface and 500-m depth are presented in Figs. 8.13(a) and 8.13(b). The features of the density field at a given depth reflect the combined effects of the temperature and salinity distributions at that depth. In general, at the surface the temperature contributions to the density field seem to be more important than those of the salinity except at high latitudes. Some useful order of magnitude estimates are that a change in density of $\Delta\rho = 0.2$ kg m^{-3} is approximately equivalent with a change in temperature of $\Delta T \approx 1$ °C or a change in salinity of $\Delta S \approx 0.2$‰. The location of the maximum meridional gradients of surface density near 30° latitude in both hemispheres compared to that of surface temperature near 40° latitude must be a consequence of the salinity maxima in subtropical latitudes. The density minimum in the equatorial regions clearly results from the high values of temperature combined with low salinities. The influence of the outflow of fresh water by the large rivers is also evident in the equatorial regions with the lowest values of density observed near the continents.

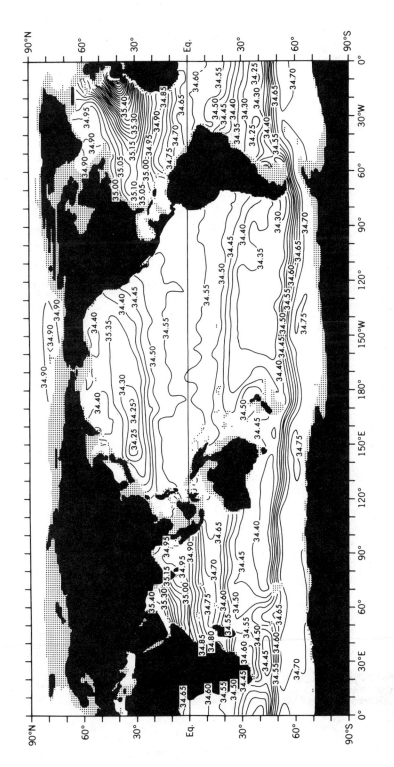

FIGURE 8.11. Global distribution of the annual-mean salinity in ‰ at 1000-m depth (from Levitus, 1982).

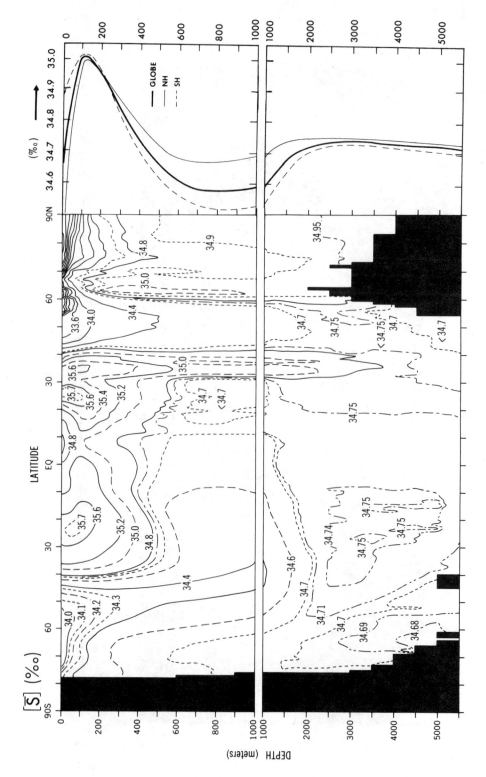

FIGURE 8.12. Zonal-mean cross section of salinity for the top ocean layer 0–1000 m and the layer below 1000-m depth in ‰ for annual-mean conditions. Vertical profiles of the hemispheric and global mean salinity are shown on the right (after Levitus 1982).

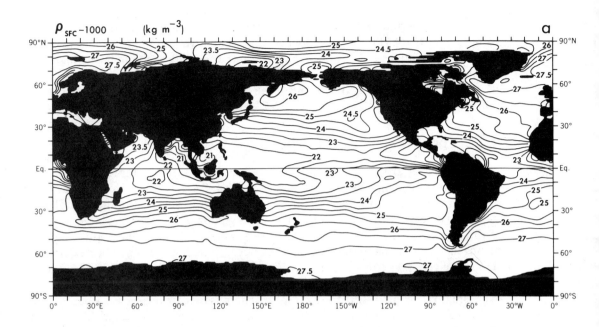

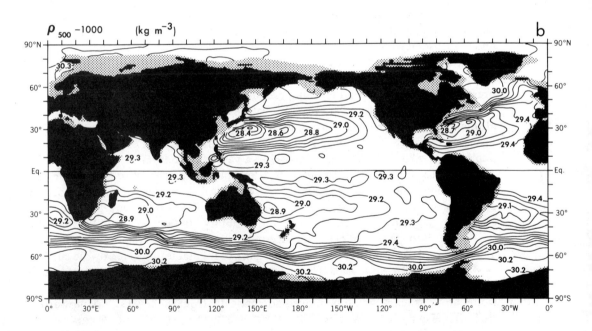

FIGURE 8.13. Global distributions of the annual-mean density (actually shown $\bar{\rho} - \rho_0$, where $\rho_0 = 1000$ kg m^{-3}) in units of kg m^{-3} at the sea surface (a) and at 500-m depth (b).

The distribution of density at 500-m depth, shown in Fig. 8.13(b), is more uniform than at the surface. Nevertheless, it presents two minimum belts in the subtropical midlatitudes in both hemispheres alternating with maxima in the high and equatorial latitudes. The influence of temperature is now prevalent since the salinity has a smaller spatial variability at this level.

8.3.2 Vertical structure of the density

The vertical distribution of the density can, in principle, be inferred from the cross sections of temperature and salinity in Figs. 8.4 and 8.12. Nevertheless, we present in Fig. 8.14 a cross section of σ_t and in Fig. 8.15 some latitudinal profiles of the actual density at various depths. The cross section of σ_t shows the large stability of the upper ocean above 1000 m equatorward of about 60° latitude and the more homogeneous, less stable waters below that level. The few areas with a weak inversion in the vertical gradient of σ_t (i.e., $\partial \sigma_t / \partial z > 0$) below 3000-m depth do not indicate real instabilities. These inversions are an artifact of the reduction of the density over a huge range in pressure from these great depths to sea level [see Eq. (3.69)]. The latitudinal profiles summarize the global structure of density as evaluated *in situ* at various depths. It is clear that the latitudinal gradients decrease strongly with depth.

Due to their importance for the dynamics of the oceans, we present in Fig. 8.16 vertical profiles of the global mean *in situ* density $\tilde{\rho}(z)$ and various measures of the stability. The mean density curve shows a rapid increase with depth in the first few 100 m's associated with the decrease in temperature and the increase in salinity, and then a somewhat slower increase of density below about 700 m where the ocean temperature and salinity are more uniform (see Figs. 8.4 and 8.12). The almost linear increase in density with increasing depth below 1000 m is primarily due to the compression of the water with increasing pressure of the ocean column above it.

Let us next discuss the curve of the vertical derivative of the density $-d\tilde{\rho}/dz$, although it is not a true measure of the static stability as we have seen before in the discussions of Sec. 3.5.2. The curve shows a peak of about 0.017 kg m^{-4} near 75-m depth. Below 75 m, it decreases rapidly to a value of about 0.005 kg m^{-4} at about 700-m depth, and it decreases only very slowly below that level to a near-constant value of about 0.004 kg m^{-4} at 3000 m. As we have discussed before in Sec. 3.5.2 this curve does not take into account the effects of compressibility and thus does not constitute an adequate measure of the static stability. In fact, below 700 m the value of $-d\tilde{\rho}/dz$ is too large where the actual equilibrium cannot be so stable and should be almost neutral (see Fig. 8.14). A better measure of the stability is given by the curve of $-\delta\tilde{\rho}/dz$ where $\delta\rho \equiv \rho - \rho_{\theta L}$. The last term represents the difference between the *in situ* density ρ at a level z and the local potential density $\rho_{\theta L}$ that a parcel of water would attain when it would be moved adiabatically from the level $z - \delta z$ upward or from the level $z + \delta z$ downward over a distance δz to the level z. This curve shows very small values of the local stability below about 700 m [see Eq. (3.76)]. On the right-hand side of Fig. 8.16 is also plotted a curve of the square of the Brunt–Väisälä frequency (N^2) for annual-mean global conditions since, as discussed in Sec. 3.5.2, it constitutes another important measure of the static stability. The largest values of N occur at the pycnocline level near 75-m depth with a value of about 0.011 rad/s corresponding with a period of oscillation of about 10 min. The Brunt–Väisälä frequency decreases with depth to values on the order of 0.0009 rad/s or a period of about 2 h in the deep ocean, indicating almost neutral conditions at these levels. The equivalent values for the atmosphere are on the order of 10 min.

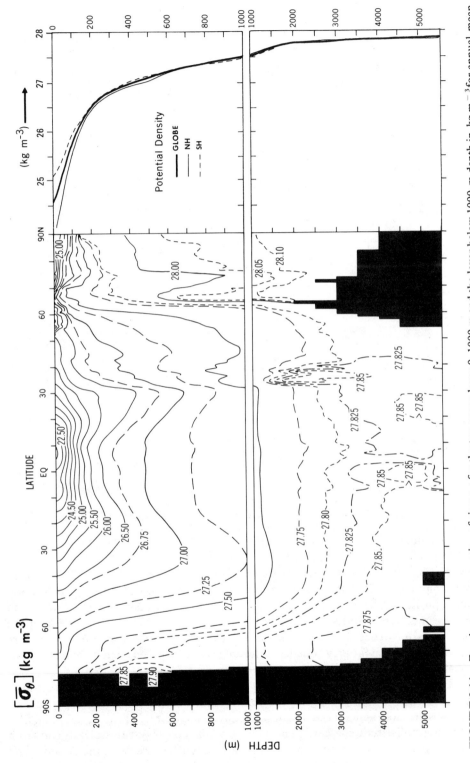

FIGURE 8.14. Zonal-mean cross section of sigma-t for the top ocean layer 0–1000 m and the layer below 1000-m depth in kg m^{-3} for annual-mean conditions (after Levitus, 1982). Vertical profiles of the hemispheric and global mean sigma-t are shown on the right.

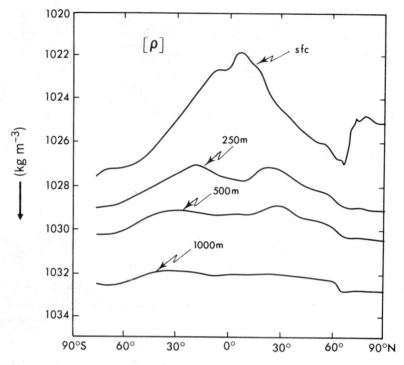

FIGURE 8.15. Meridional profiles of the zonal-mean value of the *in situ* density in kg m^{-3} at the sea surface and at different depths. Note that the density is plotted increasing downward.

8.4 MEAN OCEAN CIRCULATION

8.4.1 Global distribution of the surface circulation

The global distributions of the surface currents resulting from the atmospheric wind stresses are shown in Figs. 8.17(a) and 8.17(b) (see also Fig. 8.1). These maps are based on historical ship drift observations (Stidd, 1975; Meehl, 1980). The ship drift was computed as the difference between the actual position of the ship and its predicted position calculated 24 h earlier. Although the drift is affected by both wind and ocean currents, Stidd (1975) estimates that it is about 25% due to wind drag and about 75% due to ocean currents. In spite of these limitations, the ship drift data are very useful since so far they constitute the only set of observed current velocities on a global scale.

The maps in Fig. 8.17 depict the long-term mean winter and summer circulations. Synoptic maps would look very different since various kinds of eddies would be superposed on the general oceanic flow, mainly in the vicinity of the equator and near the strong midlatitude currents. The polar and subtropical gyres are clearly shown on both maps as well as the equatorial (eastward) countercurrents in the Atlantic and Pacific Oceans. Over the northern part of the Indian Ocean a pronounced seasonal change takes place connected with the Asian monsoon circulation leading to the reversal of the Somali current. The currents tend to be in a similar direction as the overlying mean atmospheric flow (shown before in Fig. 7.1) and the wind stress (to be

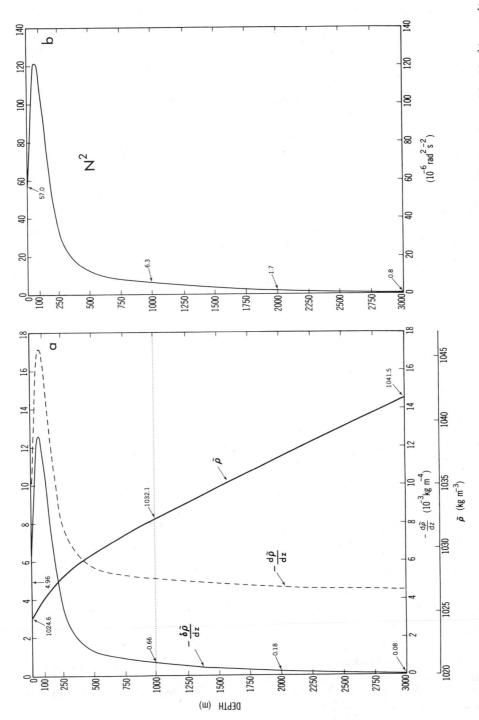

FIGURE 8.16. Vertical profiles of the global-mean density in kg m^{-3} and two measures of the mean static stability (see the text) in 10^{-3} kg m^{-4} (a), and a vertical profile of the global mean Brunt–Väisälä frequency squared (b) in 10^{-6} rad^2 s^{-2}.

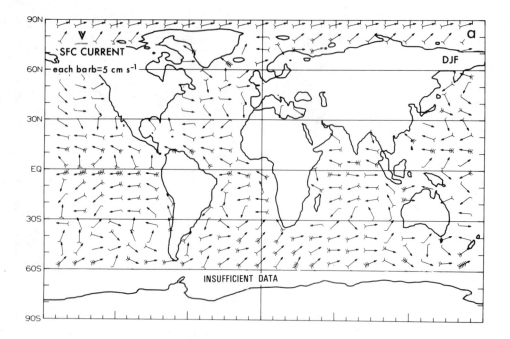

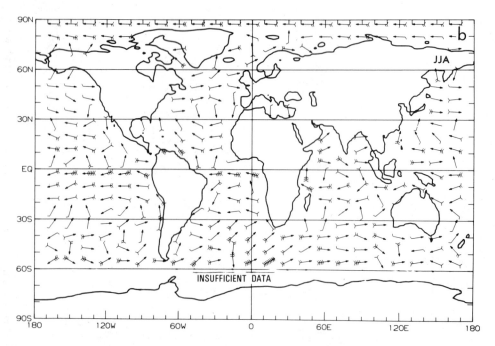

FIGURE 8.17. Global distribution of the ocean velocity vectors for DJF (a) and JJA (b) mean surface conditions. Each barb on the tail of an arrow represents a current speed of 5 cm s^{-1}. In spite of the sparseness of ocean current data south of 40 °S there is evidence of a strong eastward circumpolar current around Antarctica at about 60 °S [see also Fig. 8.19(a)].

shown in Fig. 10.3) but with a systematic deviation of about 45° in the direction of higher pressure due to the Ekman drift [see Eq. (3.25)].

The mass transports associated with the subtropical gyres (due to both the Ekman and geostrophic flow components) shown in Fig. 8.17 produce a zone of convergent surface water in the subtropics leading to the large dome of warm, saline water discussed before. At the same time, we find divergent surface waters in the high-latitude gyres resulting in upwelling of cold water under the belt of the atmospheric cyclones.

The gyres are also clearly visible in Fig. 8.18, which shows the relative topography of sea level with respect to the 1000-m depth level, assumed to be the level of no motion. The gradient of the sea-level topography is related to the surface circulation through the geostropic relation $\mathbf{v}_g = -(\mathbf{k} \times \text{grad } p)/\rho f$, where $p = p_0 + \rho g \Delta z$, p_0 is the atmospheric pressure, and Δz is the anomaly of the geopotential thickness ($\approx$ sea surface elevation). The actual ocean slopes near the coasts are more pronounced than is shown in Fig. 8.18 because sea level in the shallow coastal waters obviously cannot be represented using a reference level at 1000-m depth. The belt between 45 and 55 °S shows a steep slope of sea level indicating strong circumpolar flow around Antarctica.

The oceanic circulations are generally in phase with their atmospheric counterparts. However, in contrast with the fairly symmetric gyres in the atmosphere, the oceanic gyres show a marked east–west asymmetry, with an intense poleward flow in the western oceans and a weak, diffuse equatorward flow in the eastern oceans. As Stommel (1960) has pointed out, this so-called western intensification can be explained using the conservation of absolute vorticity ($\zeta + f = \text{curl } \mathbf{v} + 2\Omega \sin \phi$).

To show the western intensification we will use a simplified version of the general vorticity equation (3.39) in which the solenoidal and twisting terms are neglected:

$$\frac{d}{dt}(\zeta + f) = -(\zeta + f) \text{ div } \mathbf{v} + \mathbf{k} \cdot \text{curl } \mathbf{F}, \qquad (8.3)$$

where $\mathbf{F} = (F_x, F_y, F_z)$ is the friction force. The friction force includes the effects of the surface wind stress (τ_0), the stresses at the lateral boundaries, and the stress at the bottom of the ocean. The last stress term will be neglected here. Taking the vertical (mass-weighted) mean of Eq. (8.3) denoted by $\langle \ \rangle$ we obtain

$$\left\langle \frac{d\zeta}{dt} \right\rangle = -\beta \langle v \rangle + \frac{1}{\rho H} \text{curl}(-\tau_0) + \langle \partial F_y/\partial x - \partial F_x/\partial y \rangle, \qquad (8.4)$$

where H = depth of the ocean. In deriving Eq. (8.4), we have used

$$\frac{df}{dt} = v \frac{df}{dy} = 2\Omega \cos \phi \, v/R = \beta v$$

and we have assumed that

$$\langle (\zeta + f) \text{div } \mathbf{v} \rangle \approx (\zeta + f) \langle \text{div } \mathbf{v} \rangle = 0$$

because of conservation of mass. In using Eq. (8.4), we consider the rate of change of relative vorticity of a vertical column of water following its motion around the large ocean gyres. This change is due to the combination of three effects, namely the change of planetary vorticity (βv term), the change in vorticity due to the curl of the wind stress at the ocean surface, and the change of vorticity due to friction stresses at the side boundaries of the ocean. In a symmetric oceanic gyre on the rotating earth, the curl of the wind stress would be the same at both sides of the ocean, while the planetary effect would be of opposite sign so that there would be an imbalance on the western side of the ocean. Since both quantities have the same sign on the western side of the ocean, we need a very strong frictional torque ($\partial F_y/\partial x -$ term) on the

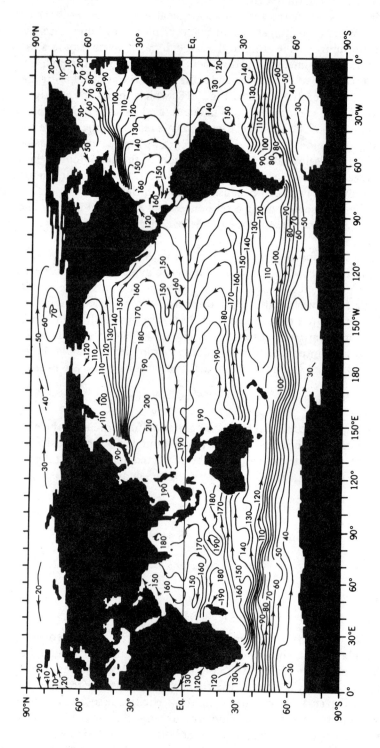

FIGURE 8.18. Global distribution of the annual-mean anomaly of the oceanic geopotential thickness in cm computed for the 0–1000 m layer (from Levitus and Oort, 1977).

OBSERVED MEAN STATE OF THE OCEANS

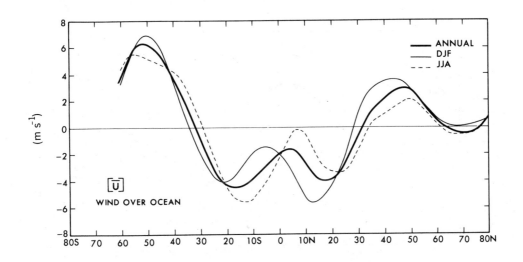

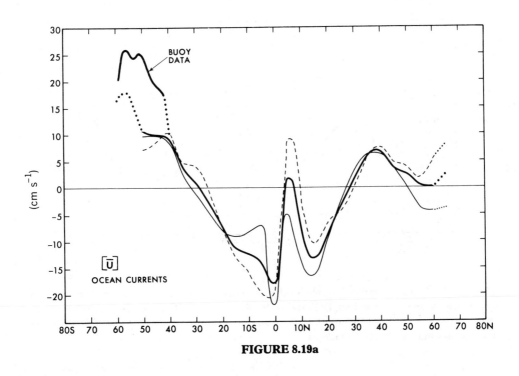

FIGURE 8.19a

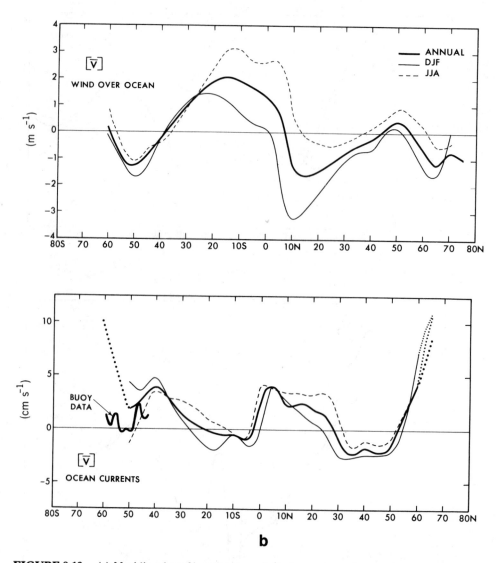

FIGURE 8.19. (a) Meridional profiles of the zonal wind component in m s^{-1} and zonal current component in cm s^{-1} for annual, DJF, and JJA mean conditions. The current velocities were estimated using ship drift data and, south of 42 °S, also using buoy data. (b) Same as (a) except for the meridional wind and current components.

western side as a source (subtropical gyre) or sink (midlatitude cyclonic gyre) of relative vorticity, if absolute vorticity is to be maintained. This implies that there has to be an asymmetry in the ocean gyres and that the ocean currents have to be much stronger on the western (e.g., $\partial F_y/\partial x \gg 0$ in the subtropical gyres) than on the eastern sides of the gyres (where $\partial F_y/\partial x \approx 0$) as confirmed by the observed circulations shown in Figs. 8.1 and 8.17.

To complete the analysis of the observed surface circulation, we also give zonal-mean profiles of the current components and of the corresponding surface winds over the oceans in Fig. 8.19. The zonal ocean current profiles are almost in phase with those in the atmosphere, although they are weaker in magnitude by roughly a factor of 30. They clearly show the main features of the ocean circulation, such as the eastward drift in mid to high latitudes which is somewhat weaker than 10 cm s^{-1} in the Northern Hemisphere and on the order of 15–20 cm s^{-1} in the Southern Hemisphere, the tropical westward flow on the order of 10 to 15 cm s^{-1}, and the narrow equatorial countercurrent centered near 6 °N.

The $[\bar{v}]$ profiles in Fig. 8.19(b) show a tendency for the opposite sign in the two media, as expected from the Ekman drift Eq. (3.24b) and the predominantly zonal nature of the winds. There is a general divergent flow (with upwelling) in the oceans at low latitudes away from the equator and a convergent flow in the subtropics, as shown by the slope of the $[\bar{v}]$ profiles. The strongest divergence is found over the equator ($\partial[\bar{v}]/\partial y > 0$) generating a narrow zone of upwelling of cold water that is consistent with the equatorial minimum in the temperature in Fig. 8.2(a).

At high southern latitudes where the ship drift data become unreliable, we have added some estimates of the ocean current velocities by drifting ocean buoys from Piola et al. (1987). These last estimates show stronger zonal and weaker meridional velocities compatible with the strong zonal westerly winds around Antarctica.

8.4.2 Vertical structure of the circulation

Existing data sets of deep water velocities are very sparse, so that our present knowledge of the subsurface circulation is based largely on indirect evidence. For example, many present representations of the deep ocean circulations are based on speculations regarding the cause of the observed vertical distributions of salinity, dissolved oxygen, and other tracers. Basically, one assumes that the deformation fields at a given depth, showing tongues of anomalous water properties, are due to advection by the prevailing mean ocean circulation, and that these fields can be used to infer the direction and magnitude of the mean flow. Some other studies have been based on computations of the dynamic height at different depths assuming geostrophic flow below the Ekman layer (see, e.g., Levitus, 1982). For extensive discussions of the predominant water masses in the oceans and the inferred thermohaline circulations see, e.g., Pickard and Emery (1982) based on observations and Bryan and Lewis (1979) based on model analyses.

Because of the difficulties in obtaining direct observations of the ocean currents at various depths we will show some results from a general circulation model of the oceans. However, these results are model dependent and therefore preliminary. The model-computed vertical circulation of the oceans as a function of depth is shown in Fig. 8.20. This cross section of the zonal mean meridional flow was derived by Bryan and Lewis (1979) using a three-dimensional world ocean model with sea surface temperature, surface salinity, and wind stress specified from observations. Their

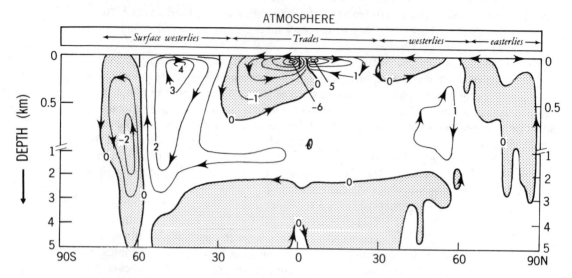

FIGURE 8.20. Zonal-mean cross section of the annual-mean flow of mass in 10^{10} kg s^{-1} for an ocean model (adapted from Bryan and Lewis, 1979).

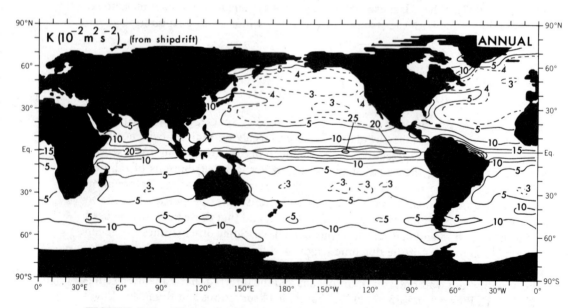

FIGURE 8.21. Global distribution of the (total) kinetic energy at the ocean surface in 10^{-2} m^2 s^{-2} for annual-mean conditions based on ship drift data.

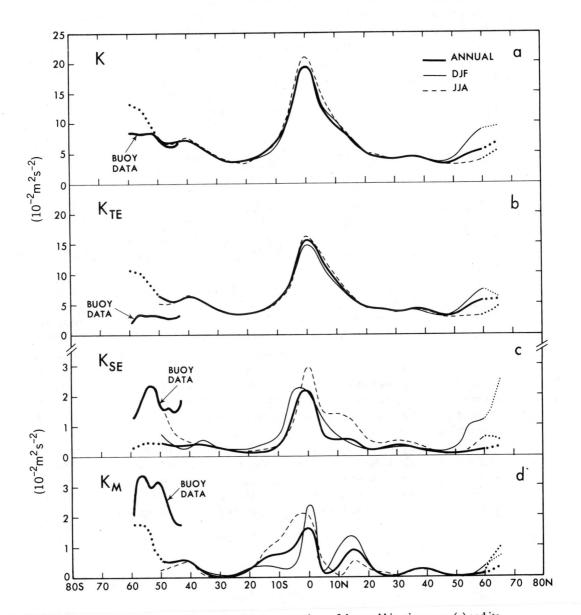

FIGURE 8.22. Meridional profiles of the zonal-mean values of the total kinetic energy (a) and its transient eddy (b), stationary eddy (c), and mean meridional circulation components (d) in 10^{-2} m² s^{-2} for annual, DJF, and JJA mean conditions based on ship drift data. South of 42 °S some results using buoy data have been added.

numerical model was driven mainly by observed wind stresses at the surface. It is interesting to compare this ocean circulation with the corresponding meridional circulation in the atmosphere as given in Fig. 7.19(a). The quantitative agreement in location and intensity of the mean meridional overturnings with those observed in the atmosphere is not accidental. In fact, the specification of observed wind stresses should, according to Eq. (3.24b), induce a similar (but opposite) ocean circulation as in the atmosphere. One should note that the depth of penetration of the cells is largely determined by the model formulation, and could therefore deviate appreciably from the unknown real circulation.

We may conclude that, to a large extent, many of the basic processes in the operation of the general circulation of the oceans are still unknown.

8.5 SURFACE KINETIC ENERGY OF THE OCEANS

The global distribution of the surface kinetic energy associated with the mean currents and their time fluctuations is shown in Fig. 8.21. Large values on the order of 2000 cm^2 s^{-2} are found over the equatorial belt where the strongest meandering occurs and values on the order of 500 to 1000 cm^2 s^{-2} in the regions of the major boundary currents.

A breakdown of the total kinetic energy in the transient eddy, stationary eddy, and zonal-mean components is given in Fig. 8.22. Besides the ship drift data we have added some presumably more reliable and representative energy estimates from floating surface buoys south of 42 °S where the ship drift data are sparse (Piola *et al.*, 1987). As we have seen before, this is an important region of the globe because of the strong Antarctic circumpolar flow. The buoy values of the total kinetic energy are not very different from the ship drift values. However, the breakdown into components is different; the buoy estimates of the transient eddy kinetic energy are smaller and those of the stationary and zonal-mean kinetic energy larger than the corresponding ship drift estimates.

It is interesting to note that the equatorial maxima in kinetic energy in Figs. 8.21 and 8.22 are mainly due to transient eddies with time scales of weeks to months. Satellite photographs have confirmed the existence of these eddies or waves in the equatorial ocean currents. We actually see that at all latitudes the transient eddies give an important contribution to the kinetic energy of the surface ocean.

CHAPTER 9

Observed Mean State of the Cryosphere

9.1 ROLE OF THE CRYOSPHERE IN THE CLIMATE

The cryosphere plays a central role in the earth's climate due to the nature and physical properties of ice and snow and due to the coupling of the cryosphere with the other components of the climatic system.

As we have mentioned in Chap. 2, the cryosphere includes the ice sheets of Greenland and Antarctica, the sea ice, and the permafrost regions of North America and Siberia. The snow, another important component of the cryosphere, extends in the northern winter into midlatitudes south of 50 °N in North America and Eurasia. The cryosphere further includes some of the major mountain glacier systems in, e.g., the Canadian Arctic, Alaska, the Alps, and Spitsbergen. Permafrost underlies extensive regions, covering 15% to 20% of the earths's land surface, mainly on the polar borders of the continents in North America and Siberia.

It is estimated that 2% of all the water on earth is frozen (see Chap. 12). This frozen water represents about 80% of all the fresh water available on earth. It is distributed in very different proportions among the various components of the cryosphere and also shows very different times of residence in each component. Figures 9.1 and 9.2 illustrate the seasonal distributions of the cryosphere in the Northern and Southern Hemispheres, respectively. Most prominent are the huge seasonal changes in snow cover over the Northern Hemisphere continents, and the seasonal changes in sea ice cover over the southern oceans around Antarctica.

The cryosphere has at all time scales a strong influence on the local climate, but its influence on the global climate is predominantly at longer time scales of years and beyond.

The distribution of ice and snow is important because they efficiently reflect shortwave solar radiation during the day and act as a nearly perfect black body radiator for the long-wave terrestrial radiation during the night. Ice and snow have a high reflectivity or albedo as compared with water and land surfaces. The cryosphere acts as an effective heat sink for the atmosphere and oceans both through its relatively high albedo and its large latent energy of melting. Variations of the global ice and snow distributions have a significant effect on the planetary albedo.

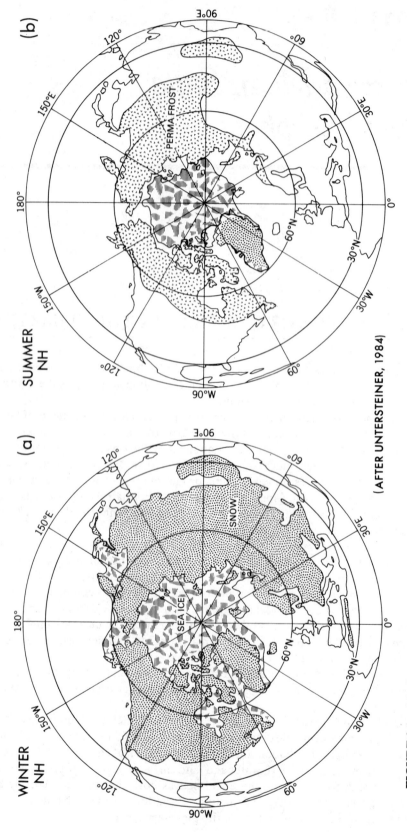

FIGURE 9.1. Maximum extent of snow and sea ice during winter (a) and minimum extent of snow and sea ice during summer (b) in the Northern Hemisphere after Untersteiner (1984). The permafrost regions are also shown in (b).

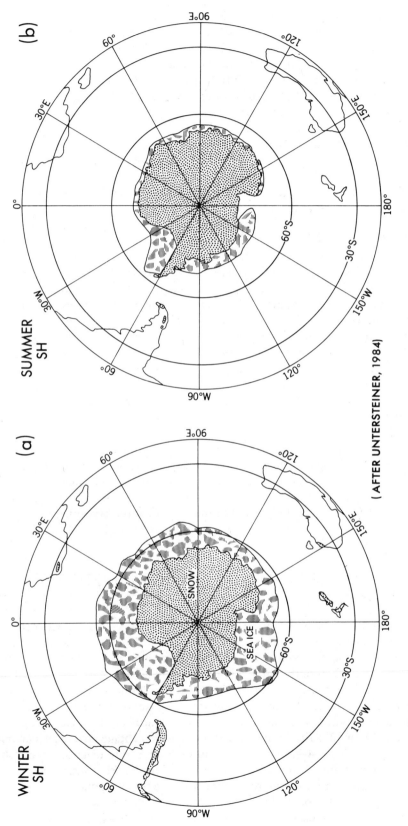

FIGURE 9.2. Maximum extent of sea ice during winter (a) and minimum extent of sea ice during summer (b) in the Southern Hemisphere after Untersteiner (1984). The ice limit was taken to be at concentrations >15%.

Because of their low thermal conductivity, ice and snow constitute excellent insulators, thus reducing the heat exchange between the land and oceans and the overlying atmosphere. Sea ice serves then as an insulating layer between the cold polar air masses and the relatively warm ocean beneath the ice.

All these effects combined tend to reinforce the cooling of the atmosphere and to induce local colder climates. The interactions between ice and snow and the atmospheric temperature could lead to a positive snow and ice cover-albedo-temperature feedback process, as mentioned in Chap. 2.

However, such a strong positive feedback process should not be considered in isolation but together with many other factors that may alter the net outcome of the feedback loop. For example, the feedback will only be effective when there is an adequate amount of precipitation (snow) available to guarantee the duration and spatial extent of the snow and ice.

The cooling of the polar atmosphere also leads to stronger meridional temperature gradients, and to an increase in the intensity of the zonal circulation in the atmosphere.

The freezing of water will tend to increase the salinity of the upper ocean by removing fresh water from the surface layers, whereas the melting of snow and ice will tend to decrease the surface salinity. The resulting changes in the vertical temperature and salinity structure will in turn affect the stability of the ocean and may lead to changes in the general regime of overturning in the world ocean as shown, e.g., in the numerical model experiments of Bryan (1986). The rejection of salt by growing sea ice and the brine drainage from sea ice constitute the initial step for the formation of bottom water masses.

The variations in the volume of ice sheets and glaciers also have an indirect influence on the climate, since they change the sea level and thus affect the area covered by the oceans.

Some other important aspects of the cryosphere, regarding the mass and energy budgets and the hydrological cycle, will be discussed later when we deal with the regional studies of the climate in the polar regions in Chaps. 12 and 13. When considered together, the present chapter and the two sections in Chaps. 12 and 13 give a fairly complete description of the physics of the climate of the cryosphere.

9.2 GENERAL FEATURES OF THE CRYOSPHERE

There are notable differences in the nature of the earth's surface between the Arctic and Antarctic regions. For example, if we compare two polar caps poleward of 70 °N and 70 °S, the ocean covers 72% of the surface area of the north polar cap whereas it covers only 22% of the surface of the south polar cap (see Table 9.1). This suggests that the role of the ocean in the exchange of such quantities as water vapor and energy with the overlying atmosphere must be very important in the north polar cap and probably of minor importance in the south polar cap. If the polar caps are extended from 70° to 60° latitude the land and ocean percentage coverages become more nearly equal. The 3–4-km high topography of the Antarctic ice sheet and the generally lower topography (except for Greenland) of the land bordering the Arctic Ocean, represent another important difference between the two polar regions.

The most important factor in determining the polar climate is the large annual cycle of insolation with almost no solar radiation during winter and an absolute maxi-

TABLE • 9.1. Some surface characteristics of the polar caps poleward of 60° and 70° latitude in the Northern and Southern Hemisphere.

	Area (10^{14} m^2)	Fraction of hemisphere	Ocean covered	Land covered
70 °N-NP	0.15	6.0%	72%	28%
70 °S-SP	0.15	6.0%	22%	78%
60 °N-NP	0.34	13.4%	46%	54%
60 °S-SP	0.34	13.4%	60%	40%

mum in solar insolation at the top of the atmosphere during summer, that is greater than anywhere else on the globe (see Fig. 6.4).

Most of our knowledge of the atmosphere in the polar regions is based on the rawinsonde network that was shown earlier in Fig. 5.4 by the distribution of rawinsonde stations in the two polar regions. The coverage in the Northern Hemisphere was found to be good. In fact, near 60 °N the zonal coverage was better than at any other latitude of the globe. On the other hand, Fig. 5.4(b) showed that in the south polar regions the network is poor, except near the rim of Antarctica. Therefore, in our later calculations of the various budgets we may expect to get reliable results for the Arctic, and only tentative results for the Antarctic regions.

9.3 ICE SHEETS AND GLACIERS

It is useful to make a distinction between the terrestrial ice bodies and the marine ice sheets (Bentley, 1984). The terrestrial ice bodies that include the Greenland ice sheet, mountain glaciers, and most of the East Antarctic ice sheet apart from the fringing ice shelves rest on ground above sea level. On the other hand, the "marine" West Antarctic ice sheet is formed by ice grounded below sea level and terminates in the two largest floating ice shelves in the Ross and Weddell seas. These differences imply that the West Antarctic ice sheet probably responds strongly to sea level changes and is therefore more affected by global climatic fluctuations than the terrestrial ice bodies. The terrestrial ice bodies respond only to changes in accumulation of snow and to changes in temperature, and at a much slower rate.

The ice sheets on Antarctica and Greenland constitute a more or less permanent feature of the earth's topography with surface elevations of about 3000 m in Greenland and over 4000 m in Antarctica.

Melting of the Greenland ice, with a mass of about 3.0×10^{18} kg, would raise mean sea level by approximately 8 m, whereas melting of the west Antarctic ice would raise it by another 8 m and melting of the east Antarctic ice by about 65 m (Barry, 1985). The energy input needed to accomplish the melting would be on the order of 10^{25} J which is twice the energy needed to warm the entire oceans by 1 K (Oerlemans and van der Veen, 1984).

Due to their high elevation combined with a high albedo the polar ice sheets act as cooling agents for the climatic system leading to very low values of the surface temperatures in winter, with values on the order of $-70\,°C$ over Antarctica and $-40\,°C$ over Greenland.

Because of the low temperatures over Antarctica there is practically no local surface melting. The mass accumulation at the surface is balanced by calving (breaking off) of icebergs mainly from the shelves and, to a lesser extent, by ablation (melting) underneath the large shelves of ice extending in the oceans around Antarctica. In Greenland there is, on the other hand, significant surface melting at the ice sheet margins and some calving of icebergs directly from the ice sheet since Greenland does not have ice shelves extending into the sea.

Glaciers are mainly found in mountainous regions at all latitudes. They represent a very small fraction of the total cryosphere and only affect the local climate. The mountain glaciers are often characterized by high rates of accumulation and ablation. Some of the earliest indirect evidence for climatic changes has come from the study of mountain glaciers, because they strongly respond to and time-integrate the climate variations.

The glacier drainage basins can range in size from a few km^2 for mountain glaciers to several $10\,000\,km^2$ for the Arctic and Antarctic glaciers. The mid- and high-latitude glaciers are of great economic importance. They retain a considerable amount of the winter snow, providing a continuous runoff for generating hydroelectric power and irrigation in the dry season.

9.4 SEA ICE

Ice covers on the average about 11% of the earth's land surface and 7% of the world ocean (Untersteiner, 1984). Sea ice acts as a barrier to the transfers of moisture, heat, and momentum between the oceans and the atmosphere, and plays an important role in the formation of deep water masses. The residence time of sea ice is on the order of months to years.

As shown in Fig. 9.1, during winter sea ice covers practically the entire Arctic Ocean and large parts of the Greenland Sea, Barents Sea, and the Sea of Okhotsk. In summer, the northern ice cover is restricted to the Greenland ice sheet and to pack ice in the Arctic Ocean.

The south polar regions in Fig. 9.2 show a much larger seasonal variation in the area covered by sea ice than the north polar regions. In winter, the sea ice distribution extends to $60\,°S$ latitude, whereas in summer the ice is confined to the regions poleward of about $70\,°S$. The distribution of sea ice near Antarctica is more symmetric and more compact than that in the Arctic.

The general pattern of ice drift in the Arctic consists of a clockwise gyre in the Beaufort Sea and a transpolar drift stream flowing from near the Siberian coast at about $100-140\,°E$ across the pole. Much of the drift ice from the Arctic is then lost through the Fram Strait between Greenland and Spitsbergen; flowing southward along the east coast of Greenland (Washington and Parkinson, 1986).

Year-to-year and regional variations in Antarctic ice cover tend to be larger than those in the Arctic because the Arctic ice is more confined by land. The maximum extent of the sea ice in the Antarctic occurs between July and October and the minimum in February–March, while in the Arctic the maximum extent occurs in January–February and the minimum in August–September (see Fig. 9.3).

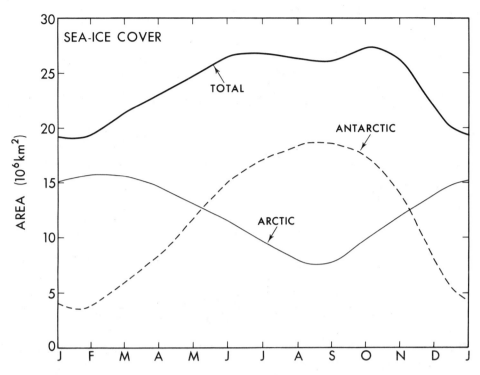

FIGURE 9.3. Annual cycle in sea ice cover in 10^6 km^2 for the Arctic, Antarctic, and the globe as a whole after Ropelewski (1989).

The boundaries between ice-covered and ice-free parts of the oceans, the so-called marginal ice zones, can move and shift in a few days in response to moving weather systems but their mean positions change at time scales of months in response to the seasonal changes in radiation. The dominant factors in the momentum equations that govern the ice dynamics, are the wind stress, the water drag, the Coriolis force, and the internal ice force which may lead to the formation of ice ridges. The relative role of wind and ocean currents on ice drift depends on the time scale. On short time scales the ice motion tends to be correlated with the winds, whereas on the long time scale the more steady ocean currents are the dominant factor.

The marginal ice zones are regions with upwelling of nutrient-rich water from the deep ocean. They are therefore areas of intense biological production, and can support large fish, mammal, and bird populations.

The annual variation of sea ice is summarized in Fig. 9.3 expressed in terms of the areal coverage. The seasonal variation in sea ice coverage for the Northern Hemisphere is very different from that for the Southern Hemisphere. The Arctic ice cover varies on the average between about 7 and 15×10^6 km^2, whereas the Antarctic ice cover varies between 3 and 18×10^6 km^2. The ice cover in the Arctic is less dense than in the Antarctic because it includes large areas of open water (Barry, 1985). Furthermore the decay in Arctic sea ice is slow and its growth relatively fast, whereas the opposite is found in the Antarctic sea ice. The reasons for these differences must be connected with the geographical differences in the land–sea distributions between the closed Arctic Ocean basin and the open southern oceans surrounding Antarctica. Once winter sets in, the Eurasian continent remains covered by snow, creating cold air masses that may lead to a rapid increase in the sea ice. In the Antarctic, the warm

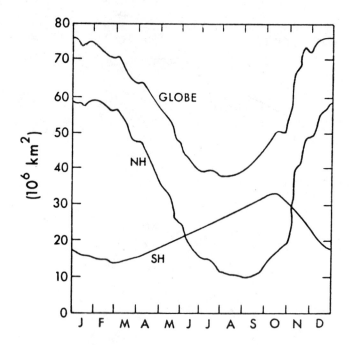

FIGURE 9.4. Annual cycle in snow and sea ice cover in 10^6 km^2 for the Northern Hemisphere, Southern Hemisphere, and the globe based on data from Kukla and Kukla (1974).

ocean waters surrounding the Antarctic continent tend to slow down the growth of sea ice in the fall and to speed up the melting of the sea ice in the spring. For the globe as a whole, the sea ice curve follows the Antarctic seasonal cycle.

9.5 SNOW

The snow cover undergoes a marked annual cycle, covering in winter about 50% of the Northern Hemisphere land surface and 10% of the oceans. However, the residence time of snow is shorter than that of sea ice and ranges only from days to months. The importance of the snow in climate stems largely from the very high values of its albedo (~80% for fresh snow).

When we consider the annual cycle of snow and ice cover combined as shown in Fig. 9.4, the Northern Hemisphere cycle shows a much larger amplitude than the Southern Hemisphere one mainly due to the vast extent of snow cover during the northern winter. Thus, for the globe as a whole the variation in snow plus ice cover must be dominated by the changes in snow cover over the Northern Hemisphere continents.

From a comparison of the Northern Hemisphere curves in Figs. 9.3 and 9.4, we can indeed infer that the variation in snow cover dominates the Northern Hemisphere curve for snow and sea ice combined. In fact, the seasonal amplitude of the Northern Hemisphere variation in snow plus sea ice cover is about 23×10^6 km^2 vs an amplitude of only 4×10^6 km^2 for sea ice cover alone. There appears to be some asymmetry in the Northern Hemisphere curve for snow and ice with a more rapid growth of the area covered by snow in the fall during the months of October through December and a slower decay or melting of the snow cover in the spring during the months of March through June. In this context, we have to also consider the depth or water equivalent of the snow pack (depth of water produced if the snow were to melt) besides its areal extent since the snow depth will affect the speed of decrease of the snow cover area in spring.

Comparing the Southern Hemisphere curves in Figs. 9.3 and 9.4 we find that they have practically the same amplitude and phase, indicating that the growth and decay of sea ice dominate the snow plus sea ice signal. There is a difference in the annual-mean values in Figs. 9.3 and 9.4 because the Antarctic snow plus sea ice curve in Fig. 9.4 includes the area of Antarctica whereas, of course, the corresponding curve in Fig. 9.3 only contains the ocean area covered by sea ice. We should add that some of the differences in the curves in Figs. 9.3 and 9.4 may also be due to differences in the data samples, data reduction, and degree of smoothing of the curves.

Snow plays a major role in the hydrological cycle for the mid- and high-latitude continents as one of the principal sources of water for runoff and soil moisture recharge. Snow is also important on economical grounds since much of the supply of water in vast regions of Europe and North America comes from snowmelt. The snowmelt and its variations are crucial factors in hydroelectric power generation, irrigation management, crop planting, protection from floods, and domestic water supply.

9.6 PERMAFROST

In the northern parts of the American and Eurasian continents, the soil temperatures are below freezing and the water is permanently frozen. Permafrost underlies about 20% of the earth's land surface. Figure 9.1(b) shows the areal extent of the permafrost regions in the Northern Hemisphere. We notice that most of the land masses of Alaska, northern Canada, and Northeast Asia are frozen. By and large, the permafrost regions are located to the north of the annual-mean -6 to $-8\,°C$ air temperature isolines (Untersteiner, 1984). Permafrost results from a delicate balance between the heat exchange with the atmosphere and the geothermal heat flux; it is affected by the water content and thermal properties of the ground. There are indications that the present distribution of continuous permafrost is of Pleistocene origin. However, in the marginal regions of thin and discontinuous permafrost, relatively rapid changes have occurred in response to climatic variations and man's activities. Permafrost prevails where the summer heating does not reach the base of the layer of frozen ground.

The permafrost prevents the movements of ground water, limits the development of vegetation, and accelerates the runoff at the surface.

CHAPTER 10

Exchange Processes Between the Earth's Surface and the Atmosphere

10.1 INTRODUCTION

As we have seen in Chap. 2, the components of the climate system are open, nonisolated subsystems. They are strongly coupled and are characterized by intense interactions on various time and space scales ranging from micro through meso to planetary scales. The fluxes across the various interfaces greatly affect the properties of the adjacent boundary regions. In the climate system the main fluxes are those associated with the transports of matter, energy, and momentum.

In this chapter, we will consider the atmosphere $\mathscr{A}$ as the central system so that we are concerned with the fluxes across the atmosphere–land surface boundary $\mathscr{A} \cap (\mathscr{L} \cup \mathscr{B})$, the atmosphere–ocean boundary $\mathscr{A} \cap \mathscr{O}$, and the atmosphere–cryosphere boundary $\mathscr{A} \cap \mathscr{C}$. In this context the land surface includes parts of the lithosphere and biosphere. The interfaces constitute highly discontinuous boundaries for the various processes that occur in the atmosphere, and have a markedly different influence on the atmosphere according to the nature of the underlying surface.

The $\mathscr{A} \cap (\mathscr{L} \cup \mathscr{B})$ interface is not a uniform boundary. It is modulated by the orography of the continents, the size of different topographic features, such as hills, valleys, and man-made structures, the nature of the soils, the vegetation coverage, etc. All these factors, their spatial variations, as well as the altitude of the surface, have a profound influence on the behavior of the lower atmosphere. They may be regarded as major factors of the regional and local climates since they introduce large variations in the momentum and energy balances and in the hydrological cycle.

An important constraint for the surface flux is the conservation of total energy. The equation for the heat budget of an interfacial layer is fairly complex since many energy transfers and transformations from one form of energy to another have to be taken into account. These processes take place continuously both at and near the earth's surface. Most of the transformations start with the absorption of solar radiation and end with the loss of infrared radiation to space.

The complete energy cycle may then be described as follows. The incoming-minus-reflected solar radiation received at the earth's surface is absorbed leading to an increase in the heat content of the ocean mixed layer or of the top layer of the soil. This thermal energy may, in turn, lead to an upward transfer of sensible, latent, and long-wave radiational energy into the atmosphere and a downward transfer of heat into the ocean below the mixed layer or deeper into the ground.

It is customary to take the flux in the first few meters of the boundary layer as representative for the flux at the actual surface. This approach is acceptable when we consider the surface to be a part of the boundary layer. It allows the use of standard meteorological observations (wind, temperature, and humidity) to compute the fluxes. The approach also avoids the conceptual and operational difficulties involved in defining the surface and in making physically meaningful measurements at such an ill-defined surface. In fact, the natural surface is highly irregular when we consider such features as sand and grass with dimensions of millimeters and less.

Near the earth's surface the atmospheric circulation is strongly affected by friction causing profound modifications in the air flow. The thickness of the modified air grows in depth until a boundary layer of 10's to 100's of meters thick is established. This planetary boundary layer is characterized by a strong wind shear since the wind must vanish right at the surface, because of the viscosity of the air. The strong wind shear together with the surface heating lead to the development of a turbulent flow in the boundary layer with eddies of various scales. The turbulent exchange or turbulent mixing constitutes a very efficient mechanism for the transfers of mass, momentum, and heat through the boundary layer both upward and downward, thus linking the earth's surface with the free atmosphere. In the surface exchange of mass we must also include the exchange of various gases, dust particles, and other aerosols.

The ocean interface $\mathscr{A} \cap \mathscr{O}$ is more uniform than that for the land. However, as mentioned in Chap. 2, the atmosphere and oceans interact strongly so that air–sea interactions take place on many scales in time and space exchanging matter, energy, and momentum. On one hand, the drag by the atmospheric winds on the ocean is responsible for the wind-driven ocean currents; on the other hand, the oceans constitute the main source of water vapor for the atmosphere feeding the atmospheric branch of the hydrological cycle. Besides, the oceans have an enormous capacity to store energy due to their very high heat capacity. The solar energy absorbed at the surface is distributed throughout the top layers of the ocean by convection and lateral transports forming a mixed layer with a thickness of about 100 m separated from the underlying cold waters by the thermocline.

The exchange processes between the cryosphere and atmosphere have a profound influence in defining the cold polar and subpolar climates. Thus, the presence of ice and snow affects the structure of the planetary boundary layer leading to a local temperature inversion in the first km of the atmosphere. Due to their high albedo, snow and ice strongly modify the radiation balance at the surface thereby leading to very low temperatures and certain feedback processes.

10.2 ENERGY BUDGET AT THE SURFACE

10.2.1 Energy fluxes at an ideal surface

An ideal surface corresponds to the geometric concept of a boundary that separates the atmosphere from the underlying medium. Physically, it can be regarded as a thin interfacial layer between the two media with no mass and heat capacity. As we saw in Sec. 6.7, the energy balance involves five modes of energy fluxes: the net radiative energy flux into or out of the surface $F_{\text{rad}}^{\text{sfc}}$, the sensible heat flux to or from the air F_{SH}, the latent (indirect) heat flux involved in evaporation F_{LH}, the sensible heat flux to or from the underlying medium (soil or water) F_G, and the latent (indirect) heat flux involved in melting or freezing F_M.

The net radiative flux $F_{\text{rad}}^{\text{sfc}}$ is the result of the radiation balance at the surface. As presented before in Eq. (6.39), the net radiational heating can be written as

$$F_{\text{rad}}^{\text{sfc}} = F_{\text{SW}}^{\downarrow}(1 - A_{\text{sfc}}) - \epsilon \sigma T_{\text{sfc}}^4 + F_{\text{LW}}^{\downarrow},$$

where T_{sfc} (in K) is the temperature, A_{sfc} the albedo, and ϵ the infrared emissivity, all taken at the surface. Further σ denotes the Stefan–Boltzmann constant, $F_{\text{LW}}^{\downarrow}$ the long-wave back radiation from the atmosphere. During the day the downward solar radiation is the prevalent mode, whereas during the night the net radiation is much weaker and directed away from the surface. Thus, the surface warms during the daytime and cools during the night, mainly under clear sky conditions.

The energy fluxes at an ideal surface obey the conservation of energy so that the balance can be expressed as

$$F_{\text{rad}}^{\text{sfc}} - F_{\text{SH}}^{\uparrow} - F_{\text{LH}}^{\uparrow} - F_G^{\downarrow} - F_M = 0,$$

which was presented before as Eq. (6.40).

The sensible heat flux $F_{\text{SH}}^{\uparrow}$ results from the difference in the temperature of the surface and the overlying air. In the first few millimeters above the surface, the transfer is mainly through molecular conduction. Above this molecular sublayer the heat exchange is accomplished by turbulent mixing and convection. The sensible heat flux is normally directed upward from the surface into the atmosphere during the daytime when the surface is warm and in the opposite direction during the evening and night hours.

The latent heat flux $F_{\text{LH}}^{\uparrow}$ results mainly from the evaporation at the surface and the later condensation within the atmosphere. Evaporation takes place from water surfaces, such as lakes, rivers, and oceans, and from moist soils and vegetation. This transfer of heat is indirect and is associated with the phase transitions of water substance, first between liquid and vapor at the surface and later between the vapor and liquid or solid phases. The latent heat flux can be expressed by the product of the latent heat of evaporation L_e, and the rate of evaporation E so that: $F_{\text{LH}} = L_e E$. The ratio of the sensible heat flux to the latent heat flux defines the so-called Bowen (1926) ratio

$$B = F_{\text{SH}}^{\uparrow}/F_{\text{LH}}^{\uparrow} = F_{\text{SH}}^{\uparrow}/L_e E. \tag{10.1}$$

The flux of energy into the subsurface medium $F_G^{\downarrow}$ is mainly due to heat conduction. The last term F_M in Eq. (6.40) is the energy used for melting snow and ice at the rate M_S and for freezing water at the rate F_S:

$$F_M = L_m(M_S - F_S). \tag{10.2}$$

Equations (6.39) and (6.40) taken together show the importance of the nature of the earth's surface for the energy budget through its albedo, infrared emissivity, thermal conductivity, and evapotranspiration characteristics.

For annual-mean conditions over land, the flux of sensible heat into the ground $F_G^\downarrow$ can be neglected in Eq. (6.40) leading to

$$F_{rad}^{sfc} - F_{SH}^\uparrow - L_e E - L_m(M_S - F_S) = 0 \tag{10.3}$$

or, using the Bowen ratio, to

$$F_{rad}^{sfc} - (1+B)L_e E - L_m(M_S - F_S) = 0. \tag{10.4}$$

10.2.2 Energy budget of a layer

We discussed in the previous section the surface energy budget for an ideal surface. For practical purposes and, in order to take into consideration the inhomogeneities of the surface, the nature of the underlying medium, the storage of energy, etc., it is better to consider the energy budget of a thin interfacial layer with a finite mass and heat capacity. We will assume that the interfacial layer is limited by horizontal planes at the top and bottom so that the fluxes to be considered are the net vertical fluxes at these boundaries of the layer.

The balance equation for this layer can be written as follows:

$$\frac{\Delta H_s}{\Delta t} = F_{rad}^{sfc} - F_{SH}^\uparrow - F_{LH}^\uparrow - F_G^\downarrow - L_m(M_S - F_S), \tag{10.5}$$

where $\Delta H_s / \Delta t$ is the rate of storage per unit area in the form of internal energy (see Fig. 10.1). It is important to note that F_{rad}^{sfc}, $F_{SH}^\uparrow$, and $F_{LH}^\uparrow$ are the vertical fluxes at the top surface, whereas F_G is the flux at the bottom surface.

The rate of sensible heat storage in the layer is given by

$$\frac{\Delta H_s}{\Delta t} = \int_{\Delta z} \frac{\Delta}{\Delta t}(\rho c T) dz, \tag{10.6}$$

where c is the specific heat and ρ and T are the density and temperature at a given depth z, respectively. The integration is done over the entire depth of the layer Δz. The heat storage is then the difference between the input and output of all forms of energy (which includes the energy in evaporation, sublimation, and melting). With a net flux convergence (input exceeds output), there is a warming of the layer, i.e., $\Delta H_s/\Delta t > 0$. In the case of a net flux divergence, the output is larger than the input, leading to a cooling of the layer, i.e., $\Delta H_s/\Delta t < 0$.

The rate of heat transfer in a given direction is proportional to the gradient of temperature according to the general Fick's law so that in the z direction

$$F_G(z,t) = -K \frac{\partial T(z,t)}{\partial z}, \tag{10.7}$$

where K is the thermal conductivity of the material. The thermal conductivity K (W m^{-1} K^{-1}) is related to the thermal diffusivity K^* (m^2 s^{-1}) by the relation $K = \rho c K^*$ and the previous equation can be written as

$$F_G / \rho c = -K^* \left(\frac{\partial T}{\partial z}\right). \tag{10.8}$$

Assuming that there are no sources or sinks of energy within the slab the net convergence of heat per unit area, $-\partial F_G/\partial z$, must be equal to the rate of increase of the internal energy. This last quantity is given by $\partial(\rho c T)/\partial t$ so that

$$\frac{\partial}{\partial t}(\rho c T) = -\frac{\partial F_G}{\partial z}. \tag{10.9}$$

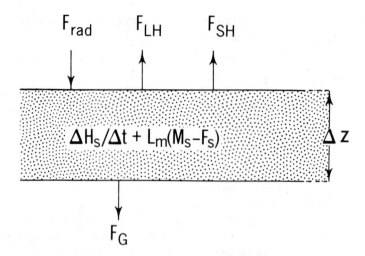

FIGURE 10.1. Important terms in the energy budget for the surface layer. The meaning of the various terms is explained in the text.

If we assume that the medium has a constant heat capacity and if we substitute Eq. (10.7) into Eq. (10.9) we obtain

$$\rho c \frac{\partial T}{\partial t} = \frac{\partial}{\partial z}\left(K \frac{\partial T}{\partial z}\right)$$

or

$$\frac{\partial T}{\partial t} = \frac{\partial}{\partial z}\left(K^* \frac{\partial T}{\partial z}\right). \tag{10.10}$$

Heat is transferred through the subsurface materials (soil, rocks, etc.) by conduction. We must note that of all materials air has the lowest thermal conductivity (see Table 10.1) but that its thermal diffusivity is very large compared to the thermal diffusivity in soils and water because of its low density. As further shown in Table 10.1, the addition of water to a dry soil increases both its thermal conductivity and its heat capacity (product of density and specific heat) since the water replaces air (a poor conductor) in the pore spaces.

Equations (10.9) and (10.10) can be used to compute the heat flux into the ground $F_G^\downarrow$ from measurements of the soil temperature at various depths (e.g., Arya, 1988). The method is based on the integration of Eq. (10.10) with respect to z.

Heat propagates into the ground according to the expression

$$T'(z) = T'(0) e^{z/d} \sin\left[\frac{2\pi}{P}(t - t_m) + z/d\right], \tag{10.11}$$

where we assume a sinusoidal variation of temperature at the surface with a period P and a phase lag $t_m(z)$. The damping depth d is given by

$$d = \sqrt{(P/\pi)K^*} \tag{10.12}$$

as can be verified by substituting Eq. (10.11) into Eq. (10.10).

At a depth $z = -3d$, the amplitude of the temperature fluctuation is reduced to about 5% of its surface value so that this depth can be taken as the effective depth of penetration (see Table 10.1). Since the heat absorbed in the ground per unit area is

TABLE 10.1. Thermal properties of some natural substances.[a]

Substance	Condition	Density ρ (10^3 kg m^{-3})	Specific heat c (10^3 J kg^{-1} K^{-1})	Heat capacity ρc (10^6 J m^{-3} K^{-1})	Thermal conductivity K (W m^{-1} K^{-1})	Thermal diffusivity K^* (10^{-6} m^2 s^{-1})	Conductive capacity c^* (10^3 J m^{-2} K^{-1} s$^{-1/2}$)	Penetration depth 3d (m) diurnal	Penetration depth annual
Air	20 °C, still	0.0012	1.00	0.0012	0.026	21.5	0.006	2.3	44
	stirred					4×10^{6} [b]	2.4	1×10^3	19×10^3
Water	20 °C, still	1.00	4.19	4.19	0.58	0.14	1.57	0.2	3.6
	stirred					130[b,c]	48	5.7	108
Ice	0 °C, pure	0.92	2.10	1.93	2.24	1.16	2.08	0.5	10.2
Snow	Fresh	0.10	2.09	0.21	0.08	0.38	0.13	0.3	6.0
Sandy soil	Dry	1.60	0.80	1.28	0.30	0.24	0.63	0.2	4.8
(40% pore space)	saturated	2.00	1.48	2.98	2.20	0.74	2.56	0.4	8.1
Clay soil	Dry	1.60	0.89	1.42	0.25	0.18	0.60	0.2	3.9
(40% pore space)	saturated	2.00	1.55	3.10	1.58	0.51	2.21	0.4	6.9
Peat soil	Dry	0.30	1.92	0.58	0.06	0.10	0.18	0.2	3.0
(80% pore space)	saturated	1.10	3.65	4.02	0.50	0.12	1.39	0.2	3.3

[a] After Oke (1987). $K^* = K/\rho c$, $c^* = \rho c \sqrt{K^*}$, $d = (PK^*/\pi)^{1/2}$.

[b] The values of K^* for stirred water and air are, of course, much greater than those for still conditions because turbulent eddy mixing is a more efficient process to transport heat vertically than molecular conduction. In spite of the large uncertainties in specifying K^* for stirred conditions, the implied annual penetration depths of about 100 m for the oceans and 19 km for the atmosphere give reasonable order of magnitude estimates if we compare them with the observed profiles in Fig. 13.20.

[c] From Munk (1966) based on geochemical data for lower thermocline ($K^* = 1.3$ cm^2 s^{-1}).

proportional to the product of the heat capacity and the penetration depth, it is useful to introduce a quantity called conductive capacity $c^* = \rho c \sqrt{K^*}$. The same expression can be used for different materials, such as water, ice, and various types of soils, in contact with the air (see Table 10.1). These materials show a different response to the same energy input $F^\downarrow_G$. Thus, when we compare the values of the conductive capacity for soil and stirred water, we can understand that a land surface heats and cools faster than a water surface and that it undergoes a greater variation of the surface temperature.

10.3 DEVELOPMENT OF THE PLANETARY BOUNDARY LAYER

As was said earlier, the planetary boundary layer is the region of the atmosphere that is directly affected by the underlying subsystems through the fluxes of matter, energy, and momentum.

10.3.1 Some characteristic features of the planetary boundary layer

The thickness of the planetary boundary layer varies considerably from some tens of m's to one or two km's, depending on the roughness of the ground, the topography, the nature of the vegetation coverage, the intensity of the wind, the rate of heating or cooling of the surface, the advection of heat and moisture, the vertical motions, etc. Over land, the depth of the planetary boundary layer changes with the time of day in response to the diurnal heating and cooling cycles. Following sunrise on a clear day, the heating of the land surface and the consequent vertical mixing lead to an increase in the thickness of the boundary layer, reaching a maximum in the afternoon. In the evening and throughout the night radiative cooling of the surface inhibits thermal mixing thereby substantially reducing the depth of the planetary boundary layer.

The effects of viscosity and turbulence are very important in the formation and maintenance of the planetary boundary layer. Viscosity is a molecular property of a fluid. It is a measure of the internal resistance of the fluid to deformation. One effect of viscosity is that the fluid particles will adhere to a solid surface as they come in contact with it. Therefore, there is no relative motion between the fluid and the solid surface. This is the well-known no-slip boundary condition. The internal resistance is associated with the shearing motion (variation of velocity) between layers. Newton found that the shear stress τ is proportional to the gradient of the velocity (rate of strain):

$$\tau/\rho = -\nu\left(\frac{\partial u}{\partial z}\right), \qquad (10.13)$$

where ν is the kinematic viscosity.

Viscous flows can be either laminar or turbulent. Flows which are originally laminar may become turbulent at large values of the Reynolds number. This number is defined as the ratio of the inertial force to the viscous force [see Eqs. (3.9) and (3.10)]:

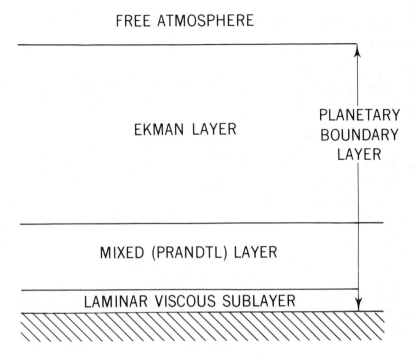

FIGURE 10.2. Schematic diagram of the structure of the planetary boundary layer.

$$R_e = \frac{uL}{\nu}, \qquad (10.14)$$

where u and L are characteristic velocity and length scales of the flow. This nondimensional number is a measure of the relative importance of inertial and viscous effects in the flow. In general, atmospheric boundary layers are characterized by very high Reynolds numbers on the order of $R_e \approx 10^6 - 10^9$.

In the case of turbulent flow, turbulence is a property of the flow and not of the fluid. In contrast to laminar flow, turbulent flows tend to be chaotic, highly irregular, almost random, dissipative, and very diffusive.

The planetary boundary layer may be idealized as a multilayered structure (see Fig. 10.2). Very close to the surface a laminar molecular sublayer develops, less than a few mm thick, where viscosity effects dominate and where there is no turbulent mixing. This layer links the earth's surface with a mixed layer some tens of m's thick, sometimes called the constant flux or Prandtl layer, where fully developed turbulence occurs. Away from the equator, above the mixed layer the Coriolis force starts to become important and both the effects of the earth's rotation and turbulence have to be included defining the Ekman layer (see Secs. 3.2.2 and 3.2.3). At the top of this layer, the so-called geostrophic level, the "free" atmosphere begins.

The atmospheric turbulence can be of a mechanical origin (forced convection) related to rapid changes in the mean wind from one point to another, or it can be of a thermal origin (free convection). Mechanical turbulence depends on the wind velocity and is especially strong over rough terrain. During the daytime, surface heating as well as strong wind shears near the ground will lead to the continuous formation of turbulent eddies. These eddies are largely responsible for the efficient mixing and exchanges of mass, heat, and momentum through the planetary boundary layer.

10.3.2 Generation and maintenance of atmospheric turbulence

It is not easy to define turbulence in precise terms. However, it may be helpful to mention some of the important properties of turbulent flow. Turbulent flow is characterized by very irregular and chaotic motions with a wide range of eddy scales so that randomness is an essential feature of such a flow. Since the velocity field is very variable in time and space, it also has very high values of the vorticity.

The large diffusivity of turbulent flows implies a high ability to mix properties efficiently, which is probably one of the most important characteristics of turbulent flow. With more organized convection, such as in cumulus clouds, turbulent exchange is responsible for the transfers of momentum, heat, water vapor, CO_2, and various pollutants through the atmospheric boundary layers. The high diffusivity of turbulence is also the cause of an increased frictional drag near the surface.

Turbulent flows are always dissipative. Since viscous shear stresses perform deformation work, kinetic energy is continuously dissipated into internal energy. Thus, in order to prevent the rapid decay of the turbulence, energy has to be supplied almost continuously. The supply of energy for the turbulent fluctuations will take place through a conversion from potential energy (buoyancy effects), through the direct transfer of energy from the mean flow, or through an indirect transfer from large-scale eddies by a cascade process in which the energy is transferred progressively to smaller scale eddies. The turbulent fluctuations will, in turn, pass their energy down to smaller scales of motion until molecular viscosity becomes dominant. Thus, we can write the kinetic energy balance for the turbulent motions in the symbolic form

$$\rho \frac{d}{dt} K_t = S + B - D + T, \tag{10.15}$$

where the left-hand side gives the rate of change of the turbulent kinetic energy (K_t). On the right-hand side of the equation

$$S = - \bar{\rho} \; \overline{w'\mathbf{v'}} \cdot \frac{\partial \bar{\mathbf{v}}}{\partial z} \tag{10.16}$$

represents the rate of production of K_t at the cost of the mean wind shear,

$$B = - \overline{w'\rho'} g \tag{10.17}$$

is the rate of production of K_t by buoyancy forces,

$$D = - \bar{\tau} \cdot \frac{\partial \bar{\mathbf{v}}}{\partial z} \tag{10.18}$$

is the rate of viscous dissipation of K_t, and T is the rate of transfer of kinetic energy from eddies with time and space scales that are not included in K_t (see Appendix A4.3).

10.3.3 Effects of stability

The stability of the environment strongly influences the vertical exchange of energy and momentum and thereby the vertical wind distribution in the boundary layer. In the presence of wind shear it is more appropriate to use a dynamic stability parameter, e.g., the Richardson number, rather than a static stability measure, such as $(1/\theta)(\partial \theta / \partial z)$ used before in Sec. 3.4.2.

The Richardson number R_i can be defined as the ratio of the destruction of turbulent kinetic energy by buoyancy forces ($-B$) and the production from the shear flow (S):

$$R_i = -B/S$$

$$= \overline{w'\rho'} g \left(-\bar{\rho}\, \overline{w'\mathbf{v}'} \cdot \frac{\partial \bar{\mathbf{v}}}{\partial z} \right)^{-1}$$

or in terms of potential temperature θ:

$$R_i \simeq \overline{w'\theta'} g \left(\bar{\theta}\, \overline{w'\mathbf{v}'} \cdot \frac{\partial \bar{\mathbf{v}}}{\partial z} \right)^{-1}. \tag{10.19}$$

Assuming a simple flux-gradient relationship (see the next section), we can write

$$\overline{w'\theta'} = -K_H \frac{\partial \bar{\theta}}{\partial z} \tag{10.20}$$

and

$$\overline{w'\mathbf{v}'} = -K_m \frac{\partial \bar{\mathbf{v}}}{\partial z}, \tag{10.21}$$

where K_H is the eddy diffusion coefficient of heat and K_M the eddy diffusion coefficient of momentum. Substituting Eqs. (10.20) and (10.21) into Eq. (10.19) and assuming that $K_H = K_M$, we find the more common expression for the Richardson number:

$$R_i = \frac{g}{\bar{\theta}} \frac{\partial \bar{\theta}/\partial z}{(\partial \bar{\mathbf{v}}/\partial z)^2}. \tag{10.22}$$

In ocean studies the alternate form

$$R_i = -g \left(\frac{1}{\bar{\rho}} \frac{\partial \bar{\rho}}{\partial z} + \frac{g}{c_s^2} \right) \left(\frac{\partial \bar{\mathbf{v}}}{\partial z} \right)^{-2} \tag{10.23}$$

is often used [see Eq. (3.76)].

The Richardson number has the same sign as the static stability, but it is a better measure of the intensity of turbulence. It also provides a criterion for the existence or nonexistence of turbulence in the case of stably stratified flow ($\partial \bar{\theta}/\partial z > 0$). A simple requirement for maintaining turbulence under statically stable conditions is that the rate of production by the wind shear (S) be equal to or greater than the rate of destruction by buoyancy forces ($-B$), i.e.,

$$R_i \leqslant 1.$$

However, in deriving this condition the viscous dissipation of turbulent energy was not taken into account. If we include viscous effects the critical value of R_i, which marks the transition from a laminar to a turbulent regime, is smaller than 1. According to observations, the critical Richardson number R_{ic} equals about 0.25. When $R_i < 0$ the flow is clearly turbulent; whereas for large positive values of R_i the turbulence tends to be weak and decaying.

10.4 EXCHANGE OF MOMENTUM

10.4.1 Eddy correlation approach

As we have seen before, the vertical flux of momentum is given by $\bar{\tau} = \rho\,\overline{v'w'}$. In principle, this expression can be used with three-dimensional wind measurements to evaluate the vertical flux in the turbulent boundary layer. This eddy correlation method has the advantage of measuring the turbulent exchange directly without restrictive assumptions as to the nature of the surface and the transfer mechanisms involved. However, the determination of the flux components requires fast-response instrumentation and a high sensitivity for measuring the wind components. The closer to the surface the more stringent the requirements for the instrumentation have to be. Above about 10 m, vanes and propellers can be used because the characteristic size of the eddies increases with height. These difficulties have prevented the wide use of the eddy correlation method in spite of its directness.

10.4.2 Gradient-flux approach

In the laminar molecular sublayer, it is acceptable to use a gradient formulation to express the fluxes by a Fickian type law (e.g., Newton, Fourier, Darcy laws) as mentioned in Sec. 10.3.1. On the molecular scale the gradient formulation has a sound physical basis in the kinetic theory of gases.

In the mixed boundary layer, the turbulent shear stress in the direction of the flow may thus be expressed as

$$\tau = -K_M \rho \frac{\partial \bar{\mathbf{v}}}{\partial z}, \tag{10.24}$$

where K_M is the eddy diffusion coefficient of momentum or the eddy viscosity. The quantity K_M is analogous to the kinematic molecular viscosity ν. The two components of Eq. (10.24) are then

$$\tau_{zx} = -K_M \rho \frac{\partial \bar{u}}{\partial z} \tag{10.25a}$$

and

$$\tau_{zy} = -K_M \rho \frac{\partial \bar{v}}{\partial z}. \tag{10.25b}$$

The value of the eddy diffusion coefficient K_M is many orders of magnitude greater than ν, showing the predominance of the turbulent eddy exchanges over the molecular ones.

Equation (10.24) expresses what one would expect intuitively, i.e., that momentum flows from regions of higher to lower momentum down the mean gradient of momentum. However, it is important to note that this gradient-flux relationship is based only on an analogy between the behavior of turbulent eddies and the behavior of molecules, and not on any sound physical laws. Furthermore, the eddy diffusivities are not fluid properties but depend on the nature of the turbulent flow which varies from one flow to another and from one place to another.

10.4.3 Mixing-length approach and wind profiles

In a further extension of the molecular analogy, Prandtl devised a physical mechanism for the turbulent exchange of momentum. This approach leads to the mixing length model of turbulence. In this model, it is assumed that a fluid element at the level z moves through an average distance l taking its velocity with it so that at height $z + l$ it is reabsorbed losing all traces of its original motions. Therefore, the mixing length l plays somewhat the same role as the mean free path in kinetic theory. Thus,

$$u' = \bar{u}(z) - \bar{u}(z + l) = -l \frac{\partial \bar{u}}{\partial z}.$$

Assuming that the turbulence is isotropic, i.e., the magnitude of the velocity fluctuations is the same in all directions,

$$|w'| \simeq |u'| = l \left| \frac{\partial \bar{u}}{\partial z} \right|$$

so that

$$\overline{u'w'} = -l^2 \left| \frac{\partial \bar{u}}{\partial z} \right| \frac{\partial \bar{u}}{\partial z}$$

and

$$\tau_{zx} = -\rho l^2 \left| \frac{\partial \bar{u}}{\partial z} \right| \frac{\partial \bar{u}}{\partial z}. \tag{10.26}$$

Thus, the relationship with the eddy diffusion coefficient comparing Eq. (10.26) with Eq. (10.25a) is

$$K_M = l^2 \left| \frac{\partial \bar{u}}{\partial z} \right|. \tag{10.27}$$

We will now consider the idealized case of a uniform surface under a boundary layer which is in neutral equilibrium so that the momentum flux can be assumed to be constant with height ($\tau_{zx} = \tau_0$). Equation (10.26) suggests then that

$$\left| \frac{\partial \bar{u}}{\partial z} \right| = \frac{1}{l} \sqrt{\frac{|\tau_0|}{\rho}}$$

or

$$\left| \frac{\partial \bar{u}}{\partial z} \right| = \frac{u_*}{l},$$

where

$$u_* = \sqrt{|\tau_0|/\rho} \tag{10.28}$$

is the so-called friction velocity with a typical value of ~ 0.2 to 0.4 m s^{-1}. Let us further assume that the scale of the mixing elements is proportional to the available space ($\approx$ the distance to the boundary) so that l changes proportionally with height, $l = kz$, where $k \cong 0.4$ is the von Kármán constant. Integration of the previous equation with respect to z then leads to a logarithmic profile for the variation of the wind with height

$$|u(z)| = \frac{1}{k} u_* \ln \frac{z}{z_0}, \tag{10.29}$$

where the integration constant z_0 is the height above the ground where $u = 0$. The constant z_0 is called the roughness length. The value of the roughness length depends on the nature of the terrain, and must be determined empirically. It is sometimes taken as 15% of the average height of the roughness elements. Estimates of z_0 range from much less than 1 cm for relatively smooth land and ocean surfaces to meters for forests and urban areas (Arya, 1988).

10.4.4 Bulk aerodynamic method

The surface stress can also be obtained in terms of a drag coefficient C_D through the expression

$$\tau_0 = -\rho C_D |\mathsf{v}(z)| \mathsf{v}(z). \tag{10.30}$$

In particular, for neutral stability conditions, the nondimensional drag coefficient C_D at a reference level z_r (usually 10-m height) is given by

$$C_D = k^2/(\ln^2 z_r/z_0) \tag{10.31}$$

as results from Eq. (10.29) noting that $u_* = \sqrt{|\tau_0|/\rho}$. In the more general case, C_D is not only a function of z_r/z_0 but also of the stability. The values for C_D range from 1.2×10^{-3} for large lake and ocean surfaces at moderate wind speeds to 7.5×10^{-3} for tall crops and fairly rough surfaces (Arya, 1988).

Equation (10.30) allows the evaluation of the drag of the underlying surface on the atmosphere using observed surface winds. The typical oceanic value of $C_D = 0.0013$ is only valid under neutral conditions, and should also be modified under strong wind conditions (e.g., Large and Pond, 1981).

Using the historical ship data covering the world oceans, maps of the wind stress on the oceans have been computed using Eq. (10.30) with a constant $C_D = 0.0013$. The results are shown in Figs. 10.3(a) and 10.3(b) for the December–February and June–August seasons. The patterns of the wind stress are very similar to those of the surface winds (Fig. 7.1), but are more pronounced over the North Atlantic and Northwest Pacific Oceans and in the westerly wind belt in the Southern Hemisphere high and midlatitudes. The stresses tend to be strongest in the winter hemisphere, particularly at midlatitudes. However, the largest seasonal changes are found in the Indian Ocean off the Somali coast. In fact, the largest stress values anywhere on the globe are observed in this area during the summer monsoon. The winter strengthening is clearly shown in the zonal averages given in Fig. 10.4.

As expected, the zonal-mean stress profiles $[-\bar{\tau}_{ox}]$ and $[-\bar{\tau}_{oy}]$ have a shape that is very similar to the shape of the zonal-mean wind profiles $[\bar{u}]$ and $[\bar{v}]$ given earlier in Figs. 8.19(a) and 8.19(b). However, the magnitude of the stress values is larger than one would compute on the basis of Eq. (10.30) using estimates of a typical mean wind speed $|\bar{\mathsf{v}}|$ and the $[\bar{u}]$ or $[\bar{v}]$ values in Fig. 8.19 because of eddy contributions, such as $[\overline{|\mathsf{v}|'u'} + |\bar{\mathsf{v}}|^*\bar{u}^*]$ in $[\bar{\tau}_{ox}]$ and $[\overline{|\mathsf{v}|'v'} + |\bar{\mathsf{v}}|^*\bar{v}^*]$ in $[\bar{\tau}_{oy}]$. The maximum eastward stress on the oceans in midlatitudes is for annual-mean conditions on the order of 0.1 Pa or 1 dyn cm^{-2} and the maximum westward stress in the tradewind region is about 0.06 Pa. The zonal-mean meridional stress values are smaller but not negligible especially in the tropics. Since $[-\bar{\tau}_{ox}]/f$ represents the southward Ekman mass transport [see Eq. (3.24) and Fig. (8.20)], its derivative $\partial([-\bar{\tau}_{ox}]\cos\phi/f)/R\cos\phi\partial\phi$ is a measure of the upwelling or downwelling in the upper ocean. Thus, from the annual-mean curve in Fig. 10.4(a), we may expect strong upwelling in the equatorial region equatorward of about 20° latitude, sinking between about 20° and 45° latitude, and again upwelling between about 50° and 70° latitude. Over the intertropical region the direct meridional stress values in Fig. 10.4(b) tend to reinforce the

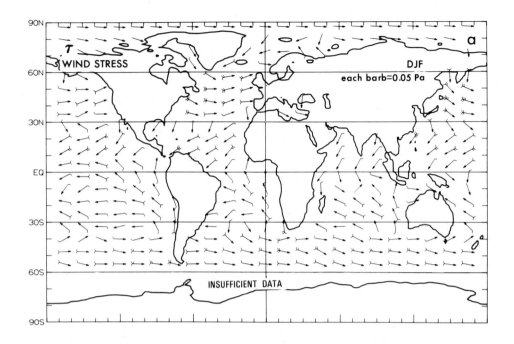
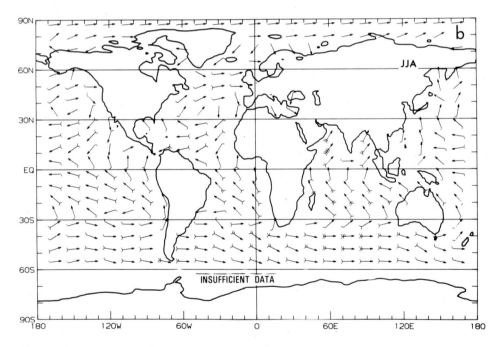

FIGURE 10.3. Global distribution of the surface wind stress ($-\tau_0$) on the oceans for DJF (a) and JJA (b) mean conditions. Each barb on the tail of an arrow represents a wind stress of 0.05 Pa (1 Pa = 10 dyn cm^{-2}).

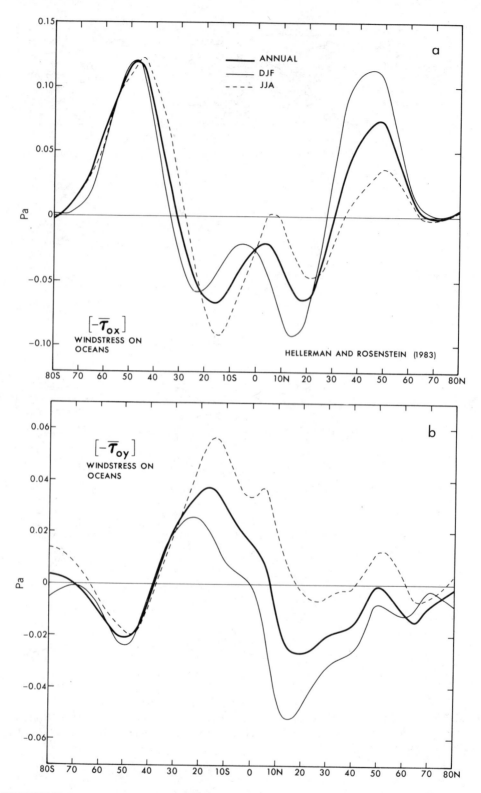

FIGURE 10.4. Meridional profiles of the zonal-mean value of the zonal (a) and meridional (b) wind stress components over the oceans in units of Pa (1 Pa = 10 dyn cm^{-2}) for annual, DJF, and JJA mean conditions (based on data from Hellerman and Rosenstein, 1983).

upwelling near the surface. It is important to note that the wind stress is the main factor responsible for driving the ocean currents through the exchange of momentum at the interface.

10.5 TRANSFER OF MECHANICAL ENERGY INTO THE OCEANS

The action of the wind on the ocean can be demonstrated when we superpose the maps of the wind stress (Fig. 10.3) on the corresponding maps of the ocean currents (Fig. 8.17). The energy transferred by the winds into the oceans can then be evaluated from the product $-\tau_0 \cdot \mathbf{v}$. It is important to note that we use here the velocity of the ocean currents and not the wind velocity since our present interest is to evaluate the energy received by the oceans. However, this expression for the transfer of energy into the oceans is formally equivalent with a term in the general energy equations to be discussed in Chap. 13, that involves the product of the wind stress and the wind velocity. This last product includes the frictional dissipation of kinetic energy in the surface boundary layer. Quantitatively, the two products are very different because the transfer of mechanical energy into the ocean is about two orders of magnitude smaller than the frictional dissipation in the atmosphere–ocean boundary layer.

A horizontal map of the energy transfer is shown in Fig. 10.5 for annual-mean conditions. The transfer is found to be most pronounced, with values up to 0.02 to 0.04 W m^{-2}, in the regions of the persistent trade winds, as well as in the 40°–60° belt of strong westerly winds in the Southern Hemisphere. The wind generation of the currents is also evident in the regions of the Kuroshio and Gulf Stream systems. However, the largest values of the energy input are found in the Tropics along the

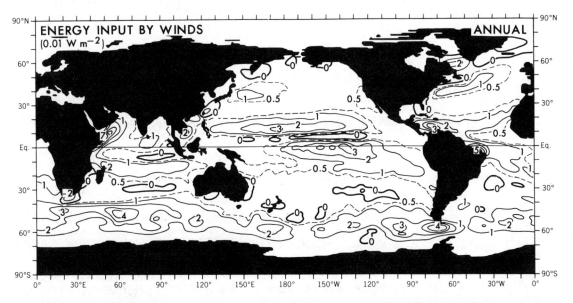

FIGURE 10.5. Global distribution of the kinetic energy input into the ocean by the winds in units of 10^{-2} W m^{-2} for annual-mean conditions.

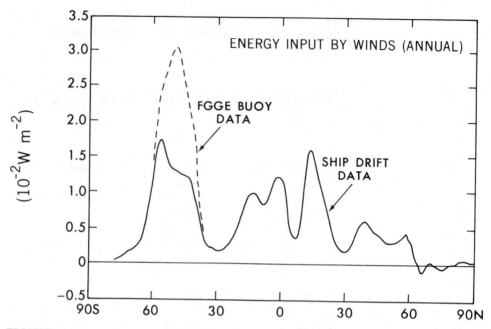

FIGURE 10.6. Meridional profile of the zonal-mean input of kinetic energy into the ocean by the winds in units of 10^{-2} W m^{-2} for annual-mean conditions. Estimates using buoy data rather than ship drift data south of 42 °S are also added as a dashed line.

north coast of South America and, especially, in the Indian Ocean near the Somali coast connected with the summer monsoon.

The zonal-mean profile of the energy input for the year in Fig. 10.6 synthesizes the earlier findings showing clearly the role of the trade winds and the midlatitude westerly wind belt in the Southern Hemisphere. The globally averaged input of mechanical energy into the world ocean is about 0.007 W m^{-2} for annual-mean conditions.

10.6 EXCHANGE OF SENSIBLE HEAT

Basically the methods for computing the heat transfer are not different from those used in the evaluation of the momentum transfer. Thus, we can use the eddy correlation approach based on the expression

$$F_{\text{SH}}(z) = \rho c_p \, \overline{w'\theta'} \qquad (10.32\text{a})$$

or

$$F_{\text{SH}}(z) = \rho c_p \, \overline{w'T'}, \qquad (10.32\text{b})$$

since near the surface $T' \simeq \theta'$. Thus, in principle, the vertical fluxes can be evaluated from direct measurements of the vertical velocity and temperature. However, the difficulties in measuring w' and T' are similar to those for momentum and have prevented a wide use of this method.

Using the same arguments as we did for momentum we can write for the turbulent flux of heat on the basis of the Prandtl assumption

$$F_{SH}(z) = \rho c_p \overline{w'\theta'} = -K_H \rho c_p \frac{\partial \overline{\theta}}{\partial z}, \qquad (10.33)$$

where K_H is the eddy diffusion coefficient for heat. This equation implies that the vertical flux of heat is determined by the mean vertical potential temperature gradient. It also corresponds to the intuitive idea that heat flows from warm to cold levels proportionally to the gradient of the mean potential temperature. Equation (10.33) gives the foundation for the gradient method of calculating the turbulent vertical heat fluxes.

In analogy with Eq. (10.30) for momentum in neutral stability conditions, we can express the sensible heat flux in terms of a dimensionless heat transfer coefficient C_H as

$$F_{SH} = -\rho c_p C_H |\mathbf{v}(z)| \{\theta(z) - \theta(0)\}. \qquad (10.34)$$

In practice over the oceans, C_H and C_D are often taken as constants and of the same order of magnitude $C_H \approx C_D \approx 0.0013$. Over land the values of C_D and C_H vary over a wide range but constant values are also usually assigned for specific applications. In general, C_D and C_H should be taken as complex functions of the roughness length, stability, and wind speed (see Large and Pond, 1982). Equation (10.34) is the foundation for the bulk aerodynamic method. It is very useful for computing the heat fluxes over the oceans.

The air–sea temperature difference $T_A - T_s \approx \theta(10 \text{ m}) - \theta(0)$ in Eq. (10.34) is one of the important factors in determining the flux of sensible heat. Figures 10.7(a) and 10.7(b) show the observed temperature difference $T_A - T_s$ for January and July. Near the equator the air temperature tends to be slightly lower than the ocean surface temperature. This difference increases with latitude in the winter hemisphere. The largest air–water temperature differences up to $-8\,°C$ are observed in northern mid and high latitudes just east of the continents. In summer, the northern mid and high latitudes show positive temperature differences up to $2\,°C$.

The turbulent flux of sensible heat between the earth's surface and the atmosphere has been determined notably by Budyko (1963, 1986). His map for annual-mean conditions is shown in Fig. 10.8. In general, the sensible heat flux is directed from the surface into the atmosphere. However, there are exceptions showing a downward (negative) surface heat flux, such as over Greenland, much of the Arctic and Antarctic, some ocean areas south of 40°S, and the cold currents of the equatorial Pacific Ocean and to the west of California. Except for the western boundary current regions of the Gulf Stream and Kuroshio, the sensible heat flux values over land are generally much larger than those over the oceans. Over the continents the heat flux shows the highest upward values in the arid regions with maxima of 60–70 W m^{-2} over the deserts. Over the oceans, the turbulent heat flux attains maximum values of 50–60 W m^{-2} over the warm ocean currents. Over the equatorial regions where the temperature differences are small the turbulent heat flux is also small.

10.7 EXCHANGE OF WATER VAPOR, EVAPORATION

Water is the most important substance that is continually transferred across the earth's surface in both directions. As we have seen in Sec. 7.6, the oceans provide the

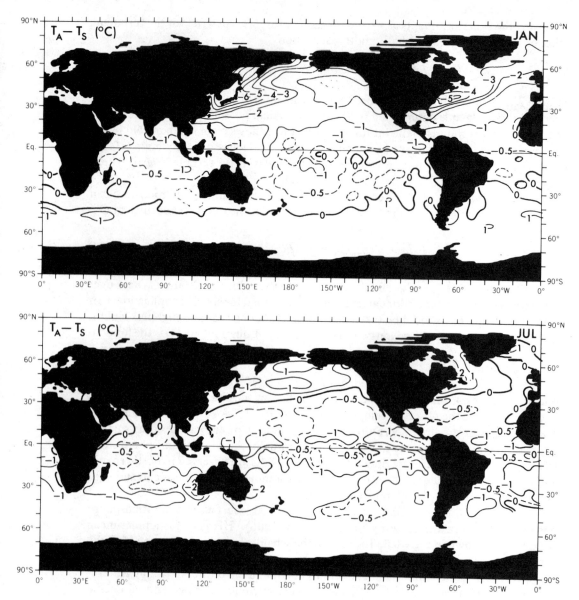

FIGURE 10.7. Global distribution of the air–sea temperature difference in °C for January and July.

atmosphere with a large amount of water in the vapor phase that, after condensation, returns to the earth's surface as precipitation over the oceans and continents. Evaporation is important not only because it is one of the fundamental elements of the hydrological cycle but also because of the large amounts of energy that are involved in the phase transition.

Evaporation occurs continuously from the earth's water surfaces, the soil and the snow and ice fields, the last process is usually called sublimation. The liquid water contained in the soil can be extracted by plants and released as water vapor in the atmosphere through transpiration. The combined process of evaporation from the soil and transpiration from the vegetation is called evapotranspiration (see Fig. 10.9).

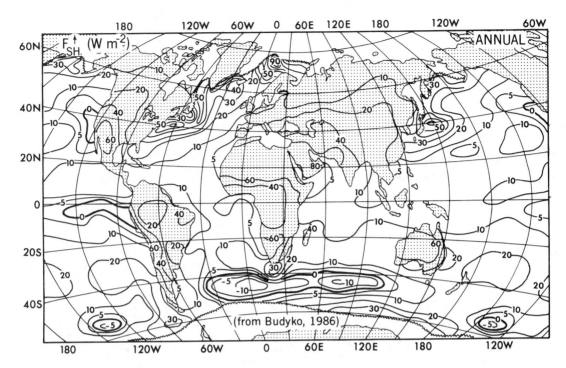

FIGURE 10.8. Global distribution of the sensible heat flux from the earth's surface into the atmosphere in W m^{-2} for annual-mean conditions after Budyko (1986).

When the soil is fully saturated with water, the water content is not a limiting factor for the evaporation. Then the soil is at field capacity. The maximum amount of water that can be lost by evaporation from a saturated surface is called the potential evaporation. Usually, the evaporation from the soil is lower than the potential evaporation because the soil surface is not always completely wet and the plants are not always transpiring at the maximum rate. However, the potential evaporation is a useful concept for agricultural and hydrological purposes.

As mentioned before in Sec. 7.6.2, evaporation is often measured with a shallow circular pan but the measured values are strongly influenced by local conditions. The water that evaporates from the pan tends to be carried away by the wind and replaced by fresh moist air evaporated from the pan. This will lead to an overestimate of the local evaporation. In reality, the surface air over the oceans will be in close contact with the water and will attain some equilibrium state with the sea surface conditions so that the rate of evaporation will be limited. The many uncertainties in this method and in other methods as well explain the wide range of evaporation estimates that are found in the literature (Sellers, 1965). They also suggest that evaporation may be better computed by indirect means.

In the next subsection, we will give some micrometeorological methods for determining the evaporation.

10.7.1 Eddy correlation, gradient-flux, and bulk transfer methods

According to Fick's law the net vertical flux of water vapor across the earth's surface is proportional to the specific humidity (concentration) gradient, in a similar

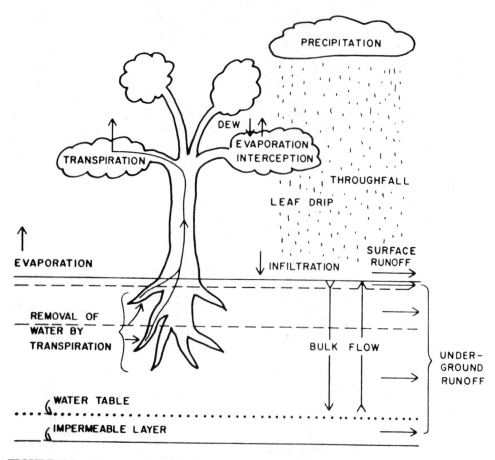

FIGURE 10.9. Schematic diagram of the various hydrological processes that appear important in understanding the surface hydrology over a land surface covered by vegetation (from Abramopoulos et al., 1988).

manner as for momentum and heat. The proportionality coefficient for the vapor flux α_w is the molecular diffusivity for water vapor:

$$E = -\rho \alpha_w \frac{\partial \overline{q}}{\partial z}. \qquad (10.35)$$

This expression is only valid in the laminar viscous sublayer, where molecular exchange is the prevalent transfer mechanism. This mechanism operates at the air–water, air–soil, and air–vegetation interfaces.

The lower region of the surface boundary layer is almost always turbulent so that in this layer the water is transferred away from the molecular sublayer by turbulence. The vertical flux is then given by

$$E = \rho \,\overline{q'w'}. \qquad (10.36)$$

The transport in this mixed layer depends on surface roughness, wind shear, and thermal stratification. Therefore, the rate of transfer of water vapor (evaporation) depends on all these factors as well as on the gradient of specific humidity. We can then accept a gradient-flux relation

$$E = -\rho K_w \frac{\partial \overline{q}}{\partial z}, \qquad (10.37)$$

where by analogy the eddy diffusivity for water vapor K_W replaces the molecular diffusivity α_W.

In a similar way as in the bulk transfer approach for determining the heat flux (10.34), we can also write an equivalent expression for the evaporation as

$$E = -\rho C_W |\mathbf{v}(z)| \{q(z) - q(0)\}, \tag{10.38}$$

where C_W is the bulk transfer coefficient for water vapor. Over land, Eq. (10.38) gives only a measure of the potential not necessarily the actual evaporation. The value of C_W is not well established particularly at high wind speeds, such as for $|\mathbf{v}| > 15$ m s^{-1}, for which there are as yet no direct estimates. Sometimes for convenience $C_W = C_D$ and $C_H = C_D$ are used although they have a somewhat different behavior depending on the stability and wind shear or on the two factors combined in terms of the Richardson number (10.22). The main difficulty in applying this last method is that the specific humidity at the surface $q(0)$ is not easy to measure or estimate over land.

10.7.2 Energy balance method

The energy equation at the surface (6.40) can be written as

$$F_{\text{rad}}^{\text{sfc}} = F_{\text{SH}}^{\uparrow} + L_e E + F_G^{\downarrow} + L_m(M_S - F_S).$$

Solving for E we find

$$E = \{F_{\text{rad}}^{\text{sfc}} - F_G^{\downarrow} - L_m(M_S - F_S)\}/(1 + B)L_e, \tag{10.39}$$

where B is the Bowen ratio [Eq. (10.1)]. Knowing the Bowen ratio and the other fluxes, Eq. (10.39) can be used to compute the evapotranspiration.

The Bowen ratio can be computed from the gradient-flux relationships for heat (10.33) and water vapor (10.37), assuming that $K_H = K_W$:

$$B = \frac{c_p}{L_e} \frac{\partial \theta/\partial z}{\partial q/\partial z} \tag{10.40}$$

or

$$B = \frac{c_p}{L_e} \partial \theta/\partial q \simeq \frac{c_p}{L_e} \frac{\Delta T}{\Delta q}. \tag{10.41}$$

Thus, to evaluate B one needs to measure the temperature difference $\Delta T = T_2 - T_1$ and the specific humidity difference $\Delta q = q_2 - q_1$ between two levels in the surface mixed layer.

Noting that $q = 0.622\, e/(p - e) \approx 0.622(e/p)$, we can express B in terms of the vapor pressure e instead of q, so that

$$B = \gamma \frac{\partial \theta}{\partial e} \simeq \gamma \frac{T(z) - T(0)}{e(z) - e(0)}, \tag{10.42}$$

where

$$\gamma = \frac{c_p}{L_e} \frac{p}{0.622} \tag{10.43}$$

is the Bowen constant (1926) which is the well-known psychrometer constant with a mean value of $\gamma \approx 0.61$ mb °C^{-1} for $p = 1000$ mb.

10.7.3 Combined approaches, Penman formula

The combination of the energy balance method with the bulk aerodynamic method or with the flux-gradient method may lead to other useful relationships to evaluate the evaporation. Equation (10.39) constitutes the foundation of the combined approaches.

Inserting Eq. (10.42) in Eq. (10.39) we obtain the expression

$$E = \{F_{\text{rad}}^{\text{sfc}} - F_{\text{G}}^{\downarrow} - L_{\text{m}}(M_{\text{S}} - F_{\text{S}})\}\left\{L_{\text{e}}\left(1 + \gamma\frac{T(z) - T(0)}{e(z) - e(0)}\right)\right\}^{-1}. \quad (10.44)$$

Equation (10.44) was used by Budyko (1958) to compute global distributions of the evaporation rate E from data for the net radiational flux $F_{\text{rad}}^{\text{sfc}}$ and for the temperature and vapor pressure both at the earth's surface and in the atmosphere just above it.

Let us now consider Penman's approach. Penman (1948) derived an expression to estimate the evaporation from open water or water-saturated soil surfaces. His method has been used widely because it requires mainly routine observations. Penman's equation is essentially based on a combination of the energy balance and bulk aerodynamic methods.

Let us consider Eq. (10.38). If we add and subtract the same quantity $q_s(z)$, the saturated specific humidity at level z, to the expression in parentheses in Eq. (10.38), we obtain

$$E = -\rho C_{\text{W}}|\mathbf{v}(z)|\{q(z) - q_s(z) + q_s(z) - q(0)\}$$

or

$$E = \rho C_{\text{W}}|\mathbf{v}(z)|\{q(0) - q_s(z)\} + E_{\text{a}}, \quad (10.45)$$

where

$$E_{\text{a}} = \rho C_{\text{W}}|\mathbf{v}(z)|\{q_s(z) - q(z)\}. \quad (10.46)$$

The last expression is proportional to the saturation deficit of the air $q_s - q$ as measured at level z and to the wind velocity. It can be regarded as a version of Dalton's evaporation law. This explains why E_{a} is called the drying or evaporation power of the air or is sometimes designated as the aerodynamic term.

When the measurement height is on the order of a few m's the bulk heat transfer relation (10.34) is a good measure of the heat flux at the surface. Assuming that $C_{\text{H}} = C_{\text{W}}$ we can use Eqs. (10.45) and (10.34) to obtain

$$\frac{L_{\text{e}}E}{F_{\text{SH}}} = \frac{L_{\text{e}}}{c_p}\left\{\frac{q(0) - q_s(z)}{T(0) - T(z)}\right\} + \frac{L_{\text{e}}E_{\text{a}}}{F_{\text{SH}}}$$

or, in terms of water vapor pressure and the Bowen ratio,

$$\frac{1}{B} = \frac{0.622 L_{\text{e}}}{\rho c_p}\left\{\frac{e(0) - e_s(z)}{T(0) - T(z)}\right\} + \frac{1}{B}\frac{E_{\text{a}}}{E}.$$

Solving for B, we find

$$B = \frac{\gamma}{\Delta}\left\{\frac{E - E_{\text{a}}}{E}\right\}, \quad (10.47)$$

where γ is the psychrometer constant (10.43) and Δ a parameter given by

$$\Delta = \frac{e(0) - e_s(z)}{T(0) - T(z)}. \quad (10.48)$$

Over the oceans, a wet surface, or a water-saturated soil the atmosphere in contact with the surface must also be saturated, i.e., $e(0) = e_s(0)$ at the surface temperature $T(0)$. Under these conditions,

$$\Delta \simeq \frac{de_s}{dT}, \tag{10.49}$$

so that Δ is approximately equal to the slope of the saturation water vapor curve vs temperature [Clausius–Clapeyron equation (3.65b)] at temperature T.

Combining expression (10.47) for the Bowen ratio with the general energy balance equation (10.39) leads to the well-known Penman formula. Thus, rewriting Eq. (10.39) in the form

$$E(1+B) = \frac{A}{L_e},$$

where

$$A = F_{\text{rad}}^{\text{sfc}} - F_G^{\downarrow} - L_m(M_S - F_S) \tag{10.50}$$

and using Eq. (10.47) we obtain

$$E + \frac{\gamma}{\Delta}(E - E_a) = \frac{A}{L_e},$$

and finally

$$E = \frac{\Delta}{\Delta + \gamma} \frac{A}{L_e} + \frac{\gamma}{\Delta + \gamma} E_a. \tag{10.51}$$

The last expression (10.51) shows that the available radiation energy (i.e., the portion not used to heat the subsurface layers or to melt snow and ice) contributes to the evaporation with the weight $\Delta/(\Delta + \gamma)$ and that the evaporation power contributes with the weight $\gamma/(\Delta + \gamma)$. The relative importance of the two factors is given by the dimensionless ratio Δ/γ, which increases rapidly with air temperature, being 1.3 at 10 °C, 2.3 at 20 °C, and 3.1 at 26 °C. The aerodynamic term E_a determines the evaporation at low-temperature conditions found in winter while the first term determines the evaporation at the high-temperature conditions found in summer or in the tropics.

The maximum amount of water vapor that, under given meteorological conditions, can be discharged from a large, fully wet surface or from a surface covered by green vegetation is the potential evapotranspiration or the maximum possible evaporation. It is given by the same expressions as Eqs. (10.38) and (10.46) for an ocean surface.

As mentioned at the beginning of this chapter, real evaporation E over soil is usually lower than the potential evapotranspiration E_0, so that

$$E = f E_0,$$

where f is a parameter that depends on the nature of the surface and on the type of crop covering the soil. The annual values in midlatitudes range from 0.80 to 0.93 but may be much lower for dry soils. As Eq. (10.51) shows the evapotranspiration from a vegetation-covered land surface depends primarily on the radiant energy supply but can also be limited by the evaporation power of the air. Some of the complexity of the role played by vegetation in the hydrological cycle can be appreciated by examining Fig. 10.9. The figure shows the influence of vegetation on the surface water budget through infiltration, runoff, percolation, evapotranspiration, and the interception of precipitation as we will discuss further in Chap. 12.

10.8 FORMATION OF ATMOSPHERIC AEROSOL

Another important aspect of the exchange of mass between the earth's surface and the atmosphere is the formation of liquid and solid particles suspended in the air, forming the so-called atmospheric aerosol (see also Sec. 6.3.3). Besides the Aitken nuclei (with a radius of about 10^{-3} to 10^{-1} μm; Houghton, 1985) that act as condensation nuclei for cloud drops, we must consider a much broader spectrum of atmospheric particles in gaseous, liquid, and solid form, ranging from 10^{-4} μm to 10's of μm. The Aitken nuclei originate mainly from combustion processes but they are also present in appreciable concentrations in unpolluted continental and maritime air, indicating the existence of other sources besides combustion. Over the oceans, giant aerosol particles are abundant due to the breaking of waves and the bursting of water bubbles. The giant particles consist mainly of salt particles (NaCl) that come about through evaporation of the water droplets ejected by ocean surface waves into the air. These particles have diameters of around 0.3 μm.

As mentioned before, we must consider not only the production of aerosols from natural phenomena but also from human activities which contribute mainly to urban pollution. On the average, aerosols are removed from the atmosphere at the same rate as they enter. The Aitken nuclei are removed mainly by precipitation but there are also other processes such as gravitational removal (dry fall out) that operates when the particles exceed a critical dimension. Over land, the wind can remove dust, clay, sand, and quartz particles from the surface. In addition, we must consider volcanic eruptions that may release large amounts of particles, as well as SO_2, CO_2, and other gases into the atmosphere. These gases may chemically react to form particles that can also have some influence on the climate.

CHAPTER 11

Angular Momentum Cycle

Angular momentum is a basic physical quantity for any rotating system. It is a conservative property for a closed system. Since we may regard the planet earth as an almost closed system for mass and external forces, the total angular momentum of the earth including the atmosphere, oceans, lithosphere, and cryosphere remains constant in time apart from small effects due to tidal friction. Any changes in the angular momentum of one component of the climate system must be balanced by a corresponding change in the angular momentum of the other components of the system. In this chapter we will discuss how these changes and exchanges occur, such as the exchange of angular momentum between the atmosphere and the underlying earth through the surface torques. If the net atmospheric torque acts in the direction of rotation the solid earth will increase its angular momentum at the cost of that of the atmosphere and will therefore rotate faster, whereas a net torque in the opposite direction would tend to slow the solid earth down and increase the angular momentum of the atmosphere.

11.1 BALANCE EQUATIONS FOR ANGULAR MOMENTUM

11.1.1 Introduction

Because of the rotation of the earth around the polar axis, it is most appropriate to use a spherical coordinate system, and to study angular momentum about the earth's axis rather than linear momentum. We must then deal with body torques instead of body forces. Furthermore, we have to consider two components of angular momentum on the rotating earth, the first one associated with the earth's rotation and the second one with the movement of the air relative to the rotating earth.

To illustrate the basic angular momentum balance in the atmosphere we will often consider volumes enclosed vertically by surfaces of constant height or of constant pressure and horizontally by conceptual, vertical walls along latitude circles. A polar cap represents such a volume.

Application of the conservation of angular momentum to the climate system will lead to some general conclusions regarding the transports of angular momentum within the atmosphere and the mechanisms involved.

The angular momentum **M** of a parcel of unit mass about a point 0, which moves with a velocity c_A, is defined as $\mathbf{M} = \mathbf{r} \times \mathbf{c}_A$, where **r** is the radius vector from the point 0 to the parcel. Moreover,

$$\frac{d\mathbf{M}}{dt} = \mathbf{r} \times \mathbf{F}, \qquad (11.1)$$

where **F** is the sum of all external forces applied to the parcel. The right-hand side defines the moment of force or the torque with respect to the point 0. If the total torque vanishes,

$$\frac{d\mathbf{M}}{dt} = 0$$

and the angular momentum **M** is conserved in time. The resulting forces applied to the total climatic system obey this condition because the external torques, which are mainly due to the solar wind and electromagnetic forces on the upper atmosphere, are very small and can be disregarded in this context. Therefore, the total angular momentum of planet earth does not vary in time, except for a slow secular decrease in angular momentum associated with the gravitational torque exerted by the moon and other planets.

Since the earth moves around its axis with a mean angular velocity Ω, the component of angular momentum about this axis is given by $\mathbf{M} \cdot \mathbf{n}$, where **n** is the unit vector in the direction of $\Omega (= \Omega \mathbf{n})$. For a unit parcel in the atmosphere the absolute angular momentum (about the earth's axis) is then $M = \mathbf{M} \cdot \mathbf{n} = (\mathbf{r} \times \mathbf{c}_A) \cdot \mathbf{n}$.

For the rotating atmosphere we find

$$M = [\mathbf{r} \times (\Omega \times \mathbf{r} + \mathbf{c})] \cdot \mathbf{n}.$$

After some elementary vector calculus, we obtain

$$M = \Omega r^2 \cos^2 \phi + ur \cos \phi, \qquad (11.2)$$

where $u = \mathbf{c} \cdot \mathbf{i}$ is the zonal component of the wind, as defined before in the equations of motion.

Since the thickness of the atmosphere is very small and the variations in the earth's radius are also relatively small, we can substitute $r \approx R$ in Eq. (11.2), where R is the mean radius of the earth. Then we find that the total or absolute angular momentum (M) about the axis of rotation for a unit of atmospheric mass is composed of two terms, the Ω angular momentum (M_Ω) and the relative angular momentum (M_r). The first term represents the angular momentum of the atmosphere if it were in solid rotation with the earth, and the second term the angular momentum relative to the rotating earth:

$$M = M_\Omega + M_r,$$

where

$$M_\Omega = \Omega R^2 \cos^2 \phi, \qquad (11.3a)$$

$$M_r = uR \cos \phi. \qquad (11.3b)$$

This breakdown is illustrated in Fig. 11.1.

Let us now assess the influence of the rotation on the circulation using the following example. Without any other forces acting on the particle, a particle that moves poleward will acquire a zonal component from the west through the action of the Coriolis force [see Eq. (3.8)] in order to compensate for the decrease in distance to the earth's axis, whereas a particle moving equatorward will acquire an easterly component to compensate for the increase in distance. For example, for a particle initially at

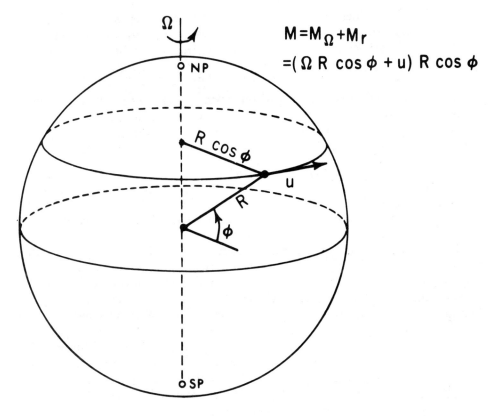

FIGURE 11.1. Schematic diagram of the angular momentum component around the earth's axis of rotation. NP = North Pole; SP = South Pole.

rest over the equator relative to the earth ($u = 0$, $v = 0$) its zonal momentum per unit mass is $\Omega R = 464$ m s^{-1}, and its angular momentum ΩR^2. If the particle moves poleward and reaches a latitude ϕ it will have acquired a zonal velocity u such that

$$\Omega R^2 = \Omega (R \cos \phi)^2 + uR \cos \phi$$

or

$$u = \Omega R(1 - \cos^2 \phi)/\cos \phi.$$

11.1.2 Angular momentum in the climatic system

The angular momentum of the atmosphere and oceans is very small compared to that of the solid earth. In fact, assuming a perfect sphere and a uniform density for the solid earth, its total angular momentum is given by

$$I_e \Omega = \tfrac{2}{5} m_e R^2 \Omega,$$

where $m_e \simeq 5.98 \times 10^{24}$ kg and $\Omega \simeq 7.29 \times 10^{-5}$ rad s^{-1}. Using a more realistic model of the earth (e.g., Smith and Dahlen, 1981) one finds for the moment of inertia of the earth $I_e \approx 8.04 \times 10^{37}$ kg m^2 and for its angular momentum a value of 5.86×10^{33}

kg m² s⁻¹. On the other hand, the total angular momentum of the atmosphere considered as a rigid, rotating body is given by $I_a \Omega$, where I_a represents the moment of inertia of the atmosphere. If the atmosphere is regarded as a spherical shell, we find that

$$I_a \Omega \approx \tfrac{2}{3} m_a R^2 \Omega.$$

The ratio of the moment of inertia of the earth to the moment of inertia of the atmosphere rotating as a rigid body is thus on the order $m_e/m_a \approx 10^6$.

The relative angular momentum can be computed from the observed zonal wind component using the expression

$$\int M_r \, dm = \int uR \cos \phi \, dm.$$

The integrals of M_r for the two hemispheres and the globe, as computed from observed winds and from NMC wind analyses, are given in Table 11.1. The table shows that the annual-mean value for the globe is on the order of 13×10^{25} kg m² s⁻¹. It is small (about 1%) compared to the corresponding value for the Ω angular momentum M_Ω which is 1.01×10^{28} kg m² s⁻¹ (i.e., when we assume that the atmosphere is in solid rotation with the underlying surface). The different values of M_r in Table 11.1 show that there is a seasonal variation in the relative angular momentum of the global atmosphere. The reason is that the seasonal variation in the winds is much larger in the Northern Hemisphere than in the Southern Hemisphere as one may expect from the more continental climate in the Northern Hemisphere. In the Northern Hemisphere, we find a decrease from a strong westerly circulation in winter to a very weak westerly circulation in summer, while in the Southern Hemisphere the winter-to-summer reduction in the westerlies is only on the order of 50%.

Because there are no reliable global measurements of ocean currents, we will make a rough estimate of the relative angular momentum in the oceans in the following way. Let us assume a strong westward current in the Atlantic and Pacific Oceans of 100 Sv (1 Sverdrup = 10^6 m³ s⁻¹) or 10^{11} kg s⁻¹ at the equator and an equivalent return flow toward the east at 40° latitude. This would give an oceanic contribution of about

$$-10^{11} \times \frac{2\pi R^2}{2}\left(1 - \cos^2 \frac{40\pi}{180}\right)$$
$$= -0.5 \times 10^{25} \quad \text{kg m}^2 \text{ s}^{-1},$$

where the longitudinal span of the Atlantic and Pacific Oceans is taken care of by the factor 1/2. Because temporal variations in ocean circulation are presumably weaker than the mean circulation itself, the oceanic contributions to the changes in angular momentum must be an order of magnitude smaller than those for the atmosphere as shown, e.g., in Fig. 11.2.

TABLE • 11.1. Hemispheric and global integrals of the relative angular momentum M_r in the atmosphere in units of 10^{25} kg m² s⁻¹ from Oort and Peixoto (1983) for the years 1963–1973, and, in parentheses, from Rosen and Salstein (1983; updated by Salstein, personal communication) based on NMC analyses for the years 1976–1987.

	Year	DJF	JJA	DJF-JJA
NH	5.3 (5.7)	9.6 (9.4)	0.2 (1.4)	9.4 (8.0)
SH	7.6 (8.4)	4.8 (6.4)	9.5 (10.0)	−4.6 (−3.6)
GLOBE	12.9 (14.1)	14.4 (15.8)	9.7 (11.4)	4.7 (4.5)

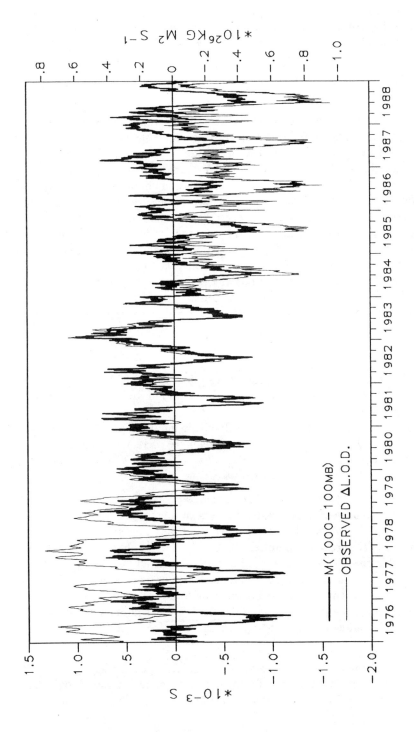

FIGURE 11.2. Time series of daily values of the relative westerly angular momentum M_r of the global atmosphere between 1000 and 100 mb based on NMC analyses in units of 10^{26} kg m^2 s^{-1} (heavy solid line; scale on right) and values of the length of day (LOD) in units of 10^{-3} s (thin solid line; scale on left) for the years 1976–1988. The daily LOD data are based on an optimal combination of observing techniques prior to April 1985 with approximately 5-day effective resolution and on VLBI (very long base line interferometry) observations with approximately daily resolution from April 1985 (based on data from Rosen et al., 1990, updated by Salstein).

Let us now discuss how important systematic latitudinal shifts in atmospheric and oceanic mass can be for the Ω component of the earth's angular momentum. Evidence for shifts in oceanic mass are found in the changes of sea level height. For example, a uniform shift of water, $\Delta z = 2$ cm, from middle and high latitudes across 30° latitude into the tropics occurring in both hemispheres would give a change in angular momentum of

$$\Delta M_\Omega = 2\pi\rho\Omega R^4 \Delta z \left\{ \int_{-\pi/6}^{\pi/6} \cos^3 \phi \, d\phi - 2 \int_{\pi/6}^{\pi/2} \cos^3 \phi \, d\phi \right\} = \pi\rho\Omega R^4 \Delta z$$

$$= 0.8 \times 10^{25} \quad \text{kg m}^2 \text{ s}^{-1}.$$

This shift in water level would be equivalent with a shift in atmospheric mass of 2 mb. (The observed monthly mean climatological shifts in atmospheric mass between the tropics and high latitudes tend to be at most a few mb.)

In summary, the observed changes in the angular momentum of the atmosphere are on the order of 5×10^{25} kg m^2 s^{-1} and the corresponding changes in the oceans are estimated to be less than 1×10^{25} kg m^2 s^{-1}. These values are extremely small (on the order of 10^{-8}) compared with the total angular momentum of the globe (5.86×10^{33} kg m^2 s^{-1}). However, small as they are, they do have profound geophysical and astronomical implications. We have seen before that the angular momentum of the earth–atmosphere–ocean system remains constant throughout the year because no appreciable external torques affect the system. Thus, in view of the observed seasonal variations in total atmospheric angular momentum, the solid earth has to adjust its rotation rate seasonally in order to keep the total angular momentum invariant:

$$I_e \Omega + \int_{V_{\text{atm}}} \rho \, ur \cos \phi \, dV = \text{const.}$$

Specifically, in terms of the seasonal differences given in Table 11.1, the angular momentum of the solid earth (or in other words the rotation rate Ω) has to be greater in July than in January.

To relate the change in rotation rate or the change in length of day (ΔLOD) to the change in relative angular momentum of the atmosphere (ΔM_r) we can use the following relation (Rosen et al., 1987):

$$\Delta \text{LOD} = 0.168 \; \Delta M_r,$$

where ΔLOD is in ms and ΔM_r in units of 10^{25} kg m^2 s^{-1}. In deriving this relation a value of 7.04×10^{37} kg m^2 was used for the moment of inertia of the earth, which is smaller than the value of 8.04×10^{37} kg m^2 given before. The smaller value only includes the effects of the crust and mantle, whereas the larger value also includes the effects of the earth's core. The reason for choosing the smaller value is that there is some evidence that the liquid core of the earth may at times be rotating at a different rate from the crust and mantle. We shall come back to this question later.

We can now estimate the necessary seasonal change from January to July in the length of day using a value of $\Delta M_r = 5 \times 10^{25}$ kg m^2 s^{-1} from Table 11.1:

$$\Delta \text{LOD} \approx 0.8 \text{ ms.}$$

Astronomical measurements show the expected lengthening of the day in northern winter and the shortening in summer. Expressed as a change in the global-mean zonal wind speed, $\Delta M_r = 5 \times 10^{25}$ kg m^2 s^{-1} would correspond with

$$\Delta \tilde{u} = \int (\Delta u) R \cos \phi \, dm \bigg/ \int R \cos \phi \, dm$$

$$= 2.0 \text{ m s}^{-1}.$$

Rosen et al. (1990) have compared direct daily estimates of the length of day with estimates of the atmospheric angular momentum during a 13-year period. Their comparison is shown in Fig. 11.2. The two independent data sets beautifully confirm the expected relationship at all time scales, except for a slow downward trend in the LOD data and some occasional unexplained differences. These last discrepancies may be caused by the neglected storage of angular momentum in the oceans, ice caps, and snow masses or may be due to inaccuracies in the determination of M_r and LOD (Rosen et al., 1990).

Although it is not easily noticeable in the curves in Fig. 11.2, spectacular changes in the record have occurred, such as the sudden drop in M_r between May 20 and June 1, 1979, corresponding with a mean decrease in zonal wind of almost 1.5 m s^{-1}, while the length of day was found to decrease by about 0.6 ms.

The downward trend in LOD in Fig. 11.2, as well as the other large interdecadal variations on the order of several ms shown in Fig. 11.3, are too large to be associated with changes in the angular momentum of the atmosphere. They are thought to be connected with slow exchanges of angular momentum between the earth's crust and mantle and the underlying fluid core (Hide et al., 1980; see Fig. 11.4). In fact, from records such as those shown in Fig. 11.3, we may learn more about the liquid core of the earth after removal of those variations that can be explained by exchanges with the atmosphere. The secular increase in the length of day due to tidal friction with the moon is much smaller than the interdecadal and seasonal variations and is only on the order of 2 ms/century (Christodoulidis et al., 1988).

11.1.3 Angular momentum in the atmosphere

The angular momentum with respect to the earth's axis of rotation is one of the fundamental parameters used to characterize the general circulation of the atmosphere and the climate. Much of the historic development of modern meteorology is connected with the study of the transport of angular momentum in the atmosphere and of its exchange with the oceans and solid earth. For further discussions of the angular momentum balance see Starr (1968) and Newell et al. (1972).

Recently attention has also been given to the components of angular momentum about axes located in the equatorial plane at right angles to the earth's axis of rotation in order to study, e.g., the earth's wobble (see Barnes et al., 1983). However we will not pursue this here any further.

As we stated earlier, in our rotating coordinate system, the time rate of change of total angular momentum for a unit volume equals the sum of all torques acting on it. From Eq. (3.7) we obtain

$$\rho \frac{dM}{dt} = -\frac{\partial p}{\partial \lambda} + \rho F_\lambda R \cos \phi + \text{tidal and other extraterrestrial torques},$$

where

$$\frac{d}{dt} = \frac{\partial}{\partial t} + u \frac{\partial}{R \cos \phi \partial \lambda} + v \frac{\partial}{R \partial \phi} + w \frac{\partial}{\partial z}.$$

For the bulk of the atmosphere only the first two terms on the right-hand side of the equation are important, being the pressure and frictional torques, respectively,

$$\rho \frac{dM}{dt} = -\frac{\partial p}{\partial \lambda} + \rho F_\lambda R \cos \phi. \tag{11.4}$$

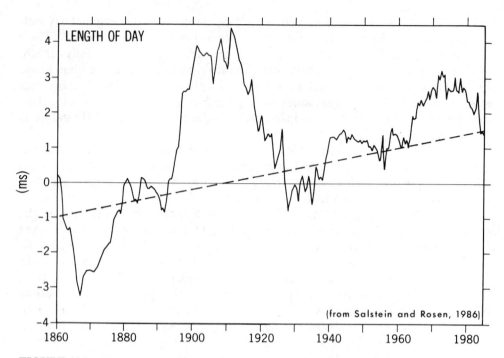

FIGURE 11.3. Time series of semiannual values of the length of day during 1860–1985, taken from the work by McCarthy and Babcock (1986). A mean annual signal has been subtracted. The dashed line indicates the secular trend from tidal friction estimates (adapted from Salstein and Rosen, 1986).

An alternative form for the time rate of change of relative instead of absolute angular momentum can be obtained by splitting $M = M_\Omega + M_r$ using Eq. (11.3). First of all, we find that

$$\rho \frac{dM_\Omega}{dt} = \rho \frac{d}{dt}(\Omega R^2 \cos^2 \phi) = -2\rho v R\Omega \cos \phi \sin \phi$$
$$+ \rho w 2R\Omega \cos^2 \phi$$
$$= -\rho v f R \cos \phi + \rho f' w R \cos \phi,$$

where

$$f = 2\Omega \sin \phi \text{ and } f' = 2\Omega \cos \phi.$$

Thus, Eq. (11.4) can be written

$$\rho \frac{dM_r}{dt} = -\frac{\partial p}{\partial \lambda} + \rho F_\lambda R \cos \phi$$
$$+ \rho f v R \cos \phi - \rho f' w R \cos \phi. \tag{11.5}$$

The last two terms cannot be reduced to boundary integrals when one integrates over a certain volume. They indicate sources or sinks of relative angular momentum. These terms did not appear in the equation for the absolute angular momentum (11.4), since absolute angular momentum is conserved, and cannot be destroyed or created. However, if one splits the absolute into relative and Ω angular momentum, internal conversions of the type $\rho f v R \cos \phi$ and $\rho f' w R \cos \phi$ may take place. As discussed before, a northward or downward flow of the volume unit considered will

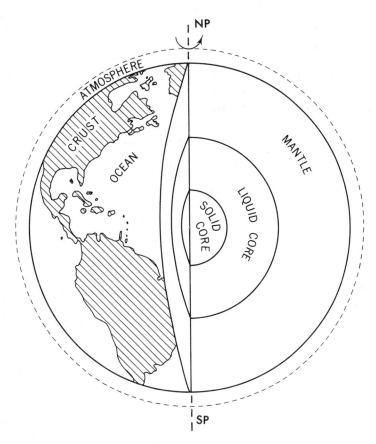

FIGURE 11.4. Schematic diagram of the atmosphere, oceans, crust, mantle, and core of the earth. Note that the vertical scale of the atmosphere (≈ 20 km) is grossly exaggerated (from Oort, 1989).

lead to an increase in its relative westerly angular momentum, equal in magnitude to the decrease in its Ω angular momentum, and vice versa. In general, the term $\rho f'wR \cos \phi$ can be neglected compared to $\rho fvR \cos \phi$ because of the small vertical extent of the atmosphere. In fact, because $\partial u/\partial x$ and $\partial v/\partial y$ almost compensate each other and thus $\partial u/\partial x + \partial v/\partial y \approx 0$, the ratio

$$w/v \approx 0.1 (\text{depth scale/length scale}) \approx 0.1 \times 10 \text{ km}/1000 \text{ km} = 10^{-3}$$

from scale analysis of large-scale motions. Equation (11.5) can be easily reduced to the zonal equation of motion (3.9a) discussed earlier, since the two equations are equivalent.

The first two terms on the right-hand side of Eq. (11.5) can be written as boundary terms after integration over a certain volume. They indicate the exchange of angular momentum with the environment. For a discussion of the frictional effects see Sec. 3.2.1.

11.1.4 Volume integrals

Let us now consider the balance of angular momentum for a unit volume. Thus, using the general balance equation (3.80) we may express Eq. (11.4) in the form

$$\frac{\partial \rho M}{\partial t} = -\text{div } \rho M\mathbf{c} - \frac{\partial p}{\partial \lambda} + \rho F_\lambda R \cos \phi, \tag{11.6}$$

where $F_\lambda = -\alpha \partial \tau_{zx}/\partial z$. In addition to the torques which constitute the sources and sinks we have a divergence term in Eq. (11.6), which has a simple physical interpretation; i.e., when integrated over a specific volume the divergence term represents the transport of angular momentum across the boundaries of the volume. If we want to evaluate a volume integral, it is convenient to choose the volume in such a way that the boundaries are fixed in the rotating coordinate system, so that the limits of integration do not vary with time. The total volume of the atmosphere or the volume of a polar cap approximately satisfy the condition of invariance. For a volume V we find

$$\int_V \frac{\partial \rho M}{\partial t} dV = -\int_S \rho M c_n \, ds - \int_V \frac{\partial p}{\partial \lambda} dV + \int_V \rho F_\lambda R \cos \phi \, dV,$$

where in the first term on the right-hand side we have made use of the Gauss divergence theorem. Here c_n denotes the normal outward component of **c** across the boundary S that limits V.

Using expression (3.10) for F_λ we obtain for the atmosphere as a whole the following, important result:

$$\frac{\partial}{\partial t} \int \rho M \, dV = -\int \frac{\partial p}{\partial \lambda} dV$$

$$+ \iint_{\text{sfc}} \tau_0 R^2 \cos^2 \phi \, d\lambda \, R \, d\phi$$

$$= \mathscr{P} + \mathscr{T} \qquad (11.7)$$

because all other components are identically zero.

The sign convention used here is that the pressure and friction torques $\mathscr{P}$ and $\mathscr{T}$ are counted positive when they tend to increase the eastward angular momentum of the atmosphere. This equation shows that the total angular momentum of the atmosphere can only change through the action of topographic features and friction. In the presence of mountains, the pressure or mountain torque $\mathscr{P}$ will generally not vanish, and it will contribute to the angular momentum cycle through east–west pressure differences across the various mountain ranges on earth. This can be seen by carrying out the integration in λ over the air above the mountains (see Fig. 11.5):

$$\mathscr{P} = -\iint \left\{ \int \left(\frac{\partial p}{\partial \lambda}\right) d\lambda \right\} dz \, R^2 \cos \phi \, d\phi \qquad (11.8a)$$

or

$$\mathscr{P} = \int_{-\pi/2}^{\pi/2} \int_{z_0}^{\infty} \sum_i (p_E^i - p_W^i) dz \, R^2 \cos \phi \, d\phi, \qquad (11.8b)$$

where the summation is made over all mountain ranges along a latitude circle, z_0 is the elevation of the earth's surface, and p_E^i and p_W^i are the pressures at the east and west sides of the ith mountain range, respectively. In the (λ, ϕ, p, t) system, the mountain torque is given by

$$\mathscr{P} = \int_{-\pi/2}^{\pi/2} \int_0^{p_{\text{sfc}}} \sum_i (z_E^i - z_W^i) dp \, R^2 \cos \phi \, d\phi. \qquad (11.8c)$$

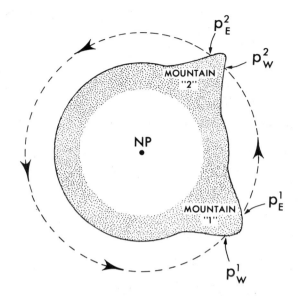

COMPUTATION MOUNTAIN TORQUE

FIGURE 11.5. Schematic diagram for the computation of the mountain torque.

An alternative method to evaluate the mountain torque is from surface data alone. This result can be derived from Eq. (11.8) through integration by parts:

$$\mathscr{P} = -\iint \left\{ \frac{\partial}{\partial \lambda}\left(\int_{z_0}^{\infty} p \, dz\right) + p \frac{\partial z_0}{\partial \lambda} \right\} d\lambda \, R^2 \cos \phi \, d\phi$$

$$= -\iint_{\text{sfc}} \left(p \frac{\partial z_0}{\partial \lambda} \right) d\lambda \, R^2 \cos \phi \, d\phi. \tag{11.8d}$$

If there were no mountains, the mountain torque would obviously be identically zero.

Since the surface stress τ_0 is directed against the surface winds, it will be counted positive when the winds are from the east and negative when they are from the west. Therefore, friction is always adding (westerly) angular momentum to the atmosphere in the regions of surface easterlies (e.g., the trade winds), where the globe rotates faster than the atmosphere, and friction is always removing angular momentum in middle and high latitudes, where surface westerlies prevail. As a result, the tropics act as the principal source of absolute angular momentum for the atmosphere, and the midlatitudes as the principal sink. We may conclude that, in order to conserve angular momentum, the easterlies and westerlies at the earth's surface have to balance in the long run. Indeed, if surface westerlies (easterlies) would predominate the angular momentum of the atmosphere would decrease (increase) in time.

Let us now consider a polar cap with a latitudinal wall that separates the surface easterlies from the surface westerlies, approximately at a latitude of 30° ($\phi = \pi/6$). This polar cap is bounded by the earth's surface, the latitudinal wall, and the top of the atmosphere. Based on our previous discussion, we conclude that there must be a transport of angular momentum from the source regions in the tropics to the sink regions in the westerly belt. This idea will allow us to gain some insight in the transports of angular momentum within the atmosphere. To show the angular momentum budget in the polar cap we can write

$$\frac{\partial}{\partial t} \int_V \rho M \, dV = -\int_V \operatorname{div} \rho M\mathbf{c} \, dV - \int_V \frac{\partial p}{\partial \lambda} dV$$

$$+ \iint_{\text{sfc}} \tau_0 R^3 \cos^2 \phi \, d\lambda \, d\phi.$$

Using Gauss' divergence theorem, we find

$$\frac{\partial}{\partial t} \int \rho M \, dV = \int_{\text{vert wall}} \rho M v \, ds + \mathscr{P}^* + \mathscr{T}^*, \qquad (11.9)$$

where

$$\mathscr{P}^* = \int_0^\infty \int_{\pi/6}^{\pi/2} \sum_i (p_E^i - p_W^i) R^2 \cos \phi \, d\phi \, dz,$$

$$\mathscr{T}^* = \iint_{\text{sfc polar cap}} \tau_0 R^3 \cos^2 \phi \, d\lambda \, d\phi,$$

and $\rho \mathbf{c} \cdot \mathbf{n} = -\rho v$ for the latitudinal wall, and $\rho \mathbf{c} \cdot \mathbf{n} = \rho w = 0$ at the top surface of the atmosphere.

This expression shows that fluctuations of angular momentum in the polar cap are the net result of the meridional flux of angular momentum across the vertical wall and of the action of the mountain ($\mathscr{P}^*$) and friction torques ($\mathscr{T}^*$) inside the cap. In a steady state, the meridional transport has to compensate exactly for the surface losses due to the friction and mountain torques.

11.1.5 Modes of transport

The time derivative in Eq. (11.6) is in general different from zero. However, it changes sign frequently so that the average over a long period tends to vanish. For example, for a year it is very small compared with the other terms.

Thus, it seems appropriate to analyze the time-averaged equation. Using expansion (4.3), Eq.(11.6) becomes

$$\frac{\overline{\partial \rho M}}{\partial t} = - \operatorname{div} \overline{M} \, \overline{\rho \mathbf{c}} - \operatorname{div} \overline{M'(\rho \mathbf{c})'}$$

$$- \frac{\overline{\partial p}}{\partial \lambda} - R \cos \phi \, \frac{\partial \overline{\tau}_{zx}}{\partial z}.$$

This expression can be rewritten in the (λ, ϕ, p, t) system in the form

$$\frac{\overline{\partial M}}{\partial t} = - \frac{1}{R \cos \phi} \left(\frac{\partial}{\partial \lambda} \overline{\mathscr{J}}_\lambda + \frac{\partial \overline{\mathscr{J}}_\phi \cos \phi}{\partial \phi} \right) - \frac{\partial}{\partial p} (\overline{\mathscr{J}}_p)$$

$$- g \frac{\partial \overline{z}}{\partial \lambda} - R \cos \phi \, \frac{\partial \overline{\tau}_{p\lambda}}{\partial p}, \qquad (11.10)$$

where $\mathbf{J} = (\mathscr{J}_\lambda, \mathscr{J}_\phi, \mathscr{J}_p)$ is the angular momentum transport density vector defined by

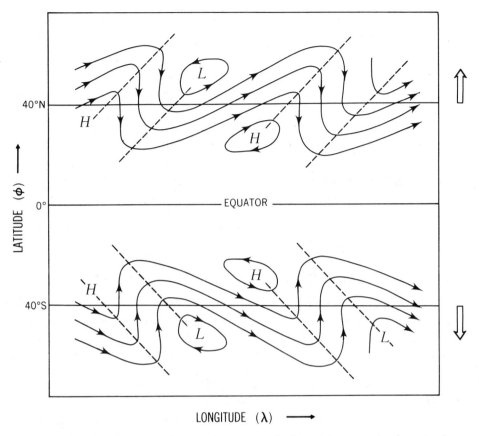

FIGURE 11.6. Schematic picture of the dominant mechanism of the upper air transport of westerly angular momentum by quasihorizontal eddies in midlatitudes. Note the preferred SW–NE tilt of the streamlines in the Northern Hemisphere, and the SE–NW tilt in the Southern Hemisphere leading to a net poleward transport of angular momentum in both hemispheres.

$$\left.\begin{aligned}
\overline{\mathcal{F}}_\lambda &= R^2 \cos^2 \phi \, \Omega \bar{u} + R \cos \phi (\bar{u}\bar{u} + \overline{u'u'}) \\
&= \overline{\mathcal{F}}_{\Omega\lambda} + \overline{\mathcal{F}}_{r\lambda} \\
\overline{\mathcal{F}}_\phi &= R^2 \cos^2 \phi \, \Omega \bar{v} + R \cos \phi (\bar{u}\bar{v} + \overline{u'v'}) \\
&= \overline{\mathcal{F}}_{\Omega\phi} + \overline{\mathcal{F}}_{r\phi} \\
\overline{\mathcal{F}}_p &= R^2 \cos^2 \phi \, \Omega \bar{\omega} + R \cos \phi (\bar{u}\bar{\omega} + \overline{u'\omega'}) \\
&= \overline{\mathcal{F}}_{\Omega p} + \overline{\mathcal{F}}_{rp}.
\end{aligned}\right\} \quad (11.11)$$

The density flux vector is then represented by the sum of the flux of Ω angular momentum by the time-mean motions and fluxes of relative angular momentum by mean motions and perturbations. The horizontal components of the friction torque were disregarded in Eq. (11.10) because in the free atmosphere they are very small compared with the torques associated with the large-scale horizontal motions. In the surface boundary layer they are small compared with the vertical component of the friction torque.

The zonally averaged form of the previous equation is expressed by

$$\overline{\frac{\partial}{\partial t}}[M] = -\frac{1}{R\cos\phi}\frac{\partial}{\partial\phi}([\overline{\mathcal{F}}_\phi]\cos\phi) - \frac{\partial}{\partial p}[\overline{\mathcal{F}}_p]$$

$$-g\left[\overline{\frac{\partial \bar{z}}{\partial \lambda}}\right] - R\cos\phi\left[\frac{\partial \bar{\tau}_{p\lambda}}{\partial p}\right] \quad (11.12)$$

since

$$\left[\frac{\partial}{\partial \lambda}\mathcal{F}_\lambda\right] = \frac{1}{2\pi}\oint \frac{\partial \mathcal{F}_\lambda}{\partial \lambda}\, d\lambda = 0.$$

The flux terms can be written explicitly using expansion (4.9) as

$$[\overline{\mathcal{F}}_\phi] = R^2 \cos^2\phi\, \Omega[\bar{v}] + R\cos\phi([\bar{u}][\bar{v}]$$
$$+ [\bar{u}^*\bar{v}^*] + [\overline{u'v'}])$$
$$= [\overline{\mathcal{F}}_{\Omega\phi}] + [\overline{\mathcal{F}}_{r\phi}],$$

$$[\overline{\mathcal{F}}_p] = R^2 \cos^2\phi\, \Omega[\bar{\omega}] + R\cos\phi([\bar{u}][\bar{\omega}] + [\bar{u}^*\bar{\omega}^*] + [\overline{u'\omega'}])$$
$$= [\overline{\mathcal{F}}_{\Omega p}] + [\overline{\mathcal{F}}_{rp}]. \quad (11.13)$$

These expressions show that the Ω angular momentum components $[\overline{\mathcal{F}}_{\Omega\phi}]$ and $[\overline{\mathcal{F}}_{\Omega p}]$ represent the transports by mean meridional circulations, $[\bar{v}]$ and $[\bar{\omega}]$, i.e., the overturnings in the (ϕ, p) plane. These transports are thus proportional to those of the mass.

The expressions also show that the meridional transports of relative angular momentum $[\overline{\mathcal{F}}_{r\phi}]$ and $[\overline{\mathcal{F}}_{rp}]$ are accomplished by three different modes of exchange resulting from the expansions of $[\overline{uv}]$ and $[\overline{u\omega}]$, respectively. Let us further analyze the terms in $[\overline{\mathcal{F}}_{r\phi}]$. The first term represents the contributions by the time-mean meridional cells. The contributions to the northward transport of momentum are positive if $[\bar{u}]$ and $[\bar{v}]$ are positively correlated. For example, if in the Northern Hemisphere the poleward branch $([\bar{v}] > 0)$ in a meridional cell is associated with a westerly wind $([\bar{u}] > 0)$ and the equatorward branch $([v] < 0)$ with an easterly wind $([\bar{u}] < 0)$ there is a net poleward contribution to the transport. The second term in $[\overline{\mathcal{F}}_{r\phi}]$ represents the contribution to the poleward transport by the stationary perturbations. The poleward transport is positive when $\bar{u}^*$ and $\bar{v}^*$ are positively (negatively) correlated in the Northern (Southern) Hemisphere. These perturbations are those shown on time-mean maps. The third term gives the contribution to the transport of angular momentum by the transient eddies, which results from the instantaneous correlation of u and v along the latitude circle. The transient eddies are mainly the perturbations shown on daily weather maps, such as moving low- and high-pressure systems.

Let us now consider the vertical transport of angular momentum. As mentioned before, $[\overline{\mathcal{F}}_{\Omega p}]$ represents the transport of angular momentum by the vertical component of the mean meridional circulations. This transport is different from zero when there is a correlation between $[\bar{\omega}]$ and ϕ, i.e., if there is an upward flow of mass at one latitude and a downward flow of an equal amount of mass at another latitude. The first terms in $[\overline{\mathcal{F}}_{rp}]$ represent the effects of mean meridional motions in transporting relative angular momentum; the other two terms give the vertical transports of relative angular momentum by the stationary and transient eddies.

The estimation of these vertical terms requires measuring ω in space and time, which cannot be done directly. The terms involving $[\bar{\omega}]$ can be evaluated indirectly from $[\bar{v}]$. However, the $[\bar{v}]$ are, by themselves, somewhat uncertain. The terms involving $\bar{\omega}^*$ can be approximately estimated from the grid point values of $\bar{u}$ and $\bar{v}$ using the continuity equation. The terms containing ω' are the most difficult to estimate, because they require an instantaneous evaluation of ω.

Starr (1948, 1953) was able to explain how the transport of momentum was achieved by horizontal motions. He showed that for the eddies to transport momentum poleward the streamlines should exhibit a preferential tilt from SW to NE in the Northern Hemisphere (as sketched in Fig. 11.6) and from NW to SE in the Southern Hemisphere. In the Northern Hemisphere, with this configuration, the winds from the south have a much larger eastward component than the winds from the north, i.e., the components u and v are positively correlated along a given latitude circle. Such a pattern is indeed observed to occur frequently in the upper air flow.

11.2 OBSERVED CYCLE OF ANGULAR MOMENTUM

11.2.1 Angular momentum in the atmosphere

11.2.1.1 Distribution of angular momentum

As we have shown before, the Ω angular momentum in the atmosphere is much larger than the relative angular momentum since the angular velocity of rotation in middle latitudes ($\Omega R \cos \phi$) is of the order of 250 m s^{-1} compared with a typical wind speed of about 10 m s^{-1}. Its horizontal distribution is only a function of latitude ($\sim \cos^2 \phi$) and it does not vary in time because the mass distribution stays relatively constant. On the other hand, seasonal variations of relative angular momentum for each hemisphere and the globe are very pronounced, as was shown before in Table 11.1.

The vertical structure of the relative angular momentum is portrayed by the zonal wind cross sections $[\bar{u}]$ in Fig. 7.15 when multiplied by $R \cos \phi$.

11.2.1.2 Meridional transport of angular momentum

In order to investigate the mechanisms of horizontal transport of angular momentum we will use expansion (11.13):

$$[\overline{\mathcal{F}_\phi}] = R^2 \cos^2 \phi \Omega [\bar{v}] + R \cos \phi ([\bar{u}][\bar{v}]$$
$$+ [\bar{u}^*\bar{v}^*] + [\overline{u'v'}])$$
$$= [\overline{\mathcal{F}_{\Omega\phi}}] + [\overline{\mathcal{F}_{r\phi}}].$$

Since the Ω angular momentum is much larger than the relative angular momentum one might expect the required north–south transport of absolute momentum to be dominated by the meridional transport of the Ω component, $[\overline{\mathcal{F}_{\Omega\phi}}]$. However, conservation of mass requires in the long-term mean that the vertical integral of the first term vanishes so that

$$\int_0^{p_0} [\overline{\mathcal{F}_\phi}] \frac{dp}{g} = \int_0^{p_0} [\overline{\mathcal{F}_{r\phi}}] \frac{dp}{g}.$$

Therefore, it is the flux of relative angular momentum that must be responsible for the horizontal transport of absolute angular momentum from the source regions in the tropics to the sink regions in the middle latitudes. As discussed before, the terms in the expression for $[\mathcal{F}_{r\phi}]$ show the various mechanisms that contribute to the total northward flux.

The vertical distribution of the zonal-mean flux of momentum and a breakdown in its various components is shown in the cross sections of Fig. 11.7 for the year (see also Starr et al., 1970a). The first qualitative impressions are that there is an overall symmetry with respect to the equator (except in the case of the stationary eddies) and that transient eddies dominate the total transport in the upper levels. The lack of symmetry in the case of the stationary eddy transport $[\bar{u}^*\bar{v}^*]$ reflects the more zonal

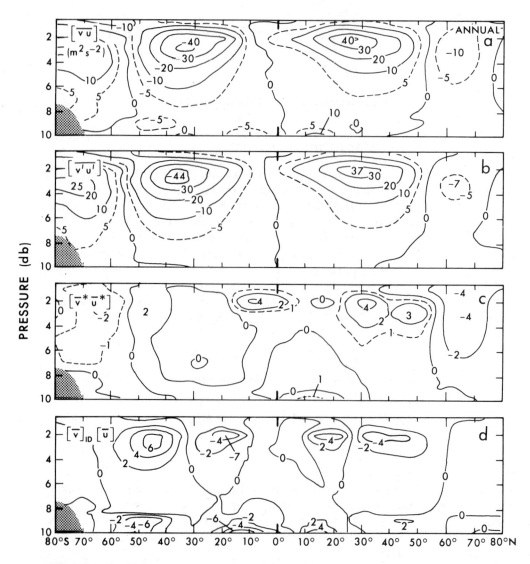

FIGURE 11.7. Zonal-mean cross sections of the northward flux of momentum by all motions (a), transient eddies (b), stationary eddies (c), and mean meridional circulations (d) in m² s⁻² for annual-mean conditions (from Oort and Peixoto, 1983).

character of the time-mean flow south of 20 °S, where the $[\bar{u}^*\bar{v}^*]$ values are unreliable. The mean meridional flux of momentum $[\bar{u}][\bar{v}]$ shows the effects of the three-cell structure in each hemisphere. Its contribution to the total transport is relatively small compared with the contribution by the transient eddies, except in the surface boundary layer.

Looking more closely, some interesting quantitative differences between the two hemispheres can be pointed out. For example, one finds the maximum poleward flux near 35 °S at the 300-mbar level, whereas the Northern Hemisphere maximum occurs near 25 °N at the 200-mbar level. At polar latitudes the maximum equatorward flux appears stronger in the Southern Hemisphere than in the Northern Hemisphere. These differences are practically all associated with the differences in the transient eddy flux.

Meridional profiles of the vertical-mean momentum fluxes for the year and for the DJF and JJA seasons are shown in Fig. 11.8. To obtain the total flux of angular momentum, one must multiply the values of the momentum transport by a factor $F = 2\pi R^2 \cos^2\phi p_0/g = 2.56 \times 10^{18} \cos^2\phi$ kg, assuming $p_0 = 1000$ mb. Although the stationary eddies are not important in the Southern Hemisphere, the total flux of angular momentum by transient, stationary, and mean meridional circulations together is somewhat greater in that hemisphere. Even for the seasons the bulk of the momentum transport is accomplished by transient eddies (i.e., eddies with time scales less than three months). For all seasons the transient fluxes are considerably stronger in the Southern Hemisphere than those during the corresponding season in the Northern Hemisphere. Throughout the year the Antarctic appears to be an important source of momentum (i.e., there is a divergence of angular momentum), probably due to the effects of topography. There appears to be a tendency for cross-equatorial transports from the winter to the summer hemisphere. In winter, the stationary eddies in the Northern Hemisphere midlatitudes are major contributors to the total flux of momentum.

11.2.1.3 Vertical transport of angular momentum

To analyze the vertical transport of angular momentum we will start with the balance equation for total angular momentum (11.12) in the form

$$\frac{\overline{\partial[M]}}{\partial t} = -\partial\{[\overline{\mathcal{F}}_\phi]\cos\phi\}\bigg/R\cos\phi\partial\phi - \frac{\partial}{\partial p}[\overline{\mathcal{F}}_p]$$
$$-R\cos\phi\left[\frac{\partial \bar{\tau}_{p\lambda}^*}{\partial p}\right], \qquad (11.14)$$

where $-(\partial \bar{\tau}_{p\lambda}^*/\partial p)$ represents a "gross" frictional force. This force is introduced for simplicity, and includes the mountain force

$$\frac{\partial \bar{\tau}_{p\lambda}^*}{\partial p} = \frac{g}{R\cos\phi}\frac{\partial \bar{z}}{\partial \lambda} + \frac{\partial \bar{\tau}_{p\lambda}}{\partial p}. \qquad (11.15)$$

It is produced by at least three factors. These are friction, subgrid-scale momentum exchange, and pressure differences across topographic features ranging from the smallest hills to the largest mountains. Since time variations of the angular momentum are usually small, the left-hand side of Eq. (11.14) can be set equal to zero for long

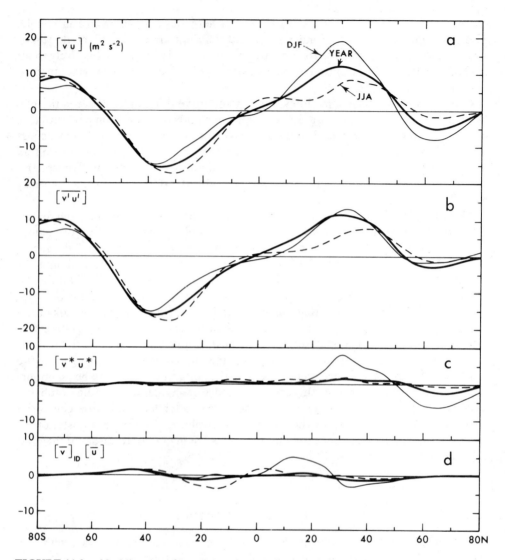

FIGURE 11.8. Meridional profiles of the vertical- and zonal-mean northward transport of momentum by all motions (a), transient eddies (b), stationary eddies (c), and mean meridional circulations (d) in units of m² s⁻² for annual, DJF, and JJA mean conditions [to convert to angular momentum transport units of 10^{18} kg m² s⁻¹ multiply values by $2\pi R^2 \cos^2 \phi (p_0/g) = 2.56 \cos^2 \phi$, where $2\pi R \cos \phi$ = length of latitude circle, $R \cos \phi$ = distance to rotation axis, and p_0/g = mass per unit area $\approx 10^4$ kg m⁻²; from Oort and Peixoto, 1983].

averaging periods. Under this condition we can define a stream function ψ_M for the angular momentum of a zonal ring of air. Following Starr *et al.* (1970b), we may write

$$2\pi R \cos \phi [\overline{\mathcal{F}}_\phi] = -\frac{\partial \psi_M}{\partial p}, \qquad (11.16a)$$

$$2\pi R \cos \phi \{[\overline{\mathcal{F}}_p] + R \cos \phi [\overline{\tau}^*_{p\lambda}]\} = \frac{\partial \psi_M}{R \partial \phi}. \qquad (11.16b)$$

We can now evaluate $[\overline{\mathcal{F}}_\phi]$ from the vertical cross sections of $[\overline{v}]$ in Fig. 7.17 and $[\overline{vu}]$ in Fig. 11.7(a) for the year. The ψ_M distribution can then be obtained for all

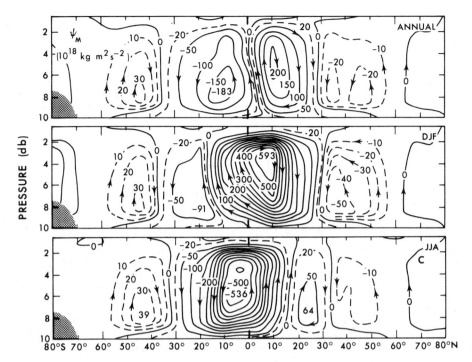

FIGURE 11.9. Streamlines of the zonal-mean transport of absolute angular momentum in the atmosphere for annual, DJF, and JJA mean conditions in 10^{18} kg m^2 s^{-2} (from Oort and Peixoto, 1983).

points in the (ϕ,p) plane by integrating Eq. (11.16a) downward starting with $\psi_M = 0$ at the top level (here chosen as $p = 25$ mb). The vertical transport $[\overline{\mathcal{J}_p}] + R \cos \phi [\overline{\tau}_{p\lambda}^*]$ can then be determined using Eq. (11.16b) (see also Hantel, 1974; Hantel and Hacker, 1978).

The resulting flow of total angular momentum for the annual, DJF, and JJA conditions is shown in Fig. 11.9. Besides recirculation of the angular momentum in the mean meridional cells, the figure depicts the interactions with the earth's surface, where the sources and sinks of angular momentum for the atmosphere are found. By and large, the mean yearly circulation of angular momentum $\{[\overline{\mathcal{J}_\phi}],[\overline{\mathcal{J}_p}] + R \cos \phi [\overline{\tau}_{p\lambda}^*]\}$ is symmetric with respect to the equator. It shows a dominant flow from the source regions in the subtropics to the sink regions in middle latitudes through the intersections of the ψ_M lines with the earth's surface. There is an apparent seasonal intensification of the tropical circulations associated with the strengthening of the Hadley cell and an increased uptake of angular momentum in the winter hemisphere. Further implications of the vertical transports and a better representation will be discussed later in the broader context of the angular momentum cycle in the climatic system.

11.2.2 Angular momentum exchange between the atmosphere and the underlying surface

As was shown before in Eq. (11.7) through integration of the balance equation of angular momentum over the entire global atmosphere, the sum of the friction and mountain torques at the earth's surface must vanish in the long run (except for the very small tidal deceleration), since

$$\overline{\frac{\partial}{\partial t} \int \rho M \, dV} \simeq 0.$$

Thus, we may write

$$\oiint \overline{\tau}_0 R^2 \cos^2 \phi \, d\lambda \, R \, d\phi$$

$$+ \oiint \sum_i (z_E^i - z_W^i) R^2 \cos \phi \, d\phi \, dp = 0,$$

where the $\oiint$ symbol indicates that the integrals are taken over all points of the earth's surface. As will be seen later, the mountains generally reinforce the friction effects, although they are of less importance (White, 1949; Newton, 1971a and 1971b). This equation leads to the general requirement that, from a global perspective, the surface area covered by easterly trade winds ($\tau_0 > 0$) should be roughly the same as the surface area covered by midlatitude westerlies ($\tau_0 < 0$) (Lorenz, 1967).

We can investigate these questions more deeply through a study of the horizontal distribution of the torque associated with the sum of the vertically integrated horizontal divergence of angular momentum and the pressure torque in Eq. (11.6), which in the long run ($\overline{\partial M/\partial t} \approx 0$) should be equal to the surface friction torque:

$$\overline{\tau}_0 R \cos \phi = \int_0^{p_0} \mathrm{div}_2 \, \overline{Mv} \frac{dp}{g} + \int_0^{p_0} \left(\frac{\partial \overline{z}}{\partial \lambda}\right) dp. \tag{11.17}$$

This expression can be rewritten by expanding the first term on the right-hand side:

$$\overline{\tau}_0 R \cos \phi = \int_0^{p_0} \mathrm{div}_2 (\overline{u'v'} R \cos \phi) \frac{dp}{g}$$

$$+ \int_0^{p_0} \mathrm{div}_2 (\overline{u} \, \overline{v} R \cos \phi) \frac{dp}{g}$$

$$- \int_0^{p_0} f \overline{v}_{\mathrm{ag}} R \cos \phi \frac{dp}{g}, \tag{11.18}$$

where

$$\overline{v}_{\mathrm{ag}} = \overline{v} - \frac{g}{f R \cos \phi} \frac{\partial \overline{z}}{\partial \lambda}.$$

The last term is difficult to measure because it represents the sensitive balance between wind and geopotential height. In order to expect reasonable results, as pointed out by Holopainen (1982), the ageostrophic wind component should be determined from samples of independent v and z data that are as complete as possible. There is some evidence, although indirect and tentative, of an approximate balance between the second and third terms on the right-hand side of Eq. (11.18) (Holopainen and Oort, 1981). This possible balance implies that the transient eddy term may provide the essential contribution to the surface stress. A map of this last quantity is shown in Fig. 11.10 for annual-mean conditions. The distribution generally shows divergence in the tropical latitudes (source region) and convergence in the middle latitudes (sink region), as one would expect. One finds a tendency for the largest values (about 2 dyn cm^{-2} or 0.2 Pa) to occur over the oceans. The narrow convergence zone over the central Sahara in between two divergence centers is probably unreliable and due to data problems. The problem of spatial data gaps was discussed before in Sec. 5.3.2,

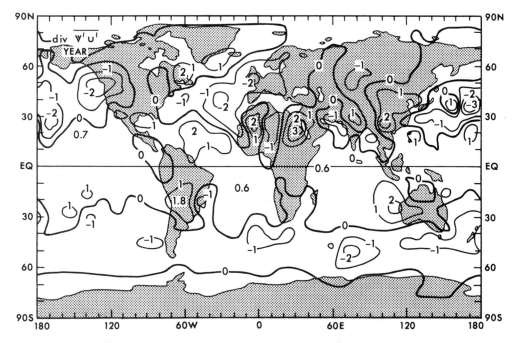

FIGURE 11.10. Global distribution of the vertical-mean divergence of the momentum flux by transient eddies for annual-mean conditions in dyn cm^{-2} (from Oort and Peixoto, 1983).

where the spatial patterns of $\overline{v'u'}$ at individual levels were shown to be unreliable. However, Fig. 11.10 gives the divergence of the vertically integrated values which appears to be more reliable.

The exchange of momentum between the atmosphere and the oceans is also shown in terms of zonal-mean ocean profiles by the dashed curves in Fig. 11.11 [see also Fig. 10.4(a)]. These dashed curves can then be compared with the ocean-plus-land stress values for the entire zonal belt as inferred from the vertically integrated momentum-flux divergence in the atmosphere using Eq. (11.17). The surface stress values over the oceans are usually computed from ship reports with the bulk aerodynamic formula (10.30), as was pioneered by Priestley (1951),

$$\tau_0 = -\rho C_D |\mathbf{v}| u,$$

where C_D represents the drag coefficient. In spite of the uncertainties in the formulation and the fact that only ocean values were included in the dashed curves, the quantitative agreement between the zonal-mean land-plus-ocean values and the all-ocean values in Fig. 11.11 is reasonably good for both the annual-mean and the extreme seasons. However, such a direct comparison is only valid if the (as yet unproven) assumption is made that in a latitude belt the average stress over the continents due to friction and mountains is equal to the average stress as computed over the ocean longitudes alone. Because of the small ocean–land fraction, one should disregard the comparison north of about 50 °N. We should further mention that Hellerman's (1967) stress values around 50 °S are probably too large.

Figure 11.12(a) shows meridional profiles of the zonal-mean divergence of angular momentum integrated over 5° latitude-wide belts in Hadley units (1 Hadley = 10^{18} kg m^2 s^{-2}). The relative contributions by the large-scale mountains are given on the right-hand side of the figure. Comparing these mountain torque curves with the friction plus mountain torque curves on the left-hand side of the figure, one finds

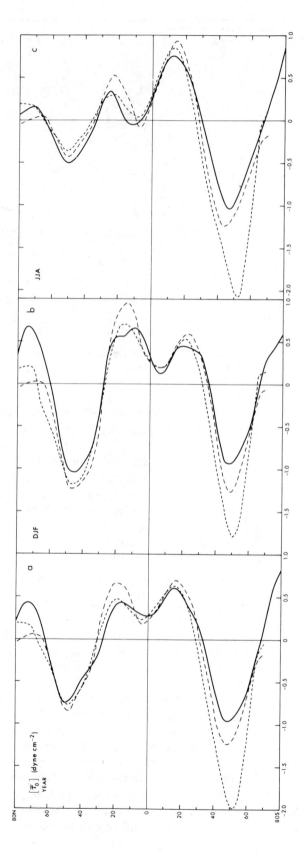

FIGURE 11.11. A comparison of meridional profiles of the zonal-mean surface stress for land plus ocean from Oort and Peixoto (1983; solid curve) with similar profiles for the oceans only from Hellerman (1967; short dash curve) and Hellerman and Rosenstein (1983; long-dash curve) for annual (a), DJF (b), and JJA (c) mean conditions. Units are in dyn cm^{-2} = 0.1 Pa. [To convert to torque units of 10^{18} kg m^2 s^{-2} for 5° latitude belts multiply stress values by $2\pi R^3 \cos^2\phi \, (\pi/36) = 14.2 \cos^2\phi$, where $2\pi R \cos\phi$ = length of latitude circle, $R \cos\phi$ = distance to rotation axis, and $(\pi/36) \, R$ = 5° latitude distance.]

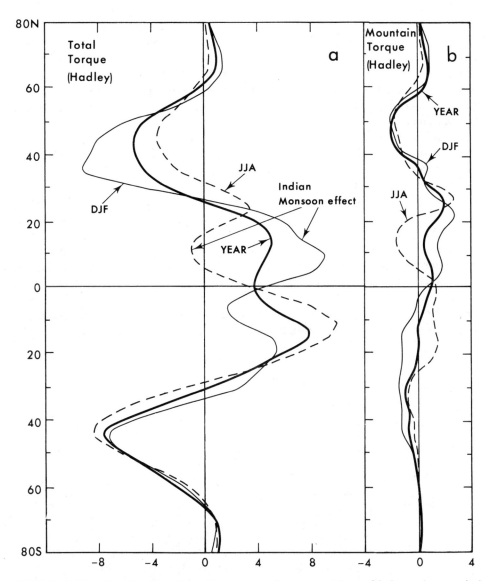

FIGURE 11.12. Meridional profiles of the mean surface torque (due to friction and mountains) (a) integrated over 5° latitude belts in Hadleys (1 Hadley = 10^{18} kg m^2 s^{-2}) and the mean mountain torque (b) after Newton (1971a) in the same units (after Oort and Peixoto, 1983).

qualitative similarity between the two sets of curves, at least in the Northern Hemisphere where the major mountain ranges are located. The strong seasonal variations in the mountain torque in the tropics of both hemispheres are interesting. In general, we may conclude that the mountain torque tends to work in the same direction as the surface friction torque.

11.2.3 Cycle of angular momentum in the climatic system

After studying the flow of angular momentum in the atmosphere and its interactions with the oceans and solid earth, we will consider the cyclic nature of the flow of angular momentum in the climatic system. In order to better show the surface source and sink regions of angular momentum for the atmosphere than was done before, we will use here the nondivergent component of the flow of relative angular momentum. This component was obtained by neglecting the $\Omega R \cos \phi [\bar{v}]$ term in $[\overline{\mathcal{F}_\phi}]$ (i.e., the internal source term of relative angular momentum) in integrating Eqs. (11.16a) and (11.16b) for ψ_M. The outcome is presented in Fig. 11.13. Through these operations the "free-wheeling" circulations of the earth's angular momentum associated with the mean meridional overturnings which do not yield a net transport across latitude circles are eliminated. At the earth's surface Figs. 11.9 and 11.13 should be identical. Figure 11.13 gives a clearer picture of the overall cycle of angular momentum in the atmosphere, better defining the sources and sinks of momentum for the atmosphere.

In the annual-mean picture the circulation of angular momentum is symmetric with respect to the equator with a dominant flow from the source regions in the tropics to the sink regions in midlatitudes. The atmosphere gains westerly momentum in the region of surface easterlies between 30 °S and 25 °N with maxima near 15 °S and 15 °N, where the easterlies are strongest (see Fig. 7.15). Here, in the free atmosphere the upward transport seems to be largely accomplished by the Hadley cells through the term $[\omega]\Omega R^2 \cos^2 \phi$ in Eq. (11.13) because of the $\cos^2 \phi$ dependence. Similarly, the downward transports into the midlatitude sink regions, in the area of surface westerlies, also seem to be accomplished by the mean meridional circulations (i.e., the Ferrel cells). On the other hand, the poleward transports across about 30°

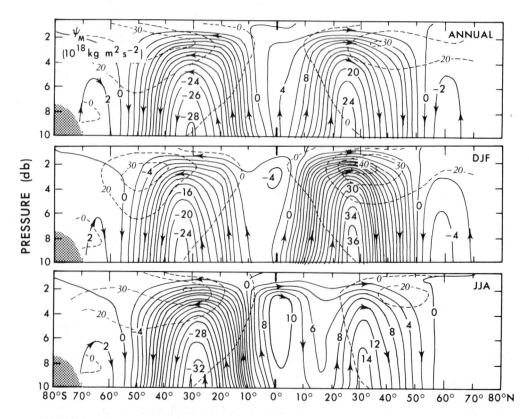

FIGURE 11.13. Streamlines of the nondivergent component of the zonal-mean transport of relative angular momentum in the atmosphere for annual, DJF, and JJA mean conditions in 10^{18} kg m^2 s^{-2}. Added are some dashed contours of $[u]/\cos \phi$ in units of m s^{-1}, which show the countergradient nature of the eddy transports (from Oort and Peixoto, 1983).

latitude in the upper troposphere are due mainly to the large-scale eddies. The polar surface easterlies also provide a weak source of westerly angular momentum for midlatitudes.

Let us next consider the seasonal diagrams. The main source of westerly angular momentum is found in the subtropics of the winter hemisphere, where the surface easterlies are strongest. Most of this angular momentum is transported toward midlatitudes of the same hemisphere, where it is returned to the earth's surface. Only a small fraction is transported across the equator into the summer hemisphere.

In Fig. 11.13, some isolines of $[\bar{u}]/\cos\phi$, which is a measure of the relative angular velocity of the atmosphere, are superposed on the cross section of the stream function, showing a maximum at jet stream levels in both hemispheres. Inspection of the figure shows that there is a flow of angular momentum toward the regions of maximum relative angular velocity. This flow against the gradient is a manifestation of what Starr (1968) called "negative viscosity" phenomena.

As we can see from Figs. 11.7 and 11.13 there is a strong interaction between the eddies and the mean motions in the atmosphere. In fact, the eddies are found to play an important role in forcing and maintaining the zonal-mean circulation fields $[\bar{u}]$, $[\bar{v}]$, and $[\bar{\omega}]$. We must keep in mind that the eddies also transport heat. For example, the existence of the indirect Ferrel cells in midlatitudes can only be explained when these combined forcings are taken into account, as will be discussed in Sec. 14.5.

11.2.4 Angular momentum exchange between the oceans and the lithosphere

To close the cycle of angular momentum, it is necessary that the angular momentum return from middle to low latitudes within the oceans and/or continents. Thus, there has to be an equatorward flow of angular momentum, as shown in Fig. 11.14, within either the oceans or the solid earth (or both) from middle to low latitudes, equal in value but with opposite sign to the transport in the atmosphere. Following Oort (1985, 1989), the first problem is now to determine what the relative roles of the oceans and the solid earth are in this process. To answer this question we will make a

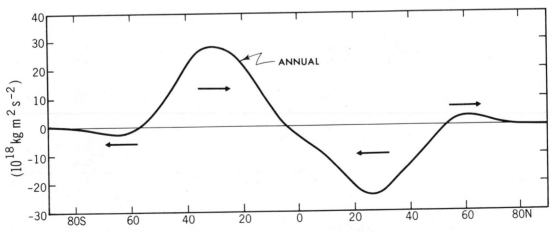

FIGURE 11.14. The annual-mean northward transport of angular momentum in the oceans and/or land in 10^{18} kg m^2 s^{-2} required to close the angular momentum cycle (from Oort, 1985).

rough comparison between the maximum meridional transports to be expected in the oceans and those measured in the atmosphere at middle latitudes. Typical wind velocities in the atmosphere are on the order of 10 m s^{-1} and the directly observed (and therefore reliable) vertical and zonal mean values of the northward flux of momentum in the atmosphere $[\overline{vu}]$ in the middle latitudes are about 10 m^2 s^{-2}. On the other hand, in the oceans typical current velocities are much weaker than in the atmosphere, i.e., on the order of 0.01 to 0.1 m s^{-1}. Therefore, we may expect oceanic $[\overline{vu}]$ values of about 0.001 m^2 s^{-2} or, in other words, values a factor of 10 000 smaller than in the atmosphere. Taking into account the greater mass of the oceans, i.e., about 1000 m of water in the oceans vs 10 m of equivalent water mass in the atmosphere, we find that the oceanic transports are too weak by about a factor of 100. Thus, a different mechanism for the equatorward transport of angular momentum must be found.

What may happen is that the oceans transport the angular momentum *laterally* within each latitude belt, and not across latitude circles. In fact, the oceans must transfer the angular momentum to the continents in the same belt thereby acting as intermediary or hand-over agents between the atmosphere and the continents. This, we speculate, takes place through "continental torques" exerted by the oceans through raised or lowered sea levels along the continental margins in a fashion comparable to the pressure torques across the mountains in the atmosphere. This idea is sketched in Figs. 11.15 and 11.16. In mathematical terms, the continental torque is given by an expression similar to the mountain torque (11.8b):

$$\mathscr{C} = \int_{-D}^{\eta} \int_{\phi_1}^{\phi_2} \sum_i (p_E^i - p_W^i) R^2 \cos \phi \, d\phi \, dz, \qquad (11.19)$$

which represents an area integral over the entire ocean bottom in the latitude belt between ϕ_1 and ϕ_2. In Eq. (11.19), $p_E^i - p_W^i$ represents the difference in bottom pressure between the east and west sides of the ith continent or ith submarine mountain ridge, and is only a function of z. The integration is carried out with respect to z along the boundaries of all continents and marine ridges in the belt between D, the greatest depth of the ocean in the belt, and η, the greatest height of sea level above the geoid. The main contribution to this pressure torque is probably due to the difference in sea level across each continent.

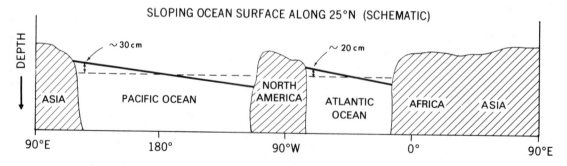

FIGURE 11.15. Schematic diagram of the observed east–west sloping of sea level along the 25 °N latitude circle. The resulting pressure differences across the low-latitude continents together with similar differences (but of opposite sign) across the midlatitude continents may lead to continental torques needed to satisfy global angular momentum constraints (from Oort, 1985).

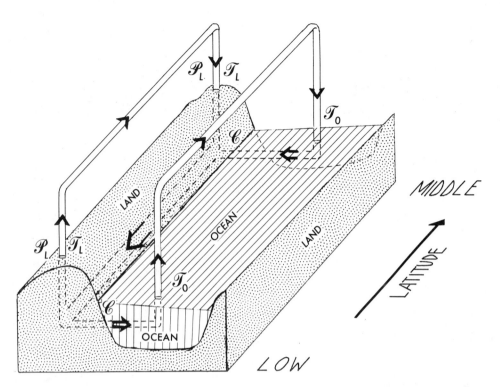

FIGURE 11.16. Schematic diagram of the cycle of angular momentum in the atmosphere–ocean–solid earth system. In the atmosphere, there is a continuous poleward flow of westerly angular momentum with sources in low latitudes through the mountain and friction torques over land $\mathscr{P}_L$ and $\mathscr{T}_L$, and through the friction torque over the oceans $\mathscr{T}_O$. The corresponding sinks of westerly angular momentum are found in the middle and high latitudes. When we consider the atmosphere plus the oceans as the total fluid envelope of the earth, the low-latitude sources and midlatitude sinks are given by the three terms $\mathscr{P}_L, \mathscr{T}_L,$ and the continental torque $\mathscr{C}$. The return equatorward flow of angular momentum must occur entirely in the solid earth (land) (from Oort, 1985).

Because the actual height of sea level is not known globally we have to make some simplifying assumptions. First of all, we only include torque contributions from levels above 1000-m depth. This is equivalent with assuming that $p_E^i = p_W^i$ below 1000-m depth. We then assume that the value of sea level (or dynamic height) η computed at the point where the 1000-m depth level intersects the continent represents the height of sea level for all points between the intersection point and the nearest coast (taken along the same latitude circle). This is a drastic assumption but it enables us to compute the actual bottom pressure p_D from the hydrostatic equation

$$p_D = \rho_0 g \eta + \int_{-D}^{0} \rho_0 g \, dz, \qquad (11.20)$$

using the observed density of sea water ρ_0 which may vary with depth and geographical location.

A cursory look at a map of mean sea level (see Fig. 8.18), as estimated from dynamic height calculations, seems to support our ideas. Differences in sea level between the west and east coasts of the major continents are found to be on the order of 50 cm. It appears that the trade winds pile up the water on the east coasts, leading to a westward torque on the low-latitude continents. Similarly, an eastward torque

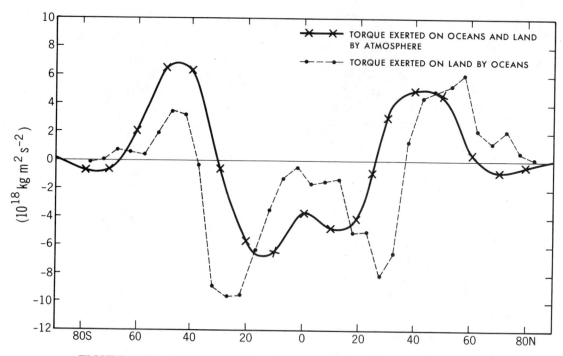

FIGURE 11.17. Meridional profiles of the surface torque exerted by the atmosphere on the oceans and land (solid line), $\mathscr{P}_L + \mathscr{T}_L + \mathscr{T}_O$, based on the data in Fig. 11.14, and the torque exerted by the oceans (due to sloping sea level) on the continents (dashed line) $\mathscr{C}$, based on the oceanographic data in Fig. 8.18 with some further assumptions (see the text). If there were no net surface torques over land ($\mathscr{P}_L + \mathscr{T}_L = 0$), and if we disregard observational inaccuracies, the two curves should almost overlap ($\mathscr{C} \approx \mathscr{T}_O$; see the text). The values represent integrals over 5° latitude belts in Hadley units of 10^{18} kg m² s⁻² (from Oort, 1985).

must be exerted in middle latitudes where westerly winds dominate. The computed meridional profile of the continental torque $\mathscr{C}$ is presented by the dashed curve in Fig. 11.17.

The total required surface torque associated with the mountain torque and the friction torques over land and oceans, $\mathscr{P}_L + \mathscr{T}_L + \mathscr{T}_O$, based on atmospheric data only, and computed by taking the derivative of the transport curve in Fig. 11.14, is also presented in Fig. 11.17 as a solid curve. The agreement between the two curves is remarkable because (1) drastic simplifications were made in computing the oceanic curve, and (2) $\mathscr{P}_L$ and $\mathscr{T}_L$ were neglected in the comparison. It gives support to the hypothesis of a dominant lateral exchange of angular momentum between the oceans and continents as advanced above, or in other words, $\mathscr{T}_O \approx \mathscr{C}$.

Summarizing the previous discussions we have shown that the oceanic stresses $\mathscr{T}_O$ are almost completely applied on the continents at the same latitudes through the $\mathscr{C}$ term, and that, therefore, the total torques on the continents consist of the usual mountain and friction torques at the atmosphere–land interface combined with the new continental pressure torque at the ocean–land interface. However, we have still left unanswered the question of how the required equatorward transports of angular momentum actually take place.

One fact is clear by now, namely that the required return flow of angular momentum from middle to low latitude has to occur almost completely *within* the solid earth.

A possible mechanism that suggests itself is by a preferred tilting of the motions along faults in the continents (Oort, 1985, 1989).

If these considerations are found to be correct, the implications could be important not only in a practical sense, but also in a philosophical sense. The continental torques would link, in a clear manner, the motions in the atmosphere and oceans with some of the motions in the solid earth's crust. Their future study might lead to a better understanding, and perhaps also prediction, of how certain stress patterns may build up in the earth's crust, and how these stresses may eventually be released, perhaps leading to the intermittent occurrence of earthquakes along the fault zones.

CHAPTER 12

Water Cycle

12.1 FORMULATION OF THE HYDROLOGICAL CYCLE

12.1.1 Introduction

The global hydrosphere consisting of various reservoirs (subsystems) connected by the transfers of water in the various phases plays a central role in the climatic system of the earth. In decreasing order of water amount held in storage, the five reservoirs of the hydrosphere are: the world oceans, the ice masses and snow deposits, the terrestrial waters, the atmosphere, and, finally, the biosphere.

Vast quantities of water are continuously on the move in the climatic system. Under the direct or indirect influence of solar energy, water evaporates from the oceans and continents and is transpired by plants and animals into the atmosphere. In the atmosphere, water is transported in the condensed phase (liquid water and ice crystals) as clouds, or in the vapor phase (water vapor). It falls on the continents and oceans in the form of rain, snow, dew, or hail, or other forms of precipitation. The water then returns to the atmosphere through evaporation and evapotranspiration, infiltrates into the ground or runs off over or under the ground to the rivers and streams, which carry it back to the oceans and seas (see Figs. 12.1 and 12.2).

This gigantic and complex system of transport of water substance in its many forms and through many stages constitutes the hydrological cycle, and is a consequence of the conservation of water substance. Not all water of the climatic system takes part in this continuous cycle. Some remains for shorter or longer periods of time in the atmosphere, in the biosphere, as snow accumulations in the cryosphere, in the oceans, river valleys, reservoirs, and lakes, or as connate waters chemically or physically bound to the lithosphere (soil and rocks), etc.

The hydrologic cycle has two major branches—the terrestrial and the atmospheric branches. The terrestrial branch consists of the inflow, outflow, and storage of water in its various forms on and in the continents and in the oceans, while the atmospheric branch consists of the atmospheric transports of water, mainly in the vapor phase. The two branches of the hydrologic cycle join at the interface between the atmosphere and the earth's surface (including the ocean). Figures 12.1 and 12.2 schematically show the distribution and the flow of water in the climatic system.

WATER CYCLE 271

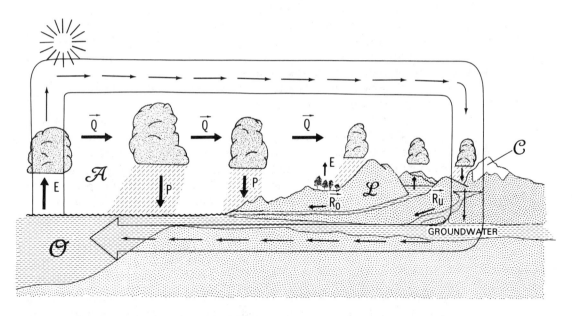

FIGURE 12.1. Schematic diagram of the atmospheric and terrestrial branches of the hydrological cycle showing the importance of evaporation E, advection of water vapor in the atmosphere $\mathbf{Q}$, precipitation P, river runoff $\mathbf{R}_0$, and underground runoff $\mathbf{R}_u$.

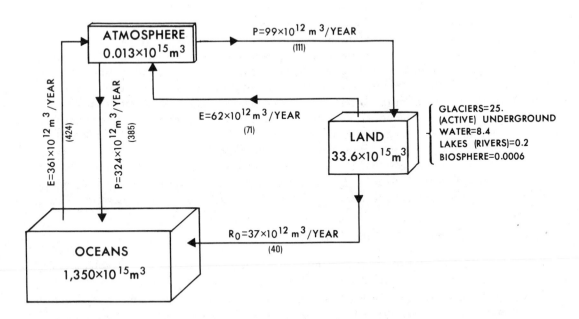

FIGURE 12.2. The amounts of water stored in the oceans, land, and atmosphere, and the amounts exchanged annually between the different reservoirs through evaporation, precipitation, and runoff (estimates are from Peixoto and Kettani, 1973, and, in parentheses, from Baumgartner and Reichel, 1975).

The loss or "output" of water from the earth's surface through evaporation and evapotranspiration is the "input" of water for the atmospheric branch, whereas precipitation, the atmospheric output, may be regarded as a gain for the terrestrial branch of the hydrological cycle. Water is thus one of the crucial links between the various components of the climatic system. It is clear that the atmospheric and terrestrial branches of the hydrological cycle have to be taken as a whole, bringing together the sister disciplines of meteorology and hydrology.

As we have seen in Sec. 7.6, the distributions of precipitation and evaporation over the globe show that there is an excess of precipitation over evaporation in the equatorial region associated with the intertropical convergence zone and a similar excess in midlatitudes associated with baroclinic perturbations along the polar front, whereas in the subtropical and polar regions evaporation exceeds precipitation. Thus, the water vapor released mainly over the subtropical oceans is continuously transported both equatorward and poleward to maintain the moisture supply for the observed precipitation belts. This shows the fundamental role played by the atmosphere and its general circulation as a forcing factor in maintaining the hydrological cycle. This fact has long been recognized by climatologists, hydrologists, and glaciologists. However, quantitative studies of the gaseous hydrosphere and of its aerial runoff have only been possible in recent decades since the development of an adequate network of aerological stations. Traditionally, hydrologists have studied the hydrological cycle in a piecemeal way, considering only the earth-bound branch. However, progress along this line has been kept back by the difficulties in obtaining reliable values of evaporation, change of water storage, and precipitation. Estimates of evaporation, based on the diffusion theory or on the energy-budget approach, are uncertain and may differ widely. Precipitation data over the oceans are still rather scarce and difficult to interpret, especially if they are taken on islands.

Water can exist in the climatic system in various phases, and there are considerable amounts of thermodynamic energy involved in the phase transitions, namely in the form of various latent heats. For example, condensation and freezing release considerable amounts of heat, while evaporation and melting require similar amounts of heat. Evaporation occurs mainly at the surface of the oceans and continents leading to a transfer of latent heat into the atmosphere, that is realized as sensible heat where the water vapor condenses. Furthermore, water vapor is an important selective absorber of solar and terrestrial radiation. These facts, as well as the radiative effects of clouds, make water a crucial factor in the energetics of the climatic system.

Water vapor can sometimes be used profitably as a tracer for the circulation in the atmosphere through an analysis of the various modes of transport. Such an analysis may help to identify the important processes that take place in the general circulation.

12.1.2 Water in the climatic system

The amount of water substance on earth does not vary appreciably, although some new water is being produced by volcanoes and hot springs. However, most steam ejected by volcanoes is actually either rainwater that in earlier times had saturated the upper layers of rock, or seawater that was trapped at the time the marine sediments were deposited. A certain amount of water vapor may be destroyed in the upper atmosphere through photodissociation by solar radiation. All these effects are negligible on the time scales we are concerned with here so that the total amount of water vapor on earth can indeed be considered to be constant.

The main reservoirs of water on the earth, namely the oceans, cryosphere, terrestrial waters, and atmosphere, are shown schematically in Fig. 12.2. The interior of the earth also contains an appreciable amount of water, either dissolved or chemically combined with solid or molten rock, but there is no satisfactory estimate of the amount of water thus locked up.

As shown in Fig. 12.2 about $1,350 \times 10^{15}$ m^3 of water (about 97% of all the water in the hydrosphere) is contained in the oceans; while 33.6×10^{15} m^3 (the remaining 2.4%) is found on and in the continents, mostly in the glaciers of the Arctic and the Antarctic. The atmosphere contains 0.013×10^{15} m^3 or only one hundred-thousandth of the total water present on the earth. The transfer rates between the various reservoirs are also shown in Fig. 12.2. They are associated mainly with precipitation, evaporation, and runoff which were discussed extensively in Sec. 7.6.

The water on the continents is distributed in several reservoirs, namely, in glaciers (25×10^{15} m^3), ground water (8.4×10^{15} m^3), lakes and rivers (0.2×10^{15} m^3), and in the living matter of the biosphere (0.0006×10^{15} m^3). The amount of water locked in the polar ice is impressively large, totaling some 1.8% of all the water in the hydrosphere. Of the total amount of underground water, vadose water (water present in soils) accounts for only 0.066×10^{15} m^3. The remainder is almost evenly divided between reservoirs deeper than 800 m and reservoirs shallower than that level (for an extensive review of the world water resources, see L'vovich and Ovtchinnikov, 1964, and L'vovich, 1979).

The distribution of terrestrial water has slowly changed with time. For instance, continental ice caps have periodically grown and melted during the past two million years. Such fluctuations must, of course, have been accompanied by profound changes in the hydrological cycle.

The residence time of water in the various reservoirs can be deduced from the ratio of the amount of water in the particular reservoir and the accumulation or depletion rate as presented in Fig. 12.2. It is found to vary from about ten days for atmospheric water vapor to thousands of years for the polar ice and the oceans.

12.2 EQUATIONS OF HYDROLOGY

12.2.1 Classic equation of hydrology

The classic equation of hydrology is obtained as a water balance requirement for the terrestrial branch of the hydrological cycle. Applying the principle of continuity to a specific region, the balance equation for the terrestrial branch may be written as

$$S = P - E - R_0 - R_u, \qquad (12.1)$$

where

S = rate of storage of water,
P = precipitation rate (in liquid and solid phase),
E = evaporation rate (which includes evapotranspiration over land and sublimation over snow and ice),
R_0 = surface runoff, and
R_u = subterranean runoff.

For a large land region, the net subterranean runoff is usually small so that the classic equation can be simplified to the form

$$\{\overline{S}\} = \{\overline{P} - \overline{E}\} - \{\overline{R}_0\}, \qquad (12.2)$$

where $\overline{(\ )}$ denotes a time average and $\{\ \}$ a space average over the region of area A, $\{\overline{S}\}$ is the rate of change of total surface and subterranean water storage, $\{\overline{P} - \overline{E}\}$ is the average rate of precipitation minus evaporation per unit area, and $\{\overline{R}_0\}$ is the average rate of runoff. For long periods of time and large areas $\{\overline{S}\}$ tends to be small compared to the other terms, and Eq. (12.2) can be reduced to

$$\{\overline{E}\} = \{\overline{P}\} - \{\overline{R}_0\}. \tag{12.3}$$

Traditionally, the quantity of major interest for hydrology has been runoff which can be measured fairly accurately through gauging the extensive network of streams. Precipitation is the primary cause for runoff. As we have seen in Tables 7.1 and 7.2, these two quantities, R_0 and P, are measured and recorded over many continental areas of the globe.

Measurements of evaporation, evapotranspiration, and the change in water storage are very difficult to make, although some semi-empirical estimates of these quantities are available for local areas. Since there is an increasing need for large-scale estimates of evapotranspiration and soil moisture storage for the planning of more efficient irrigation systems, for various water resource projects and, in general, for the optimum use of the available water, hydrologists are now beginning to study the atmospheric branch of the water cycle.

12.2.2 Balance equation for water vapor

The formulation of the atmospheric branch of the hydrological cycle is based on the balance requirements of water vapor in the atmosphere. We will follow the general discussions of Peixoto (1973) and Peixoto and Oort (1983).

The amount of water vapor contained in a unit area column of air which extends from the earth's surface to the top of the atmosphere is given by the expression

$$W(\lambda,\phi,t) = \int_0^{p_0} q \frac{dp}{g}, \tag{12.4}$$

where q is the specific humidity. The term W can be called the amount of precipitable water in the atmosphere. It represents the amount of liquid water that would result if all the water vapor in the unit column of the atmosphere were condensed. It is usually expressed in units of g cm^{-2} or cm. By integrating the horizontal transport of water vapor with respect to pressure the "aerial runoff" $\mathbf{Q}$ is obtained:

$$\mathbf{Q}(\lambda,\phi,t) = \int_0^{p_0} q\mathbf{v} \frac{dp}{g} = Q_\lambda \mathbf{i} + Q_\phi \mathbf{j}. \tag{12.5}$$

The zonal and meridional components of $\mathbf{Q}$ are given by

$$Q_\lambda = \int_0^{p_0} qu \frac{dp}{g} = \langle qu \rangle p_0/g,$$

$$Q_\phi = \int_0^{p_0} qv \frac{dp}{g} = \langle qv \rangle p_0/g.$$

The term Q_ϕ represents the flux of water vapor across a unit strip of a latitudinal wall (ϕ = constant) and Q_λ the flux across a unit strip of a meridional wall (λ = const). The previous equations may be averaged in time leading to the corresponding mean values $\overline{W}$, $\overline{\mathbf{Q}}$, $\overline{Q}_\lambda$, and $\overline{Q}_\phi$.

When the basic balance equation for water vapor (3.79) and the mass continuity equation (3.4) in the (x, y, p, t) system are combined, the budget of water vapor in the atmosphere can be expressed by the following balance equation:

$$\frac{\partial q}{\partial t} + \text{div } q\mathbf{v} + \frac{\partial(q\omega)}{\partial p} = s(q) + D, \qquad (12.6)$$

where the source-sink term $s(q)$ equals the rate of generation or destruction of water vapor within a unit mass of air associated with phase changes, and $D = -\alpha \text{ div } \mathbf{J}_q^D$ the molecular and turbulent eddy diffusion through the boundaries. The main sources and sinks of water vapor in the atmosphere are primarily evaporation and condensation and, to a lesser extent (except near the earth's surface), diffusion from the surroundings. Thus, $s(q) = e - c$, where e is the rate of evaporation (plus sublimation) and c the rate of condensation per unit mass (in units of g of water per kg moist air per second).

Similarly as for water vapor, we can write a balance equation for the condensed phase (liquid plus solid) q_c, noting that the rate of formation in clouds is given by $s(q_c) = -s(q)$, so that

$$\frac{\partial q_c}{\partial t} + \text{div } q_c\mathbf{v} + \frac{\partial q_c \omega_c}{\partial p} = -(e - c). \qquad (12.7)$$

In this equation, ω_c is the vertical velocity of the water droplets or snow and solid ice particles and $q_c \omega_c$ the net vertical transport of the condensed water.

A balance equation for the total water content is obtained by adding Eqs. (12.6) and (12.7):

$$\frac{\partial q}{\partial t} + \text{div } q\mathbf{v} + \frac{\partial(q\omega)}{\partial p} + \frac{\partial q_c}{\partial t}$$
$$+ \text{div } q_c\mathbf{v} + \frac{\partial q_c \omega_c}{\partial p} = D. \qquad (12.8)$$

Vertical integration of this equation with respect to pressure between the earth's surface and the top of the atmosphere gives the balance equation for the total water substance in the atmosphere:

$$\frac{\partial W}{\partial t} + \text{div } \mathbf{Q} + \frac{\partial W_c}{\partial t} + \text{div } \mathbf{Q}_c + P = E, \qquad (12.9)$$

where W_c is the amount of condensed water in a unit area column of air:

$$W_c(\lambda, \phi, t) = \int_0^{p_0} q_c \, dp/g \qquad (12.10)$$

and $\mathbf{Q}_c$ the horizontal transport vector of condensed water:

$$\mathbf{Q}_c(\lambda, \phi, t) = \int_0^{p_0} q_c \mathbf{v} \, dp/g = Q_{c\lambda}\mathbf{i} + Q_{c\phi}\mathbf{j}. \qquad (12.11)$$

Usually $\partial W_c/\partial t \ll \partial W/\partial t$ and $\mathbf{Q}_c \ll \mathbf{Q}$ so that both the time rate of change of the liquid and solid water in clouds and their horizontal transports can be neglected in Eq. (12.9). Thus, we can simplify the general balance equation after averaging in time to

$$\frac{\overline{\partial W}}{\partial t} + \text{div } \overline{\mathbf{Q}} = \overline{E} - \overline{P}. \qquad (12.12)$$

A possible exception is the case of the formation of dense cumulonimbus clouds in the tropics or over warm ocean currents, such as the Gulf Stream and Kuroshio currents during winter (Peixoto, 1973).

Equation (12.12) shows that the excess of evaporation over precipitation at the earth's surface is balanced by the local rate of change of water vapor storage $\partial W/\partial t$ and by the inflow or outflow of water vapor, div $\mathbf{Q}$. Averaging of Eq. (12.12) in space over a region bounded by a conceptual vertical wall (e.g., a river-drainage basin or an interior sea) leads to another form of Eq. (12.12):

$$\left\{\overline{\frac{\partial W}{\partial t}}\right\} + \{\text{div } \overline{\mathbf{Q}}\} = \{\overline{E} - \overline{P}\}. \tag{12.13}$$

Using the Gauss theorem, Eq. (12.13) may be rewritten in a form, which is often more useful for regional studies:

$$\left\{\overline{\frac{\partial W}{\partial t}}\right\} + (1/A)\oint (\overline{\mathbf{Q}}\cdot\mathbf{n})d\gamma = \{\overline{E} - \overline{P}\}, \tag{12.14}$$

where A denotes the area of the region and $\mathbf{n}$ the unit vector directed outward, normal to the boundary of the region.

Equations (12.13) and (12.14) describe the atmospheric branch of the hydrological cycle. Except in the case of severe storms and for short intervals of time, the rate of change of precipitable water $\partial W/\partial t$ is very small compared with the other terms. Thus, for sufficiently long periods of time, divergence of water vapor is found over those regions of the globe where evaporation exceeds precipitation, whereas convergence is found where precipitation is greater than evaporation.

The term $\{\overline{E} - \overline{P}\}$ is common to Eqs. (12.2) and (12.13) and establishes the connection between the terrestrial and atmospheric branches of the hydrological cycle. Elimination of $\{\overline{E} - \overline{P}\}$ between these two equations yields

$$\{\overline{R}_0\} + \{\overline{S}\} = -\{\text{div } \overline{\mathbf{Q}}\} - \left\{\overline{\frac{\partial W}{\partial t}}\right\}, \tag{12.15}$$

which shows how the two branches of the hydrological cycle are linked together.

If, besides the aerological terms, $\{\overline{R}_0\}$ and $\{\overline{P}\}$ are known over a certain catchment basin from stream flow and precipitation data, one can estimate the rate of change in ground water and the rate of evaporation. Thus, using finite differences as is usually done in hydrology, Eq. (12.15) may be written

$$\{\overline{S}\} = -\{\overline{\Delta W/\Delta t}\} - (1/A)\oint (\overline{\mathbf{Q}}\cdot\mathbf{n})d\gamma - \{\overline{R}_0\}. \tag{12.16}$$

Similarly the mean evaporation can be obtained from Eq. (12.14):

$$\{E\} = \{\overline{\Delta W/\Delta t}\} + (1/A)\oint (\overline{\mathbf{Q}}\cdot\mathbf{n})d\gamma + \{\overline{P}\}. \tag{12.17}$$

Over long periods of time, such as a year, changes in storage in the land and in the atmosphere become small so that, e.g., for a continent the surface and subsurface runoff have to be exactly balanced by the aerial "runoff" into the continent from the surrounding ocean areas.

When the entire global atmosphere is considered over a long period of time, all transport and storage terms vanish, and we can conclude from Eq. (12.12) that the global-mean evaporation has to be equal to the global-mean precipitation.

12.2.3 Modes of water vapor transport

The water balance equation (12.12) for a zone bounded by the latitudes ϕ and $\phi + \Delta\phi$ can be written as

$$\overline{\frac{\partial [W]}{\partial t}} + \Delta [\overline{Q}_\phi] \cos \phi / R \cos \phi \Delta\phi = [\overline{E} - \overline{P}], \qquad (12.18)$$

where $Q_\phi = \int_0^{p_0} qv \, dp/g$. To get a better understanding of the various mechanisms responsible for the global transports of atmospheric water vapor we will expand the zonally averaged transports in the space-time domain (Starr and Peixoto, 1971), similarly as was done before in the case of angular momentum.

For example, for the mean northward transport of water vapor across a specific latitude circle at a certain pressure level we may use expansion (4.9):

$$[\, \overline{qv} \,] = [\overline{q}][\overline{v}] + [\overline{q}^*\overline{v}^*] + [\, \overline{q'v'} \,]. \qquad (12.19)$$

Here, the term $[\overline{q}][\overline{v}]$ represents the transport of water vapor by the mean meridional circulation which dominates in the tropics. The term $[\overline{q}^*\overline{v}^*]$ is associated with the mean stationary eddies of the general circulation, such as the semipermanent subtropical anticyclones and the semipermanent cyclones prevailing in high latitudes. Finally, the term $[\, \overline{q'v'} \,]$ represents the northward transport of moisture due to the transient perturbations which develop along the polar front and in the intertropical convergence zone.

Integration of Eq. (12.19) in the vertical then leads to the expansion of the total meridional transport in terms of the various modes of transfer:

$$[\overline{Q}_\phi] = (1/g) \int [\overline{q}][\overline{v}] dp$$

$$+ (1/g) \int [\overline{q}^*\overline{v}^*] dp + (1/g) \int [\, \overline{q'v'} \,] dp. \qquad (12.20)$$

Similarly we can consider the vertical transport of water vapor $q\omega$ and expand its zonal and time mean into the various components

$$[\, \overline{q\omega} \,] = [\overline{q}][\overline{\omega}] + [\overline{q}^*\overline{\omega}^*] + [\, \overline{q'\omega'} \,]. \qquad (12.21)$$

The vertical transport of moisture constitutes a link between the earth's surface and the free atmosphere, providing water vapor and latent heat to the upper levels of the atmosphere. The transient eddy term $[\, \overline{q'\omega'} \,]$ describes the vertical transfer of water vapor due to turbulent diffusion, cumulus convection, and other mesoscale convective phenomena. A similar expression as Eq. (12.21) can be derived for the condensed phase. In that case the term $[\, \overline{q_c \omega_c} \,]$ represents the vertical transport of water droplets or solid ice particles, which, in general, is downward when $[\, \overline{q\omega} \,]$ is upward. It represents the rate of precipitation at a given pressure level.

12.3 OBSERVED ATMOSPHERIC BRANCH OF THE HYDROLOGICAL CYCLE

12.3.1 Water vapor in the atmosphere

12.3.1.1 Precipitable water

The annual-mean distribution of the specific humidity near the surface expressed in g's of water vapor per kg of moist air is shown in Fig. 12.3(a). As expected, the highest humidities of 18 to 19 g kg^{-1} are found in the equatorial regions. There is a continuous decrease of humidity with latitude down to very low values on the order of 1 g kg^{-1} or less over the polar regions. The global pattern of humidity resembles the temperature patterns in Fig. 7.4(a) since the capacity of the atmosphere to retain water vapor depends strongly on temperature. Of course, there are exceptions in the major desert regions where the surface air is very dry compared to the zonally averaged humidity in spite of its relatively high temperature.

The January minus July difference map of the surface specific humidity in Fig. 12.3(b) shows that the largest annual variations occur in relatively low latitudes over the continents. The variations associated with the Asian monsoon are also very pronounced. Comparing this difference map of specific humidity with the corresponding map of surface temperature in Fig. 7.4(c) we recognize that similar features appear on both maps. However, in the case of humidity the centers are displaced equatorward by about 15° latitude over North Africa, 30° latitude over Asia and North America, and 10° latitude over the Southern Hemisphere continents.

The spatial distribution of the annual-mean precipitable water content $\overline{W}$ is represented in Fig. 12.3(c). With only a few exceptions, the analysis shows a continuous decrease of precipitable water from the equatorial regions, where it attains the highest values, to the north and south poles.

The departures from zonal symmetry are associated with the physiography of the earth's surface, and are apparent in both hemispheres. As a general rule, the precipitable water is higher over the oceans than over the continents. The deflection of the isolines near the western and eastern coasts of the continents is reinforced by the topography and the presence of warm and cold ocean currents. The distribution over the Southern Hemisphere is practically zonal, since the ocean coverage exceeds by far that of the continents. As expected, the lowest values of $\overline{W}$ (< 5 kg m^{-2}) occur over subpolar and polar regions.

The precipitable water over the desert areas is considerably smaller than the corresponding zonal average, mainly due to strong subsidence. This effect is pronounced in the eastern portions of the large semipermanent anticyclones of the subtropics. In addition, the effects of high terrain on the precipitable water distribution are illustrated by relatively dry areas (often $\overline{W} < 10$ kg m^{-2}) over the major mountain ranges, such as the Rockies, Himalayas, highlands of Ethiopia, and the Andes. The effects of topography and the land–sea contrast in the Southern Hemisphere are shown by the dipping of the 20 kg m^{-2} isoline towards lower latitudes.

It is also useful to consider water vapor in terms of the relative humidity of the air U, given by the ratio of the actual vapor pressure and the saturation vapor pressure [see Eq. (3.62)]. Figure 12.3(d) shows the horizontal distribution of relative humidity in the lower troposphere for annual-mean conditions. It clearly shows the humid

WATER CYCLE 279

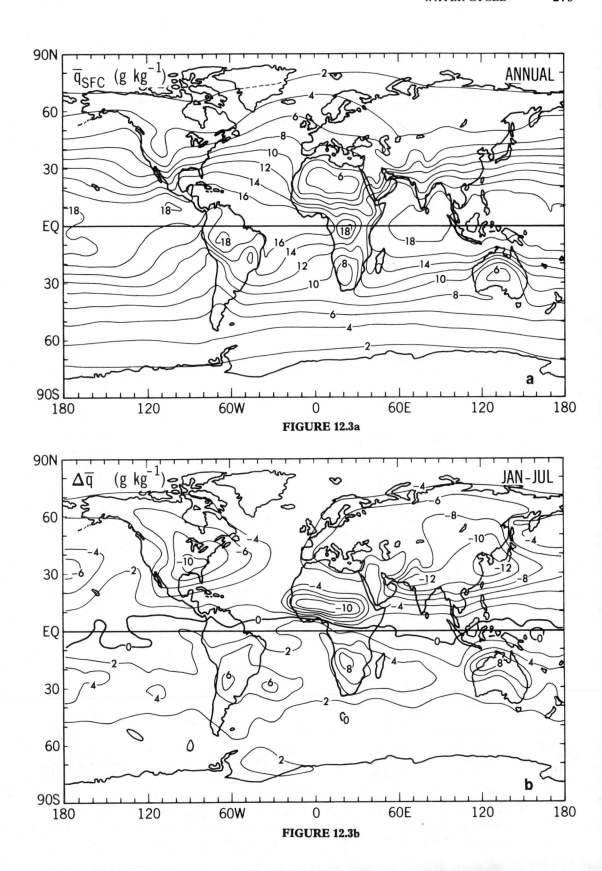

FIGURE 12.3a

FIGURE 12.3b

280   PHYSICS OF CLIMATE

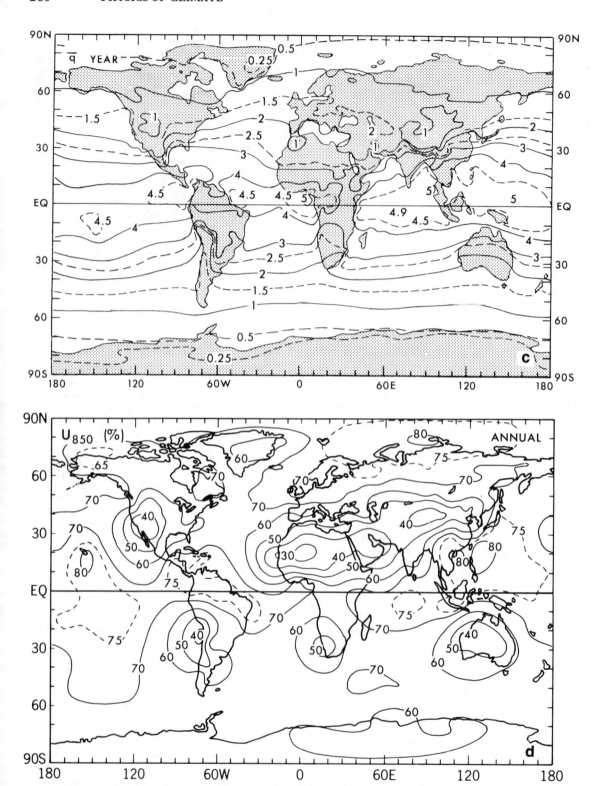

FIGURE 12.3. Global distributions of the surface specific humidity for annual-mean (a) and January-minus-July conditions (b), of the vertical-mean specific humidity (c) in g kg^{-1}, and of the relative humidity (d) at 850 mb in %. [When multiplied by p_0/g (≈ 1 over the oceans and low-level land areas) the third field gives the precipitable water in the atmosphere in units of 10 kg m^{-2}.]

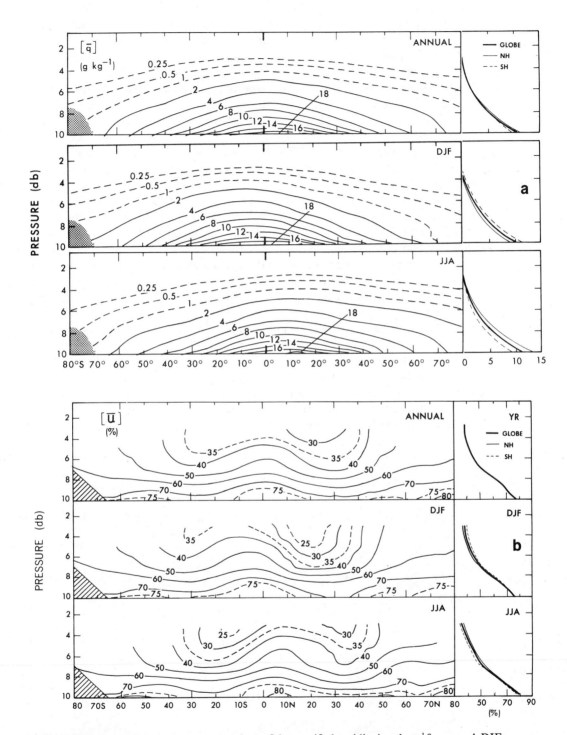

FIGURE 12.4. (a) Zonal-mean cross sections of the specific humidity in g kg^{-1} for annual, DJF, and JJA mean conditions. Vertical profiles of the hemispheric and global mean values are shown on the right. (b) Zonal-mean cross sections of the relative humidity in % for annual, DJF, and JJA mean conditions.

equatorial regions with values on the order of 75% and the dry subtropical regions with minimum values of 30%–40% over the continental deserts. As expected, the relative humidity is higher, in general, over the oceans than over land.

The vertical structure of water vapor is shown in Figs. 12.4(a) and 12.4(b) in the form of zonal-mean cross sections of the mean specific humidity $[\bar{q}]$, and the mean relative humidity $[\bar{U}]$, respectively. The specific humidity decreases rapidly with height, almost following an exponential law as shown in the vertical profiles on the right-hand side of Fig. 12.4(a). It also decreases with latitude. More than 50% of the water vapor is concentrated below the 850-mb surface, while more than 90% is confined to the layer below 500 mb. The seasonal variations are more intense in the Northern than in the Southern Hemisphere, as expected from the corresponding temperature variations.

The cross sections of relative humidity in Fig. 12.4(b) also show a decrease with height but weaker than in the case of specific humidity. The relative humidity ranges from values on the order of 70% to 80% near the surface to 30%–50% near the 400-mb level. The latitudinal gradients tend to increase with height, reflecting the influence of the rising and sinking branches of the mean meridional circulation (see Figs. 7.18 and 7.19). Thus, in the annual mean, strong sinking motions near 20°–30° latitude lead to a minimum in relative humidity at those latitudes in each hemisphere, whereas the rising motions in the ITCZ give rise to the maximum just north of the equator. The seasonal differences in intensity and location of the Hadley cells are clearly reflected in the latitudinal shifts from about 25° latitude in winter to 35° latitude in summer. The seasonal range has a maximum of $\Delta U \simeq 15\%$–20% near 15° latitude in each hemisphere at the 500-mb level.

The variability in time and space of the vertical moisture distribution is portrayed in Figs. 12.5(a)–12.5(d) for annual-mean and seasonal conditions. Figures 12.5(a) and 12.5(c) represent the intra-annual (seasonal and synoptic) temporal variability

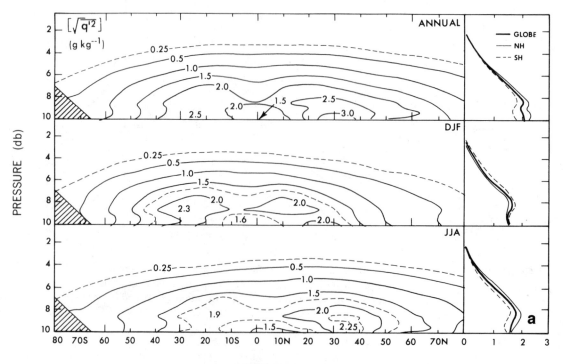

FIGURE 12.5a

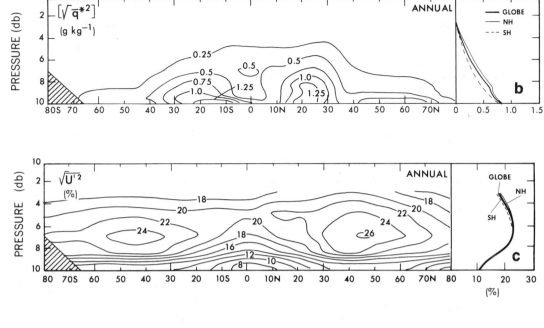

FIGURE 12.5. (a) Zonal-mean cross sections of the day-to-day (transient eddy) standard deviation of the specific humidity in g kg^{-1} for annual, DJF, and JJA mean conditions. Vertical profiles of the hemispheric and global-mean values are shown on the right. (b) Zonal-mean cross section of the east–west standard deviation of the annual-mean specific humidity field in g kg^{-1}. (c) Zonal-mean cross section of the day-to-day (transient eddy) standard deviation of the relative humidity in % for annual-mean conditions. (d) Zonal-mean cross section of the east–west standard deviation of the annual-mean relative humidity field in %.

and Figs. 12.5(b) and 12.5(d) the east–west spatial variability. The variability in specific humidity is most pronounced near the surface showing a bimodal distribution with maxima in the subtropics of each hemisphere. The temporal variability in relative humidity in Fig. 12.5(c) increases with height with a maximum of about 25% at 700 mb at midlatitudes. The spatial variability in relative humidity in Fig. 12.5(d) is much weaker with maxima on the order of 10% near 850 mb in the subtropics.

Using the grid point values of the $\overline{W}$ maps, the zonally averaged amount of water in the atmosphere $[\overline{W}]$ can be evaluated. The results are shown in Fig. 12.6(a). The values correspond to the vertical integrals of the values presented in the previous cross sections. These profiles give the gross distribution of water vapor in the atmosphere. The seasonal profiles are almost symmetrical with respect to the annual curve. They show a maximum in the equatorial zone, with a slight seasonal migration

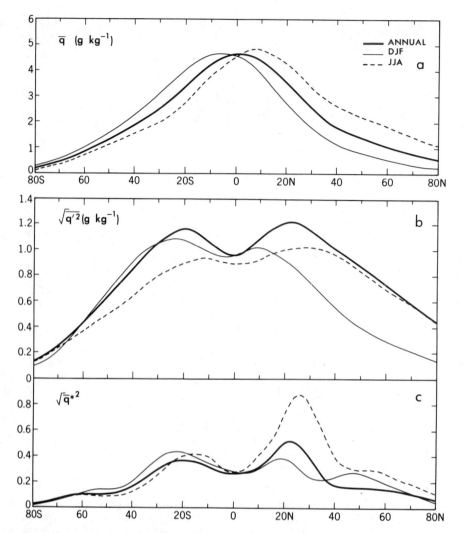

FIGURE 12.6. Meridional profiles of the vertical- and zonal-mean values of the time-mean specific humidity (a), the day-to-day standard deviation of the specific humidity (b), and the east–west standard deviation of the time-mean specific humidity (c) in g kg^{-1} for annual, DJF, and JJA mean conditions.

into the summer hemisphere, and a monotonic decrease to polar latitudes with the steepest gradients in the subtropics.

The global water content in the atmosphere is on the order of 13.1×10^{15} kg which is equivalent with a uniform layer of about 2.5 cm of water covering the globe. Seasonal changes in the hemispheric water content are more pronounced in the Northern than in the Southern Hemisphere with summer minus winter values of 3.0 and 1.8×10^{15} kg (or 1.2 and 0.7 cm of water covering the hemisphere), respectively.

Assuming a mean annual precipitation value over the globe of 1.0 m, the ratio of the amount of water in the air and the precipitation rate leads to a residence time of water in the atmosphere of (0.025/1.00) yr or about 9 days. This indicates that the water vapor in the atmosphere is replenished about 40 times a year.

As expected, the temporal variability in Fig. 12.6(b) is largest for the year since it includes the interseasonal variability, which is most pronounced in the subtropics [see

Fig. 12.3(b)]. The spatial variability as given in Fig. 12.6(c) shows a maximum around 20 °N with a large difference between the winter and summer values. This maximum is associated with the large amount of water vapor over the western borders of the subtropical highs during summer and over the southeast Asian monsoon area relative to the zonal average.

12.3.2 Transport of water vapor

12.3.2.1 Zonal transport of water vapor

A map of the field of total zonal transport of water vapor $\overline{Q}_\lambda$ for yearly conditions is presented in Fig. 12.7(a). The flow of water vapor reflects the planetary behavior of the general circulation in the lower half of the atmosphere, since the specific humidity q acts as a weighting factor for the wind field. Thus, the general pattern of the $\overline{Q}_\lambda$ maps is consistent with the distribution of the mean zonal flow, i.e., westerlies in midlatitudes and easterlies in the tropical belt (see Fig. 7.1). The zero isoline coincides with the mean location of the centers of the subtropical anticyclones in both hemispheres.

The zonal belt of easterly transport is interrupted over some continental areas, such as over India, associated with the summer-monsoon circulation. The westerly transport in midlatitudes is associated with the circulations along the polar fronts and along the polar borders of the subtropical anticyclones.

The field of transient eddy zonal transport of water vapor in Fig. 12.7(b), although weaker than the time-mean transport, presents a considerable degree of east–west detail with centers of eastward (positive) transport alternating with centers of westward (negative) eddy transport in both hemispheres. It may be noted that the map of the zonal transient eddy transport of water vapor shows very little similarity with the map of the total zonal flux. In fact, the eddy field is much less orderly and the intensities are at least one order of magnitude smaller than those for the total flux. The distribution of the centers over the globe is associated with the land–sea contrast, and with the invasion of moist air masses of maritime origin into the continents with synoptic time scales of a few days to one or two weeks. The map reveals the strong influence of the continental character of the Northern Hemisphere. In this hemisphere, the moist zonal eddy transport is predominantly westward (negative), with an extensive longitudinal belt of eastward transport at 10 °N. In the Southern Hemisphere, the flow of water vapor is mostly eastward, except over Australia, South Africa, South America, and Antarctica where the flow is westward. A negative value of the zonal eddy transport indicates that the air is relatively dry when the wind is from the west and relatively humid when the wind is from the east.

The vertical structure of the mean zonal flux is depicted in the cross section of $[\overline{qu}]$ given in Fig. 12.8. Within the tropics, there is a very strong westward transport below 800 mb with one main center at the surface near 15° latitude in each hemisphere. Some weak westward (negative) transport is observed in subpolar latitudes. In middle latitudes, the figure shows two well-developed centers of eastward flow attaining their maximum values just above the 850-mb level. The maximum in the Southern Hemisphere is larger than the corresponding maximum in the Northern Hemisphere. The flow is still fairly large at 400 and 300 mb in both hemispheres, and cannot be disregarded.

Meridional profiles of the total zonal flow of water vapor are presented in Fig. 12.9. These profiles summarize the main characteristics of the zonal transport of moisture

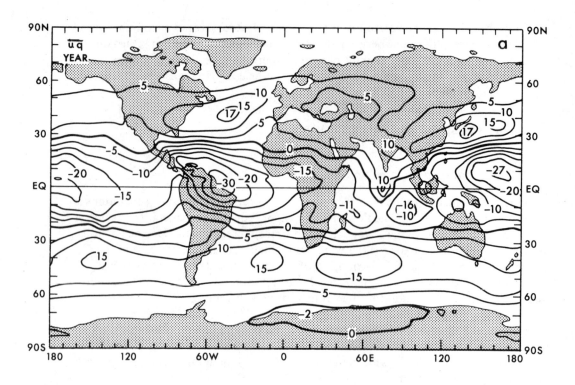
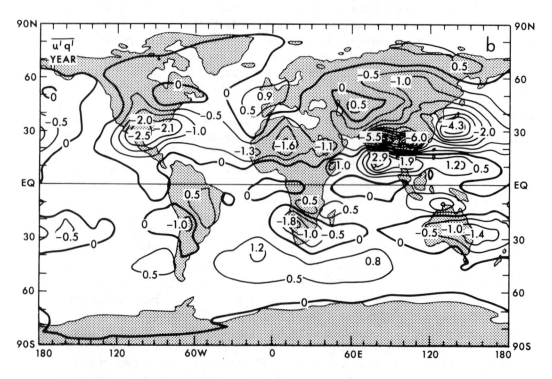

FIGURE 12.7. Global distributions of the vertical-mean water vapor transport in the zonal direction by all motions (a) and transient eddies (b) in m s^{-1} g kg^{-1} for annual-mean conditions. Positive values indicate eastward transport of moisture. The vertically integrated zonal transports in units of 10 kg m^{-1} s^{-1} can be obtained by multiplying the fields by p_0/g (≈ 1 over oceans and low-level land).

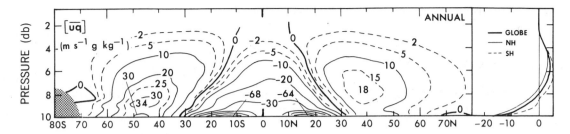

FIGURE 12.8. Zonal-mean cross section of the zonal (eastward) transport of specific humidity by all motions in m s^{-1} g kg^{-1}. Vertical profiles of the hemispheric and global mean values are shown on the right.

in the atmosphere. In each hemisphere, the flow is from the west in mid to high latitudes, with a maximum around 40° latitude. The Southern Hemisphere maxima exceed those of the Northern Hemisphere. In tropical and equatorial latitudes, the flux is westward with a bimodal distribution for the seasons, the dominant maximum being in the winter hemisphere. The bimodal configuration of the $[\overline{Q}_\lambda]$ is associated with the light and variable winds over the equatorial trough (see Fig. 7.1). The seasonal curves in Fig. 12.9 show that the maximum in eastward transport in each hemisphere moves to higher latitudes during summer, and that its intensity changes differently in the two hemispheres. In fact, the maximum value in the Southern Hemisphere is higher during summer than during winter, whereas they are about the same in the Northern Hemisphere. The curves for the year cross the zero line at about 23° latitude.

The global-mean transport values of $\overline{Q}_\lambda$ as inferred from the three curves in Fig. 12.9 show that the gaseous hydrosphere as a whole moves eastward, faster than the earth, with relative values of 0.56, 0.83, and 0.49 m s^{-1} g kg^{-1} for the year, and DJF and JJA seasons, respectively. However, the behavior of the mean zonal fluxes for the two individual hemispheres is very different. In the Northern Hemisphere, the gaseous hydrosphere moves slower than the Earth (-0.54, -0.35, and -0.40 m s^{-1} g kg^{-1} for the year, DJF and JJA seasons, respectively) whereas it moves more rapidly in the Southern Hemisphere (1.66, 2.01, and 1.37 m s^{-1} g kg^{-1} for the year, DJF and JJA seasons, respectively) (see Peixoto and Oort, 1983).

12.3.2.2 Meridional transport of water vapor

The planetary distribution of the $\overline{Q}_\phi$ field is presented in Fig. 12.10(a). The patterns reveal considerable detail and are not as simple as those of the $\overline{Q}_\lambda$ field. The most intense centers of the $\overline{Q}_\phi$ fields are observed over the oceans and over the fringes of the continents. Although the $\overline{Q}_\phi$ values are a factor of 2 to 3 smaller than the corresponding $\overline{Q}_\lambda$ values, they do play an essential role in maintaining the global water balance as well as in the energetics of the atmosphere.

The meridional flux of moisture over midlatitudes varies considerably with the seasons but is predominantly poleward throughout the year. The transport is mainly accomplished by baroclinic lows associated with the polar front and by stationary eddies, such as subpolar lows and subtropical anticyclones, together with their transient pulsations. The largest variability during the year is associated with the movement and changes in strength of the Hadley cells. The lower branches of these cells are very effective in transporting moisture into the ITCZ.

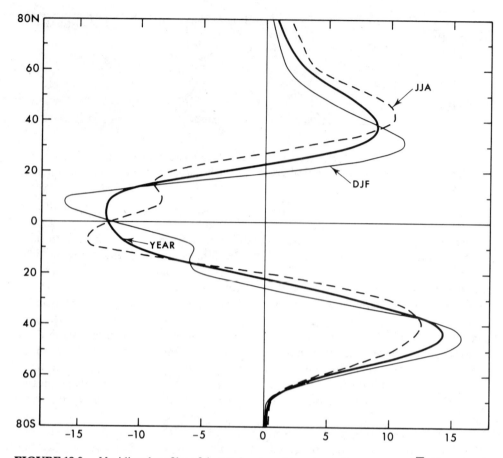

FIGURE 12.9. Meridional profiles of the total eastward transport of water vapor $[\overline{Q}_\lambda]$ in units of $10 \text{ kg m}^{-1} \text{s}^{-1}$ for annual, DJF, and JJA mean conditions (after Peixoto and Oort, 1983).

The vertically integrated transient-eddy northward flux of water vapor is presented in Fig. 12.10(b). In contrast to the zonal eddy flow, the present patterns are better defined, with predominantly poleward flow in each hemisphere throughout the year. We also notice that the transient eddies are responsible for a much greater fraction of the total transport of water vapor in the meridional than in the zonal direction. In fact, the configurations of the transient eddy and total meridional transport fields are very similar in middle latitudes.

The analysis of the transient eddy field shows a belt of strong poleward flux in middle latitudes of both hemispheres. In the polar regions and in the tropics, the meridional eddy transports are weak, almost vanishing over the equator. The influence of the continents is evident in the location of the maxima just east of the continents, especially in the Northern Hemisphere due to the poleward flow of relatively humid, tropical air and the equatorward flow of relatively dry, continental air. The maxima in the Southern Hemisphere are less intense but more zonally uniform, so that the zonal-mean poleward fluxes are of the same magnitude in both hemispheres.

The $[\overline{qv}]$ cross sections in Fig. 12.11 show the vertical structure of the meridional transfer of water vapor, and its breakdown into various components. The largest values occur generally closer to the surface than in the case of the zonal transport.

WATER CYCLE 289

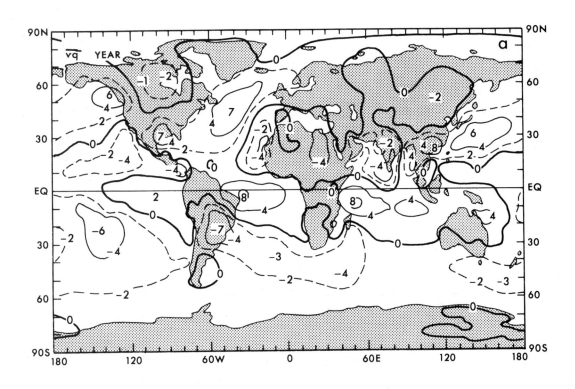

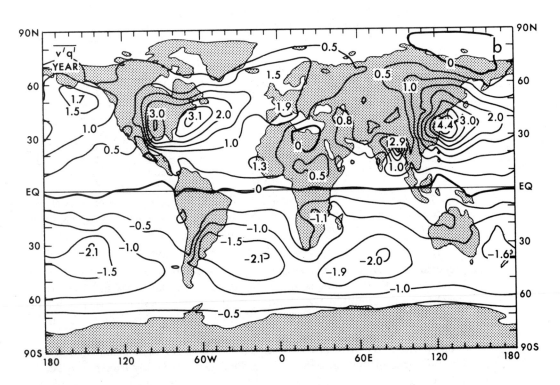

FIGURE 12.10. Global distributions of the vertical-mean meridional (northward) transport of water vapor by all motions (a) and transient eddies (b) in m s^{-1} g kg^{-1} for annual-mean conditions. To convert to $\overline{Q}_\phi$ in units of 10 kg m^{-1} s^{-1}, multiply by p_0/g (≈ 1 over oceans and low-level land).

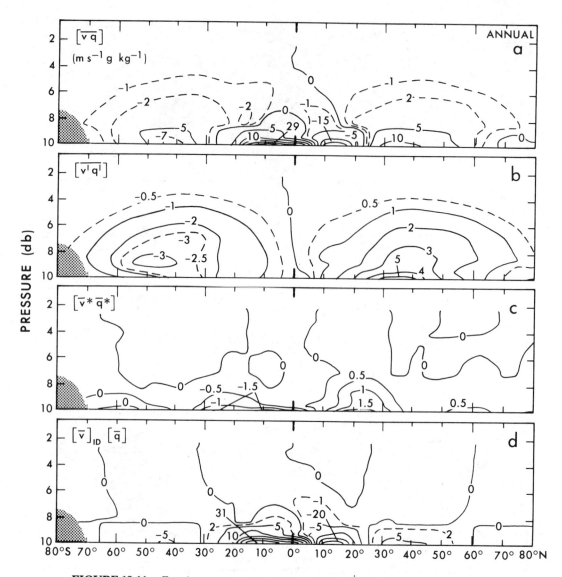

FIGURE 12.11. Zonal-mean cross sections of the northward transport of water vapor by all motions (a), transient eddies (b), stationary eddies (c), and mean meridional circulations (d) in m s^{-1} g kg^{-1} for annual-mean conditions.

The vertical structure of the transient eddy meridional transport of water vapor in Fig. 12.11(b) shows that the flow is directed poleward at all latitudes and levels, and that it reaches maximum intensity in middle latitudes below 850 mb. Inspection of the various cross sections confirms that the vertical distribution of the total flux in middle latitudes is mainly determined by the transient eddy term.

The contributions by the stationary eddies in Fig. 12.11(c) are smaller and less coherent than those of the transient eddies. The stationary eddy transports are, of course, associated with the quasistationary features of the general circulation, in particular with the subtropical anticyclones and the semipermanent lows at mid to high latitudes.

The contributions by the mean meridional circulations in Fig. 12.11(d) reveal a three-cell circulation structure in both hemispheres. The low-level branches of the

Hadley cells give the largest contributions to the total meridional flux of water vapor over the tropics due to the large values of $[\bar{v}]$ combined with high values of $[\bar{q}]$. The upper branches of the cells do not contribute much because of the low humidities at those levels. In contrast with the Hadley cells, the midlatitude Ferrel cells and the polar cells are unimportant compared with the transient and stationary modes of transfer.

The zonal- and vertical-mean profiles in Fig. 12.12 synthesize the general behavior of the meridional flux fields. In midlatitudes, the meridional transports are poleward in both hemispheres, with maxima near 40° latitude and with small seasonal variations. In the tropical zone, the mean annual transports are positive south of the equator and negative to the north of it. There is evidence for a strong interaction between the two hemispheres as shown by the seasonal curves in Figs. 12.12(a) and 12.12(d). In fact, the cross-equatorial flow in the Hadley cells changes direction with the seasons leading to a water vapor flux into the Northern Hemisphere during JJA of about 18.8×10^8 kg s^{-1} and a flow into the Southern Hemisphere during DJF of about -13.6×10^8 kg s^{-1}. For the year as a whole, there is a net flux into the Northern Hemisphere of 3.2×10^8 kg s^{-1}. In other words, on an annual basis, the Southern Hemisphere supplies a considerable amount of water vapor to the Northern Hemisphere confirming our earlier conclusions in Sec. 7.6. The cross-equatorial flow of water vapor implies an annual excess of precipitation over evaporation in the Northern Hemisphere of 39 mm yr^{-1} and an excess of 58 mm for the three months of the JJA season. The water vapor exported by the Northern Hemisphere during the winter season corresponds to an excess of evaporation over precipitation in that hemisphere of about 42 mm for the three months of the DJF season. The annual value of 39 mm yr^{-1} can be compared with the corresponding, less reliable value of 73 mm yr^{-1} shown in Table 7.1 which was based on individual estimates of precipitation and evaporation over the Northern Hemisphere.

12.3.2.3 Vertical transport of water vapor

The vertical transport of water vapor in the atmosphere plays an essential role in the hydrological cycle, since it links the terrestrial and atmospheric branches. Figure 12.13 shows the vertical flux associated with the time-mean flow at the 850-mb level where it attains the highest values. Because of the difficulties involved in computing the vertical velocity field $\bar{\omega}$ the analysis of $\bar{\omega}\,\bar{q}$ is tentative. Nevertheless, the different features are reasonable with an upward flux ($\bar{\omega}\,\bar{q} < 0$) over the equatorial region and at high latitudes, and a downward flux ($\bar{\omega}\,\bar{q} > 0$) in the subtropics. The belt of maximum upward transport over the equatorial region is, of course, associated with the ascending branches of the Hadley cells, whereas the upward flux in middle and high latitudes must be connected with the quasistationary low-pressure systems. The centers of maximum downward flux occur mainly in the eastern parts of the subtropical anticyclones over the oceans where general subsidence prevails.

The meridional profiles in Fig. 12.14 show the relative contributions by the transient eddies, stationary eddies, and mean meridional circulations to the total vertical transports. They show that the vertical flux by the transient eddies [Fig. 12.14(b)], ranging in scale from organized convection to synoptic disturbances, must be at least as important for the total vertical transport as the mean meridional circulations in the Hadley cells [Fig. 12.14(d)]. However, the latter seems to oppose the transient eddy effects in the subtropics.

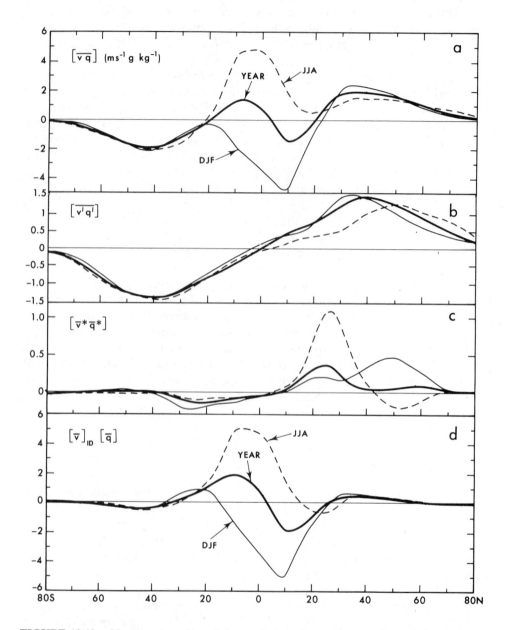

FIGURE 12.12. Meridional profiles of the vertical- and zonal-mean values of the northward transport of water vapor by all motions (a), transient eddies (b), stationary eddies (c), and mean meridional circulations (d) in m s^{-1} g kg^{-1} for annual, DJF, and JJA mean conditions. [To convert to total transport estimates multiply values by $10^{-3} 2\pi R \cos\phi\, p_0/g = 4\cos\phi$ to find values in units of 10^8 kg s^{-1} or by $12.6 \cos\phi$ to find units in 10^{15} kg yr^{-1}, where $2\pi R \cos\phi$ = length of latitude circle and $p_0/g = 10^4$ kg m^{-2} the total atmospheric mass per unit area.] (After Peixoto and Oort, 1983).

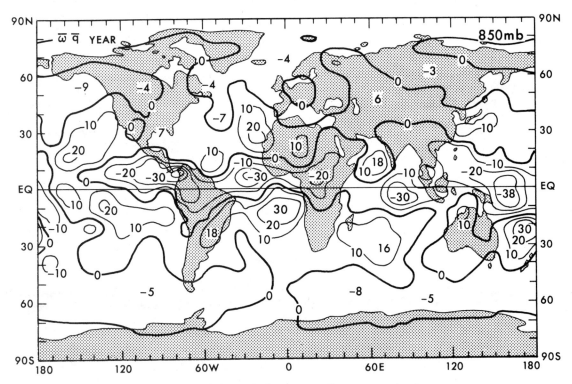

FIGURE 12.13. Global distribution of the vertical transport of water vapor by time-mean motions at the 850-mb level, $\overline{\omega}\,\overline{q}$, in p-velocity units of 10^{-4} g kg^{-1} mb s^{-1} or in actual flux units of 10^{-6} kg m^{-2} s^{-1} for annual-mean conditions. Negative values indicate upward transports.

12.3.3 Divergence of water vapor

The importance of the divergence field of water vapor in the atmosphere depends on its relationship with the mean difference of evaporation and precipitation $(\overline{E}-\overline{P})$, as shown earlier in Eqs. (12.12) and (12.13). Thus, divergence maps are of great interest for the study of the planetary water balance since regions of mean positive divergence $(\overline{E}-\overline{P}>0)$ constitute source regions of water vapor, whereas the regions of convergence $(\overline{E}-\overline{P}<0)$ are sink regions of water vapor for the atmosphere. Using the grid point values of $\overline{Q}_\lambda$ and $\overline{Q}_\phi$ shown earlier for the year in Figs. 12.7(a) and 12.10(a), divergence maps of the water vapor flux can be evaluated. The resulting maps for both the mean yearly and seasonal conditions are presented in Fig. 12.15.

Convergence generally prevails over the equatorial and mid- to high-latitude zones, while divergence predominates in the subtropics. The convergence and divergence centers are, as a rule, more intense over the ocean than over the land. The equatorial convergence of water vapor is associated with the ITCZ. Water vapor is carried towards the region of mean rising motion by the lower branches of the Hadley cells, leading to heavy precipitation confirming the earlier distributions of P given in Fig. 7.24. The belt of convergence of water vapor consists of various centers located over the headwaters and drainage basins of large river systems, such as the Amazon in South America, the Ubangi, Congo, Senegal, and Blue Nile in Africa, and the Indus, Ganges, Mekong, and Yangtze in Southeast Asia. The seasonal shifts of the equatorial belt of convergence, to the north in the JJA season and to the south in the DJF season, are related to the migration of the rising branches of the Hadley cells. The

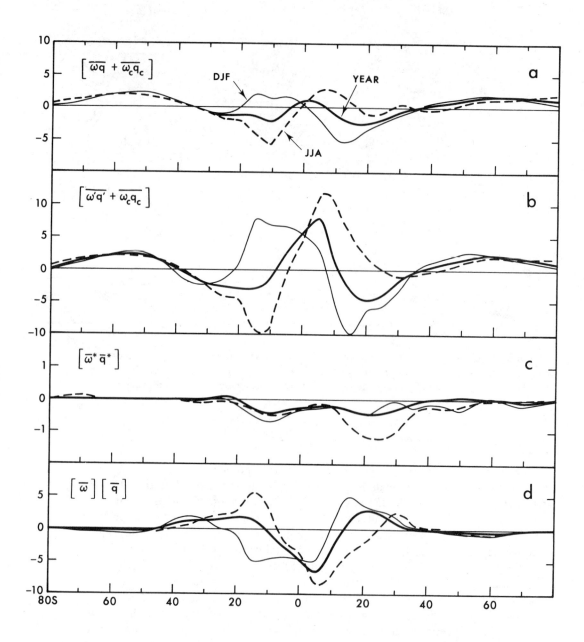

FIGURE 12.14. Meridional profiles of the vertical- and zonal-mean values of the vertical transport of water vapor in units of 10^{-4} g kg^{-1} mb s^{-1} or 10^{-6} kg m^{-2} s^{-1} by all motions (a), transient eddies (b), stationary eddies (c), and mean meridional circulations (d) for annual, DJF, and JJA mean conditions. The total curves in (a) and the residual curves in (b) include also the total vertical transport of water in the condensed form. Negative values indicate upward transports (from Peixoto and Oort, 1983).

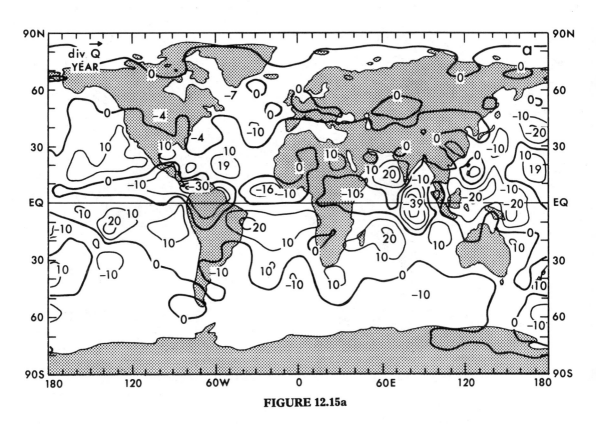

FIGURE 12.15a

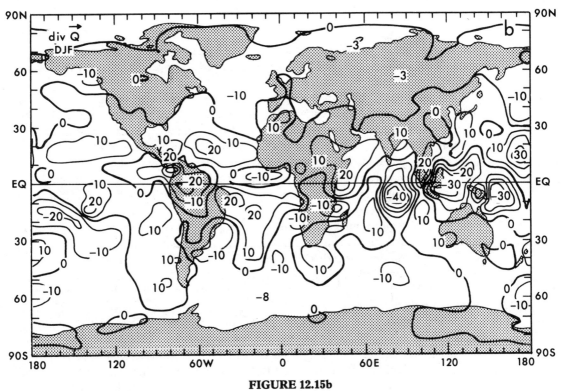

FIGURE 12.15b

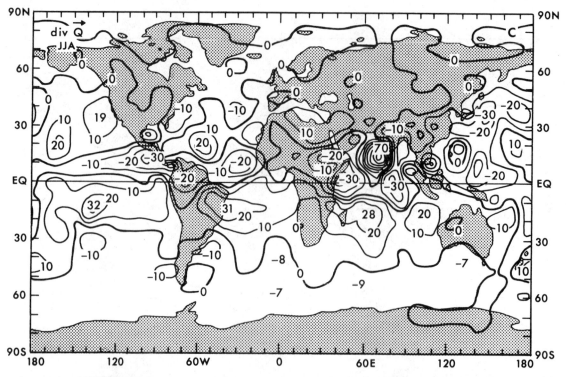

FIGURE 12.15. Global distributions of the horizonal divergence of the total water vapor transport div **Q** in units of 0.1 m yr^{-1} for annual (a), DJF (b), and JJA (c) mean conditions. Negative values indicate areas of convergence of moisture or excess of precipitation over evaporation (from Peixoto and Oort, 1983).

JJA map shows an intense center of divergence just west of India that may be connected with large evaporation over the Arabian Sea, supplying part of the moisture for the Indian summer monsoon.

The subtropical belts of divergence coincide largely with the arid zones of the globe. They are associated with strong evaporation over the oceans and with subsidence that prevails over the large subtropical anticyclones. Their latitudinal oscillation follows the yearly movement of the subtropical anticyclones. The existence of areas of divergence over the oceans is easy to understand since there is always water available for evaporation and the prevailing ocean currents may advect the necessary fresh water to maintain equilibrium. However, the situation is more difficult to explain when divergence occurs over land. In this case, surface and underground flows from less arid regions must supply the water required to counterbalance the observed excess of evaporation over precipitation (Starr and Peixoto, 1958).

The mid- to high-latitude convergence in both hemispheres is mainly associated with the transient baroclinic lows that accompany the polar front. Over the polar regions, there are some indications of a slight divergence, especially in the vicinity of the north pole.

As was shown before in Fig. 7.26, the evaporation is very large over the oceanic regions with strong divergence of water vapor (see Fig. 12.15) and local relative maxima in surface salinity are to be expected. On the other hand, in regions of convergence, such as in equatorial latitudes, the excess of fresh water from rain will dilute the ocean water, leading to lower salinity values. Indeed, the general configuration of the salinity field (see Fig. 8.9) reveals a high correlation with the mean annual divergence map of water vapor in Fig. 12.15.

Zonally averaged profiles of the div $\mathbf{Q}$ ($\approx E - P$) fields are shown in Fig. 12.16 and are also tabulated in Table 12.1. The profiles synthesize the main features of the divergence maps already discussed. In all cases, they indicate a strong convergence with an excess of mean precipitation over mean evaporation, [div $\overline{\mathbf{Q}}$] $\approx [\overline{E - P}] < 0$, over the equatorial region and strong divergence in the subtropics. The divergence intensifies during the winter season. As can be seen from Figs. 7.25 and 7.27, precipitation has a minimum and evaporation a maximum in each of the subtropical zones.

It is instructive to superimpose the $[\overline{Q}_\phi]$ and [div $\overline{\mathbf{Q}}$] profiles from Figs. 12.12(a) and 12.16, since the latter profiles are the derivatives of the $[\overline{Q}_\phi]$ curves. They show a large export of water vapor from the zones of divergence and a strong import of moisture into the belts of convergence.

12.4 SYNTHESIS OF THE WATER BALANCE

The role of the general circulation in the hydrological cycle can be shown well through maps of the vertically integrated atmospheric moisture flow in terms of the $\overline{\mathbf{Q}}$ vector field, the so-called aerial runoff (see Fig. 12.17). In addition to the vectors, streamlines are drawn. In a hypothetical steady state, the streamlines would show the prevailing paths of water vapor in the atmosphere after its release from the source regions at the earth's surface.

TABLE • 12.1. Zonal-mean divergence of the vertically integrated atmospheric water vapor flux, $\Delta([Q_\phi]\cos\phi)/R\cos\phi\,\Delta\phi(=E-P)$, in units of cm yr^{-1} for annual-mean and DJF and JJA seasonal conditions from Peixóto and Oort (1983).

	YEAR	DJF	JJA		YEAR	DJF	JJA
85 °N	−9.0	−8.5	−18.5	−85 °S	−2.9	−1.9	−3.6
80 °N	−9.8	−7.4	−20.3	−80 °S	−3.8	−2.2	−5.8
75 °N	−11.7	−8.1	−20.8	−75 °S	−8.6	−5.5	−11.8
70 °N	−16.4	−14.3	−17.4	−70 °S	−18.1	−14.2	−20.0
65 °N	−23.1	−20.4	−16.2	−65 °S	−25.8	−23.3	−25.1
60 °N	−27.1	−26.1	−18.9	−60 °S	−28.4	−29.2	−26.5
55 °N	−26.0	−31.6	−18.6	−55 °S	−28.5	−32.4	−27.4
50 °N	−21.1	−30.5	−13.8	−50 °S	−25.8	−30.4	−25.5
45 °N	−15.0	−22.9	−8.0	−45 °S	−16.8	−18.8	−18.3
40 °N	−11.5	−16.6	−6.3	−40 °S	−0.7	4.8	−6.2
35 °N	−6.2	−3.7	2.6	−35 °S	14.0	24.5	6.7
30 °N	14.7	35.9	14.9	−30 °S	23.7	24.7	27.4
25 °N	48.8	73.0	15.7	−25 °S	30.9	25.4	48.8
20 °N	61.8	96.1	0.9	−20 °S	39.3	−2.0	76.0
15 °N	43.2	116.2	−37.8	−15 °S	40.4	−54.3	111.8
10 °N	−19.3	48.6	−80.4	−10 °S	16.7	−55.5	79.7
5 °N	−64.1	−37.5	−79.4	−5 °S	−15.3	−40.0	13.7
0 °N	−43.9	−46.4	−32.6				
NH	−3.9	16.6	−23.2	SH	3.9	−16.6	23.2

298 PHYSICS OF CLIMATE

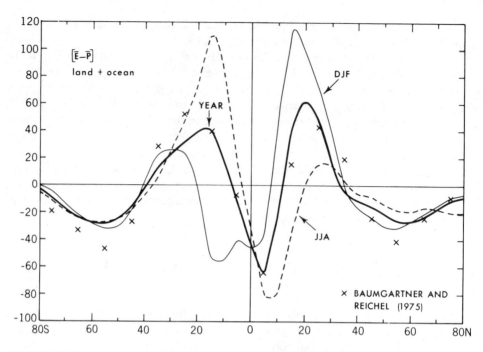

FIGURE 12.16. Meridional profiles of the zonal-mean divergence of the total water vapor transport $[\text{div }\mathbf{Q}] \approx [E - P]$ in 0.01 m yr^{-1} for annual, DJF, and JJA mean conditions. Some annual-mean estimates of $E - P$ by Baumgartner and Reichel (1975) are added for comparison (see also Table 7.1).

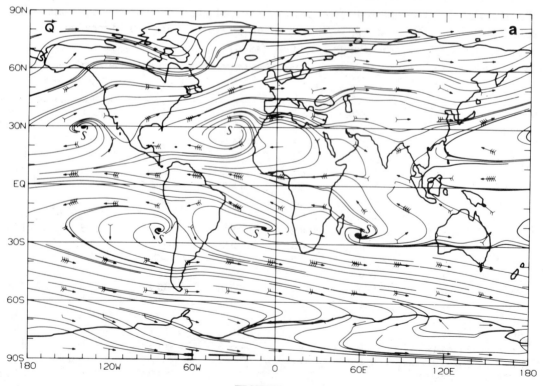

FIGURE 12.17a

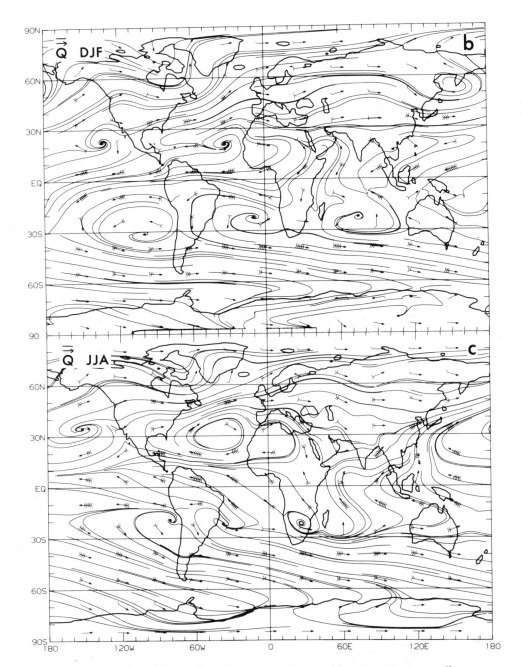

FIGURE 12.17. Global distributions of the total aerial runoff **Q** and some corresponding streamlines for annual (a), DJF (b), and JJA (c) mean conditions. Each barb on the shaft of an arrow indicates a value of 2 m s^{-1} g kg^{-1} (from Peixoto and Oort, 1983).

The $\overline{\mathbf{Q}}$ maps provide a good indication of the prevailing movements of the main moist air masses in the atmosphere and the sites of their formation. They show again that the main sources of water vapor for the atmosphere are located over the subtropical oceans and that most of the water vapor necessary for precipitation over the continents comes from the oceans. In steady-state conditions, this net inflow of moisture to the continents must be compensated by runoff of the rivers into the oceans [see Eq. (12.15)]. The air masses also receive some moisture over the continents due to evapotranspiration and evaporation from lakes, soil, etc. However, this last moisture supply constitutes only a small fraction of the water that falls locally as precipitation over land. These results confirm that the *in situ* theory (i.e., local evaporation would provide the water vapor required for the local precipitation) cannot be accepted, as a general rule.

Since the water vapor transport occurs mainly in the lower troposphere, it is clearly affected by the earth's topography. Indeed, the absence of large mountains along the Atlantic coast in Europe favors the deep penetration of moisture from the Atlantic Ocean into the Eurasian continent and the Mediterranean region. On the other hand, the existence of the Rocky Mountains parallel to the west coast of North America does not allow moisture from the Pacific Ocean to penetrate deeply into the American continent. Most of the moisture falling as precipitation over North America seems to be supplied by water vapor originating over the warm waters of the Gulf of Mexico, with a deep northward intrusion of water vapor in all seasons. An analogous situation occurs in South America with respect to the Andes. In fact, most of the moisture over South America comes from the Atlantic Ocean.

When the hydrosphere of the climatic system is considered as a whole, there must be, on the average, a compensating meridional flux of water substance in the terrestrial hydrosphere opposite to that which is observed in the gaseous hydrosphere. Thus, balance considerations require a small net southward transport of water by ocean currents and rivers in midlatitudes of the Northern Hemisphere, as well as across the equator in the subequatorial zone. At other latitudes, the net meridional flux of liquid water must be northward. Preliminary results (Hellerman, 1981) indicate that there is a detectable meridional runoff by major rivers in the required direction, although it is much smaller than the net contribution by ocean flows.

In order to complement the above discussion of the horizontal flow of water vapor, we also present zonal-mean cross sections of the stream flow in the vertical-meridional plane. As before in the angular momentum transport, we can define a stream function in the meridional-vertical plane, by taking the zonal average of Eq. (12.6), which leads to

$$[\overline{\partial q/\partial t}] + \{\partial [\overline{qv}]\cos\phi/(R\cos\phi\partial\phi)\} + \partial[\overline{q\omega}]/\partial p = [\overline{s(q)}]. \qquad (12.22)$$

A similar equation can be written for the condensed form, noting that $s(q_c) = -(e - c)$. As mentioned before, the local rate of change can be neglected for the long-term mean. Since the horizontal transports in the condensed form are very small, the final balance equation becomes

$$\{\partial[\overline{qv}]\cos\phi/(R\cos\phi\partial\phi)\} + \partial[\overline{q\omega} + \overline{q_c\omega_c}]/\partial p = 0, \qquad (12.23)$$

where ω_c is the vertical velocity of the condensed phase. Under these conditions, a stream function ψ_q can be defined by the equations

$$\frac{\partial \psi_q}{\partial p} = (2\pi R/g)\cos\phi [\overline{qv}], \qquad (12.24a)$$

$$-\frac{\partial \psi_q}{R\partial\phi} = (2\pi R/g)\cos\phi [\overline{q\omega} + \overline{q_c\omega_c}] \qquad (12.24b)$$

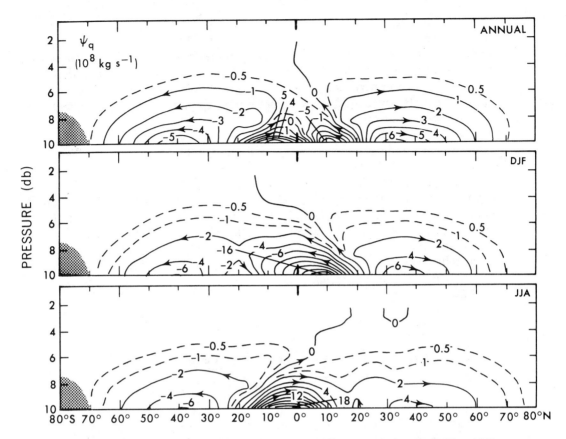

FIGURE 12.18. Streamlines of the zonal-mean transport of water vapor for annual, DJF, and JJA mean conditions in 10^8 kg s^{-1} (from Peixoto and Oort, 1983).

satisfying the boundary condition $\psi_q = 0$ at the top of the atmosphere. The integration of these equations leads to the field ψ_q in Fig. 12.18 which portrays the general features of the zonal-mean meridional flow of water substance in a vertical plane.

Inspection of the ψ_q cross sections in Fig. 12.18 confirms that the water vapor circulates in the lower troposphere, and that the maximum transports occur in the planetary boundary layer. The streamlines that begin at the earth's surface are due to the excess of evaporation over precipitation, while those that end at the earth's surface define the sink regions for atmospheric water. The yearly ψ_q cross section shows that there is an export of water vapor from the zones, between 12 and 35 °N, and 10 and 35 °S. In the Northern Hemisphere, a considerable portion of this water vapor (to the right of the zero streamline) is exported poleward throughout the lower half of the atmosphere, while the remainder, confined to much lower levels, is directed to the equatorial regions, mainly to feed the ITCZ. In the Southern Hemisphere low latitudes, a large part of the water vapor is exported to the north in a shallow layer near the earth's surface, some of it crossing the equator and falling as precipitation in the ITCZ. The other part is transported southwards at much higher levels, up to 400 mb. Polewards of 40° latitude, in both hemispheres, the streamlines terminate at the surface, indicating the existence of an excess of precipitation over evaporation.

The seasonal cross sections show simpler patterns of the sources and sinks of moisture for the atmosphere. In the DJF season, there is an export of moisture, produced over subtropical regions, from the Northern into the Southern Hemisphere. Most of this moisture, circulating in the lowest layers, falls over the southern subequatorial zone, while some of it is transported at higher altitudes further to the south. From the moisture released in this season, a small fraction (northward of the zero streamline) flows to the north. In the JJA season, the main source of water vapor is located in the Southern Hemisphere between 10 and 32 °S, invading practically the entire globe. However, most of the water vapor is transported across the equator into the Northern Hemisphere.

The interactions between the atmospheric and terrestrial branches are essential to achieve and maintain the state of quasi-equilbrium of the climatic system. Occasionally, however, imbalances may persist for an extended period of time, leading to drought or flood conditions.

Of course, solar radiation provides the bulk of the energy that sets the water circulation, as a whole, in motion. Part of this solar energy is redistributed within the climatic system by the aerological branch of the hydrological cycle when the phase transitions of the water substance occur. If we add the role of water vapor and clouds in the reflection, absorption, and emission of both solar and terrestrial radiation, we may conclude that the hydrological cycle is one of the crucial factors in the dynamics of climate.

12.5 HYDROLOGY OF THE POLAR REGIONS

Due to their importance for the global climate, we will select the polar regions poleward of 70° latitude as special regional applications of the general principles developed earlier in this chapter. Such a study offers the opportunity to also extend our earlier analyses of the water vapor to include the snow and ice, and to focus directly on the interactions between the atmosphere, oceans, and cryosphere. Some comments on the general characteristics of the cryosphere were made earlier in Chap. 9, while the role of the polar regions as principal heat sinks of the global atmospheric heat engine will be discussed in the next chapter.

A comparison of the conditions in the Arctic and Antarctic regions is of interest because of the very different physiography of the earth's surface. As clearly shown in Figs. 5.4, 9.1, and 9.2 the north polar region consists of the relatively flat, generally ice-covered Arctic ocean, whereas the south polar region contains a land mass with the pronounced Antarctic ice sheet reaching up into the atmosphere to above the 700-mb level. Some basic characteristics of the polar regions were also shown earlier in Table 9.1.

12.5.1 Equations of hydrology including vapor, liquid, and solid water substance

In considering the hydrology of the polar regions we must study the water in the vapor, liquid, and solid forms and the interactions among the three forms. As a first step, we will develop the balance equations for each component separately.

For an atmospheric polar cap bounded by a hypothetical latitudinal wall at 70° latitude the balance equation for water vapor can be written as

$$\frac{\partial \{W_v\}}{\partial t} = F_v + S(W_v), \qquad (12.25)$$

where the braces indicate a horizontal area integral over the polar cap and $\partial W_v/\partial t$ the rate of change of water vapor in the atmosphere. Further

$$F_v = \pm \iint_{\substack{\text{atm}\\\text{wall}}} \rho q v \, dx \, dz \qquad (12.26)$$

represents the meridional flux of water vapor (the positive sign refers to the north polar cap, 70 °N-NP, and the negative sign to the south polar cap, 70 °S-SP) across the vertical wall, and

$$S(W_v) = \{E_1 - P_1\} + \{E_s - P_s\} \qquad (12.27)$$

the surface source of water vapor for the atmosphere. It contains the surface evaporation minus the surface precipitation in the liquid form $\{E_1 - P_1\}$ and the surface sublimation minus the surface precipitation in the form of snow or ice pellets $\{E_s - P_s\}$.

Similarly the balance equation for water in the liquid form can be written as

$$\frac{\partial \{W_1\}}{\partial t} = F_1 + S(W_1), \qquad (12.28)$$

where $\partial W_1/\partial t$ is the rate of change with time of the polar water in the liquid form.

The flux term

$$F_1 = \pm \iint_{\substack{\text{atm}\\\text{wall}}} \rho_1 v_1 \, dx \, dz \pm \iint_{\substack{\text{ocean}\\\text{wall}}} \rho_1 v_1 \, dx \, dz$$

$$\pm \iint_{\text{rivers}} \rho_1 v_1 \, dx \, dz \qquad (12.29)$$

represents the net flow of liquid water across the latitudinal wall in the atmosphere (as clouds), the ocean and rivers, and subterranean runoff. Again the positive signs in the expression for F_1 refer to the north polar cap, 70 °N-NP, and the negative signs to the south polar cap, 70 °S-SP. Usually the first term is negligible since most of the water in the atmosphere is observed to occur in the vapor phase. Further,

$$S(W_1) = -\{E_1 - P_1\} + \{M_s - F_s\} \qquad (12.30)$$

is the net source of liquid water through precipitation in the form of rain minus the evaporation and sublimation, $\{P_1 - E_1\}$, and through melting of snow and ice minus freezing of water, $\{M_s - F_s\}$. The melting of ice may include ablation to the side of and underneath the polar ice sheets.

The balance for water in the solid form (snow and ice) in a polar cap is expressed by

$$\frac{\partial \{W_s\}}{\partial t} = F_s + S(W_s), \qquad (12.31)$$

where $\partial W_s/\partial t$ is the rate of change with time of the snow and ice contained in the polar cap. Furthermore,

$$F_s = \pm \iint \rho_s v_s \, dx \, dz \qquad (12.32)$$

represents the flow of snow and ice, mainly in the form of ice flow and calved icebergs, across the polar wall, and

$$S(W_s) = -\{E_s - P_s\} - \{M_s - F_s\} \qquad (12.33)$$

the source of snow and ice. This source includes the precipitation in the form of snow and ice pellets minus the sublimation, $\{P_s - E_s\}$, and the freezing of water minus the melting of snow and ice, $\{F_s - M_s\}$.

For the total water substance we can derive a balance equation by adding Eqs. (12.25), (12.28), and (12.31). Since the source terms cancel we obtain a conservation equation without sources and sinks:

$$\frac{\partial W_v}{\partial t} + \frac{\partial W_l}{\partial t} + \frac{\partial W_s}{\partial t} = F_v + F_l + F_s. \qquad (12.34)$$

The various relations between the three water components in the vapor, liquid, and solid forms are shown schematically in Fig. 12.19.

Under steady (e.g., annual-mean) conditions the time derivatives tend to zero, and the equations reduce to

$$F_v + S(W_v) = 0, \qquad (12.35)$$
$$F_l + S(W_l) = 0, \qquad (12.36)$$
$$F_s + S(W_s) = 0, \qquad (12.37)$$
$$F_v + F_l + F_s = 0. \qquad (12.38)$$

12.5.2 Observed water budget of the polar regions

We will start our discussion with the water budget in the vapor phase. Figure 12.20 illustrates the annual variation of the precipitable water in the two polar caps poleward of 70° latitude. The two curves are shifted with respect to each other by 6 months so that a direct interhemispheric comparison can be made for the corresponding seasons.

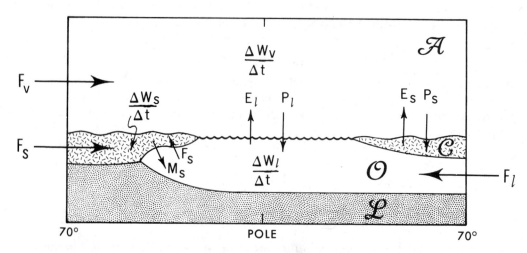

FIGURE 12.19. Schematic diagram of the water budget for a polar cap showing the fluxes across the latitudinal wall, the storage terms, and the exchange processes across the interfaces between the atmosphere, oceans, and cryosphere. The exchange processes are associated with phase changes, such as evaporation, precipitation, melting, and freezing.

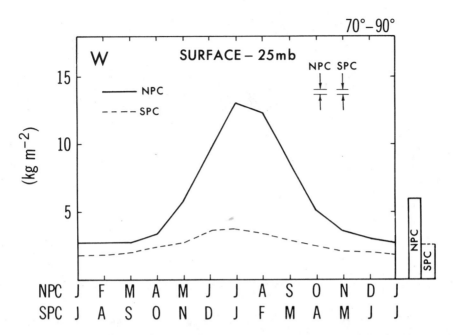

FIGURE 12.20. Annual cycle in the mean precipitable water for the north (solid) and south (dashed) polar caps in units of kg m^{-2}.

As expected, the moisture content is less in the south polar cap (SPC) than in the north polar cap (NPC) due to the higher altitudes and lower temperatures over Antarctica. The NPC–SPC differences in the precipitable water range from 1.1 kg m^{-2} in midwinter to 9.8 kg m^{-2} in midsummer. The rate of change of precipitable water in the NPC is large in the late spring and early fall with values on the order of 3.5 kg m^{-2} months^{-1}, whereas the rate of change in the SPC is at most 0.5 kg m^{-2} months^{-1}.

The seasonal variation of the water vapor transport across the latitudinal wall at 70 °N is shown in Fig. 12.21. A maximum in the poleward transport of moisture of 2.5×10^{15} kg yr^{-1} is found in July and a minimum of 1.3×10^{15} kg yr^{-1} in late winter. The corresponding curve for the transport across 70 °S is not shown because the present transport data are not reliable enough to determine the seasonal variation at those latitudes. The yearly inflow of water vapor into the NPC is estimated to be 1.8×10^{15} kg yr^{-1}. In terms of $P - E$ [see Eq. (12.25)] this net inflow would correspond to an excess of precipitation over evaporation of about 12 cm yr^{-1} (see also Table 7.1 for comparison). An estimate of the annual transport of water vapor across 70 °S was obtained from Bromwich's (1988) analysis of the annual accumulation rates of snow over Antarctica. When his values are somewhat reduced to take into account more recent lower accumulation estimates over the ice caps, sublimation and some other effects a similar value of about 1.8×10^{15} kg yr^{-1} for the SPC is obtained as for the NPC. In the Antarctic region sublimation plays a significant role besides evaporation, whereas in the Arctic it is much smaller than the evaporation.

The transport of liquid water into the polar caps as given by Eq. (12.29) comes about through the net poleward flow of water in the oceans and the northward component of the runoff by rivers. The oceanic transport cannot be measured directly because it represents a small residual value of large inflows and outflows of water. Therefore, we will estimate its magnitude by using the balance equation (12.38) as we shall see later.

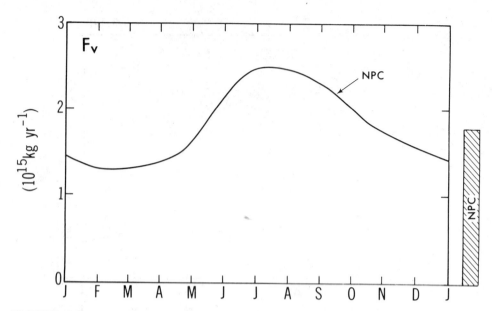

FIGURE 12.21. Annual cycle in the northward transport of water vapor across 70 °N into the north polar cap in units of 10^{15} kg yr^{-1}.

The runoff into the Arctic Ocean by the rivers of Scandinavia, Russia, Siberia, Canada, and Alaska is estimated to be on the order of $F_{riv} \approx 4 \times 10^{15}$ kg yr^{-1} (Vowinckel and Orvig, 1970; Lockwood, 1979; from Table 7.2, we would infer a value of 2.6×10^{15} kg yr^{-1}).

Regarding the ice transport F_i the annual-mean export of ice from the Arctic through the Denmark Strait is fairly well known and is on the order of 3×10^{15} kg yr^{-1} or about 10^5 m^3 s^{-1}. For the mean width of the channel this would correspond to an annual average outflow of ice of 3-m thickness at the speed of 5 cm s^{-1}. The annual export of ice from Antarctica is very uncertain and estimates by various authors vary widely. A very tentative estimate of the ice export from the SPC in the form of icebergs and sea ice can be obtained from the seasonal variation in the total area extent of sea ice around Antarctica shown before in Fig. 9.3. Thus, we will assume that one-third of the summer–winter difference of 15×10^6 km^2 in sea ice extent is produced poleward of 70 °S in the Weddell and Ross Seas and that all this ice is exported from the SPC. If we take the thickness of the ice to be on the average 1 m, we find a total export across 70 °S of 5×10^{15} kg yr^{-1}. Jacobs and Fairbanks (1985) have used data on salinity and oxygen isotope ratios to estimate the formation of icebergs at the rate of 1.1×10^{15} kg yr^{-1} and of sea ice at the rate of 2.4×10^{15} kg yr^{-1} (i.e., on the average 95-cm sea ice formation over a continental shelf area of 2.5×10^6 km^2) leading to a total value of 3.5×10^{15} kg yr^{-1} for the ice export across 70 °S. A third method by Radok *et al.* (1986) based on precipitation data over Antarctica and the Antarctic ice balance equation assuming steady-state conditions gives a value of about 2×10^{15} kg yr^{-1} for the ice export. We will use a value of $F_s = 3.5 \times 10^{15}$ kg yr^{-1} as being representative at 70 °S.

Going back now to the equation for the total water balance (12.34) and assuming steady conditions, the net influx of water in the vapor, liquid, and solid phases has to vanish [Eq. (12.38)]:

$$F_v + F_l + F_s \approx 0.$$

Combining the previous results for the north polar cap, namely the water vapor transport $F_v = 2 \times 10^{15}$ kg yr^{-1}, the river transport $F_{riv} = 4 \times 10^{15}$ kg yr^{-1}, and the ice transport $F_s = -3 \times 10^{15}$ kg yr^{-1}, we find a net outflow of ocean water across 70 °N with a value of

$$F_0 = -3 \times 10^{15} \text{ kg yr}^{-1}.$$

In the south polar cap, the water transport by rivers is, of course, negligible. The previous results of a water vapor inflow $F_v = 1.8 \times 10^{15}$ kg yr^{-1} and an ice outflow $F_s = -3.5 \times 10^{15}$ kg yr^{-1} would require a net inflow of ocean water across 70 °S with a value of about

$$F_0 = 1.7 \times 10^{15} \text{ kg yr}^{-1}.$$

Although crude, the previous analyses show the relative importance of the different water transports in establishing the hydrological balance in the polar regions. The main terms in the arctic balance are the river runoff and the atmospheric inflow of water vapor that must result in a net export of ocean water from the NPC. On the other hand, in the Antarctic polar cap, the main terms appear to be the atmospheric inflow of water vapor and the export of icebergs and sea ice across 70 °S that must be balanced by a net import of ocean water from middle latitudes.

CHAPTER 13

Energetics

13.1 BASIC FORMS OF ENERGY

The main intrinsic forms of energy in the atmosphere are internal energy I, gravitational-potential energy Φ, kinetic energy K, and latent energy associated with phase transitions of water, LH (evaporation and condensation of water vapor, freezing of water, and melting and sublimation of snow and ice). Expressions for these energy forms per unit mass are

$$I = c_v T, \tag{13.1a}$$

$$\Phi = gz, \tag{13.1b}$$

$$K = \tfrac{1}{2}\mathbf{c}\cdot\mathbf{c} = \tfrac{1}{2}(u^2 + v^2 + w^2)$$

$$\approx \tfrac{1}{2}(u^2 + v^2). \tag{13.1c}$$

In case of the latent energy, we will use the general form

$$\text{LH} = Lq. \tag{13.1d}$$

However, it depends on the particular phase transition how much latent energy is actually realized as sensible heat and whether we should use the latent heat of evaporation L_e ($=2501$ J g^{-1}) or the latent heat of sublimation L_s ($=2835$ J g^{-1}) for L in Eq. (13.1d). In most situations, the release of latent heat will occur in rain and $L = L_e$ will be the appropriate value to use. We will further assume that L is a constant.

Finally, we have for the total energy

$$E = I + \Phi + \text{LH} + K. \tag{13.1e}$$

The main relevant form of external energy is the solar radiation. Radiation is the only form of energy which can alter the global energy of the climate system. The direct effect of solar radiation is to heat the atmosphere and the underlying surface (ground and oceans) by the absorption of the short-wave, incoming radiation. We may ignore geothermal, electrical, and nuclear forms of energy since they are negligible for the energetics of the atmosphere. From the thermodynamic point of view, latent heat is a form of internal energy but we will consider it separately.

In a hydrostatic system, the internal energy for a unit area column of the atmosphere which extends from the earth's surface to the top of the atmosphere is given by

$$\int_0^\infty \rho I \, dz = \int_0^\infty \rho c_v T \, dz = \int_0^{p_0} c_v T \frac{dp}{g},$$

and the potential energy by
$$\int_0^\infty \rho\Phi\,dz = \int_0^\infty \rho gz\,dz = \int_0^{p_0} z\,dp.$$

Integration by parts results in
$$\int_0^\infty \rho\Phi\,dz = \int_0^\infty p\,dz + zp\Big|_0^{p_0}$$
$$= \int_0^\infty p\,dz = \int_0^\infty \rho R_d T\,dz.$$

At the limits of integration, $zp = 0$.

The potential and internal energy in the atmosphere are not independent forms of energy. In fact, in hydrostatic equilibrium they are proportional to each other with the ratio $R_d/c_v \approx 2/5$, and it is convenient to consider them together as one form of energy, the so-called total potential energy. Thus

$$\int_0^\infty \rho(\Phi + I)\,dz = \int_0^\infty \rho c_p T\,dz = \int_0^{p_0} c_p T \frac{dp}{g}, \qquad (13.2)$$

where $c_p = c_v + R_d$.

This sum represents the enthalpy in the atmospheric column. For our later discussions of the energy transformations in the atmosphere it is also useful to rewrite Eq. (13.2) in the (λ,ϕ,θ,t) system.

Introducing the potential temperature $\theta = T(p_0/p)^\kappa$, the total potential energy becomes

$$\int_0^\infty \rho(\Phi + I)\,dz = \frac{c_p}{g(1+\kappa)p_0^\kappa} \int_0^\infty \theta\,dp^{\kappa+1}. \qquad (13.3a)$$

Integration by parts gives

$$\int_0^\infty \rho(\Phi + I)\,dz$$
$$= \frac{c_p}{g(1+\kappa)p_0^\kappa} \int_0^\infty p^{\kappa+1}\,d\theta + \frac{c_p}{g(1+\kappa)p_0^\kappa} \theta p^{\kappa+1}\Big|_0^\infty$$
$$= \frac{c_p}{g(1+\kappa)p_0^\kappa} \int_0^\infty p^{\kappa+1}\,d\theta. \qquad (13.3b)$$

In Eq. (13.3b) we have used an extended definition of $p(\theta)$ below the earth's surface ($p > p_0$), in which $p(\theta) = p_0$ for all subsurface $\theta \leqslant \theta_0$ so that we can integrate from $\theta = 0$ to ∞ rather than from θ_0 to ∞. The second term on the right-hand side of Eq. (13.3b) can then be discarded because it vanishes at the surface ($p = p_0$) where $\theta = 0$ as well as at the top ($p \to 0$) where $\theta p^{\kappa+1} \to 0$. The previous integrals (13.3a) and (13.3b) are dual with respect to the differential operator d, i.e., the d operates on $p^{\kappa+1}$ or θ.

Knowing that the speed of sound c_s is given by

$$c_s^2 = \left(\frac{\partial p}{\partial \rho}\right)_{\theta=\text{const}} = \frac{c_p}{c_v} R_d T,$$

the expression of total potential energy (13.2) can also be written in the form

$$\int_0^\infty (\Phi + I)\frac{dp}{g} = \frac{c_v}{gR_d} \int_0^\infty c_s^2\,dp. \qquad (13.4)$$

The total kinetic energy for the unit area column is

$$\int_0^\infty K \frac{dp}{g} = \frac{1}{2g} \int_0^\infty c^2 \, dp,$$

where c is the wind speed. Comparing these expressions for $c = 15$ m s^{-1} and $c_s = 300$ m s^{-1}, we find that the ratio $K/(\Phi + I) \approx 10^{-3}$, and that the kinetic energy is a very small fraction of the total energy in the atmosphere. However, if we consider the total energy available for conversion into kinetic energy the ratio is not that small (see estimates in Table 14.1).

For the oceans, we would obtain similar expressions for the forms of energy with the exception of latent heat. Obviously for the oceans the specific heats are different.

13.2 ENERGY BALANCE EQUATIONS

13.2.1 Introduction

The rate of change of the potential energy per unit mass is given by

$$\frac{d\Phi}{dt} = gw \tag{13.5}$$

and the rate of change of internal energy, as can be seen from Eqs. (3.42) and (3.2), by

$$\frac{dI}{dt} = Q - p\alpha \, \text{div } \mathbf{c}. \tag{13.6}$$

The diabatic heating term Q in Eq. (13.6) can be written explicitly in the form

$$Q = Q_h + Q_f,$$

where

$$Q_h = -\alpha \, \text{div } \mathbf{F}_{\text{rad}} - L(e - c) - \alpha \, \text{div } \mathbf{J}_H^D \tag{13.7a}$$

and

$$Q_f = -\alpha \tau : \text{grad } \mathbf{c}. \tag{13.7b}$$

Thus, we have decomposed the total heating Q into a first part Q_h containing radiational heating, latent heating, and heating due to conduction, and a second part Q_f associated with frictional dissipation. The heat flux due to conduction is denoted by $\mathbf{J}_H^D$.

For the total potential energy we have

$$\frac{d}{dt}(\Phi + I) = gw + Q - p\alpha \, \text{div } \mathbf{c} \tag{13.8a}$$

or

$$\frac{d}{dt}(\Phi + I) = gw + Q + \alpha \, \mathbf{c} \cdot \text{grad } p - \alpha \, \text{div } p\mathbf{c}. \tag{13.8b}$$

The kinetic energy can be obtained immediately from Eq. (3.8) by using the common procedure of taking the scalar product with $\mathbf{c}$:

$$\frac{dK}{dt} = -gw - \alpha \, \mathbf{c} \cdot \text{grad } p - \alpha \, \mathbf{c} \cdot \text{div } \tau \tag{13.9a}$$

or in an alternative form

$$\frac{dK}{dt} = -gw + p\alpha \text{ div } \mathbf{c} - \alpha \text{ div}(p\mathbf{c} + \boldsymbol{\tau}\cdot\mathbf{c})$$
$$+ \alpha\boldsymbol{\tau}\text{: grad } \mathbf{c}. \quad (13.9\text{b})$$

In hydrostatic equilibrium, we have

$$-gw - \alpha \mathbf{c}\cdot\text{grad } p = -\alpha \mathbf{v}\cdot\text{grad } p,$$

where $\mathbf{v}$ is the horizontal, two-dimensional wind vector.

Inspection of the previous energy equations shows that they are not independent of each other since they are linked by common terms. In fact, Eq. (13.5) shows that the rate of change of potential energy comes from the work done against the force of gravity gw. This same term appears with the opposite sign in Eq. (13.9) which suggests that there is a conversion of potential into kinetic energy, or vice versa.

As shown by Eq. (13.6), the sources or sinks of internal energy are the rate of heating Q and the rate of work performed by compression against the pressure field $-p\alpha \text{ div } \mathbf{c}$. This last term together with the frictional heating part Q_f also occurs in Eq. (13.9b) but with the opposite sign, which shows the link between the kinetic and internal energy.

The term $-\alpha \text{ div}(p\mathbf{c} + \boldsymbol{\tau}\cdot\mathbf{c})$ in Eq. (13.9b) indicates the work at the boundaries by pressure and friction forces. This term plays an important role in transferring energy from the atmosphere into the oceans; it generates the wind-driven ocean currents. It is important to note that the only processes that can produce or destroy kinetic energy are those involving real forces. Motions of the atmosphere or oceans with or against the gravity force, i.e., downward or upward motions, convert potential into or from kinetic energy as shown by the common term gw in Eqs. (13.8b) and (13.9a). Analogously, motions of the atmosphere with or against the pressure gradient force across isobars will also convert potential energy into or from kinetic energy. These processes are adiabatic (reversible) because they can proceed in both directions. However, frictional dissipation is an irreversible process.

For the latent heat we have from Eq. (12.6):

$$L\frac{dq}{dt} = L(e - c) - \alpha L \text{ div } \mathbf{J}_q^D. \quad (13.10)$$

The last term is associated with molecular diffusion of water vapor across the boundaries and can usually be neglected.

Let us next consider the total energy of the atmosphere. We must then explicitly take into account the radiation balance which results from the incoming solar radiation and the outgoing infrared radiation emitted by the earth. Thus, for total energy E we obtain by adding Eqs. (13.8b), (13.9a), and (13.10)

$$\frac{dE}{dt} = -\alpha \text{ div } p\mathbf{c} - \alpha \mathbf{c}\cdot\text{div } \boldsymbol{\tau} + L(e-c) + Q - \alpha L \text{ div } \mathbf{J}_q^D. \quad (13.11\text{a})$$

Using Eq. (13.7), we can rewrite Eq. (13.11a) in the form

$$\frac{dE}{dt} = -\alpha \text{ div } \mathbf{F}_{\text{rad}} - \alpha \text{ div } p\mathbf{c}$$
$$- \alpha \text{ div } (\mathbf{J}_H^D + L\mathbf{J}_q^D + \boldsymbol{\tau}\cdot\mathbf{c}). \quad (13.11\text{b})$$

To illustrate our interpretation of the previous energy equations (13.5) through (13.11), and to show the meaning of the various source, conversion and sink terms of energy a schematic diagram of the energy cycle is given in Fig. 13.1.

13.2.2 The climate equations

Since the fields that characterize the state of the atmosphere are quite variable in time, it is important to study the mean state of the atmosphere and its temporal fluctuations. Therefore, the equations will be averaged in time. Furthermore, using the procedures discussed in Sec. 3.7, we will write the balance equations of the various energy components in a local Eulerian framework.

For potential energy we obtain from Eq. (13.5):

$$\frac{\overline{\partial \rho \Phi}}{\partial t} = - \operatorname{div} \overline{\Phi} \, \overline{\rho \mathbf{c}} + g \overline{\rho w}, \tag{13.12}$$

for internal energy from Eq. (13.6):

$$\frac{\overline{\partial \rho I}}{\partial t} = - \operatorname{div} \overline{I} \, \overline{\rho \mathbf{c}} - \operatorname{div} \overline{I'(\rho \mathbf{c})'} + \overline{\rho Q} - \overline{p \operatorname{div} \mathbf{c}}, \tag{13.13}$$

for kinetic energy from Eq. (13.9a):

$$\frac{\overline{\partial \rho K}}{\partial t} = - \operatorname{div} \overline{K} \, \overline{\rho \mathbf{c}} - \operatorname{div} \overline{K'(\rho \mathbf{c})'}$$
$$- \overline{\mathbf{c} \cdot \operatorname{grad} p} - \overline{\mathbf{c} \cdot \operatorname{div} \tau} - g \overline{\rho w}, \tag{13.14}$$

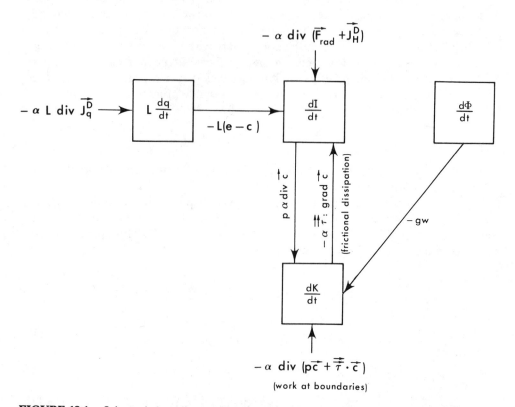

FIGURE 13.1. Schematic box diagram showing the terms which connect the various forms of energy in the atmosphere. Sometimes the conversion from internal into kinetic energy, $p\alpha \operatorname{div} \mathbf{c}$, is written in an alternate form as $-\alpha \mathbf{c} \cdot \operatorname{grad} p$, which includes the pressure work at the boundaries $-\alpha \operatorname{div} p\mathbf{c}$ (after Peixoto and Oort, 1984).

for latent heat from Eq. (13.10):

$$L \overline{\frac{\partial \rho q}{\partial t}} = -L \operatorname{div}(\overline{q}\ \overline{\rho \mathbf{c}}) - L \operatorname{div}\overline{q'(\rho \mathbf{c})'}$$
$$+ \rho L \overline{(e-c)} - L \operatorname{div}\overline{\mathbf{J}_q^D}, \qquad (13.15)$$

and for total energy from Eq. (13.11b):

$$\overline{\frac{\partial \rho E}{\partial t}} = -\operatorname{div}(\overline{E}\ \overline{\rho \mathbf{c}})$$
$$- \operatorname{div}\overline{E'(\rho \mathbf{c})'} - \operatorname{div}(\overline{\mathbf{F}_{\text{rad}} + p\mathbf{c} + \mathbf{J}_H^D}$$
$$+ L\overline{\mathbf{J}_q^D} + \overline{\boldsymbol{\tau}\cdot\mathbf{c}}). \qquad (13.16)$$

The third term on the right-hand side of Eq. (13.16) represents the combined effects of radiation fluxes, pressure work, diffusion of sensible and latent heat, and frictional stresses across the boundaries of the volume.

The terms involving products of primed variables are the transient eddy terms. These terms can be very important in the overall balances. The observed atmospheric parameters and related climatic quantities are evaluated usually in the (λ,ϕ,p,t) system so that it is advantageous to rewrite the equations in this coordinate system.

We will combine the equations for potential and internal energy of the dry atmosphere, considering then the total potential energy in hydrostatic equilibrium and using Eq. (3.43):

$$\overline{\frac{\partial}{\partial t} c_p T} = -\frac{1}{R \cos\phi}\left(\frac{\partial}{\partial \lambda}\overline{\mathcal{F}}_{H\lambda} + \frac{\partial \overline{\mathcal{F}}_{H\phi}\cos\phi}{\partial \phi}\right)$$
$$- \frac{\partial}{\partial p}\overline{\mathcal{F}}_{Hp} + Q + \overline{\alpha\omega}, \qquad (13.17)$$

where $\mathbf{J}_H = (\mathcal{F}_{H\lambda},\mathcal{F}_{H\phi},\mathcal{F}_{Hp})$ = the enthalpy transport density vector,

$$\begin{cases} \mathcal{F}_{H\lambda} = c_p T u \\ \mathcal{F}_{H\phi} = c_p T v \\ \mathcal{F}_{Hp} = c_p T \omega \end{cases}$$

and Q is given by Eq. (13.7).

For horizontal kinetic energy, $K_H = (u^2 + v^2)/2$, we obtain from Eq. (3.20) after multiplying by u and v:

$$\overline{\frac{\partial}{\partial t} K_H} = -\frac{1}{R \cos\phi}\left\{\frac{\partial}{\partial \lambda}\overline{\mathcal{F}}_{K\lambda} + \frac{\partial(\overline{\mathcal{F}}_{K\phi}\cos\phi)}{\partial \phi}\right\} - \frac{\partial}{\partial p}\overline{\mathcal{F}}_{Kp}$$
$$- \overline{\alpha\omega} + \overline{uF_\lambda} + \overline{vF_\phi}, \qquad (13.18)$$

where $\mathbf{J}_K = (\mathcal{F}_{K\lambda},\mathcal{F}_{K\phi},\mathcal{F}_{Kp})$ = the mechanical energy transport density vector

and $\begin{cases} \mathcal{F}_{K\lambda} = \tfrac{1}{2}(u^2 + v^2)u + gzu \\ \mathcal{F}_{K\phi} = \tfrac{1}{2}(u^2 + v^2)v + gzv \\ \mathcal{F}_{Kp} = \tfrac{1}{2}(u^2 + v^2)\omega + gz\omega \end{cases}$.

In these expressions, in the (λ,ϕ,p,t) system the pressure work term is formally equivalent with the flux of potential energy across the boundaries of the unit mass.

In deriving Eq. (13.18) we may note that

$$-gu\frac{\partial z}{R\cos\phi\partial\lambda} - gv\frac{\partial z}{R\partial\phi}$$

$$= -g\left(\frac{\partial uz}{R\cos\phi\partial\lambda} + \frac{\partial vz\cos\phi}{R\cos\phi\partial\phi} + \frac{\partial \omega z}{\partial p}\right)$$

$$+ gz\left(\frac{\partial u}{R\cos\phi\partial\lambda} + \frac{\partial v\cos\phi}{R\cos\phi\partial\phi} + \frac{\partial \omega}{\partial p}\right) + g\omega\frac{\partial z}{\partial p}$$

$$= -g\left(\frac{\partial uz}{R\cos\phi\partial\lambda} + \frac{\partial vz\cos\phi}{R\cos\phi\partial\phi} + \frac{\partial \omega z}{\partial p}\right) - \omega\alpha, \quad (13.19)$$

assuming hydrostatic equilibrium and using the continuity equation. For latent heat we obtain from Eq. (13.15):

$$\overline{\frac{\partial Lq}{\partial t}} = -\frac{L}{R\cos\phi}\left(\frac{\partial}{\partial\lambda}\overline{\mathcal{J}}_{q\lambda} + \frac{\partial\overline{\mathcal{J}}_{q\phi}\cos\phi}{\partial\phi}\right)$$

$$- L\frac{\partial}{\partial p}(\overline{\mathcal{J}}_{qp}) + L(\overline{e-c}) - Lg\frac{\partial}{\partial p}(\overline{\mathbf{J}^D_q}), \quad (13.20)$$

where $\mathbf{J}_q = (\mathcal{J}_{q\lambda}, \mathcal{J}_{q\phi}, \mathcal{J}_{qp}) =$ the water vapor transport density vector and

$$\begin{cases} \mathcal{J}_{q\lambda} = qu \\ \mathcal{J}_{q\phi} = qv \\ \mathcal{J}_{qp} = q\omega \end{cases}$$

In the diffusion term we have only included the diffusion along the vertical since it is the most important contribution near the earth's surface.

Finally, we obtain for total energy by adding the three equations (13.17), (13.18), and (13.20):

$$\overline{\frac{\partial}{\partial t}E} = -\frac{1}{R\cos\phi}\left(\frac{\partial\overline{\mathcal{J}}_{E\lambda}}{\partial\lambda} + \frac{\partial\overline{\mathcal{J}}_{E\phi}\cos\phi}{\partial\phi}\right) - \frac{\partial}{\partial p}(\overline{\mathcal{J}}_{Ep})$$

$$- g\frac{\partial}{\partial p}(\overline{F}_{rad} + \overline{\mathcal{J}^D_H} + \overline{\mathcal{J}^D_q} + \overline{\tau\cdot\mathbf{c}}), \quad (13.21)$$

where $\mathbf{J}_E = (\mathcal{J}_{E\lambda}, \mathcal{J}_{E\phi}, \mathcal{J}_{Ep}) =$ the total energy transport density factor and

$$\begin{cases} \mathcal{J}_{E\lambda} = [c_pT + gz + Lq + \tfrac{1}{2}(u^2+v^2)]u \\ \mathcal{J}_{E\phi} = [c_pT + gz + Lq + \tfrac{1}{2}(u^2+v^2)]v \\ \mathcal{J}_{Ep} = [c_pT + gz + Lq + \tfrac{1}{2}(u^2+v^2)]\omega \end{cases}$$

In the last term of Eq. (13.21), we have again only included the vertical component since it is the primary contributor in the radiation and diffusion fluxes.

The introduction of the transport density vectors $\mathbf{J}_x$ for a property x can bring some clarity to the physical interpretation of the equations since these vectors often appear associated with the divergence operator. The equations are particular versions of the general equation (2.1). Then by the Gauss divergence theorem, the transport vectors represent the net fluxes of the property x across the boundaries. The most important remaining terms in the equations are now internal generation or destruction terms besides some usually smaller diffusion terms across the boundaries.

Some of these source or sink terms are common to two or more of the equations with opposite sign. Whenever such a term can be identified with a real physical process, it will represent a conversion $C(A,B)$ of an energy form A into an energy form B.

Due to the importance of the zonal symmetry of many of the climatic parameters, it is often appropriate to consider the zonally averaged climate equations.

For total potential energy we have from Eq. (13.17):

$$\overline{\frac{\partial}{\partial t}[c_p T]} = -\frac{1}{R\cos\phi}\frac{\partial}{\partial \phi}([\overline{\mathcal{F}}_{H\phi}]\cos\phi)$$

$$-\frac{\partial}{\partial p}[\overline{\mathcal{F}}_{Hp}] + [\overline{Q}] + [\overline{\alpha\omega}], \quad (13.22)$$

where according to expansion (4.9):

$$[\overline{\mathcal{F}}_{H\phi}] = c_p[\overline{T}][\overline{v}] + c_p[\overline{T^*\overline{v}^*}] + c_p[\overline{T'v'}]$$

and

$$[\overline{\mathcal{F}}_{Hp}] = c_p[\overline{T}][\overline{\omega}] + c_p[\overline{T^*\overline{\omega}^*}] + c_p[\overline{T'\omega'}].$$

This clearly shows the contributions by the mean meridional cells, stationary eddies, and transient eddies.

For the horizontal kinetic energy we obtain from Eq. (13.18):

$$\overline{\frac{\partial}{\partial t}[K_H]} = -\frac{1}{R\cos\phi}\frac{\partial}{\partial \phi}([\overline{\mathcal{F}}_{K\phi}]\cos\phi)$$

$$-\frac{\partial}{\partial p}[\overline{\mathcal{F}}_{Kp}]$$

$$-[\overline{\alpha\omega}] + [\overline{uF_\lambda}] + [\overline{vF_\phi}], \quad (13.23)$$

where

$$[\overline{\mathcal{F}}_{K\phi}] = [\overline{Kv}] + g[\overline{zv}]$$

and

$$[\overline{\mathcal{F}}_{Kp}] = [\overline{K\omega}] + g[\overline{z\omega}].$$

For the total energy E we obtain from Eq. (13.21):

$$\overline{\frac{\partial}{\partial t}[E]} = -\frac{1}{R\cos\phi}\frac{\partial}{\partial \phi}([\overline{\mathcal{F}}_{E\phi}]\cos\phi) - \frac{\partial}{\partial p}[\overline{\mathcal{F}}_{Ep}]$$

$$-g\frac{\partial}{\partial p}[\overline{F}_{\text{rad}} + \overline{\mathcal{F}}_{Hp}^D + L\,\overline{\mathcal{F}}_{qp}^D + \overline{\tau\cdot\mathbf{c}}], \quad (13.24)$$

where

$$[\overline{\mathcal{F}}_{E\phi}] = c_p[\overline{Tv}] + g[\overline{zv}] + L[\overline{vq}] + \tfrac{1}{2}[\overline{(u^2+v^2)v}]$$

and a similar expression holds for $[\overline{\mathcal{F}}_{Ep}]$.

13.2.3 Volume integrals

For regional or global energy budget studies the previous equations must be integrated over a given volume V with a mass M. This volume must be fixed in space, as explained before. Let us consider then a polar cap limited by a vertical wall at latitude ϕ.

Equation (13.17) now becomes

$$\frac{\partial}{\partial t}\overline{\int_M c_p T\, dm} = -\int_{\text{boundary}} \overline{\mathbf{J}_H}\cdot \mathbf{n}\, ds$$
$$+ \int_M \overline{Q}\, dm + \int_M \overline{\alpha\omega}\, dm. \qquad (13.25)$$

The boundaries are formed by the earth's surface, the top of the atmosphere, and the latitudinal wall. At the top and bottom boundaries $\mathbf{J}_H\cdot\mathbf{n}$ vanishes, so that

$$-\int_{\text{boundary}} \overline{\mathbf{J}_H}\cdot \mathbf{n}\, ds = \int_{\text{wall}} \overline{\mathcal{J}}_{H\phi}\, ds$$
$$= 2\pi R \cos\phi \int_0^{p_0} [\overline{\mathcal{J}}_{H\phi}]\frac{dp}{g}.$$

This shows that the total meridional transport can be obtained by vertical integration of the mean zonal values. Using the expansions shown before in the case of Eq. (13.22), various terms appear each representing a different mode of transport of enthalpy. The second and third integrals at the right-hand side of Eq. (13.25) represent the sources and sinks of enthalpy through nonadiabatic heating and diffusion across the boundaries (Q term) and conversion from potential into kinetic energy ($-\alpha\omega$ term).

Similarly for kinetic energy, we obtain from Eq. (13.18):

$$\frac{\partial}{\partial t}\overline{\int_M K_H\, dm} = 2\pi R \cos\phi \int_0^{p_0} [\overline{\mathcal{J}}_{K\phi}]\frac{dp}{g}$$
$$- \int_{\text{sfc}} \overline{\omega z}\, ds - \int_M \overline{\omega\alpha}\, dm$$
$$+ \int_M (\overline{uF_\lambda} + \overline{vF_\phi})dm. \qquad (13.26)$$

For latent heat, we find using Eq. (13.20):

$$\frac{\partial}{\partial t}\overline{\int_M Lq\, dm} = 2\pi R \cos\phi \int_0^{p_0} [\overline{\mathcal{J}}_{q\phi}]\frac{dp}{g}$$
$$+ L\int_M \overline{(e-c)}dm + \int_{\text{sfc}} \overline{F}^\uparrow_{\text{LH}}\, ds, \qquad (13.27)$$

where $F^\uparrow_{\text{LH}} = L\mathcal{J}_q^D$ is the upward flux of latent heat at the earth's surface.

For total energy we obtain using Eq. (13.21):

$$\overline{\frac{\partial}{\partial t} \int_M E\,dm} = 2\pi R \cos\phi \int_0^{p_0} [\overline{\mathcal{J}}_{E\phi}]\frac{dp}{g} + \int_{\text{top}} \overline{F}^{\downarrow}_{\text{rad}}\,ds$$

$$+ \int_{\text{sfc}}(-\overline{F}^{\downarrow}_{\text{rad}} + \overline{F}^{\uparrow}_{\text{SH}} + \overline{F}^{\uparrow}_{\text{LH}}$$

$$- \overline{\omega z} + \overline{\tau_0 \cdot \mathbf{c}})\,ds. \tag{13.28}$$

The sign convention used here is to count $F^{\downarrow}_{\text{rad}}$ positive if downward, and $F^{\uparrow}_{\text{SH}}(=\mathcal{J}^D_H)$ and $F^{\uparrow}_{\text{LH}}$ positive if directed upward.

If we consider the latent heat release as an internal energy source, we can write the energy equation for the atmosphere in a different form as

$$\overline{\frac{\partial}{\partial t}\int(c_p T + K_H)\,dm} = 2\pi R\cos\phi \int_0^{p_0}([\overline{\mathcal{J}}_{H\phi}] + [\overline{\mathcal{J}}_{K\phi}])\frac{dp}{g}$$

$$+ \int_{\text{top}} \overline{F}^{\downarrow}_{\text{rad}}\,ds$$

$$+ \int_{\text{sfc}}(-\overline{F}^{\downarrow}_{\text{rad}} + \overline{F}^{\uparrow}_{\text{SH}} - \overline{\omega z} + \overline{\tau_0 \cdot \mathbf{c}})\,ds$$

$$+ \int_{\text{sfc}} L\overline{P}\,ds. \tag{13.29}$$

The last term in Eq. (13.29) containing the precipitation rate P at the earth's surface is now the only effect of water in this alternative energy equation.

13.2.4 Globally averaged climate equations

Let us now consider the global atmosphere. Integration of Eqs. (13.25) through (13.29) over the entire volume of the atmosphere will bring out more clearly the internal sources and sinks, as well as the interactions with the external system through the boundaries at the top of the atmosphere and at the earth's surface. Thus, we find for the various quantities:

$$\overline{\frac{\partial}{\partial t}\int_M c_p T\,dm} = \int_M \overline{Q}\,dm + \int_M \overline{\alpha\omega}\,dm, \tag{13.30}$$

$$\overline{\frac{\partial}{\partial t}\int_M K_H\,dm} = -\int_{\text{sfc}} \overline{\omega z}\,ds - \int_M \overline{\omega\alpha}\,dm$$

$$+ \int_M (\overline{uF_\lambda} + \overline{vF_\phi})\,dm, \tag{13.31}$$

$$\overline{\frac{\partial}{\partial t}\int_M Lq\,dm} = \int_M L(\overline{e-c})\,dm + \int_{\text{sfc}} \overline{F}^{\uparrow}_{\text{LH}}\,ds$$

$$= \int_{\text{sfc}} L(\overline{E} - \overline{P})\,ds, \tag{13.32}$$

where E denotes the evaporation (or sublimation) rate at the earth's surface. Because the contributions of the surface pressure work ($\overline{\omega z}$) and the surface friction ($\overline{\tau_0 \cdot \mathbf{c}}$) are very small compared to the other surface terms in Eqs. (13.28) and (13.29) we can write for the total energy equations:

$$\overline{\frac{\partial}{\partial t} \int_M E \, dm} = \int_{\text{top}} \overline{F}_{\text{rad}}^{\downarrow} \, ds + \int_{\text{sfc}} (-\overline{F}_{\text{rad}}^{\downarrow} + \overline{F}_{\text{SH}}^{\uparrow} + \overline{F}_{\text{LH}}^{\uparrow}) \, ds \qquad (13.33)$$

and

$$\overline{\frac{\partial}{\partial t} \int (c_p T + K_H) \, dm} = \int_{\text{top}} \overline{F}_{\text{rad}}^{\downarrow} \, ds + \int_{\text{sfc}} (-\overline{F}_{\text{rad}}^{\downarrow} + \overline{F}_{\text{SH}}^{\uparrow} + L\overline{P}) \, ds. \qquad (13.34)$$

From Eq. (13.30) we conclude that the rate of change of total potential energy for the entire atmosphere is equal to the total heating of the atmosphere minus the conversion $C(P,K)$ of potential into kinetic energy. This last quantity can be written in various alternative forms for the entire atmosphere:

$$C(P,K) = -\int_M \alpha \omega \, dm \qquad (13.35a)$$

$$= -\int_M R_d \frac{T\omega}{p} \, dm \qquad (13.35b)$$

$$= -\int_M g \, z \, \text{div}_2 \, \mathbf{v} \, dm \qquad (13.35c)$$

$$= -\int_M g \, \mathbf{v} \cdot \text{grad} \, z \, dm. \qquad (13.35d)$$

All these expressions contain the ageostrophic motions (motions across isobars) and are connected with the divergent component of the wind. However, they can be interpreted in rather different ways. Since the work done by the geostrophic wind is identically zero, the net work results from the ageostrophic flow against the pressure gradient force (13.35d). This is associated with horizontal divergence or convergence mainly near the earth's surface and at jet stream levels [Eq. (13.35c)]. Moreover, in order to have a generation of kinetic energy the resulting vertical motions are such that there must be an expansion (rising) of lighter air and a compression (sinking) of denser air as shown by expressions (13.35a) and (13.35b).

As can be seen from Eq. (13.31), the time rate of change of horizontal kinetic energy in the global atmosphere is equal to the conversion from total potential energy $C(P, K)$ minus the frictional dissipation of kinetic energy in the atmosphere D_A and the mechanical energy loss to the oceans D_{oc}. The latter quantity is ultimately dissipated by friction inside the oceans.

Symbolically the equations for total potential and kinetic energy can be written as

$$\frac{\partial P}{\partial t} = -C(P,K) + H, \qquad (13.36)$$

$$\frac{\partial K}{\partial t} = C(P,K) - D_A - D_{\text{oc}}, \qquad (13.37)$$

where

$$H = \int_M Q \, dm$$

or using Eq. (13.7):

$$H = -\int_M \alpha \operatorname{div} \mathbf{F}_{\text{rad}}\, dm - \int_M L(e-c)\, dm + \int_{\text{sfc}} F^{\uparrow}_{\text{SH}}\, ds - \int_M \alpha\tau{:}\operatorname{grad} \mathbf{c}\, dm,$$

$$D_A = -\int_M \alpha\tau{:}\operatorname{grad} \mathbf{c}\, dm$$

$$= \text{atmospheric dissipation rate,}$$

and

$$D_{\text{oc}} = \int_{\text{sfc}} (\omega z - \tau_0 \cdot \mathbf{c})\, ds.$$

For a long period of time $\overline{\partial P/\partial t}$ and $\overline{\partial K/\partial t}$ vanish, and we have $\overline{H} = \overline{C} = \overline{D}_A + \overline{D}_{\text{oc}}$. Because there is always dissipation of kinetic energy $\overline{D}_A > 0$, $\overline{D}_{\text{oc}} > 0$ and, consequently, $\overline{H} > 0$.

The previous climate equations also demonstrate the importance of the interactions between the atmosphere and the other subsystems through the $F^{\downarrow}_{\text{rad}}, F^{\uparrow}_{\text{SH}}, F^{\uparrow}_{\text{LH}}$, and pressure work terms in the energy equation, through the friction and pressure work terms in the kinetic energy equation, and through the $\overline{E}$ and $\overline{P}$ terms in the latent heat equation.

13.3 OBSERVED ENERGY BALANCE

13.3.1 Diabatic heating in the atmosphere

When we presented the thermodynamic energy equation (3.42), we mentioned how important the diabatic processes were. In fact, they constitute the sources and sinks of internal energy for the atmosphere. Thus, they play a crucial role in most meteorological phenomena (such as convection, fronts, and synoptic weather systems) and, on a global scale, in the generation and destruction of available potential energy as we will show later.

To get an idea of the order of magnitude of the various components of the diabatic heating in Eq. (13.7), we present tentative zonal-mean cross sections of them as given by Newell *et al.* (1970) for the northern winter season in Fig. 13.2.

The net radiative heating is generally negative on the order of -0.5 to -2.0 °C/day, as expected (see Sec. 6.6). In the intertropical high stratosphere it is slightly positive mainly due to the strong absorption of solar radiation by ozone.

The latent heat release was estimated from the precipitation (see Fig. 7.25) assuming a given vertical distribution of the cloud-forming processes (Newell *et al.*, 1974). Three maxima on the order of 2 °C/day occur in the troposphere. They are associated with strong convective precipitation in the ITCZ and with the polar frontal precipitation in midlatitudes of both hemispheres.

The boundary layer heating is confined mainly to the lower troposphere below 800 mb with the largest values near the ground. They are due to small-scale turbulent motions in the surface boundary layer.

The total diabatic heating resulting from the superposition of the three effects is also presented in Fig. 13.2. The release of latent heat dominates in the tropics where it leads to net heating. At high latitudes and at high altitudes the dominant factor is the radiative cooling.

13.3.2 Energy in the atmosphere

As we have seen before, energy can be stored in the atmosphere in various forms, namely as internal energy, potential energy, latent heat, and kinetic energy. The total energy per unit mass E is given by expression (13.1):

$$E = c_v T + gz + Lq + \tfrac{1}{2}(u^2 + v^2),$$

where we will use $L = L_e$ in the actual calculations. This expression shows that the evaluation of the various forms of energy requires the three-dimensional distributions of the fields of $T, z, q, u,$ and v. Various maps of these quantities were already discussed in Chap. 7. Because the vertical-mean internal and potential energy are linearly related to each other through the relation $\Phi = (c_p/c_v - 1)I$, only maps of the vertically averaged temperature $\langle \overline{T} \rangle$ and specific humidity $\langle \overline{q} \rangle$ are needed to describe the geographical distribution of the energy.

In Fig. 13.3, we present meridional profiles of the vertically averaged $\overline{T}$, $\overline{z} - z_{SA}$, $\overline{q}$, and, finally, the total energy $\overline{E}$. These profiles synthesize the general behavior of the corresponding forms of energy. The contribution from kinetic energy is negligible. The latent heat shows a strong seasonal variation mainly in the lower latitudes, in contrast with the internal energy and the potential energy for which the seasonal variations occur mainly in the high latitudes. The profiles of total energy reflect these facts. The profiles also show that the Southern Hemisphere high latitudes contain less internal and potential energy, as well as less latent heat, than the Northern Hemisphere high latitudes contain. It is worth mentioning that mainly due to the influence of topography on the value of the potential energy, the profiles of $\langle \overline{E} \rangle$ show an artificial maximum poleward of 70 °S.

The mean values of the various forms of energy for the two hemispheres and the entire globe are presented in Table 13.1. The inspection of the table shows that the most important forms of energy are the internal energy (70.4% for global annual mean), potential energy (27.1%), and latent heat (2.5%). The kinetic energy is only a minute fraction (0.05%) of the total energy of the atmosphere. However, as we will show later, the kinetic energy represents a considerable fraction of the energy *available* for the general circulation of the atmosphere and plays a very important role in the energetics of the general circulation. Of course the winds are also crucial in redistributing the energy over the globe. The amplitude of the annual cycle in the Northern Hemisphere is almost twice as large as that in the Southern Hemisphere which is mostly due to the differences in land–sea distribution in the two hemispheres.

13.3.3 Transport of atmospheric energy

It is well known that the motions in the atmosphere and oceans play an important role in carrying energy from the regions of net incoming radiation to the regions of net outgoing radiation. Thereby the atmospheric and oceanic currents have a moderat-

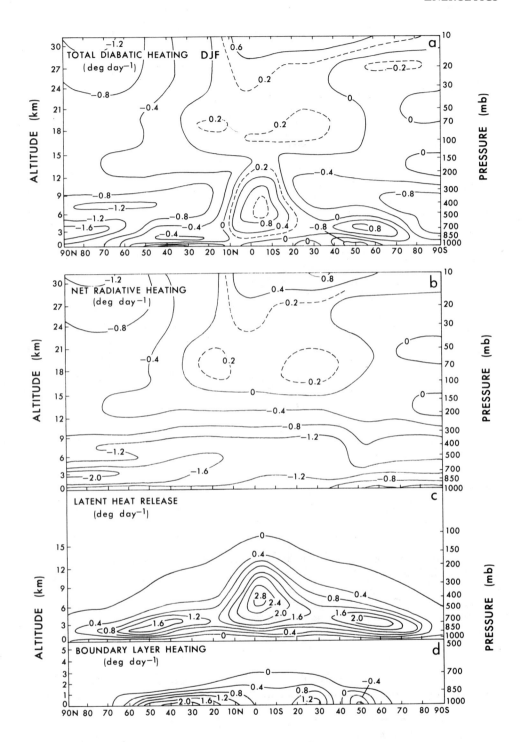

FIGURE 13.2. The total diabatic heating rate for the atmosphere for December–February (a) and its components, i.e., the net radiative heating (b), the latent heat release (c), and the sensible heating in the boundary layer (d) in units of °C day^{-1}. To obtain equivalent energy input units multiply values by c_p $(\Delta p/g)$ or a factor of 11.6 to convert from °C day^{-1} to W m^{-2} for a 100-mb $(\Delta p/g = 10^3$ kg m$^{-2})$ thick layer (from Newell *et al.*, 1970).

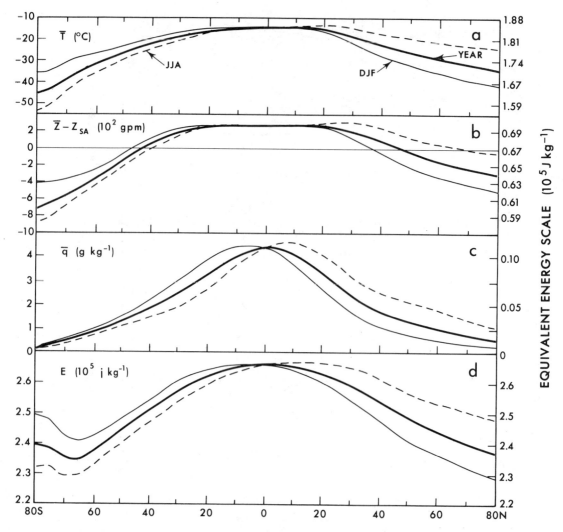

FIGURE 13.3. Meridional profiles of the zonal- and vertical-mean temperature (a), geopotential height departure from the NMC standard atmosphere (b), specific humidity (c), and total energy (d) for annual, DJF, and JJA mean conditions. An equivalent energy scale is given on the right-hand side of the figure. The kinetic energy profiles shown before in Fig. 7.23 give much smaller values on the order of 150 J kg^{-1} (from Oort and Peixoto, 1983).

ing influence in the formation of the climate. After some simplifications, we may write Eq. (13.28) for the energy balance in a polar cap as

$$\overline{\frac{\partial}{\partial t} \int (c_v T + gz + Lq)\, dm}$$

$$= \iint_{\text{wall}} \overline{(c_p T + gz + Lq)v}\, \frac{dx\, dp}{g} + \overline{F}^{\downarrow}_{\text{TA}} - \overline{F}^{\downarrow}_{\text{BA}}, \qquad (13.38)$$

where the term F_{TA} represents the net downward energy flux at the top of the atmosphere, and F_{BA} the net downward flux of energy at the bottom.

TABLE ● 13.1. Hemispheric and global integrals of the atmospheric energy[a] per unit surface area in units of 10^7 J m^{-2} from Oort and Peixoto (1983).

	Year			DJF			JJA			DJF-JJA		
	NH	SH	Globe	NH	SH	Globe	NH	SH	Globe	NH	SH	Globe
I	180.6	180.0	180.3	178.2	181.6	179.9	183.4	178.4	180.9	−5.2	3.2	−1.0
Φ	70.0	68.7	69.3	69.0	69.3	69.2	71.0	68.3	69.6	−2.0	1.0	−0.4
LH	6.48	6.28	6.38	5.15	7.16	6.16	8.07	5.38	6.72	−2.92	1.78	−0.56
K	0.116	0.131	0.123	0.168	0.100	0.134	0.072	0.156	0.114	0.096	−0.056	0.020
E	257.2	255.1	256.1	252.5	258.2	255.4	262.5	252.3	257.4	−10.0	5.9	−2.0
K/E (%)	0.05	0.05	0.05	0.07	0.04	0.05	0.03	0.06	0.04			
LH/E (%)	2.52	2.46	2.49	2.04	2.77	2.41	3.07	2.13	2.61			

[a] I = internal energy; Φ = potential energy; LH = latent heat; K = kinetic energy; and E = total energy.

Equation (13.38) states that the change of energy in a polar cap is the net result of the exchanges of energy with the rest of the atmosphere, outer space, and the underlying surface. The relative importance of the various forms of energy in transporting energy across the three boundaries can be quite different. In fact, the exchange through the vertical, latitudinal wall takes place in all three forms of energy, whereas across the upper boundary (F_{TA}) the exchange occurs in the form of radiation only, and through the bottom (F_{BA}) mainly in the form of radiation, sensible heat, and latent heat.

The transport across the wall can be decomposed into the contributions by transient eddies, stationary eddies, and mean meridional circulations in order to get a better understanding of the physical mechanisms involved:

$$[\overline{(c_p T + gz + Lq)v}]$$
$$= c_p [\overline{v'T'}] + c_p [\overline{v}^* \overline{T}^*] + c_p [\overline{v}][\overline{T}]$$
$$+ g[\overline{v'z'}] + g[\overline{v}^* \overline{z}^*] + g[\overline{v}][\overline{z}]$$
$$+ L[\overline{v'q'}] + L[\overline{v}^* \overline{q}^*] + L[\overline{v}][\overline{q}]. \quad (13.39)$$

The transport of water vapor (latent heat) was discussed before in Chap. 12, and the transports of the other components will be considered next.

13.3.3.1 Meridional transport of sensible heat

Due to the importance of large-scale turbulent processes in the atmosphere, maps of the vertical-mean poleward flux of sensible heat by transient eddies are shown in Figs. 13.4(a)–13.4(c). By and large, the eddy fluxes are predominantly poleward in both hemispheres with a maximum in midlatitudes. The yearly map of $\langle \overline{v'T'} \rangle$ shows midlatitude zones of strong poleward heat flux fairly uniform in the Southern Hemisphere, but with distinct maxima in the Northern Hemisphere over North America and eastern Asia. These are clearly associated with baroclinic disturbances along the polar front. Over the equator the meridional eddy fluxes are very small, as they are over the poles. However, at some longitudes near 70° latitude they attain sizeable values directed away from the poles.

The fluxes for the seasons are most intense in the winter hemisphere. In the Northern Hemisphere the influence of the land–sea distribution is very pronounced. In the Southern Hemisphere, we find a rather uniform belt of maximum poleward heat flow around 45 °S.

Comparison with similar maps of $\langle \overline{v'T'} \rangle$ for the years 1950 (Peixoto, 1960) and 1958 (Peixoto, 1974) suggests that the annual transient eddy heat flux does not vary much from year to year and that it is a relatively stable measure of the general circulation, at least on an annual basis.

The vertical distribution of the zonally averaged values for the various modes of sensible heat transport are given in Fig. 13.5. Near 50° latitude the eddy transports exhibit two maxima in the vertical at about the 850- and 200-mbar levels, which are associated with the alternation of air masses and the fluctuations of the tropopause, respectively. The $[\overline{v'T'}]$ pattern between 20 °S and 20 °N shows that the transient eddies in the tropics transport heat toward the equator. Thereby, they act in an

ENERGETICS 325

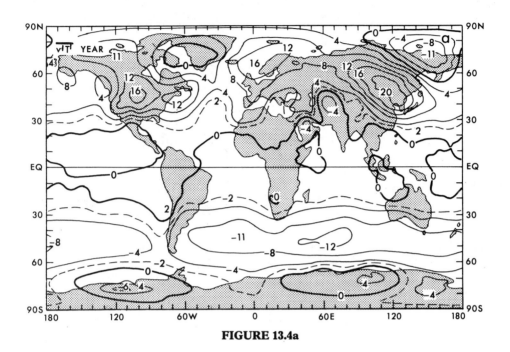

FIGURE 13.4a

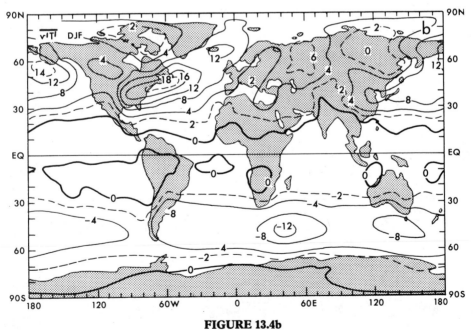

FIGURE 13.4b

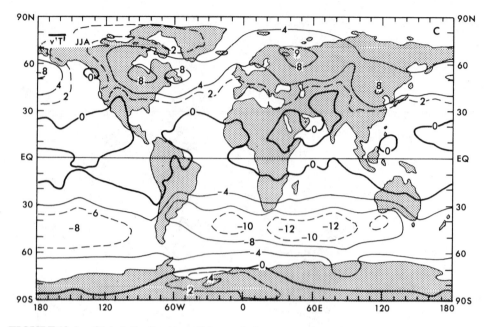

FIGURE 13.4. Global distributions of the vertical-mean northward transport of sensible heat by transient eddies in °C m s^{-1} for annual (a), DJF (b), and JJA (c) mean conditions [to convert to units of 10^7 W m^{-1} multiply values by $c_p\,(p_0/g) \approx 1$, where p_0/g = mass per unit area] (from Oort and Peixoto, 1983).

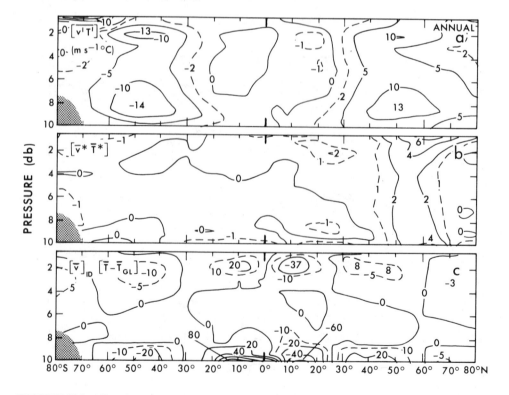

FIGURE 13.5. Zonal-mean cross sections of the northward transport of sensible heat by transient eddies (a), stationary eddies (b), and mean meridional circulations (c) in °C m s^{-1} (from Oort and Peixoto, 1983).

abnormal manner since they tend to heat rather than cool the inner tropics. This behavior resembles that of the eddies in the lower stratosphere (Starr and Wallace, 1964). The patterns for the yearly and seasonal conditions are similar, except for the mean meridional fluxes in the tropics. This fact is shown in the vertical-mean profiles in Fig. 13.6.

The stationary eddy pattern ($[\bar{v}^*\overline{T}^*]$) in Fig. 13.5 has a different, usually weaker structure with much higher values in the Northern Hemisphere poleward of 40 °N. As shown in the profiles in Fig. 13.6, the stationary eddies have a large seasonal variation in the Northern Hemisphere. In the winter their transport can exceed the transport by the transient eddies.

The mean meridional circulation transports $[\bar{v}]_{\text{ind}}[\overline{T}]$ were evaluated using the indirectly calculated values of $[\bar{v}]$. The vertical distribution in Fig. 13.5 shows that the mean meridional circulations are most active in the upper and lower levels. The vertical cross section reveals the three-cell regime in both hemispheres. Large seasonal variations occur in the tropics since the mean meridional flow of sensible heat tends to be from the winter into the summer hemisphere, associated with the shifts in the Hadley cells. This is clearly shown in the vertical mean profiles in Fig. 13.6.

13.3.3.2 Meridional transport of potential energy

The vertical-mean flux of potential energy is presented in Fig. 13.7. The eddy fluxes are very small compared with those of sensible heat. This is in agreement with the quasigeostrophic character of the eddies, since for geostrophic flow

$$[v_g z] = \frac{g}{f}\left[\frac{\partial z}{R \cos \phi \, \partial \lambda} z\right] = 0.$$

It is of interest to note that the transient eddy transport of potential energy across about 25° latitude is directed into the tropics in both hemispheres. This eddy transport is equivalent with a pressure forcing in the (λ, ϕ, p) system. Such a forcing was first suggested by Mak (1969) as a possible energy source for tropical disturbances.

By far the most important mode is the mean meridional cell flux. The patterns are similar to those of the sensible heat flow, but with the sign reversed, as can be seen from the adiabatic limit condition $g\,dz = -c_p\,dT$. The net cross-equatorial flux of potential energy is directed from the summer to the winter hemisphere. In a thermally direct circulation, as in the Hadley cells, it overcompensates for the flow of sensible heat, leading to a net residual transport of energy into the winter hemisphere.

13.3.3.3 Meridional transport of kinetic energy

To complete the analysis of the transport of energy, a discussion of the kinetic energy is given now, in spite of the smallness of its contribution to the total energy balance. The transport of kinetic energy in the various modes is shown in Figs. 13.8 and 13.9. The transport occurs mainly at the jet stream level in the upper troposphere. The cross sections show alternating positive and negative centers, leading to a well-defined convergence of total kinetic energy in middle latitudes. Among the various modes, the transient eddies attain the largest values. The stationary eddies are almost nonexistent in the Southern Hemisphere, but important in the Northern

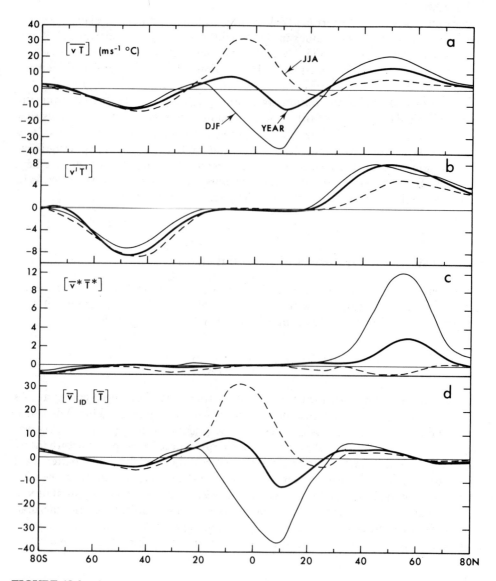

FIGURE 13.6. Meridional profiles of the zonal-and vertical-mean northward flux of sensible heat by all motions (a), transient eddies (b), stationary eddies (c), and mean meridional circulations (d) in units of °C m s^{-1} [to convert to units of 10^{15} W multiply values by $2\pi R \cos\phi \, c_p \, (p_0/g) \approx 0.4 \cos\phi$; from Oort and Peixoto, 1983].

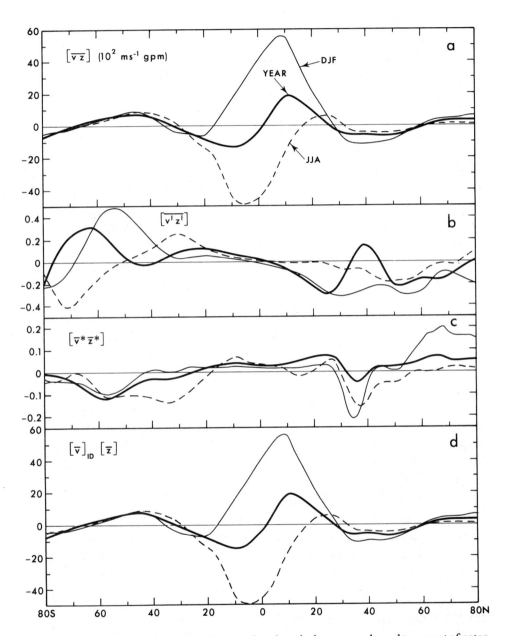

FIGURE 13.7. Meridional profiles of the zonal-and vertical-mean northward transport of potential energy by all motions (a), transient eddies (b), stationary eddies (c), and mean meridional circulations (d) in 10^2 gpm m s^{-1} [to convert to units of 10^{15} W multiply values by $2\pi R \cos\phi(p_0/g)$ or $0.4 \cos\phi$; from Oort and Peixoto, 1983].

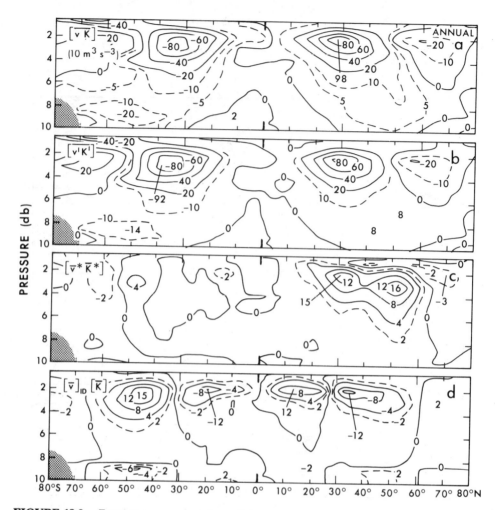

FIGURE 13.8. Zonal-mean cross sections of the northward transport of kinetic energy by all motions (a), transient eddies (b), stationary eddies (c), and mean meridional circulations (d) in $10 \, m^3 \, s^{-3}$ for annual-mean conditions (from Oort and Peixoto, 1983).

Hemisphere. As shown by the vertically integrated values in Fig. 13.9 the stationary eddies are as important as the transient eddies during the DJF season at 30 °N. The contribution by the mean meridional cells to the total transport of kinetic energy is quite small, even in the tropical Hadley cells.

13.3.3.4 Meridional transport of total energy

We will now present a more integrated picture of the energy transport in the atmosphere, by combining the previous zonal-mean cross sections and profiles all in units of °C m s^{-1}. To convert the various fluxes to the same energy flux units, e.g., to J g^{-1} m s^{-1}, one should multiply the sensible heat flux in °C m s^{-1} by 1.0, the geopotential flux values in gpm m s^{-1} by 0.01, the water vapor flux values in g kg^{-1} m s^{-1} by 2.5, and the kinetic energy flux values in m^3 s^{-3} by 0.001. Figures 13.10 and 13.11 show the resulting cross sections and profiles of the transport of total energy. We find that the most important modes of total energy transport are the

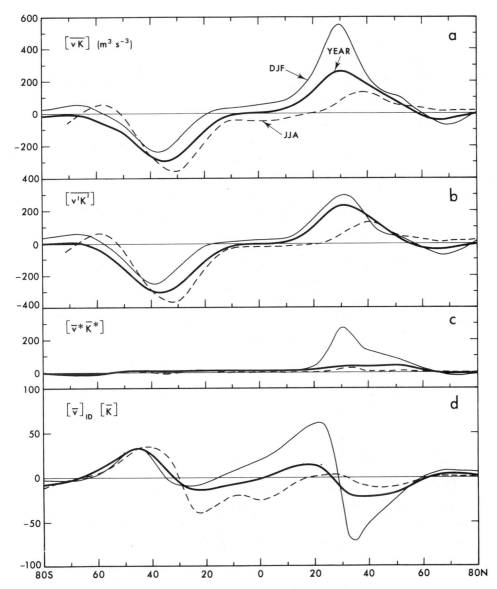

FIGURE 13.9. Meridional profiles of the zonal-and vertical-mean northward transport of kinetic energy by all motions (a), transient eddies (b), stationary eddies (c), and mean meridional circulations (d) in m^3 s^{-3} [to convert to units of 10^{15} W multiply values by $2\pi R \cos\phi(p_0/g)$ or $0.0004 \cos\phi$; from Oort and Peixoto, 1983].

transient eddies and the mean meridional circulations. In fact, between 20 °S and 20 °N the mean meridional cell circulations appear to be the important mechanism to transport energy meridionally in its various forms, whereas poleward of 30 °N the eddies play a dominant role for all energy forms except potential energy. Furthermore, for the total energy flux by the eddies the seasonal changes are large in the Northern and small in the Southern Hemisphere.

We should note that only indirectly calculated $[\bar{v}]_{\text{ind}}$ were used to evaluate the mean meridional circulation transports shown in Figs. 13.10 and 13.11 because the direct estimates of $[\bar{v}]$ are unrealistic in the Southern Hemisphere. Further, in the earlier presentation of the mean meridional circulation fluxes of sensible heat, potential energy, and total energy their mean global values, averaged both horizontally and vertically, were subtracted since they do not lead to a net transport.

The main sources and sinks of atmospheric energy are apparent when we consider the divergence of the energy transports. Their vertical-mean profiles are shown in Fig. 13.12 for annual-mean and winter and summer conditions. These profiles are the derivatives with respect to latitude of the annual curves in Fig. 13.11. As expected, the main sources of energy are in the intertropical regions showing a bimodal distribution, and the main sinks are in high latitudes.

The net effect of the transient eddies on a regional scale can be seen by studying maps of the divergence of the energy transport. Thus, we present in Fig. 13.13 global maps of the vertical-mean divergence of the transient eddy flux of energy. The centers of energy divergence, i.e., the sources of energy, are located around 30 °N and 30 °S. In the Southern Hemisphere, a zonally uniform value of nearly 50 W m^{-2} is found, whereas in the Northern Hemisphere we find two strong sources of energy over the Gulf Stream and Kuroshio regions with maximum divergence up to 150 W m^{-2}. Regions of convergence (sinks) of atmospheric energy are found poleward of about 45° latitude with centers on the order of -100 W m^{-2} over eastern Canada, northeast of Iceland, and eastern Siberia. In the Southern Hemisphere, an extensive belt of convergence is found over the oceans around the rim of the Antarctic continent.

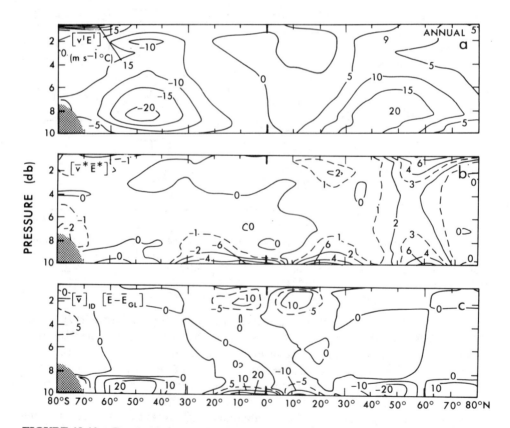

FIGURE 13.10. Zonal-mean cross sections of the northward transport of total energy by transient eddies (a), stationary eddies (b), and mean meridional circulations (c) for annual-mean conditions in °C m s^{-1} (from Oort and Peixoto, 1983).

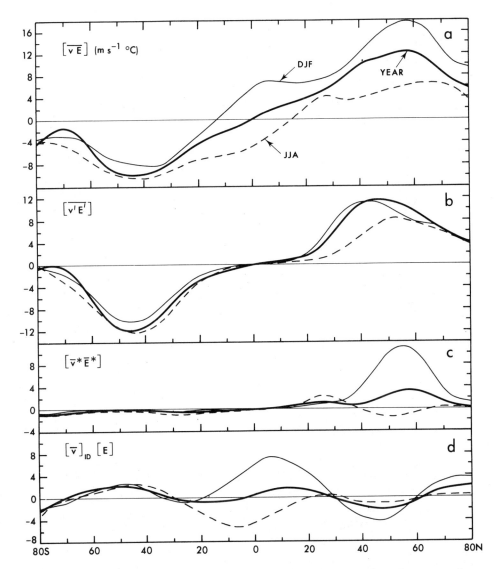

FIGURE 13.11. Meridional profiles of the zonal- and vertical-mean northward transport of energy by all motions (a), transient eddies (b), stationary eddies (c), and mean meridional circulations (d) in °C m s^{-1} [to convert to units of 10^{15} W multiply values by $2\pi R \cos\phi\, c_p (p_0/g) \approx 0.4 \cos\phi$; from Oort and Peixoto, 1983].

One should note that the field of the energy convergence by the time-mean flow is much more difficult to determine than the transient eddy field. When the time-mean flow is included, the patterns (not shown here) tend to be more noisy with several maxima and minima around each latitude circle and with an apparent poleward shift of the oceanic divergence maxima of roughly 10° latitude. The essence of the divergence of energy in the atmosphere is probably contained in the divergence of the transient eddy transports.

It is of interest to compare the yearly divergence map by transient eddies [Fig. 13.13(a)] with the earlier map of net incoming radiation [Fig. 6.11(a)]. In the comparison, two other factors should be taken into account, namely the divergence of

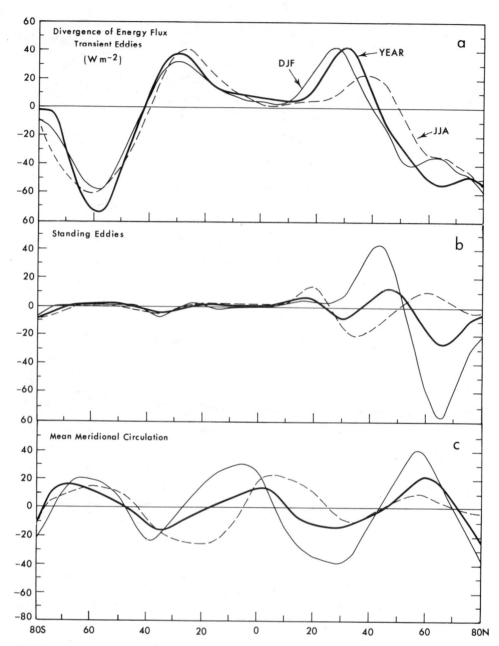

FIGURE 13.12. Meridional profiles of the zonal-mean divergence of energy by transient eddies (a), stationary eddies (b), and mean meridional circulations (c) in W m^{-2} (from Oort and Peixoto, 1983).

energy by the time-mean atmospheric flow and by ocean currents. Since both terms are probably important, but cannot be determined well at present, we can only speculate as to their relative significance. Over the continents, the differences between Figs. 13.13(a) and 6.11(a) must be entirely due to the time-mean atmospheric flow. In fact, from the comparison of the maps we would expect convergence of energy by the mean flow over western Canada and western Europe, and divergence over eastern Canada and Siberia. Over the oceans, the differences between the two maps are large

ENERGETICS 335

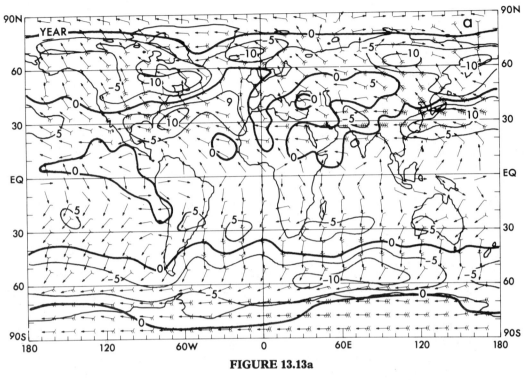

FIGURE 13.13a

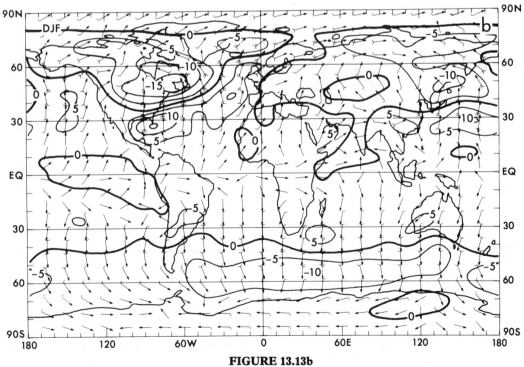

FIGURE 13.13b

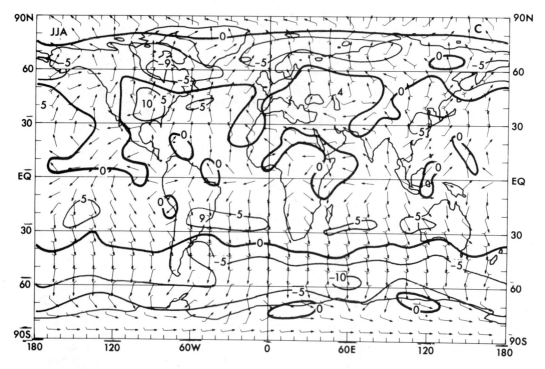

FIGURE 13.13. Global distributions of the vertical-mean horizontal flow of total energy by transient eddies for annual (a), DJF (b), and JJA (c) mean conditions as shown by arrows. Each barb on the tail of an arrow represents a transport of 5 °C m s^{-1} (or in terms of an integrated value a transport of 5×10^7 W m^{-1}). Also shown are some isolines of the transient eddy divergence in units of 10 W m^{-2} (from Oort and Peixoto, 1983).

and an important part of the imbalance must be due to oceanic heat transports. Thus, we would expect oceanic heat divergence equatorward of 20° latitude, convergence in a zonal belt around 30 °S, and strong convergence near 30 °N localized over the Gulf Stream and Kuroshio, as well as a general convergence over the eastern North Atlantic Ocean.

To compare the seasonal maps of net incoming radiation with those of the transient eddy atmospheric divergence, the storage of energy in the atmosphere and, especially, the oceans must be taken into account. In a qualitative sense, we find over the continents the tendency for a strong transient eddy convergence of energy in winter and for some divergence in summer.

13.3.3.5 Mechanisms of releasing energy

Let us now take a Lagrangian point of view. The poleward particle trajectories in the eddies must be slightly inclined with respect to the horizontal, carrying warmer air poleward and upward so that the necessary kinetic energy be released and the general circulation is kept going (see Fig. 13.14). If the trajectories had the same slope as the isentropes [case (b)] the process would be adiabatic (isentropic) and no heat would be transported or kinetic energy released. If for point B the slope were greater than the slope of the isentropes [case (c)] cooler rather than warmer particles would be trans-

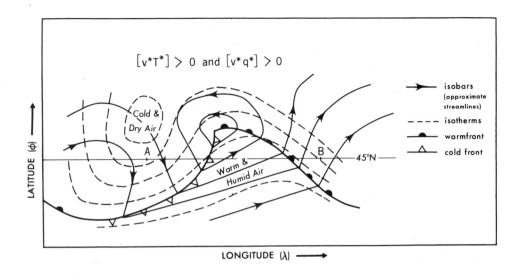

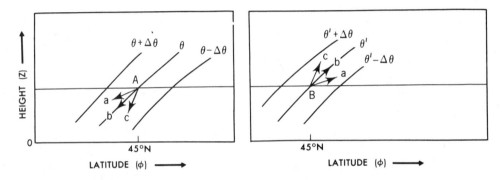

FIGURE 13.14. Schematic picture of the dominant mechanism of northward transport of sensible and latent heat by eddies in midlatitudes of the Northern Hemisphere. The transport takes place mainly in the lower troposphere. Shown are latitude–longitude and height–latitude cross sections (from Peixoto and Oort, 1984).

ported poleward, leading to a cooling instead of the required warming. On the other hand, in case (a) the motions would become dynamically unstable because in their new position the particles would be warmer than their environment and would accelerate away from their original position. For point A to release kinetic energy, the slope of the particle trajectories would again have to be smaller than the slope of the isentrope [case (a)] so that the temperature of the particle in its new position would become colder than its environment and the particle would continue to sink.

In order for these mechanisms to operate, it is obvious that the isentropic surfaces have to be inclined with respect to the pressure surfaces, and that the atmosphere be baroclinic. We therefore find that the perturbations (eddies) grow, leading to what is called baroclinic instability (Charney, 1947; Eady, 1949). The perturbations are called baroclinic because the horizontal gradient of the mean temperature plays a crucial role in their development. These processes are an essential element of the general circulation of the atmosphere, and thus of the climate. They may be regarded as the result of macroscale slant convection that is a manifestation of baroclinic instability.

13.3.3.6 Vertical transport of atmospheric energy

We can use Eq. (13.24) to estimate the vertical transport of energy. For a steady state the left-hand side of Eq. (13.24) vanishes. Neglecting the small diffusion terms inside the atmosphere, we can define a stream function for the total energy ψ_E such that

$$2\pi R \cos \phi \, [\overline{\mathcal{J}}_{E\phi}] = \frac{-\partial \psi_E}{\partial p}, \tag{13.40a}$$

$$2\pi R \cos \phi \, \{[\overline{\mathcal{J}}_{Ep}] + g \, [\overline{F}_{\text{rad}}]\} = \frac{\partial \psi_E}{R \, \partial \phi}. \tag{13.40b}$$

As was done in the case of angular momentum and water vapor, we can separate each term in $[\overline{\mathcal{J}}_{Ep}]$ into the various modes of transport, but this is not necessary for the present discussion. In fact little is known about the predominant scales at which vertical transports take place in the atmosphere. However, it is clear that from the surface upward progressively bigger and bigger eddies become important in the transfer. Dry and wet (cumulus) convective transfer is very important in the tropics. In middle latitudes the baroclinic eddies probably play a more dominant role (see Starr *et al.*, 1970b; Palmén and Newton, 1969).

At the top of the atmosphere we can neglect the first term on the left-hand side of Eq. (13.40b). Thus, using the observed net radiation fluxes as top boundary condition, we can evaluate the ψ_E distribution from the horizontal fluxes alone by starting the integration of expression (13.40a) at one of the poles, where we assume $\psi_E = 0$. Figure 13.15 shows the resulting distribution of ψ_E as a function of latitude and pressure for the year and for the DJF and JJA seasons. The streamlines of ψ_E represent the density of the flux of energy in units of 10^{15} W. It is of great interest to analyze the vertical energy fluxes inside the atmosphere as inferred from these figures using Eq. (13.40b), since they cannot be obtained directly from the present observations. To avoid the appearance of some closed streamlines induced by the tropical Hadley cells, the global mean value of the energy was subtracted in the construction of the streamlines. In the computations of the streamlines for DJF and JJA, the (small) energy storage in the atmosphere was neglected.

In the yearly diagram [Fig. 13.15], a large inflow of radiational energy occurs in the tropics, between about 30 °S and 30 °N. Roughly half of this energy passes unattenuated through the atmosphere and is absorbed mainly in the tropical oceans, where it is transported poleward by oceanic circulations. However, in the subtropics and middle latitudes most of the action seems to be due to the atmospheric circulations which transport the energy poleward until it is radiated back into space. The seasonal diagrams show the enormous direct inflow of energy into the summer hemisphere and the corresponding outflow from the winter hemisphere. The tilting of the streamlines in the vertical is due to atmospheric motions, i.e., mean meridional circulations in the tropics and horizontal eddies in middle and high latitudes. The streamlines that intersect the 1000-mbar surface represent, approximately, the amount of energy which is given to or taken away from the earth's surface. Because of the small heat capacity of the land and the small amounts of energy stored in snow and ice, the most important energy exchange occurs with the oceans. Large amounts of heat are stored in the oceans of the summer hemisphere and lost by the oceans in the winter hemisphere. Only in the annual-mean case is the rate of storage equal to zero, and are the values at the surface equal to the required oceanic heat transport.

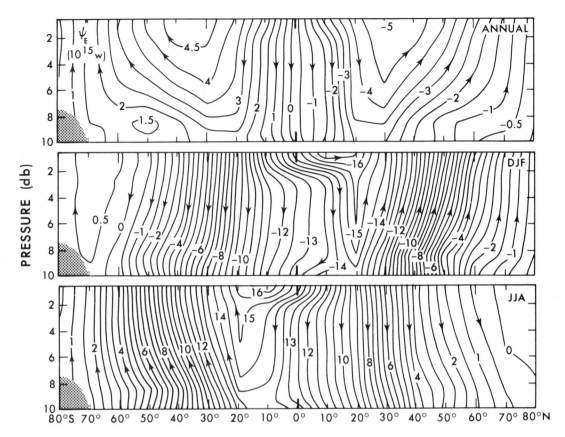

FIGURE 13.15. Streamlines of the zonal-mean transport of total energy in the atmosphere for annual, DJF, and JJA mean conditions in 10^{15} W (from Oort and Peixoto, 1983).

13.3.4 Energy in the oceans

From the distributions of temperature and density in the oceans, we can estimate the values of the internal energy $I = \int \rho c_0 T \, dV$ with respect to a reference temperature of 0 K and of the potential energy $\Phi = \int \rho g z \, dV$ with respect to the mean ocean depth. The internal and potential energy in the oceans are not as simply related as in the atmosphere. The seasonal and mean annual values obtained for the entire globe and the two hemispheres are presented in Table 13.2. Inspection of the values in Table 13.2 shows that the total internal energy is about 100 times larger than the total potential energy for the world ocean. The January–July difference values in the table show the importance of the oceans as reservoirs for storing energy. These vast amounts of internal energy storage are associated with the high thermal capacity of the ocean waters, the strong mixing in the upper 100 m, and the large ocean mass involved.

Even more than for the atmosphere, the kinetic energy in the oceans constitutes only a minute fraction of the potential and internal energy. Therefore, it will be useful to introduce the notion of "available" potential energy. Later, in Chap. 14 we will separate the reservoir of potential plus internal energy into an available and unavailable part. Only the available part would participate in the conversion into and from kinetic energy.

TABLE • 13.2. Hemispheric and global integrals of the oceanic energy[a] and their seasonal range per unit ocean surface area (inland seas, such as the Baltic Sea, Mediterranean Sea, Red Sea, Persian Gulf, and Hudson Bay have not been included) from Oort et al. (1989).

	Energy per unit area (0–5500 m)			January–July (March–September) differences Energy per unit area (0–275 m)		
	NH	SH	Globe	NH	SH	Globe
I	43.0	46.8	45.2	−46	54	13
Φ[b]			0.60	(−79)	(87)	(18)
K	0.55	0.63	0.80×10^{-7}	c	c	c
	0.78×10^{-7}	0.81×10^{-7}		d	d	d
E	43.6	47.4	45.8	−46	54	13
Units	10^{11} J m^{-2}			10^{7} J m^{-2}		

[a] I = internal energy (defined with respect to 0 K), Φ = potential energy (defined with respect to mean ocean depth), K = kinetic energy, and E = total energy.
[b] Values of the potential energy Φ for the NH, SH, and globe were computed using a mean world ocean depth of 3740 m as a reference level.
[c] Values are on the order of 10^{5} J m^{-2} or less but cannot be determined accurately from the data.
[d] Values are on the order of 10^{3} J m^{-2} but cannot be determined accurately from the data.

From the seasonal values we see that the oceans lose heat during the winter season and receive heat during the summer season. Because of the greater seasonal lag in the oceans than in the atmosphere, the March–September differences are also given in parentheses in Table 13.2 as a more representative measure of the seasonal range than the January–July differences. The seasonal range for the Southern Hemisphere is 87×10^7 J m^{-2} as compared with -79×10^7 J m^{-2} for the Northern Hemisphere. The difference in range would be larger when we take into account the larger ocean area in the Southern Hemisphere. The energy amounts involved in the solution and dissolution of salts in the ocean can be neglected in this context.

13.3.5 Transport of oceanic energy

The northward transport of energy across a latitudinal wall in the oceans can be written as

$$T_{oc} = \iint_{wall} \rho(c_0 T + gz + c^2/2 + p/\rho)\, v\, dx\, dz. \tag{13.41}$$

As before, the kinetic energy contributions can be neglected. Furthermore, for the oceans, the flux of potential energy and the pressure work term practically cancel each other, because of the near-incompressibility of ocean water. One can show this by writing

$$\iint (\rho g z + p) v\, dx\, dz$$

$$\simeq \iint \{-\rho_0 g d + (p_0 + \rho_0 g d)\} v\, dx\, dz$$

$$\simeq p_0 \iint v\, dx\, dz,$$

where at the ocean surface ($z = 0$) the pressure equals p_0, and at depth $z = -d$ the pressure equals $p_0 + \rho_0 g d$. Since we assume that there is no net mass transport in the ocean across the latitudinal wall ($\iint v\, dx\, dz \simeq 0$), expression (13.41) for the total energy transport in the oceans reduces to only the transport of heat:

$$T_{oc} = \iint_{wall} \rho\, c_0\, vT\, dx\, dz. \tag{13.42}$$

To show the breakdown of this transport into various components let us consider, for simplicity, one ocean basin. Formally, the flux across this ocean basin can be written as follows:

$$T_{oc} = \int_{-H}^{0} \left(\int_{x_1(z)}^{x_2(z)} \rho c_0 vT\, dx \right) dz \tag{13.43a}$$

or by changing the order of integration as

$$T_{oc} = \int_{X_1}^{X_2} \left(\int_{-h(x)}^{0} \rho c_0 vT\, dz \right) dx. \tag{13.43b}$$

As illustrated in Fig. 13.16 we use the following notation:

$h(x)$ = depth of basin;
H = maximum depth basin;
$x_1(z), x_2(z)$ = western and eastern boundaries of basin at depth z;
X_1, X_2 = western and eastern boundaries of basin at the surface, respectively.

The total poleward oceanic heat flux across a latitude circle may be obtained by summing the contributions from the different ocean basins. Indicating a zonal average over an ocean basin by square brackets, and a departure from this zonal average by an asterisk, one can write the time-averaged equation (13.43a) in the following form:

$$\overline{T}_{oc} = \int_{-H}^{0} \rho c_0 (x_2 - x_1) [\,\overline{vT}\,] dz, \tag{13.44a}$$

where

$$[\,\overline{vT}\,] = \underbrace{[\bar{v}][\bar{T}]}_{(i)} + \underbrace{[\bar{v}^*\bar{T}^*]}_{(ii)} + \underbrace{\overline{[v]'[T]'}}_{(iii)} + \underbrace{[\,\overline{v'^*T'^*}\,]}_{(iv)}$$

(i) = stationary meridional overturning component;
(ii) = stationary eddy (gyre) component;
(iii) = transient meridional overturning component;
(iv) = transient eddy component.

In an analogous fashion, the time-averaged equation (13.43b) can be broken down into a slightly different set of components, denoting a vertical average by $\langle\,\rangle$ and a departure from it by a double prime:

$$\overline{T}_{oc} = \int_{X_1}^{X_2} \rho c_0 h \langle\,\overline{vT}\,\rangle dx, \tag{13.44b}$$

where

$$\langle\,\overline{vT}\,\rangle = \underbrace{\langle\bar{v}\rangle\langle\bar{T}\rangle}_{(i)} + \underbrace{\langle\bar{v}''\bar{T}''\rangle}_{(ii)} + \underbrace{\overline{\langle v\rangle'\langle T\rangle'}}_{(iii)}$$

$$+ \underbrace{\overline{\langle (v')''(T')''\rangle}}_{(iv)}$$

(i) = time-mean barotropic component;
(ii) = time-mean baroclinic component;
(iii) = transient barotropic component;
(iv) = transient baroclinic component.

It is important to note that in Eq. (13.44a) both x_1 and x_2 are a function of z, and that in Eq. (13.44b) h is a function of x. In the atmosphere, surface topography is usually less important and is even frequently neglected (in other words, the earth is assumed to be a perfectly round smooth sphere) so that the various components in Eqs. (13.44a) and (13.44b) can be more easily interrelated in the atmospheric case. However, in the oceanic case the two formulations are best studied separately.

As far as the energy transports are concerned, very little direct information is available for the oceans. Various methods to estimate the poleward heat flux in the world oceans are listed in Table 13.3. One method of evaluating the transport, the so-

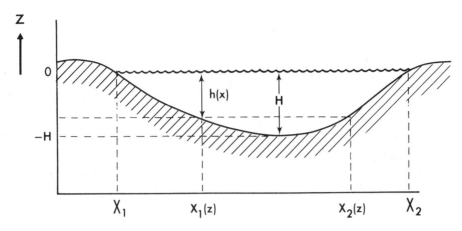

FIGURE 13.16. Schematic diagram of the various integration limits used in the computation of the northward heat flux across an ocean basin.

called planetary energy balance method (Vonder Haar and Oort, 1973; Oort and Vonder Haar, 1976; Trenberth, 1979) infers the oceanic transport from a combination of satellite-derived net radiation values at the top of the atmosphere and global atmospheric transport data. Thus, the oceanic heat transport was computed as the difference between the total energy transport required for radiative balance (see Fig. 6.15) and the observed atmospheric energy transport. The three transport profiles are plotted in Fig. 13.17. The inspection of these profiles leads to the conclusion that for the mean annual heat balance of the climate system, the oceans dominate in low latitudes with a maximum poleward heat transport of about 3.5×10^{15} W near 25 °N and of -2.7×10^{15} W near 20 °S, whereas the atmosphere is more important in middle and high latitudes of the Northern Hemisphere. In middle latitudes of the Southern Hemisphere, the atmosphere and oceans appear to be of about equal importance with values of about -2 to -3×10^{15} W near 40 °S, whereas the oceans dominate again at high southern latitudes. A stronger poleward heat flux is found near 60 °S in the vicinity of the Antarctic circumpolar current than at the same latitude in the Northern Hemisphere at 60 °N where there are mainly continents. Of course, the values of T_{oc} should vanish over land, e.g., south of 70 °S; the nonzero value obtained can be considered a measure of the error in T_{oc}.

More classical methods of evaluating the oceanic heat transport (e.g., Hall and Bryden, 1982) lead, in general, to smaller maximum values on the order of 1 to 2×10^{15} W, but they are located at about the same latitude in the tropics. The scientific debate on what the real transports are is still going on (Dobson *et al.*, 1982). It is an important issue for climate, since it touches on the relative role of the atmosphere and oceans in shaping the present "normal" climate as well as its year-to-year and interdecadal variations.

One other indirect method frequently used is to estimate the regional oceanic heat convergence or divergence needed to balance the energy loss to or gain from the atmosphere using surface ship observations. A study by Hastenrath (1982) based on this approach leads to the estimates for the transports in the three individual oceans shown in Fig. 13.18. Although very tentative, these curves give an intriguing picture of the possible large differences between the Atlantic and Pacific Oceans in transporting heat poleward. Hastenrath's total transport values agree reasonably well with those in Fig. 13.17 but the agreement may be somewhat fortuitous.

TABLE • 13.3. Various methods to estimate poleward heat flux in the world oceans.

Method[a]	Difficulties	Strength
1. Direct method From simultaneous direct *in situ* current and temperature observations	presently only done for very limited regions; inertial oscillations, transient and possibly quasistationary eddies require extensive sampling	only direct method should give definitive answer if properly planned and executed
2. Model methods (a) From *in situ* geostrophic calculations based on observed density structure in world oceans	assumption of no-motion level; transient eddy effects neglected; large uncertainty in flux calculations near continental shelves	fairly "direct" calculation
(b) From observed density structure adjusted "somewhat" to numerical model characteristics	includes only stationary component; somewhat model dependent; problems in continental shelf regions	nongeostrophic component included; dynamically consistent
(c) From a three-dimensional general circulation model specifying only surface exchange of momentum, heat, and water vapor (either as observed or from atmospheric general circulation model)	model dependent; results contaminated by *ad hoc* parameterizations (e.g., in modeling subgrid-scale eddies)	consistent method; suggests contributions by various transport mechanisms; promises possibility to study climate anomalies during recent as well as ancient climatic conditions
3. Surface heat balance method From surface heat balance estimates (net radiation, sensible and latent heat fluxes at ocean surface); the "classical" method	uncertainties in formulation air-sea exchange processes and in radiation parameterization	good global coverage by surface ship reports
4. Planetary energy balance method From heat balance estimates for entire climatic system (atmosphere, oceans, snow and ice, land) as a residual (net radiation from satellites, atmospheric storage and transports from rawinsonde network, oceanic storage from hydrographic and BT data)	data deficiencies in SH (mean meridional circulation transports in atmosphere cannot be measured reliably enough in SH); interannual variations in oceanic heat storage may affect results; errors accumulate in residual calculation	besides conservation of energy, no assumptions are needed; sum total of all components oceanic transport is obtained

[a] For a general evaluation of the various methods, see Dobson *et al.* (1982).

ENERGETICS 345

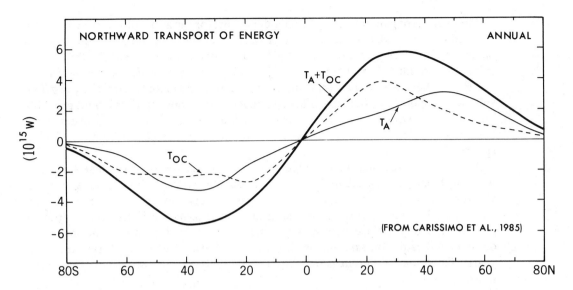

FIGURE 13.17. Meridional profiles of the northward transport of energy by the total system ($T_A + T_{OC}$), the atmosphere (T_A), and the oceans (T_{OC}) for annual-mean conditions in units of 10^{15} W (based on data from Carissimo *et al.*, 1985).

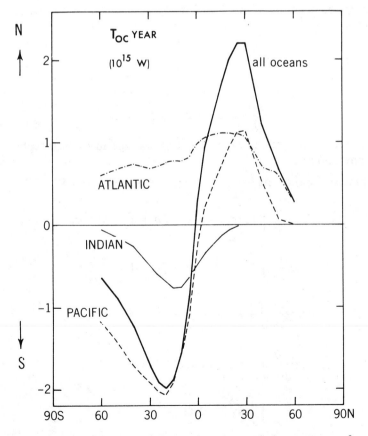

FIGURE 13.18. Meridional profiles of the northward oceanic heat transport for the various oceans in units of 10^{15} W as computed indirectly from surface heat balance conditions (adapted from Hastenrath, 1982).

The analysis of general circulation model results for the atmosphere, oceans, or the coupled system (see Chap. 17) can also provide independent estimates of the required oceanic heat flux (e.g., Miller et al., 1983; Russell et al., 1985; Masuda, 1988).

The uncertainty of the estimates of T_{oc} is still quite large, especially in the Southern Hemisphere where the radiosonde network is inadequate to measure the atmospheric transport (T_A) reliably. This deficiency in the upper air data would probably tend to underestimate the atmospheric contribution and thereby overestimate the oceanic role in the total heat transport in middle latitudes of the Southern Hemisphere.

As far as the seasonal heat budgets are concerned, the oceanic heat storage is probably less well known than the other components. Nevertheless, tentative seasonal estimates of T_{oc}, as well as of T_A and $T_A + T_{oc}$, are shown in Fig. 13.19. A very large annual variation in the required oceanic heat transport is apparent. Another possible difficulty in the estimates of T_{oc} may relate to the fact that the atmospheric, oceanic, and radiation samples used in Fig. 13.19 were not taken during the same years. Thus, some differences may be expected only because of year-to-year variations.

13.3.6 Synthesis of the energy balance

The energy balance equation for the total climatic system can be written symbolically in the form

$$\frac{\partial E}{\partial t} = F_{TA},$$

where

$$\frac{\partial E}{\partial t} \equiv S_A + S_O + S_L + S_I \tag{13.45}$$

and $S_A, S_O, S_L,$ and S_I are the rates of energy storage in the atmosphere, oceans, land, and snow and ice.

More precise expressions for the storage components are

$$S_A \equiv \frac{\partial}{\partial t} \iint \rho(c_v T + gz + Lq + c^2/2)\, dA\, dz$$

$$\simeq \frac{\partial}{\partial t} \iint \rho(c_v T + gz + Lq)\, dA\, dz, \tag{13.46a}$$

$$S_O \equiv \frac{\partial}{\partial t} \iint \rho(c_o T + gz + c^2/2)\, dA\, dz$$

$$\simeq \frac{\partial}{\partial t} \iint \rho c_o T\, dA\, dz, \tag{13.46b}$$

$$S_L \equiv \frac{\partial}{\partial t} \iint \rho c_L T\, dA\, dz, \tag{13.46c}$$

$$S_I = \frac{\partial}{\partial t} \iint \rho(c_I T - L_m)\, dA\, dz, \tag{13.46d}$$

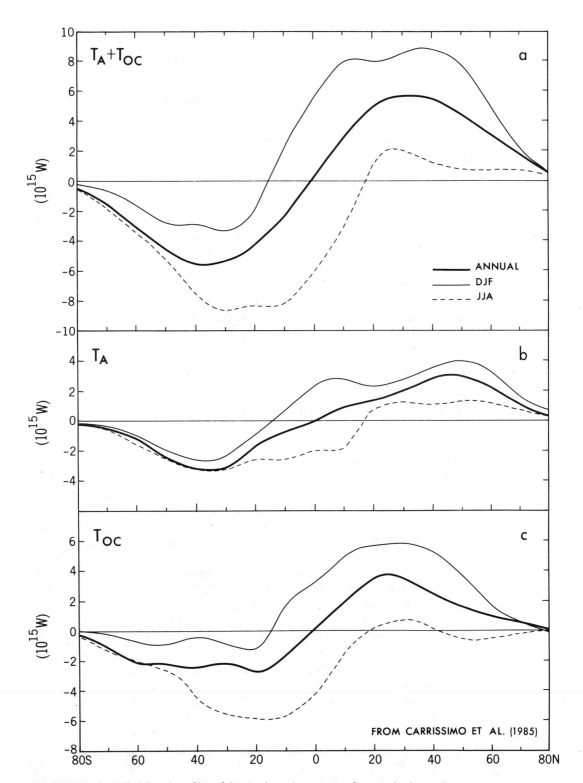

FIGURE 13.19. Meridional profiles of the northward transport of energy in the total system (a), the atmosphere (b), and the oceans (c) for annual, DJF, and JJA mean conditions in units of 10^{15} W (based on data from Carissimo *et al.*, 1985).

where

c_v, c_0, c_L, c_1 = specific heat for the atmosphere, oceans, land, snow and ice,
and

L, L_m = latent heat of evaporation (or sublimation), melting.

Changes in kinetic energy are quite small compared with those in internal energy and can be neglected. Further, changes in density of the oceans are very small so that both the second and third terms on the right-hand side of Eq. (13.46b) can be left out. All integrations are assumed to be over the entire volume taken up by the particular element. However, in practice one needs to integrate vertically only over the layer which is subject to the energy variation. The thickness of this layer depends, of course, on the period of the variation. Thus, for the annual cycle the penetration depth in the atmosphere is the full depth [Fig. 13.20(a)], in the oceans only 100 to 300 m [Fig. 13.20(b)], and in the land about 1 m. Probably about 1 m of snow and ice melts and accumulates annually in the polar regions.

Estimated values of the heat content in the atmosphere, oceans, and cryosphere are shown in Figs. 13.21(a), 13.21(b), and 9.4, respectively. For convenience, the annual-mean values have been subtracted out (see Tables 13.1 and 13.2). For the atmosphere, the global curve follows the variation of the Northern Hemisphere because of its greater continentality. For the oceans, the situation is reversed. Both the annual variation in net radiation and the ocean area are greater for the Southern Hemisphere leading to the observed global curve. The jaggedness of the curves is presumably due to data deficiencies; this is particularly clear for the month of February. For the cryosphere no global budget data are available. As a substitute the horizontal area covered by snow and ice as measured by Kukla and Kukla (1974) from satellite photographs was given before in Fig. 9.4. When a change in area occurs, we will assume that a 50-cm thick layer of snow and ice (water equivalent) melts or freezes and that no changes occur elsewhere. For example, a decrease in area of 10^7 km^2 would require an energy input of 0.17×10^{22} J. A comparison of the various curves in Figs. 13.21(a) and 13.21(b) shows the overwhelming importance of the oceans. Also of interest are the differences in the phase of the maximum, namely late July for the atmosphere, mid-August for the cryosphere, and mid-September for the oceans.

The global rates of heat storage in the atmosphere and oceans [i.e., the derivative of the curves for the globe in Figs. 13.21(a) and 13.21(b)] are presented in Fig. 13.22. The curves may be compared directly with the net radiation available at the top of the atmosphere as measured by satellite (Ellis et al., 1978) and also shown in Fig. 13.22. In spite of the deficiencies in the observing methods and networks, especially in the Southern Hemisphere, the agreement between the radiation and the ocean heat storage curves is surprisingly good, both in phase and amplitude. We note that the energy storage in the atmosphere and in the cryosphere (not shown) are relatively small. These results have been confirmed by general circulation model experiments (Manabe et al., 1979).

Going into more detail, the atmosphere's limited capacity to store heat is also shown by the latitudinal curves of the rate of energy storage (S_A) in Fig. 13.23(a). For the DJF and JJA seasons the heat storage in the atmosphere amounts to less than 10 W m^{-2}, whereas in the transition seasons, March–May and September–November (not shown), larger values on the order of about 20 W m^{-2} occur at high latitudes. Considering next the convergence due to horizontal atmospheric exchange in Fig. 13.23(b), it is found to be a much more important term in the heat balance than the

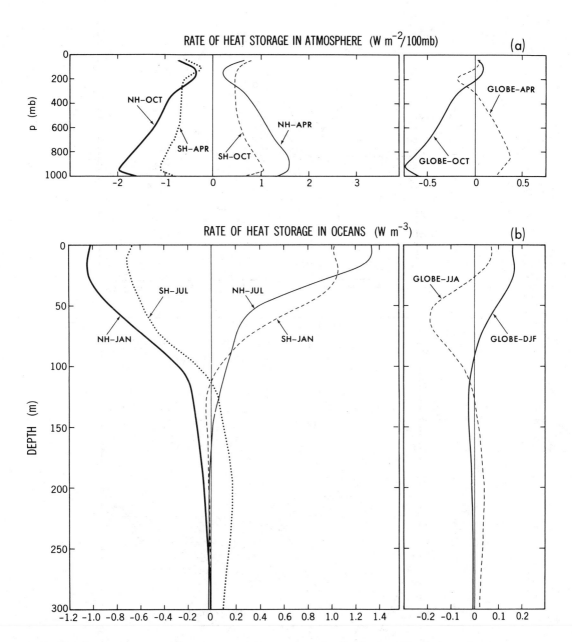

FIGURE 13.20. Vertical profiles of the rate of heat storage for the Northern Hemisphere, Southern Hemisphere, and globe during the transition months of October and April in the atmosphere (a) and during January and July as the months of largest change in the oceans (b). Energy units of W m^{-2} (100 mb^{-1}) are used for the atmosphere and W m^{-3} as units for the oceans, since 1 m of water is about equivalent to a 100-mb layer in the atmosphere.

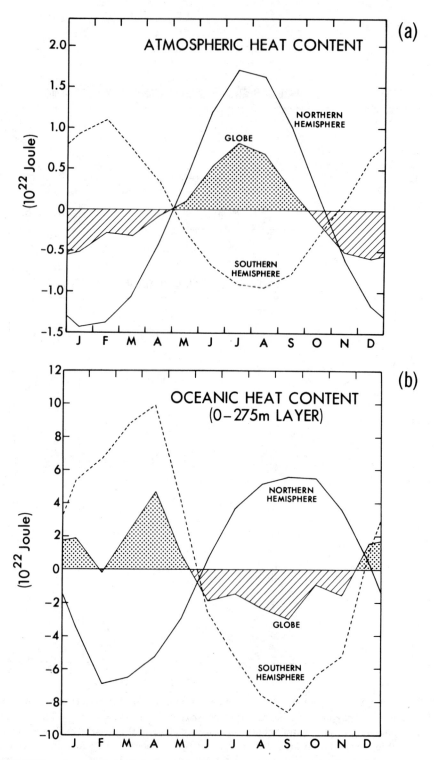

FIGURE 13.21. Annual cycle in the total energy content of the atmosphere (a) and oceans (b) for the Northern Hemisphere, Southern Hemisphere, and globe in units of 10^{22} J. Shown are departures from the annual-mean values. The atmospheric estimates are for the entire depth of the atmosphere, whereas the oceanic estimates are for the layer 0–275 m depth where most of the seasonal variation occurs (see Fig. 13.20).

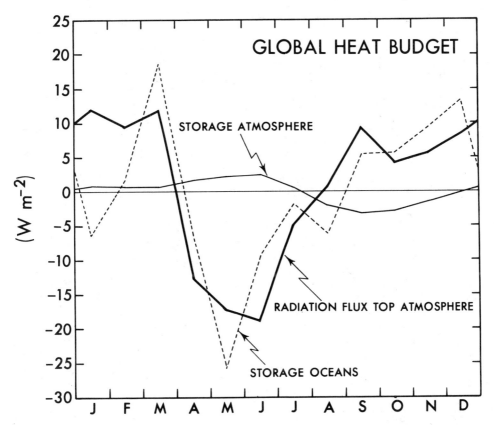

FIGURE 13.22. A comparison between the annual cycle in the global-mean net radiation flux at the top of the atmosphere as derived from satellite data and the annual cycle in the rate of energy storage in the oceans and atmosphere as computed from *in situ* data in units of W m^{-2} (after Ellis *et al.*, 1978).

atmospheric storage. On an annual basis about 20 W m^{-2} is lost between 40 °S and 40 °N, and about 50–70 W m^{-2} is gained poleward of 60°. The seasonal curves show a maximum divergence of 30–40 W m^{-2} in the tropics of the summer hemisphere, and a maximum convergence near the winter pole.

The energy that is exchanged with the underlying earth's surface (F_{BA}) can be obtained as a residual in Eq. (13.28) using the net radiation data at the top of the atmosphere [Fig. 6.14(d)] and the atmospheric data (Fig. 13.23). The estimated values are presented in Fig. 13.24. Between 30 °S and 30 °N the surface gains annually up to 40 W m^{-2}, and poleward of 30° it loses a similar amount. The seasonal variations are very large. In fact, a gain of about 90 W m^{-2} is found near 40 °S during DJF, and a gain of about 60 W m^{-2} near 40 °N during JJA. Losses of more than 100 W m^{-2} appear to occur near 50 °N during DJF and south of 40 °S during JJA. We may note that the classical method to calculate the exchange of energy between the atmosphere and the earth's surface is based on a combination of radiation calculations and empirical drag-law formulas (see, e.g., Budyko, 1963). The total exchange is then evaluated as the sum of the separate contributions from radiation, sensible, and latent heat exchange using as basic data the surface observations of cloudiness, surface wind, temperature, and humidity (see Chap. 10). Obviously, such calculations are tentative and can contain large systematic errors. The present residual calculation is an attractive alternative method, since it is based on sound physical principles and does not involve empirical relations.

352 PHYSICS OF CLIMATE

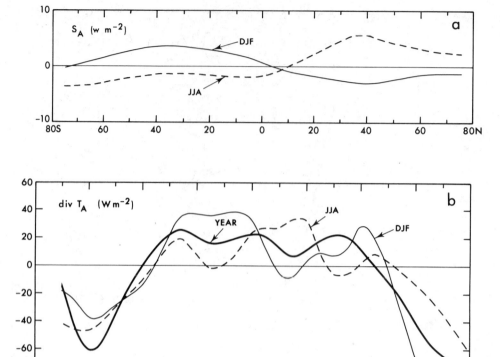

FIGURE 13.23. Meridional profiles of the zonal-mean rate of energy storage in the atmosphere (a) and the divergence of the vertically integrated atmospheric transport of energy (b) in W m^{-2} (from Oort and Peixoto, 1983).

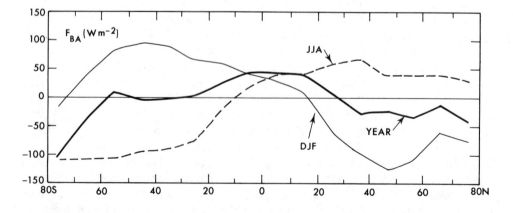

FIGURE 13.24. Meridional profiles of the zonal-mean flux of energy from the atmosphere to the earth's surface in W m^{-2} (after Oort and Peixoto, 1983).

Because the heat capacity of the continental surface is very small, its effects in the annual variation of the global heat balance can be neglected. The same holds true for snow and ice, except at high latitudes. Therefore, on a large scale, the principal exchange must occur between the atmosphere and the oceans, and the curves in Fig. 13.24 may be regarded as a measure of the energy available for the world ocean.

For time periods from months to years and probably much longer, storage of energy in the oceans may play a major role in the earth's heat budget. Figure 13.25(a) shows meridional profiles of the observed rate of storage in the oceans S_O for normal winter and summer conditions. This storage term was evaluated at each latitude from the observed variations in the vertical-mean temperature in the top layer of the oceans (0–275 m) (Levitus, 1984). In view of the large heat capacity of the oceans it is not surprising that the oceanic heat storage in midlatitudes is about a factor 20 larger than the atmospheric heat storage. In fact, a typical oceanic storage value of 100 W m^{-2} in middle latitudes is almost as large as the net seasonal radiation values at the top of the atmosphere [Fig. 6.14(d)]. Therefore, the oceans act as a huge buffer for the energy, damping the seasonal and interannual fluctuations in climate that otherwise would occur on an oceanless planet.

The relative importance of the oceanic heat storage and the transport of energy by oceanic motions can be assessed by comparing the storage and the annual (residual) flux divergence curves in Fig. 13.25(b) with the computed flux of total energy across the earth's surface F_{BA} in Fig. 13.24. It is obvious that the oceanic flux divergence dominates at low latitudes, and the oceanic storage at middle and high latitudes.

13.4 ENERGETICS OF THE POLAR REGIONS

For the same reasons invoked in studying the hydrology of the polar regions it seems appropriate to complete this chapter with the equivalent study of the energetics of the Arctic and Antarctic regions.

13.4.1 Formulation of the energy budget

Let us consider Eq. (13.28) for an atmospheric polar cap poleward of 70° latitude shown schematically in Fig. 13.26. It can be rewritten in the form

$$\frac{\partial \{E\}}{\partial t} = \{F_{TA}\} + F_{\text{wall}} - \{F_{BA}\}. \tag{13.47}$$

In this equation the braces indicate an areal average over the polar cap. Further

$$\frac{\partial \{E\}}{\partial t} \equiv \int \left(c_v \frac{\partial T}{\partial t} + g \frac{\partial z}{\partial t} + L \frac{\partial q}{\partial t}\right) dm$$

represents the rate of change of energy in the polar cap,

$$F_{TA} \equiv F^{\downarrow}_{SW} - F^{\uparrow}_{LW}$$

the net downward radiation flux (solar minus terrestrial) at the top of the atmosphere, and

$$F_{\text{wall}} \equiv \pm \iint_{\substack{\text{atm} \\ \text{wall}}} \left(c_p T + gz + Lq + \frac{1}{2}c^2\right) v \, dx \frac{dp}{g}$$

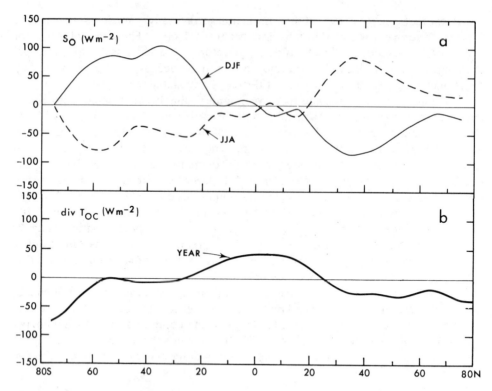

FIGURE 13.25. Meridional profiles of the zonal-mean rate of energy storage in the oceans (a) from Levitus (1982), and the divergence of the oceanic heat transport (b) in W m^{-2} (after Oort and Peixoto, 1983).

represents the total poleward transport of atmospheric energy across an imaginary wall at 70° latitude in the form of sensible heat, potential energy, latent heat, and kinetic energy. The positive sign refers to the north polar cap, 70 °N-NP, and the negative sign to the south polar cap, 70 °S-SP (v is always counted positive northward). As said before, the kinetic energy transport can be neglected. Coming back to Eq. (13.47), the last term can be written explicitly as

$$F_{BA} \equiv F^{\downarrow}_{SW} - F^{\uparrow}_{LW} - F^{\uparrow}_{SH} - F^{\downarrow}_{G} - F^{\uparrow}_{LH} - F_{M},$$

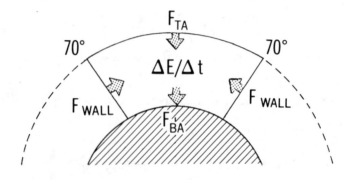

FIGURE 13.26. Schematic diagram of the energy budget for an atmospheric polar cap.

representing the net energy flux downward into the earth's surface. This equation was already analyzed in Secs. 6.7 and 10.2. The term F_{BA} is composed of the incoming short-wave minus the outgoing long-wave radiation, the sensible heat losses, and the latent heat losses. The latent heat flux is due to evaporation, evapotranspiration, and sublimation in the air–surface interface

$$F^\uparrow_{LH} \equiv L_e E_1 + L_s E_s,$$

where L_e and L_s are the latent heat of evaporation and sublimation, respectively. In Eq. (13.47) the ocean heat transport does not appear explicitly but it is contained implicitly in the term F_{BA}, as will be discussed later.

We will next consider the heat budget of an ocean–land–ice cap (to be called "oceanic" polar cap) poleward of 70° latitude, that includes all parts of the climatic system located below the atmospheric polar cap discussed so far (Fig. 13.27). For convenience we will use here the (x,y,z,t) instead of the (x,y,p,t) coordinate system. The governing equation can be written in the form

$$\frac{\partial}{\partial t}\{E^*\} = \{F_{BA}\} + F^*_{wall} + \{F_{BO}\}. \tag{13.48}$$

In this equation, the rate of change of energy in the oceanic polar cap associated with changes in heat content of the land (S_L), ocean (S_O), and cryosphere (S_I) and with net melting minus freezing (S_{LHI}) inside the cap is given by

$$\frac{\partial \{E^*\}}{\partial t} \equiv S_L + S_O + S_I + S_{LHI}$$

or

$$\frac{\partial \{E^*\}}{\partial t} \equiv \int_{land} \rho c_L \frac{\partial T}{\partial t} dV$$

$$+ \int_{ocean} \rho c_O \frac{\partial T}{\partial t} dV + \int_{cryosphere} \rho c_I \frac{\partial T}{\partial t} dV$$

$$- L_m \frac{\partial}{\partial t} \int_{cryosphere} \rho \, dV.$$

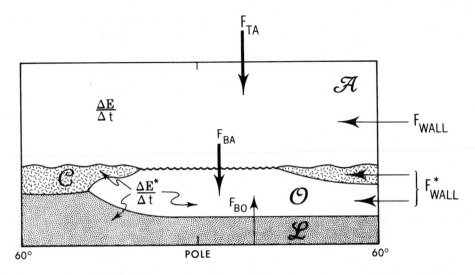

FIGURE 13.27. Schematic diagram of the energy budget for a polar cap containing atmosphere, oceans, land, snow, and sea ice.

Further,

$$F^*_{\text{wall}} = \pm \iint_{\text{ocean}} \rho c_o vT\, dx\, dz$$

$$\pm \iint_{\text{cryosphere}} \rho c_I vT\, dx\, dz$$

$$\mp \iint_{\text{cryosphere}} L_m \rho v\, dx\, dz$$

$$= \pm T_{\text{oc}} \pm T_I$$

represents the sum of the fluxes across the latitudinal wall of sensible heat in ocean currents (T_{oc}) and of sensible plus latent heat associated with the export of ice (T_I), mainly in the form of icebergs, from the polar caps. As before, the top sign in the expression for F^*_{wall} refers to the north polar cap, 70 °N-NP, and the bottom sign to the south polar cap, 70 °S-SP. Finally $\{F_{\text{BO}}\}$ represents the upward geothermal heat flux at the bottom of the oceanic polar cap, that is usually negligibly small.

For long-term mean conditions, the storage terms are also generally small so that the energy balance equations for the atmospheric polar caps reduce to

$$\{F_{\text{TA}}\} + F_{\text{wall}} - \{F_{\text{BA}}\} \approx 0, \tag{13.49}$$

for the oceanic polar caps to

$$\{F_{\text{BA}}\} + F^*_{\text{wall}} \approx 0, \tag{13.50}$$

and for the atmospheric and oceanic polar caps combined to

$$\{F_{\text{TA}}\} + F_{\text{wall}} + F^*_{\text{wall}} \approx 0. \tag{13.51}$$

For long-term mean conditions, (13.49) and (13.50) can also be written more explicitly in the form

$$(F^{\downarrow}_{\text{SW}} - F^{\uparrow}_{\text{LW}})_{\text{TA}} \pm \iint_{\substack{\text{atm}\\\text{wall}}} (c_p T + gz + Lq)v\, dx\, \frac{dp}{g}$$

$$- (F^{\downarrow}_{\text{SW}} - F^{\uparrow}_{\text{LW}} - F^{\uparrow}_{\text{SH}} - F^{\uparrow}_{\text{LH}})_{\text{BA}} = 0, \tag{13.52}$$

$$(F^{\downarrow}_{\text{SW}} - F^{\uparrow}_{\text{LW}} - F^{\uparrow}_{\text{SH}} - F^{\uparrow}_{\text{LH}})_{\text{BA}}$$

$$\pm \iint_{\text{ocean}} \rho c_o vT\, dx\, dz \pm \iint_{\text{cryosphere}} \rho c_I vT\, dx\, dz$$

$$\mp \iint_{\text{cryosphere}} L_m \rho v\, dx\, dz = 0, \tag{13.53}$$

where, as before, the top sign refers to the north polar cap, 70 °N-NP, and the bottom sign to the south polar cap, 70 °S-SP.

13.4.2 Observed energy budget in atmospheric polar caps

Following the work by Nakamura and Oort (1988) on the Arctic and Antarctic energetics, we will discuss the various terms entering in the energy budgets of the polar caps, beginning with the storage of energy in the atmospheric polar caps. The storage term $\partial\{E\}/\partial t$ appearing on the left-hand side of Eq. (13.47) can be estimated by taking the time difference of E between two neighboring months. The results are shown in Fig. 13.28 for each month of the year in terms of an equivalent vertical energy flux per unit area. For easy comparison between the corresponding seasons in the two hemispheres, the curves for the north and south polar caps have been shifted by six months with respect to each other in Fig. 13.28 as well as in later figures.

The changes in the storage rate in the north polar cap (NPC) are almost symmetric about the solstices with a maximum in spring and a minimum in fall, whereas they are slightly asymmetric for the south polar cap (SPC) related to the faster buildup of the SPC temperature during spring. The rate of storage of energy is generally much smaller than the other factors in Eq. (13.47). We may note that the integral of $\partial\{E\}/\partial t$ over a year becomes zero in both polar caps since we assume steady-state conditions for the annual mean.

The incoming short-wave radiation $F_{SW}^\downarrow$ and the outgoing long-wave radiation $F_{LW}^\uparrow$ at the top of the atmosphere are displayed in Fig. 13.29, for the two polar caps. Because of the lower air temperatures over Antarctica the long-wave radiation flux is consistently lower for the Antarctic than for the Arctic regions. The short-wave radiation flux is primarily controlled by astronomical factors, such as the tilt of the earth's axis. It reaches a maximum around the summer solstice. The values of $F_{SW}^\downarrow$ also tend to be smaller in the south polar cap except, of course, in winter when very little solar radiation is available for both polar regions. At first glance, the summer results are somewhat surprising because the earth–sun distance is smallest during

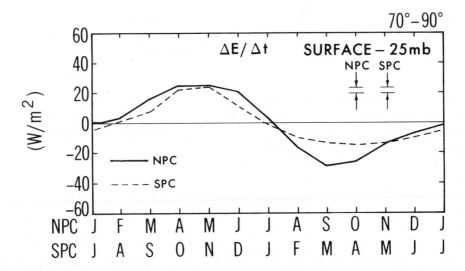

FIGURE 13.28. Annual cycle of the rate of storage of energy in the atmosphere for the north and south polar caps in terms of an equivalent vertical energy flux per unit area (W m^{-2}). The time axis is shifted by six months between the two curves to facilitate the interhemispheric comparison. Mean error bars indicating 95% confidence limits are shown on the top right-hand side of the figure (from Nakamura and Oort, 1988).

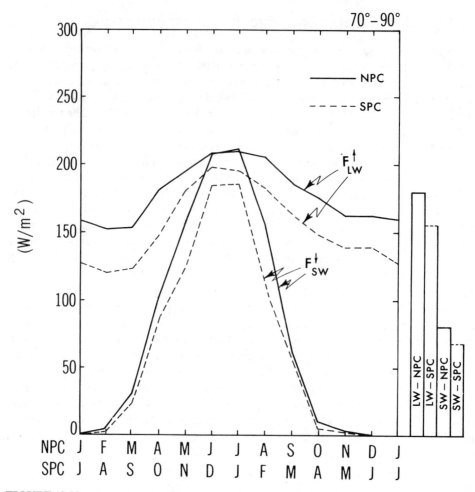

FIGURE 13.29. Annual cycle of the net incoming solar radiation $F^\downarrow_{SW}$ and the net outgoing terrestrial radiation $F^\uparrow_{LW}$ in W m^{-2} for the north and south polar caps. Annual-mean values are shown to the right of the figure (from Nakamura and Oort, 1988).

January, i.e., during summer in the Southern Hemisphere. However, the differences in albedo between the Arctic and Antarctic regions must be of overriding importance. They are mainly due to differences in the nature of the earth's surface, namely in the north polar cap a less reflective ocean surface with some sea ice that is covered by meltwater during summer, and in the south polar cap a land surface with permanent, highly reflective snow and ice.

The annual variation of the net total incoming radiation at the top of the atmosphere F_{TA} is shown in Fig. 13.30. It is clear that the earth–atmosphere system in both polar caps is losing heat by radiation throughout the year except for a short period in the summer of the NPC. During winter when there is little incoming solar radiation, F_{TA} is mostly due to the loss of long-wave terrestrial radiation that is controlled by the atmospheric temperature. The loss is smaller in the south polar cap. In summer, F_{TA} is larger over the Arctic since the reflection of the solar radiation over Antarctica is large enough to overcome the difference in $F^\uparrow_{LW}$. In summary, we note that the amplitude of the annual variation in F_{TA} is greater in the north polar cap than in the south polar cap. When averaged over a year, the north polar cap loses about 10

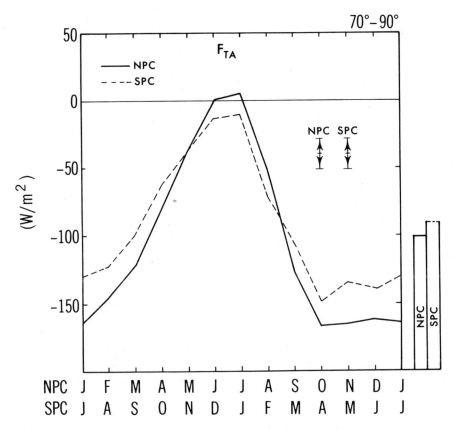

FIGURE 13.30. Annual cycle of the net incoming (solar minus terrestrial) radiation at the top of the atmosphere F_{TA} for the north and south polar caps in W m^{-2} (from Nakamura and Oort, 1988).

W m^{-2} more heat than the south polar cap mainly because of the higher atmospheric temperatures in the Arctic. Mean error bars indicating 95% confidence limits are also shown in Fig. 13.30. Error estimates for the two polar caps are on the order of 10 W m^{-2}.

Figure 13.31 shows the annual contribution of the horizontal energy flux by the atmosphere to the energy budgets of the north polar cap in units of W m^{-2}. The energy fluxes include the fluxes of sensible heat, potential energy, and latent heat [see Eq. (13.47)]. The atmosphere is more efficient during the winter semester in providing heat to the polar regions; the imported amounts across 70 °N vary between about 80 W m^{-2} in summer and 120 W m^{-2} in winter. In the Northern Hemisphere, the transport is mainly due to the transient eddies and the mean meridional circulation. The stationary eddies play an important role only in winter. No results are shown for the south polar cap because of the unreliable estimates of the mean meridional circulation transports across 70 °S.

The final term in Eq. (13.47), the flux across the earth's interface F_{BA}, can be evaluated in the north polar cap as a residual using the available atmospheric and radiation data. The results for the Arctic are shown as a solid line in Fig. 13.32. For the south polar cap, we used general circulation model results (Manabe and Broccoli, 1988, personal communication) to give a very tentative estimate of the transport of energy across the earth's surface in Antarctica, as indicated by the dotted line in Fig. 13.32. The profiles show that, especially in the north polar cap, there must be a large

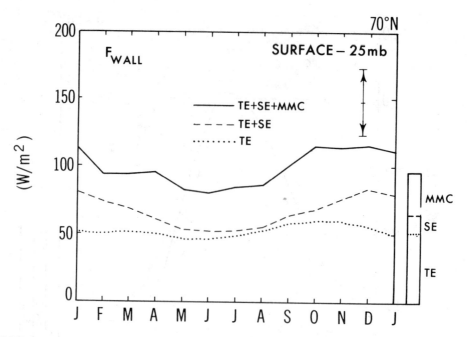

FIGURE 13.31. Annual cycle of the poleward energy flux F_{wall} at 70 °N and its breakdown into transient eddy, stationary eddy, and mean meridional circulation contributions in terms of an equivalent vertical energy flux per unit area (W m^{-2}) (from Nakamura and Oort, 1988).

upward transport of energy ($F_{\text{BA}} < 0$) from the surface into the atmosphere during the winter semester (on the order of 50 W m^{-2}) and a large downward transfer of energy ($F_{\text{BA}} > 0$) during the summer months (50–80 W m^{-2}). There is evidence from the model for a similar but weaker energy exchange in the south polar cap with an upward flux of only about 10 W m^{-2} during winter and a downward flux of about 30 W m^{-2} during midsummer. The primary reason for the large amplitude of F_{BA} in the Arctic is the very large variation in solar radiation ($F_{\text{SW}}^{\downarrow}$) with almost no radiation in winter (polar night) and continuous daylight in summer. For the south polar cap, the amplitude of $F_{\text{SW}}^{\downarrow}$ is also large but, in contrast to the situation in the Arctic, it must be compensated by a large annual variation in F_{wall} across 70 °S, leading to the much reduced annual variation of F_{BA} shown in Fig. 13.32.

13.4.3 Observed energy budget in ocean–land–ice polar caps

The presence of sea ice or open water is, of course, of great importance in the surface heat exchange in the polar caps. Therefore, before proceeding with the discussion of the other terms in the polar energy budgets, we will first discuss some of the properties of sea ice and give some sample calculations of the heat exchange for two cases: one, when sea ice is present and, the other, when there is open water. We may first note that the equilibrium temperature of sea water in contact with ice remains practically constant and is independent of the thermal state of the atmosphere and the sea ice thickness. As a consequence, the ice-covered Arctic waters tend to be much warmer in winter than the atmosphere above it, which may be as cold as -30 to -60 °C. Due to the salinity of sea water there is a depression of the freezing point so that the sea temperature can reach minimum values of about

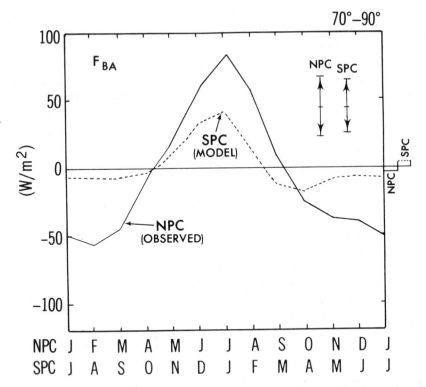

FIGURE 13.32. Annual variation of the downward surface heat flux F_{BA} for the north and south polar caps in W m^{-2}. The results for the south polar cap are taken from the output of a general circulation model rather than from real data because of data deficiencies in the Southern Hemisphere.

-1.9 °C, instead of 0 °C for fresh water. The low conductivity of sea ice ($K \approx 2.2$ W m^{-1} K^{-1}) makes sea ice a good insulator, preventing heat from escaping into the atmosphere. However, for a thin layer of ice there is always some loss as shown, to a first approximation, by Newton's law of cooling:

$$F^{\uparrow}_{SH} = -K\Delta T/\Delta z.$$

For example, for a layer $\Delta z = 2$-m thick and a temperature difference $\Delta T = 30$ °C the amount of heat transferred is on the order of 30 W m^{-2}.

The transfer of heat for exposed waters in polyn'ya and leads is of course much greater than for sea ice. It can be estimated using the parameterized expression for heat exchange at the air–sea interface [Eq. (10.34)]

$$F^{\uparrow}_{SH} = -\rho c_p C_H |\mathbf{v}| (T_A - T_S),$$

where $T_A - T_S$ is the air–sea temperature difference. For typical values of 5 m s^{-1} for the wind speed and 40 °C for the temperature difference, the sensible heat transfer into the atmosphere amounts to about 250 W m^{-2}. Using a similar expression for the transfer of latent heat and assuming a specific humidity difference of 4 g kg^{-1}, the amount of latent heat loss is about one-fourth or on the order of 60 W m^{-2}. Of course, the processes described here are especially important in the marginal ice zone.

The annual "production" of ice in the Arctic leading to an outflow of ice, mainly in the form of icebergs, has been estimated (see, e.g., Untersteiner, 1984) to be on the order of 3×10^{15} kg yr^{-1}. Averaged over the north polar cap this outflow of ice (or the inflow of latent heat of melting) is equivalent with an average annual heating of

2.1 W m^{-2} assuming that the exported ice is replaced by the same amount of water at about the same temperature. The upward surface flux in the north polar cap computed as a residual from Eq. (13.53) is only 2.4 W m^{-2} for the year (see Fig. 13.32; the right-hand side of the figure) so that we would infer a weak (equivalent with a net atmospheric heating of 0.3 W m^{-2}) poleward oceanic heat transport across 70 °N. Because we have no reliable residual values for the surface flux in the south polar cap, we cannot say much about the Antarctic conditions. Nevertheless, it is plausible that the oceanic heat flux across 70 °S is extremely small because of the predominance of land in the south polar cap (78% land).

For periods shorter than a year, the storage of energy in the oceanic polar cap given before in Eq. (13.48) must be quite important:

$$\frac{\partial \{E^*\}}{\partial t} \equiv S_L + S_O + S_I + S_{LHI}.$$

Of the four terms on the right-hand side of the equation only the rate of storage of sensible heat in the oceans (S_O) and the rate of storage of latent heat in the form of snow and ice (S_{LHI}) appear to be significant.

The rate of heat storage in the oceans (S_O) is given in Fig. 13.33. The values in the north polar cap are important, whereas those in the south polar cap are insignificant because poleward of 70 °S the ocean occupies only 22% of the surface area in the polar cap.

After subtracting the values of the oceanic heat storage from those of the surface flux in Fig. 13.32, the remaining heat lost by the surface in winter must lead to a substantial freezing of the surface waters ($S_{LHI} < 0$). In summer, the remaining heat gained must be used mainly to melt snow and sea ice. In the Arctic we find then an annual cycle with snow and ice formation between September and May, and snow and ice melting from early June through August. Time integration of the values leads to an equivalent ice formation averaged over the entire north polar cap of 1.3 m during the winter and a destruction of 1.1 m during the summer season. The annual cycle of snow and ice formation and destruction in the south polar cap must be similar to that in the Arctic but with a smaller amplitude, possibly on the order of 0.4 m averaged over the area of the south polar cap.

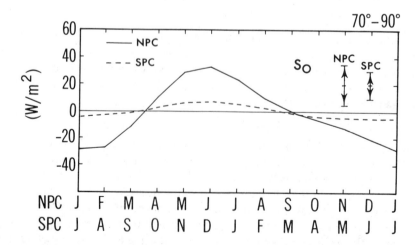

FIGURE 13.33. Annual variation of the rate of storage of sensible heat in the ocean S_O for the north and south polar caps in W m^{-2}, computed as the integrated heat storage divided by the total (land plus ocean) surface area of the polar cap (from Nakamura and Oort, 1988).

13.4.4 Synthesis of the polar energetics

In view of the importance of the polar regions in climate we summarize in Fig. 13.34 the annual-mean and typical summer/winter conditions of the energy budget for the north and south polar caps. The annual-mean budget is very similar in the two regions. The principal balance [see Eq. (13.47)] is between F_{TA} and F_{wall}, because

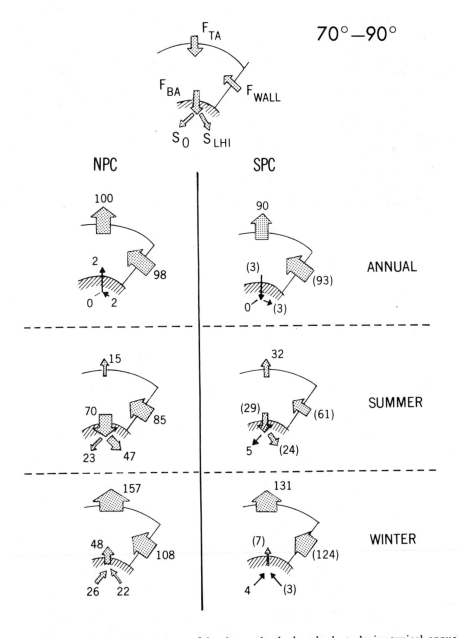

FIGURE 13.34. Schematic breakdown of the observed polar heat budgets during typical annual, summer, and winter mean conditions for the two polar caps. The values of the rate of energy storage in the atmosphere during summer are 3 W m^{-2} for the NPC and 0 W m^{-2} for the SPC, and during winter -2 W m^{-2} for the NPC and -4 W m^{-2} for the SPC. The numbers in parentheses are indirect estimates (adapted from Nakamura and Oort, 1988).

F_{BA} is small when integrated over a year. The seasonal pictures are significantly different in the two polar caps.

In summer, for the north polar cap, F_{TA} is small and the heat flowing into the atmospheric polar cap is balanced largely by storage at and below the earth's surface (melting of snow and ice and warming of the ocean). In summer, for the south polar cap, the influx of energy through the latitudinal wall appears to be much reduced (as deduced from model calculations) and is used about equally to compensate for the radiational loss at the top of the atmosphere and the melting of snow and ice mainly at the fringes of Antarctica.

In winter for the north polar cap, the radiative loss at the top of the atmosphere is very large. About two-thirds of the loss is balanced by the inflow of energy through the wall and one-third through the surface. The heat released at the surface seems to come about equally from lowering the ocean temperature and from freezing of the surface waters. For the south polar cap, the winter loss of radiation is found to be almost entirely balanced by the influx of heat through the wall, and the surface exchange seems to be negligible.

The differences in the physiography of the earth's surface in the two regions—a relatively flat, ocean-covered north polar cap and a land-glacier covered south polar cap with a high topography extending to more than 3000 m into the atmosphere— are, of course, the main reasons for the large differences in the heat budgets shown in Fig. 13.34.

CHAPTER 14

The Ocean–Atmosphere Heat Engine

As a synthesis of our discussions of the radiation and energy processes in the atmosphere, we present in this chapter an integrated picture of how the global radiation and energy balances are achieved in the mean in the climatic system. As shown in Fig. 14.1, the incoming solar radiation is not all used by the climatic system. As we have seen before, a substantial fraction (on the order of 30%, see Fig. 6.3) of the incident solar radiation is reflected by clouds, and to a lesser extent by the earth's surface, and, therefore, does not participate in the atmospheric heat engine. Out of the remaining 70%, 20% is absorbed by the atmosphere and 50% by the oceans and land. This last amount will be used partly to maintain the hydrological cycle through evaporation (24%) and hence heat the atmosphere indirectly through condensation of water, and partly to heat the atmosphere directly through the flux of sensible heat (6%). The remaining 20% is used to heat the underlying surface, and will be lost later as infrared radiation to the atmosphere (14%) and in the spectral window (8–12 μm; see Fig. 6.2) to outer space (6%).

The heat absorbed by the atmosphere is used to increase the internal and potential energy which will partially (less than 1%) be converted into kinetic energy to maintain the atmospheric and oceanic general circulations against friction. Finally, the atmosphere will radiate out to space around 64% as infrared radiation, thereby closing the energy cycle.

In the following sections, we will discuss how the energy is used to operate the ocean–atmosphere heat engine by applying the concept of availability of energy to the various subsystems of the climate system.

14.1 AVAILABILITY OF ENERGY IN THE ATMOSPHERE

We have already seen that the highest temperatures occur at low latitudes near the earth's surface and the lowest temperatures at high latitudes in the upper atmosphere.

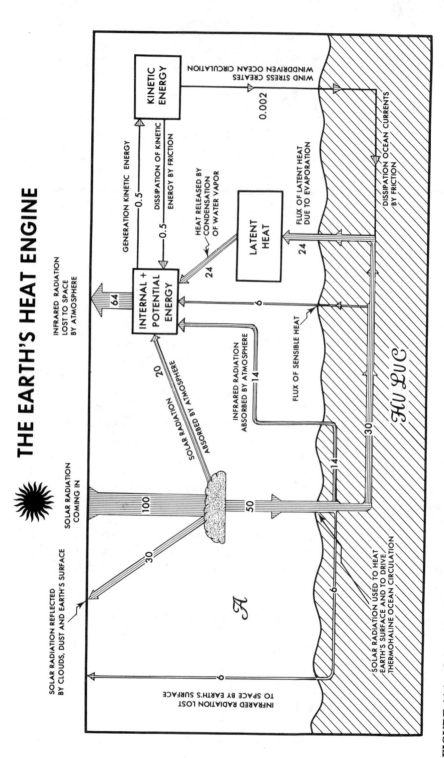

FIGURE 14.1. Schematic diagram of the flow of energy in the climatic system. A value of 100 is assigned to the incoming flux of solar energy ($=340$ W m^{-1}). All values represent annual averages for the entire atmosphere. For simplicity no separate energy boxes have been drawn for the oceans ($\mathscr{H}$), land surfaces ($\mathscr{L}$), or cryosphere ($\mathscr{C}$).

Furthermore, the atmosphere was found to act as a vehicle to transport heat poleward and upward. Thus, the atmosphere may be regarded as a heat engine with, on the whole, heat flowing from the warm sources to the cold sinks. The work performed by the atmospheric engine is used to maintain the kinetic energy of the circulations against a continuous drain of energy by friction dissipation.

Even for the ideal case of a Carnot machine, the efficiency η of the heat machine would be low since the difference between the temperature of the warm source T_W and that of the cold sink T_C is relatively small compared with the temperature of the warm source:

$$\eta = (T_W - T_C)/T_W \lesssim 10\%. \tag{14.1}$$

Later, we will present other estimates of the efficiency which, as expected, lead to lower values than the limiting ideal case. Thus, the amount of kinetic energy generated by the atmospheric heat engine has to be small compared with the total potential plus internal energy.

In our preliminary analysis of the generation of kinetic energy, we concluded that the source for kinetic energy was the total potential energy as discussed in conjunction with Fig. 13.1. However, only a fraction of the total potential plus internal energy in the atmosphere is available to be converted into kinetic energy, whereas most of the total potential energy is unusable. This same fact was already realized by Margules (1903). It is therefore necessary to refine our previous analysis of the problem when we discussed the balance equations for the energy components and their interrelations.

Although the total potential energy in a barotropic atmosphere (see Sec. 3.5.3) can be very high, the capacity to generate kinetic energy does not exist because the necessary mechanisms for conversion are not present. Indeed, in a barotropic atmosphere the horizontal pressure gradient force is zero everywhere, and moreover if the atmosphere is in hydrostatic equilibrium, there is no net vertical force since the vertical component of the pressure gradient is exactly compensated by the force of gravity. Such an atmosphere contains a large amount of potential energy but none of it is available for conversion into kinetic energy. Using the concepts of the availability theory we may say that such an atmosphere is in a "dead" state.

However, the actual atmosphere is baroclinic and in quasihydrostatic equilibrium. Thus within the real atmosphere, there exist several regions or subsystems with a nonuniform distribution of temperature on isobaric levels. Using the terminology of the availability theory we can state that there are many regional subsystems in an "alive" state, and that the atmosphere as a whole is thus in an alive state. Since the atmosphere is not a completely uniform (canonical) system it will continuously evolve through many different natural states, each having a different total potential energy. When the atmosphere goes from one natural state to another there is a change in total potential energy, and thus kinetic energy is either created or destroyed. Only the difference in total potential energy is involved in the transformation process and is thus available to be converted into kinetic energy.

The concept of available potential energy can now be defined with regard to an ideal conceptual state of the atmosphere with a minimum total potential energy. In fact, this state is founded on the conservation of mass and on its redistribution through isentropic processes. Under these conditions the sum of the internal, potential, and kinetic energy is invariant. From all the possible mass-equivalent dead states, we choose the one with the same entropy and minimum total potential energy as the reference state (indicated by a subscript r). This state is characterized by a horizontal stratification with absolute stability in both pressure, potential temperature, and height as shown in Fig. 14.2b. Thus, the availability for total potential energy is defined by

$$\Lambda = \int (\Phi + I) \, dm - \int (\Phi + I)_r \, dm \qquad (14.2)$$

which represents the maximum possible amount of total potential energy that can be converted into kinetic energy.

The actual operation of the general circulation leads to a continuous depletion of the available potential energy. Its maintenance then requires certain mechanisms for restoring the losses and, thus, ensuring the existence of alive regions in the atmosphere. To show that these mechanisms are mainly associated with the nonuniform heating of the atmosphere let us consider a dead atmosphere. If this atmosphere would be heated uniformly, its total potential energy would increase but none would be available because the atmospheric structure would remain barotropic. On the other hand, if we would keep the total amount of potential energy constant, but add and subtract heat differentially, alive subsystems would be generated and availability will be produced. In other words, although the atmosphere would initially be barotropic it would later become baroclinic so that circulations would develop (Bjerknes theorem). Therefore, to keep the general circulation of the atmosphere alive one needs to continuously generate available potential energy through heating of warm and cooling of cold regions.

Let us then consider how we can determine the reference state. For an isentropic (or adiabatic) adjustment of the mass field, the surfaces of $\theta =$ constant behave as material surfaces. We note that the total amount of mass between two isentropic surfaces remains constant during the isentropic rearrangements. In a stable atmosphere, θ increases monotonically with height ($\partial \theta / \partial z > 0$) and we can use an (x,y,θ) coordinate system. We must keep in mind that $p(x,y,\theta)$ represents the weight of the

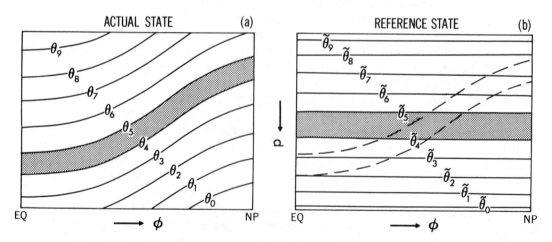

FIGURE 14.2. Schematic diagram of the actual state (a) and the reference state (b) used to compute the available potential energy in the atmosphere. The reference state is obtained by an isentropic redistribution of the atmospheric mass so that the isentropic surfaces become horizontal. During the redistribution the area (i.e., the mass) between two isentropic surfaces remains the same. Note that in a stable state the potential temperature θ increases with height.

atmosphere per unit area above the isentropic surface θ. The average pressure over an isentropic surface is given by

$$\tilde{p}(\theta) = \iint_\sigma p(x,y,\theta)\, dx\, dy \bigg/ \iint_\sigma dx\, dy,$$

where σ is the global area of $\theta =$ constant surface. Whenever a θ surface intersects the earth's surface, the surface pressure is used as a measure of the atmospheric mass above it, e.g., $p(\theta) = p_0$ for $\theta \leqslant \theta_0$.

During an isentropic rearrangement of mass, $\tilde{p}(\theta)$ does not change. Therefore, for the pressure in the reference state

$$p_r(\theta) = \tilde{p}(\theta),$$

so that the departures from the reference state (indicated by an asterisk) in isentropic coordinates are given by

$$(p^*)_\theta \equiv p - \tilde{p}(\theta),$$

and in isobaric coordinates by

$$(\theta^*)_p \equiv \theta - \tilde{\theta}(p).$$

The last two expressions are, of course, related since

$$(p^*)_\theta = -(\theta^*)_p \bigg/ \frac{\partial \tilde{\theta}}{\partial p}.$$

Taking the definition of availability (14.2) and using expressions (13.2) and (13.3a), we obtain the following equivalent expressions for the available potential energy:

$$P = \int c_p(T - T_r)\, dm = \int c_p T(1 - T_r/T)\, dm \tag{14.3a}$$

or

$$P = \int c_p \theta \frac{(p^\kappa - p_r^\kappa)}{p_0^\kappa}\, dm$$

$$= \int c_p T \left\{ 1 - \left(\frac{p_r}{p}\right)^\kappa \right\} dm. \tag{14.3b}$$

If we substitute expression (13.3b) for the total potential energy in the θ-coordinate system in Eq. (14.3a), we obtain the exact expression derived by Lorenz (1955):

$$P = \frac{c_p}{g(1+\kappa)p_0^\kappa} \iiint (p^{\kappa+1} - \tilde{p}^{\kappa+1})\, dx\, dy\, d\theta. \tag{14.3c}$$

If we compare expressions (14.3a) and (14.3b) with the corresponding expression (13.2) for $(\Phi + I)$, we see that the main differences come from the binomial factors $(1 - T_r/T)$ and $\{1 - (p_r/p)^\kappa\}$ that, within this context, can be defined as efficiency factors.

After some simplifications of Eq. (14.3c) one can derive the so-called "approximate" expression of Lorenz (1955):

$$P = \frac{c_p}{2} \int \Gamma(T^2 - \tilde{T}^2)\, dm, \tag{14.4}$$

which gives the available potential energy in terms of the two-dimensional variance of temperature on a constant pressure surface. This expression is most convenient to use in actual calculations. The quantity Γ is an inverse measure of the gross global-mean static stability [see Eq. (3.52)]:

$$\Gamma = -(\kappa\theta/pT)\left(\frac{\partial\tilde{\theta}}{\partial p}\right)^{-1}$$

$$= (\gamma_d/\tilde{T})(\gamma_d - \tilde{\gamma})^{-1}. \tag{14.5}$$

Because of the zonal character of the general circulation it is useful to partition the kinetic energy into the kinetic energy of the zonal-mean flow K_M and the kinetic energy of the perturbations K_E [see Eq. (7.9)]. We have called these components mean and eddy kinetic energy, respectively. Similarly, the available potential energy may be partitioned into the components P_M and P_E. These partitions are done using the expansion

$$[\overline{u^2}] = [\bar{u}]^2 + [\overline{u'^2}] + [\bar{u}^{*2}]$$

and similar expansions for $[\overline{v^2}]$ and $[\overline{T^2}]$. Therefore we have
(1) for kinetic energy

$$K_M = \frac{1}{2}\int ([\bar{u}]^2 + [\bar{v}]^2)\, dm, \tag{14.6}$$

$$K_E = K_{TE} + K_{SE}$$
$$= \frac{1}{2}\int ([\overline{u'^2} + \overline{v'^2}] + [\bar{u}^{*2} + \bar{v}^{*2}])\, dm; \tag{14.7}$$

(2) for available potential energy

$$P_M = \frac{c_p}{2}\int \Gamma([\bar{T}]^2 - \tilde{\tilde{T}}^2)\, dm, \tag{14.8}$$

$$P_E = P_{TE} + P_{SE} = \frac{c_p}{2}\int \Gamma([\overline{T'^2}] + [\bar{T}^{*2}])\, dm. \tag{14.9}$$

The balance equations for these basic forms of energy permit a clear interpretation of the mechanisms involved in the generation and dissipation of energy and of the transformations between the various forms of energy.

14.2 AVAILABILITY OF ENERGY IN THE OCEANS

The total energy of the oceans consists of gravitational potential energy Φ, internal energy I, and kinetic energy K. The kinetic energy and its variations are very small when compared with the corresponding values of the total energy, and can be disregarded in this analysis.

The availability of energy in the oceans can be described in terms of the potential and internal energy and their variations with respect to a reference state. We will take as reference a conceptual state of the oceans with a horizontally uniform, stably stratified density that would result after an adiabatic (or isentropic) redistribution of the mass. In the reference state the density surfaces would be also equigeopotential surfaces.

The total potential plus internal energy for the oceans is given by

$$\int \rho(gz + I)\, dV = \int\int \rho(gz + I)\, dz\, dA, \tag{14.10}$$

where I is the specific internal energy and dA is a horizontal area element. For the reference state we would have

$$\int \rho (gz + I)\, dV = \int \int \rho_r g z_r\, dz\, dA$$
$$+ \int \int \rho_r I_r\, dz\, dA. \tag{14.11}$$

The availability will be defined as

$$\int \int (\rho g z - \rho_r g z_r)\, dz\, dA + \int \int (\rho I - \rho_r I_r)\, dz\, dA, \tag{14.12}$$

where the first term represents the available gravitational potential energy and the second term the available internal energy. Integrating by parts and using the hydrostatic approximation this expression reduces to

$$\int \int (p + \rho I)\, dz\, dA - \int \int (p_r + \rho_r I_r)\, dz\, dA$$
$$= \int \int \rho h\, dz\, dA - \int \int \rho_r h_r\, dz\, dA, \tag{14.13}$$

where $h = p/\rho + I$ is the enthalpy. This expression is equivalent to Eq. (14.3a) for the atmosphere because in a dry atmosphere the enthalpy $h = c_p T$. In the integrations the terms pz at the top and bottom are the same for the actual and reference states because the total mass is conserved. Similarly as in the atmosphere, we may conclude that the availability of total energy is the difference in enthalpy between the two states.

The internal energy I as well as the enthalpy h are functions of pressure, temperature, and salinity but are not known explicitly. However, they can be discussed using the Gibbs equations for the oceans, i.e.,

$$dI = T\, ds - p\, d\alpha + \Sigma \mu_i\, dn_i$$

and

$$dh = T\, ds + \alpha\, dp + \Sigma \mu_i\, dn_i,$$

where s is the entropy, α is the specific volume, μ_i are the chemical potentials that depend on the concentrations, and n_i are the number of moles for the various dissolved components. Of course, the most important dissolved component is salt (NaCl).

Some attempts (e.g., Reid *et al.*, 1981) have been made to incorporate these equations in the evaluation of the enthalpy and internal energy of sea water but this is beyond the scope of the present book. One of the conclusions of Reid *et al.* (1981) is that the available gravitational potential energy is the dominant term and that the available internal energy can at most contribute 10% to 20% to the total. Furthermore, the contribution of the internal energy to the total available energy seems to be negative, contrary to what one would expect, and to what is observed in the atmosphere. This can be attributed to the different behavior of the compressibility as a function of temperature in the oceans and atmosphere.

To describe the energetics of the oceans we will follow the procedure of Oort *et al.* (1989). Thus, we will neglect the possible (but probably minor) contributions from the available internal energy and only present estimates of the available gravitational potential energy as given by the first part of Eq. (14.12):

$$P = \iint \rho g \, z \, dz \, dA - \iint \rho_r g \, z_r \, dz \, da. \tag{14.14}$$

After changing from the (x,y,z) to the (x,y,ρ) coordinate system, we may write

$$P = \frac{1}{2} g \iint z^2 \, d\rho \, dA - \frac{1}{2} g \iint z_r^2 \, d\rho \, dA$$

or

$$P = \frac{1}{2} g \iint (z - z_r)^2 \, d\rho \, dA.$$

The last two expressions are the same because the integral of the cross term vanishes:

$$\frac{1}{2} g \iint z_r (z - z_r) \, d\rho \, dA$$

$$= \frac{1}{2} g \, z_r \iint (z - z_r) \, d\rho \, dA = 0.$$

Substituting next the global-mean height over a constant-density surface $\tilde{z} = \tilde{z}(\rho)$ for the reference height z_r, and using the observed fact that the horizontal gradients in density are much smaller than the vertical gradients we obtain

$$P = -\frac{1}{2} g \iint (z - \tilde{z})^2 \frac{\delta \tilde{\rho}}{dz} \, dz \, dA \tag{14.15}$$

or

$$P = \frac{1}{2} \iint \tilde{\rho} N^2 (z - \tilde{z})^2 \, dz \, dA, \tag{14.16}$$

where N^2 is the Brunt–Väisälä frequency [see Eq. (3.78)].

Since we are considering vertical displacements the vertical gradient of the local potential density $-\delta \tilde{\rho}/dz$, defined earlier in Eq. (3.74), must be used instead of the vertical gradient of the *in situ* density $-d\tilde{\rho}/dz$, as a measure of the stability in the calculation of the available potential energy.

As an alternative to expression (14.15) we can use the departures of the local density from the global-mean density over a constant-height surface $\tilde{\rho} = \tilde{\rho}(z)$, and write

$$P = -\frac{1}{2} g \iint \frac{(\rho - \tilde{\rho})^2}{(\delta \tilde{\rho}/dz)} \, dz \, dA. \tag{14.17}$$

Similarly as was done for the atmosphere in terms of the variance of temperature over an isobaric surface, the integrand of the available gravitational potential energy for the oceans has now been expressed in terms of the variance of height over an isosteric surface (14.15) or in terms of the variance of density over a constant-depth surface (14.17) weighted by a mean stability factor.

We will use Eq. (14.17) to estimate the available gravitational potential energy in the oceans. Since the ocean velocities are usually very small and difficult to measure the estimates of kinetic energy in the oceans are relatively uncertain.

Similarly as for the atmosphere, we can evaluate the mean zonal, stationary eddy, and transient eddy parts of P and K in the oceans. Thus, we have

$$P = P_M + P_{SE} + P_{TE}$$
$$= -\frac{1}{2} \int \int g \frac{([\bar{\rho}] - \tilde{\rho})^2}{(\delta\tilde{\rho}/dz)} dz\, dA$$
$$-\frac{1}{2} \int \int g \frac{\bar{\rho}^{*2}}{(\delta\tilde{\rho}/dz)} dz\, dA$$
$$-\frac{1}{2} \int \int g \frac{\overline{\rho'^2}}{(\delta\tilde{\rho}/dz)} dz\, dA \qquad (14.18)$$

and

$$K = K_M + K_{SE} + K_{TE}$$
$$= \frac{1}{2} \int \int ([\bar{u}]^2 + [\bar{v}]^2)\, dz\, dA$$
$$+ \frac{1}{2} \int \int (\bar{u}^{*2} + \bar{v}^{*2})\, dz\, dA$$
$$+ \frac{1}{2} \int \int \overline{u'^2} + \overline{v'^2}\, dz\, dA, \qquad (14.19)$$

where the zonal averaging [ ] operator is defined only for the oceanic segments of the latitude circles and the ( )* operator as the departure from this zonal average.

14.3 BALANCE EQUATIONS FOR KINETIC AND AVAILABLE POTENTIAL ENERGY

14.3.1 Formal derivation of the balance equations

The derivation of the balance equations for quadratic quantities, such as the kinetic and available potential energy in the atmosphere and oceans, involves some apparent difficulties. One of these difficulties is that these equations include third-order terms describing the interactions of the eddies with the mean fields. Thus, we will give here a general derivation of this type of balance equation in the space domain.

Let us start with considering the balance equation for an arbitrary parameter A. The symbol A can represent any of the meteorological or oceanographic variables u, v, T, or ρ. Using the (λ, ϕ, p, t) coordinate system we can write

$$\frac{\partial A}{\partial t} = -\mathbf{v}\cdot\text{grad}\, A - \omega \frac{\partial A}{\partial p} + S, \qquad (14.20)$$

where S is a source or sink term. In the space domain, the parameter A can be broken down into two components:

$$A = [A] + A^*.$$

Multiplication of Eq. (14.20) by $[A]$ and using the continuity of mass equation in the form (3.4) div $\mathbf{v} + \partial \omega/\partial p = 0$ gives

$$[A]\frac{\partial A}{\partial t} = -\mathbf{v}[A]\cdot\text{grad}\,A - \omega[A]\frac{\partial A}{\partial p} + [A]S$$

$$= -\mathbf{v}[A]\cdot\text{grad}\,[A] - \omega[A]\frac{\partial[A]}{\partial p}$$

$$+ [A]S - [A]\,\text{div}(\mathbf{v}\,A^*) - [A]\frac{\partial \omega A^*}{\partial p}.$$

Next taking the zonal average:

$$[A]\frac{\partial}{\partial t}[A] = -\left[\mathbf{v}\cdot\text{grad}\,\frac{1}{2}[A]^2\right]$$

$$- [\omega]\frac{\partial}{\partial p}\frac{1}{2}[A]^2 + [A][S]$$

$$- [A][\text{div}(\mathbf{v}^*A^*)] - [A]\frac{\partial[\omega^*A^*]}{\partial p}$$

leads to a balance equation for the zonal-mean quadratic quantity

$$\underbrace{\frac{\partial}{\partial t}\frac{1}{2}[A]^2}_{(1)} = \underbrace{-\frac{\partial([v]\frac{1}{2}[A]^2\cos\phi)}{R\cos\phi\partial\phi}}_{} \underbrace{- \frac{\partial}{\partial p}\left([\omega]\frac{1}{2}[A]^2\right)}_{(2)} + \underbrace{[A][S]}_{(3)}$$

$$\underbrace{- \frac{\partial([v^*A^*][A]\cos\phi)}{R\cos\phi\partial\phi} - \frac{\partial}{\partial p}([\omega^*A^*][A])}_{(4)}$$

$$\underbrace{+ [v^*A^*]\frac{\partial[A]}{R\partial\phi} + [\omega^*A^*]\frac{\partial[A]}{\partial p}}_{(5)}. \qquad (14.21)$$

To derive a balance equation for the eddy quadratic component one multiplies Eq. (14.20) by A^*:

$$A^*\frac{\partial A}{\partial t} = -\mathbf{v}A^*\cdot\text{grad}\,A - \omega A^*\frac{\partial A}{\partial p} + A^*S$$

$$= -\mathbf{v}A^*\cdot\text{grad}\,[A] - \omega A^*\frac{\partial}{\partial p}[A] + A^*S$$

$$- \mathbf{v}A^*\cdot\text{grad}\,A^* - \omega A^*\frac{\partial A^*}{\partial p}.$$

Again, taking the zonal average gives

$$\left[A^*\frac{\partial A^*}{\partial t}\right] = -[v^*A^*]\frac{\partial[A]}{R\partial\phi}$$

$$- [\omega^*A^*]\frac{\partial[A]}{\partial p} + [A^*S^*]$$

$$- \left[\mathbf{v}\cdot\text{grad}\,\frac{1}{2}A^{*2}\right] - \left[\omega\frac{\partial}{\partial p}\frac{1}{2}A^{*2}\right]$$

and leads finally to the desired balance equation neglecting third-order terms:

$$\underbrace{\frac{\partial}{\partial t}\frac{1}{2}[A^{*2}]}_{(1')} = \underbrace{-\partial\frac{([v]\frac{1}{2}[A^{*2}]\cos\phi)}{R\cos\phi\partial\phi}}_{(2')} \underbrace{-\frac{\partial}{\partial p}\left([\omega]\frac{1}{2}[A^{*2}]\right)}_{} + \underbrace{[A^*S^*]}_{(3')}$$

$$\underbrace{-[v^*A^*]\frac{\partial[A]}{R\partial\phi} - [\omega^*A^*]\frac{\partial[A]}{\partial p}}_{(4')}. \quad (14.22)$$

The following interpretation of the various terms in Eqs. (14.21) and (14.22) may be useful:

(a) Terms (1) and (1') give the rate of change of the mean and eddy components of the quadratic quantity, respectively.

(b) Terms (2), (4), and (2') are flux divergence terms that after integration over the mass reduce to boundary terms. These boundary terms vanish in case of global integrals.

(c) Terms (3) and (3') are the source and sink terms for the mean and eddy components of the quadratic quantities.

(d) Terms (5) and (4') can be interpreted as the conversion terms between the mean and eddy components; they are of equal magnitude but have the opposite sign.

To derive analogous equations as Eqs. (14.21) and (14.22) in the time and the mixed space-time domains (see Sec. 4.1.1) similar methods can be used.

14.3.2 Balance equations for mean and eddy kinetic energy in the atmosphere

As an application of the previous methodology we will derive the balance equations for the mean and eddy kinetic energy, K_M and K_E. Starting with the two horizontal equations of motion (3.20):

$$\frac{\partial u}{\partial t} = -u\frac{\partial u}{R\cos\phi\partial\lambda} - v\frac{\partial u}{R\partial\phi} - \omega\frac{\partial u}{\partial p} + \frac{\tan\phi}{R}uv$$

$$+fv - g\frac{\partial z}{R\cos\phi\partial\lambda} + F_\lambda,$$

$$\frac{\partial v}{\partial t} = -u\frac{\partial v}{R\cos\phi\partial\lambda} - v\frac{\partial v}{R\partial\phi} - \omega\frac{\partial v}{\partial p} - \frac{\tan\phi}{R}u^2$$

$$-fu - g\frac{\partial z}{R\partial\phi} + F_\phi,$$

multiplying them by $[u]$ and $[v]$, respectively, and adding, we obtain a balance equation for K_M formally equivalent to Eq. (14.21). Integrating over the entire mass of the atmosphere, we can disregard the boundary terms so that we can write the balance equation in the following symbolic form (see Peixoto and Oort, 1974):

$$\frac{\partial}{\partial t}K_M = C(P_M, K_M) + C(K_E, K_M) - D(K_M), \quad (14.23)$$

where $C(A,B)$ indicates the rate of conversion from A to B and $D(K_M)$ the rate of dissipation of K_M. Thus,

$$C(P_M, K_M) = -\int [v] g \frac{\partial [z]}{R \partial \phi} dm$$

$$= -\int [\omega][\alpha] dm, \qquad (14.24)$$

$$C(K_E, K_M) = \int [v^* u^*] \cos\phi \frac{\partial ([u]/\cos\phi)}{R \partial \phi} dm$$

$$+ \int [\omega^* u^*] \frac{\partial [u]}{\partial p} dm$$

$$+ \int [v^{*2}] \frac{\partial [v]}{R \partial \phi} dm - \int [v] \frac{\tan\phi}{R} [u^{*2}] dm$$

$$+ \int [\omega^* v^*] \frac{\partial [v]}{\partial p} dm \qquad (14.25)$$

and

$$D(K_M) = -\int ([u][F_\lambda] + [v][F_\phi]) dm. \qquad (14.26)$$

The term $C(P_M, K_M)$ shows that kinetic energy in the zonal mean flow can be generated by horizontal cross isobaric flow down the north–south pressure gradient or by the rising of light air and the sinking of dense air in zonally symmetric meridional overturnings. The term $C(K_E, K_M)$ will lead to a conversion from eddy into mean kinetic energy as long as the north–south and up–down transports of momentum by eddies are up the gradient of the mean zonal wind ("negative" viscosity phenomenon). The rate of conversion between mean potential and mean kinetic energy can be written in either the "v grad z" or the "$\omega\alpha$" form [see Eq. (14.24)]. Provided that global integrals are considered the two formulations are equivalent. This can be shown using the expansion

$$-\mathbf{v}\cdot\text{grad } z = -\text{div } \mathbf{v}z + z \text{ div } \mathbf{v}$$

$$= -\text{div } \mathbf{v}z - z \frac{\partial \omega}{\partial p}$$

$$= -\text{div } \mathbf{v}z - \frac{\partial \omega z}{\partial p} - \omega\alpha/g.$$

Thus,

$$-\int g \mathbf{v}\cdot\text{grad } z \, dm = -\int \omega\alpha \, dm.$$

Similarly as for K_M, the balance equation for K_E can be obtained by multiplying the first and second equations of motion by u^* and v^*, respectively, adding and finally integrating over the entire atmosphere. Thus, we may write in symbolic form:

$$\frac{\partial K_E}{\partial t} = C(P_E, K_E) - C(K_E, K_M) - D(K_E). \qquad (14.27)$$

New terms are the eddy conversion and eddy dissipation terms

$$C(P_E, K_E) = -\int g[\mathbf{v}^* \cdot \text{grad } z^*] dm$$

$$= -\int [\omega^* \alpha^*] dm \tag{14.28}$$

and

$$D(K_E) = -\int [u^* F_\lambda^*] dm - \int [v^* F_\phi^*] dm. \tag{14.29}$$

The rate of conversion between eddy potential and eddy kinetic energy, $C(P_E, K_E)$, generates eddy kinetic energy when, along a latitude circle, the wind anomalies tend to be negatively correlated with the anomalies in the pressure gradient or when light (warm) air tends to rise and dense (cold) air tends to sink.

In the mixed space-time domain, transient and stationary eddy terms would appear rather than only spatial eddy terms. For example, instead of the simple term $[\omega^* \alpha^*]$ in the conversion rate from eddy potential to eddy kinetic energy in the space domain, we would have $[\overline{\omega' \alpha'}] + [\overline{\omega}^* \overline{\alpha}^*]$, explicitly showing the influence of both transient and stationary eddies.

14.3.3 Balance equations for mean and eddy available potential energy in the atmosphere

The balance equation for P_M can be derived using the first law of thermodynamics (3.49) in the form

$$\frac{\partial T}{\partial t} = -\mathbf{v} \cdot \text{grad } T - \omega \frac{T}{\theta} \frac{\partial \theta}{\partial p} + Q/c_p,$$

by first multiplying it by $c_p \Gamma([T] - \tilde{T})$, where $([T] - \tilde{T})$ is the departure of the zonal average temperature from its global mean, and then integrating it over the entire atmosphere so that the boundary terms disappear (for a discussion of the boundary terms, see Peixoto and Oort, 1974). Thus, we can write in symbolic form:

$$\frac{\partial P_M}{\partial t} = G(P_M) - C(P_M, P_E) - C(P_M, K_M). \tag{14.30}$$

Here

$$G(P_M) = \int \Gamma([T] - \tilde{T})([Q] - \tilde{Q}) dm \tag{14.31}$$

indicates the rate of generation of P_M mainly through heating of relatively warm air at low latitudes and cooling of relatively cold air at high latitudes, and

$$C(P_M, P_E) = -c_p \int \Gamma[v^* T^*] \frac{\partial [T]}{R \partial \phi} dm$$

$$- c_p \int p^{-\kappa} [\omega^* T^*] \frac{\partial}{\partial p} [\Gamma p^\kappa ([T] - \tilde{T})] dm \tag{14.32}$$

indicates the rate of conversion from mean to eddy available potential energy through eddy poleward or eddy upward heat transports down the mean zonal temperature gradient.

Using the same methodology the balance equation for P_E can be derived by multiplying Eq. (3.49) by $c_p \Gamma T^*$ and integrating over the entire atmosphere:

$$\frac{\partial P_E}{\partial t} = G(P_E) + C(P_M, P_E) - C(P_E, K_E). \tag{14.33}$$

The only new term in Eq. (14.33) is the rate of generation of P_E:

$$G(P_E) = \int \Gamma [T^* Q^*] dm. \tag{14.34}$$

This term operates through the heating of warm air anomalies and the cooling of cold air anomalies along each latitude circle, thereby increasing the east–west variance of temperature and P_E.

The balance equations for the four basic forms of energy are summarized below:

$$\frac{\partial K_M}{\partial t} = C(P_M, K_M) + C(K_E, K_M) - D(K_M),$$

$$\frac{\partial K_E}{\partial t} = C(P_E, K_E) - C(K_E, K_M) - D(K_E),$$

$$\frac{\partial P_M}{\partial t} = G(P_M) - C(P_M, K_M) - C(P_M, P_E),$$

$$\frac{\partial P_E}{\partial t} = G(P_E) - C(P_E, K_E) + C(P_M, P_E).$$

The connecting links between the various energy forms as given by these equations are illustrated schematically in Fig. 14.3 in the form of a box diagram. As we have seen before, the generation of available potential energy is proportional to the covariance of the diabatic heating and temperature. Therefore, when regions of high

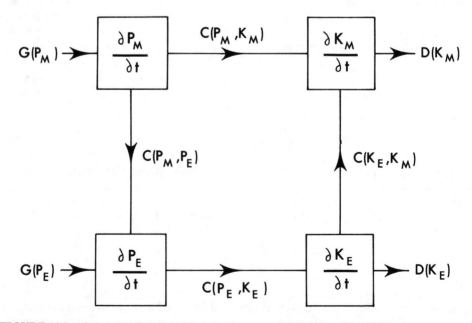

FIGURE 14.3. Schematic box diagram of the energy cycle in the atmosphere showing the rates of generation, conversion, and dissipation of the various forms of energy.

temperature are heated and regions of low temperature are cooled available potential energy will be generated. The generation of available potential energy resembles the behavior of a heat engine where fuel has to be added to heat the warm furnace and cooling has to be provided to maintain the required temperature difference. Basically, the heating of the warm regions and the cooling of the cold regions amount to decreasing the entropy of the system. Obviously, the opposite processes would lead to an increase in entropy and a destruction of available potential energy.

14.4 OBSERVED ENERGY CYCLE IN THE ATMOSPHERE

14.4.1 Spatial distributions of the energy and energy conversions

Meridional and vertical profiles of the mean and eddy energy components are shown in Figs. 14.4 and 14.5. The curves represent the integrands that contribute to the global integrals of the available potential and kinetic energy components. However, the integrands of P_M and P_E are not necessarily meaningful by themselves.

The meridional profiles in Fig. 14.4 show important differences between the Northern and Southern Hemispheres, namely:

(1) Poleward of 50° latitude the contributions to P_M are greater in the Southern than in the Northern Hemisphere for all seasons. This is because the temperatures are lower over the Antarctic continent than over the Arctic Ocean, leading to a larger meridional temperature gradient in the Southern Hemisphere.

(2) On the other hand, the eddy activity shown in P_{TE} and P_{SE} is higher in the Northern than in the Southern Hemisphere. The very large values of P_{TE} in the northern midlatitudes for the year are not due primarily to synoptic-scale eddies but to the large seasonal cycle of the temperature over the northern continents.

(3) The K_M values are, in general, smaller in the Northern than in the Southern Hemisphere, except during winter.

(4) The K_{TE} values are of comparable magnitude in both hemispheres.

(5) As we have seen in earlier sections for other parameters, the seasonal variations are, in general, larger in the Northern than in the Southern Hemisphere.

(6) The stationary eddies are very small in the Southern Hemisphere and are only important in the Northern Hemisphere.

The vertical profiles of the global averages of the various forms of energy (as illustrated in Fig. 14.5) show that there is a seasonal variation even when the global energy content of the atmosphere is considered. It seems, from the energetics point of view, that the troposphere is more active during the northern winter (DJF) than during the northern summer (JJA) for all energy components.

The contributions to the directly measurable (see the next section) global integrals of the energy conversion terms $C(P_M, P_E)$, $C(K_E, K_M)$, and $C(P_M, K_M)$ are shown in Figs. 14.6 and 14.7. The most striking differences between the two hemispheres in Fig. 14.6 are due to the differences in the amplitude of the seasonal cycle. Aside from this fact, the two hemispheres appear to behave alike.

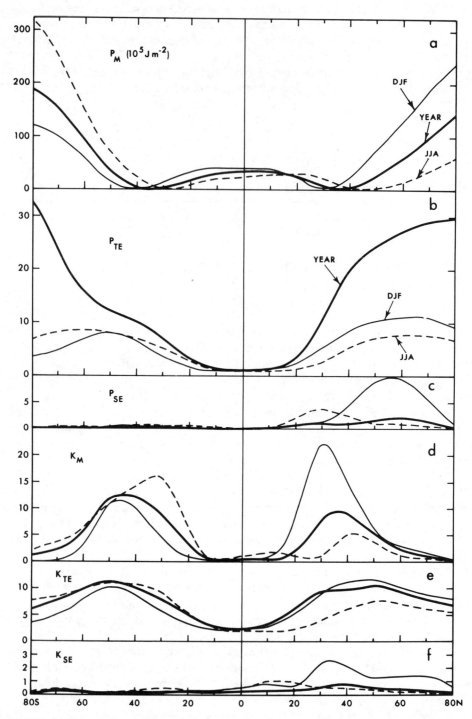

FIGURE 14.4. Meridional profiles of the contributions to the global integrals of the mean available potential energy (a), transient eddy available potential energy (b), stationary eddy available potential energy (c), mean kinetic energy (d), transient eddy kinetic energy (e), and stationary eddy kinetic energy (f) in the atmosphere in units of 10^5 J m^{-2} (from Oort and Peixoto, 1983).

THE OCEAN–ATMOSPHERE HEAT ENGINE   381

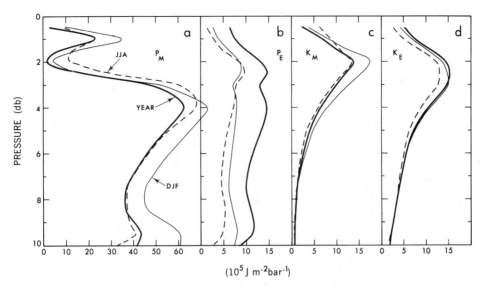

FIGURE 14.5. Vertical profiles of the contributions to the global integrals of mean available potential energy (a), eddy available potential energy (b), mean kinetic energy (c), and eddy kinetic energy (d) in the atmosphere in units of 10^5 J m^{-2} bar^{-1}. The transient and stationary eddies are combined in the eddy terms (from Oort and Peixoto, 1983).

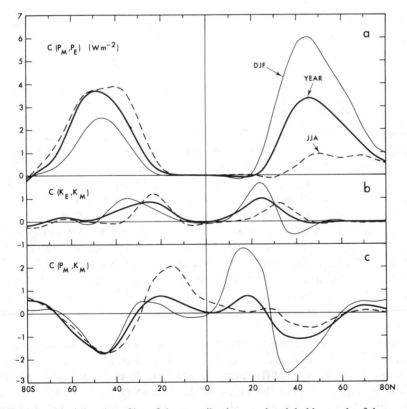

FIGURE 14.6. Meridional profiles of the contributions to the global integrals of the conversion rates from mean to eddy available potential energy (a), eddy to mean kinetic energy (b), and mean available potential to mean kinetic energy (c) in the atmosphere in units of W m^{-2} (from Oort and Peixoto, 1983).

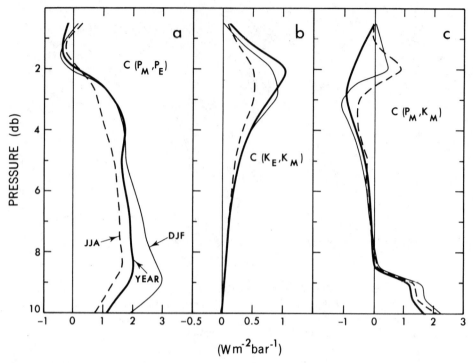

FIGURE 14.7. Vertical profiles of the contributions to the global integrals of the conversion rates from mean to eddy available potential energy (a), eddy to mean kinetic energy (b), and mean available potential to mean kinetic energy (c) in the atmosphere in units of W m^{-2} bar^{-1} (from Oort and Peixoto, 1983).

14.4.2 The energy cycle

The annual-mean energetics of the global atmosphere are summarized in Fig. 14.8 in the form of a schematic box diagram of the type shown before in Fig. 14.3. The energy amounts and three of the four conversion terms, namely $C(P_M, P_E)$, $C(P_M, K_M)$, and $C(K_E, K_M)$, can be computed directly from the available observations. However, the fourth conversion term $C(P_E, K_E)$, as well as the generation and dissipation terms, are generally estimated by indirect means since they require knowledge of the poorly known vertical velocity, diabatic heating, and frictional fields.

In the case of Fig. 14.8, the unknown quantities were evaluated as residuals in Eqs. (14.23), (14.27), (14.30), and (14.33) except for the eddy generation term $G(P_E)$. This last term was assumed on the basis of tentative evidence from earlier diagnostic and numerical model studies. A direct evaluation of $G(P_E)$ is extremely difficult because it requires the evaluation of the covariance of the diabatic heating rate and the temperature on a daily basis.

We have now reached the point where we can begin to describe the general mechanisms by which the incoming solar radiation maintains the kinetic energy of the atmosphere and oceans against frictional dissipation. As primary forcing, the radiational heating and cooling in the atmosphere tend to generate an almost zonally symmetric distribution of temperature in both hemispheres with a strong north–south gradient which is most intense in middle latitudes. The north–south variance of the zonal-mean temperature at each level is, of course, a measure of the mean zonal available potential energy P_M. However, the meridional transport of heat by the

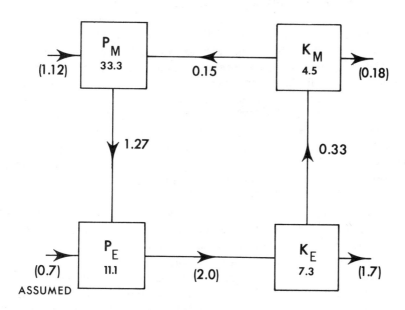

FIGURE 14.8. The observed energy cycle in the global atmosphere for annual-mean conditions. Since the rates of change of energy are negligible, the energy amounts are given instead inside each box in units of 10^5 J m^{-2}. The rates of generation, conversion, and dissipation are in W m^{-2}. Terms not directly measured are shown in parentheses (modified after Oort and Peixoto, 1983).

eddies in midlatitudes deforms the zonally symmetric distribution of temperature leading to an additional variance in the east–west direction which is proportional to the eddy available potential energy P_E.

Through mechanisms described before, such as baroclinic instability, some of the eddy available potential energy is transformed into kinetic energy of the growing disturbances K_E. These eddies also transport momentum poleward increasing the zonal motion, and thus generating zonal-mean kinetic energy K_M. A large part of the eddy kinetic energy is directly dissipated by friction or cascades into smaller and smaller eddies (turbulence) until it is finally converted into heat by molecular viscosity. This is a consequence of the second law of thermodynamics. In the absence of external forcings the system tends to evolve to states of larger and larger disorder. Of course, the maximum disorder or entropy is obtained when the ordered energy is converted into random molecular motions. The dissipation seems to be most intense in the eddy kinetic energy.

The mean meridional circulations in the tropics (Hadley cells) generate zonal angular momentum in their upper branches [see Eq. (3.9a), fv term] and thereby mean zonal kinetic energy. However, the indirect circulations in midlatitudes of each hemisphere (Ferrel cells) consume mean zonal kinetic energy at a rate slightly exceeding the production in the Hadley cells. Therefore, for the atmosphere as a whole part of the mean zonal kinetic energy is converted back into mean zonal available potential energy.

Considering the numerical estimates from Oort and Peixoto (1983) shown in Fig. 14.8, we are led to the following scheme for the energy cycle. Both (1) the net heating in the tropics by the absorption of solar radiation and by the release of latent heat and

(2) the net cooling in high latitudes by the loss of terrestrial radiation result in a generation of zonal-mean available potential energy at the rate of about 1.1 W m^{-2}. This energy is converted into eddy available potential energy by the growing baroclinic disturbances at a rate of about 1.3 W m^{-2}. The fastest growing eddies are in the range of wave number 5 to 8 about the pole. Some of the eddy available potential energy may be dissipated by a greater loss of infrared radiation to space in warm than in cold air and by a similar heat exchange with the earth's surface. However, latent heating is believed to surpass these effects leading to a small positive generation of P_E, assumed to be on the order of 0.7 W m^{-2}. The eddy available potential energy is next converted into eddy kinetic energy at the rate of about 2.0 W m^{-2} through the sinking of relatively cold air and the rising of warm air in the eddies (see Fig. 13.14). As we have seen before, there is a weak but prevalent tendency of a countergradient flow of momentum up the gradient of zonal-mean angular velocity (see Sec. 11.2.3). Therefore, some of the eddy kinetic energy is converted into kinetic energy of the mean flow in a barotropic process leading to a decascade of energy from the small into the larger scales (0.3 W m^{-2}), which can be interpreted as a negative viscosity phenomenon (Starr, 1968). However, the bulk of the kinetic energy of the large-scale eddies is dissipated by friction in a normal cascade regime (1.7 W m^{-2}).

Finally, some of the mean zonal kinetic energy is dissipated by friction and turbulence (0.2 W m^{-2}), while a small residual is converted into mean zonal available potential energy (0.15 W m^{-2}) by the combined action of the direct and indirect mean meridional circulations. If one compares the total amount of kinetic energy with the rate of dissipation of kinetic energy we find that it would take about seven days to dissipate the kinetic energy.

Schematically, we have found that the energy cycle in the atmosphere proceeds from P_M to K_M through the following scheme:

$$P_M \rightarrow P_E \rightarrow K_E \rightarrow K_M.$$

It therefore appears that the eddies play a crucial role in regulating the general circulation of the atmosphere as suggested by Starr (1948). This issue will be discussed further in the next section.

Further information on how much the two hemispheres contribute to the global integrals and how their contributions vary with the seasons is presented in Table 14.1. The amounts of energy in both the available potential and kinetic energy are found to be much higher in the winter than in the summer hemisphere. As far as the energy cycle is concerned, maximum generation of available potential energy is found in the summer hemisphere, whereas the conversion and dissipation terms seem to be higher in the winter hemisphere. In order to maintain the balance a strong cross-equatorial flux of mean available potential energy (not given in the table) must take place from the summer to the winter hemisphere.

The energy ratios are also presented in Table 14.1. They show that, for the year as a whole, the eddies (transient plus stationary) form an important fraction of the global kinetic (62%) and global available potential energy (25%). The ratio of the amount of kinetic energy to the total amount of kinetic plus available potential energy is remarkably stable with the same value of about 21% for each hemisphere independent of the season.

The intensity of the atmospheric energy cycle can be measured by estimating either the rate of conversion of available potential energy into kinetic energy $C(P,K)$, the rate of generation of available potential energy $G(P)$, or the rate of dissipation of kinetic energy $D(K)$, since in a steady state all three terms must balance. Regarded again as a heat engine, the efficiency η of the atmosphere can be taken as the ratio of

TABLE • 14.1. Estimates of the amounts of available potential and kinetic energy (10^5 J m^{-2}), some energy ratios (%), and the energy generation, conversion, and dissipation rates (W m^{-2}) for the Northern Hemisphere, Southern Hemisphere, and global atmosphere (after Oort and Peixoto, 1983; Oort *et al.*, 1989). To obtain the integrals over the total atmospheric mass the values for the Northern and Southern Hemisphere should be multiplied by 2.56×10^{14} m^2 and for the globe by 5.12×10^{14} m^2.

	Year NH	Year SH	Year Globe	DJF NH	DJF SH	DJF Globe	JJA NH	JJA SH	JJA Globe	units
P_M	31.5	35.1	33.3	51.0	32.9	41.9	22.1	48.9	35.5	10^5 J m^{-2}
P_{SE}	1.2	0.4	0.8	3.7	0.5	2.1	1.8	0.7	1.2	10^5 J m^{-2}
P_{TE} [b]	12.1	8.6	10.3	5.5	3.8	4.6	3.6	4.8	4.2	10^5 J m^{-2}
P	44.8	44.1	44.4	60.2	37.2	48.6	27.5	54.4	40.9	10^5 J m^{-2}
K_M	3.6	5.5	4.5	7.5	3.8	5.7	2.0	7.4	4.7	10^5 J m^{-2}
K_{SE}	0.6	0.3	0.4	1.7	0.4	1.0	0.8	0.5	0.6	10^5 J m^{-2}
K_{TE} [b]	6.9	6.8	6.9	7.1	5.4	6.3	4.2	7.0	5.6	10^5 J m^{-2}
K	11.1	12.5	11.8	16.3	9.6	13.0	7.0	15.0	11.0	10^5 J m^{-2}
$K/(P+K)$	20	22	21	21	21	21	20	22	21	%
K_E/K	68	57	62	54	60	56	71	50	56	%
P_E/P	30	20	25	15	12	14	20	10	13	%
$G(P_M)$	(1.26)	(0.97)	(1.12)	(0.76)	(2.33)	(1.54)	(2.03)	(0.16)	(1.10)	W m^{-2}
$G(P_E)$	0.7[a]	0.7[a]	0.7[a]	0.5[a]	0.7[a]	0.6[a]	1.0[a]	0.4[a]	0.7[a]	W m^{-2}
$C(P_M,P_E)$	1.22	1.31	1.27	2.31	0.81	1.56	0.29	1.68	0.99	W m^{-2}
$C(P_E,K_E)$	(2.0)	(2.0)	(2.0)	(2.8)	(1.5)	(2.2)	(1.3)	(2.1)	(1.7)	W m^{-2}
$C(P_M,K_M)$	−0.16	−0.14	−0.15	0.21	−0.24	−0.02	−0.02	0.24	0.11	W m^{-2}
$C(K_E,K_M)$	0.29	0.36	0.33	0.27	0.34	0.30	0.18	0.25	0.21	W m^{-2}
$D(K_M)$	(0.13)	(0.22)	(0.18)	(0.48)	(0.10)	(0.28)	(0.16)	(0.49)	(0.32)	W m^{-2}
$D(K_E)$	(1.7)	(1.6)	(1.7)	(2.5)	(1.2)	(1.9)	(1.1)	(1.9)	(1.5)	W m^{-2}

[a] The values of $G(P_E)$ are assumed (see the text).
[b] All time scales of variations less than 12 months in the case of "Year" and less than 3 months in the case of "DJF" and "JJA" are included as transient eddies in the estimates of P_{TE} and K_{TE}.

the kinetic energy dissipated by friction (2.0 W m^{-2}) and the mean incoming solar radiation (238 W m^{-2}) which leads to a value for the efficiency $\eta \approx 0.8\%$. This estimate of the efficiency is better than the one based on a Carnot cycle because it involves only the real input of energy into the system. The value for the Carnot efficiency was found earlier to be on the order of 10% [see Eq. (14.1)].

14.5 MAINTENANCE AND FORCING OF THE ZONAL-MEAN STATE OF THE ATMOSPHERE

14.5.1 Introduction

The maintenance of the zonal-mean fields and the interactions between the eddies and the zonal-mean fields have been referred to several times in earlier sections.

For example, in the preceding section the rate of production of zonal kinetic energy [Eq. (14.25)] by the Reynolds' stresses associated with the large-scale eddies was found to be of the right order of magnitude to maintain the atmospheric flow against frictional dissipation. Further, the rate of production of eddy available potential energy [Eq. (14.32)] was found to result from the interaction of the eddy flux of enthalpy with the gradient of the zonal-mean temperature. It was also shown in Chap. 11 that the angular momentum transport is primarily accomplished by the eddies so that the eddies are necessary to maintain the zonal-mean flow of the atmosphere. The heat transport by the eddies is also of the right order of magnitude to fulfill the energy balance requirements as was shown in Chap. 13. Thus, many atmospheric properties are related to the eddy heat and momentum fluxes.

The two meridional fluxes can be combined in a consistent way as was done first by Eliassen and Palm (1961) who wanted to diagnose the sources and sinks of energy for mountain lee waves. In generalizing their results a fictitious vector can be defined in the meridional plane, the so-called Eliassen-Palm flux (*E-P flux*; see Edmon et al., 1980), with the eddy flux of momentum as the meridional component and the eddy flux of sensible heat as the vertical component. As we will see later, meridional cross sections displaying the *E-P flux* by means of arrows and the *E-P flux* divergence by means of contour lines have become a valuable tool for studying the interactions between the eddies and the mean state. Such a representation is more useful than if the eddy fluxes are considered separately. The divergence of the *E-P flux* gives a measure of the total forcing of the zonal-mean state by the eddies, whereas the direction of the *E-P flux* vectors indicates the relative contributions of the eddy fluxes of heat and momentum.

Edmon et al. (1980) have also shown that the divergence of the Eliassen-Palm flux is proportional to the northward transport of quasigeostrophic potential vorticity (see Sec. 3.4.3).

14.5.2 Interactions between the eddies and the zonal-mean state

In order to analyze the nature of the intractions between the disturbances and the zonal-mean state of the atmosphere, we start with the basic equations, written in a spherical pressure coordinate system, namely

(a) the zonally averaged momentum equation (3.20a):

$$\frac{\partial [u]}{\partial t} + \frac{\partial ([u][v]\cos^2 \phi)}{R \cos^2 \phi \partial \phi} + \frac{\partial ([u][\omega])}{\partial p}$$
$$= -\frac{\partial ([u^*v^*]\cos^2 \phi)}{R \cos^2 \phi \partial \phi} - \frac{\partial [u^*\omega^*]}{\partial p} + f[v] + [F_\lambda]; \qquad (14.35)$$

(b) the zonally averaged thermodynamic equation (3.49):

$$\frac{\partial [\theta]}{\partial t} + \frac{\partial ([v][\theta]\cos \phi)}{R \cos \phi \partial \phi} + \frac{\partial ([\omega][\theta])}{\partial p}$$
$$= -\frac{\partial [\theta^*v^*]\cos \phi}{R \cos \phi \partial \phi} - \frac{\partial [\theta^*\omega^*]}{\partial p} + \left(\frac{p_0}{p}\right)^\kappa \frac{[Q]}{c_p}; \qquad (14.36)$$

(c) the zonally averaged mass continuity equation (3.6):

$$\frac{\partial [v]\cos \phi}{R \cos \phi \partial \phi} + \frac{\partial [\omega]}{\partial p} = 0 \qquad (14.37)$$

and, finally, the thermal wind equation (7.6c) rewritten in the (λ,ϕ,p,t) coordinate system

$$f\frac{\partial [u]}{\partial p} = \frac{1}{\rho[\theta]} \frac{\partial [\theta]}{R\partial \phi}. \tag{14.38}$$

Within the quasigeostrophic approximation (and using appropriate scale analysis; see Sec. 3.2) most terms involving ω in Eqs. (14.35) and (14.36) can be disregarded with the exception of the term $\partial([\omega][\theta])/\partial p$. In view of the large static stability of the atmosphere this last term can be approximated by $[\omega](\partial[\theta]/\partial p)$ or further by $[\omega](\partial\theta_s/\partial p)$ where θ_s is the constant, global-mean potential temperature at a given pressure level.

We assume further that the terms involving $[u]$, v, and θ_s are order-one quantities whereas the terms involving other quantities such as ω and $[v]$ are on the order of the Rossby number. Thus, with the quasigeostrophic approximation ($R_0 \ll 1$) the previous set of equations can be written as

$$\frac{\partial [u]}{\partial t} = f[v] + [F_\lambda] - \frac{1}{R\cos^2\phi} \frac{\partial ([u^*v^*]\cos^2\phi)}{\partial \phi}, \tag{14.39}$$

$$\frac{\partial [\theta]}{\partial t} + [\omega]\frac{\partial \theta_s}{\partial p} = \left(\frac{p_0}{p}\right)^\kappa \frac{[Q]}{c_p} - \frac{1}{R\cos\phi}\frac{\partial [v^*\theta^*]\cos\phi}{\partial \phi}, \tag{14.40}$$

$$\frac{1}{R\cos\phi}\frac{\partial [v]\cos\phi}{\partial \phi} + \frac{\partial [\omega]}{\partial p} = 0, \tag{14.41}$$

and

$$f\frac{\partial [u]}{\partial p} = \frac{1}{\rho[\theta]} \frac{\partial [\theta]}{R\partial \phi}. \tag{14.42}$$

Thus, we have four equations for four unknowns $[u]$, $[v]$, $[\omega]$, and $[\theta]$. All other terms that do not include these four unknowns are regarded as forcing terms. This set of equations shows the importance of the convergence by the eddies in forcing and maintaining the zonal-mean state of the atmoshere.

14.5.3 Eliassen-Palm flux

As an application of these equations let us consider a steady, adiabatic, nondissipative flow. The first two equations (14.39) and (14.40) then reduce to

$$f[v] = \frac{1}{R\cos^2\phi}\frac{\partial}{\partial \phi}([u^*v^*]\cos^2\phi), \tag{14.43}$$

$$[\omega] = -\frac{1}{R\cos\phi}\frac{\partial}{\partial \phi}\left(\frac{[v^*\theta^*]\cos\phi}{\partial\theta_s/\partial p}\right). \tag{14.44}$$

Thus, under these restrictive conditions, the eddy terms do not influence $[u]$ and $[\theta]$ but they do influence $[v]$ and $[\omega]$. This result leads to the nonacceleration theorem established by Charney and Drazin (1961) which states that for adiabatic, nondissipative conditions eddy fluxes do not force the zonal-mean wind and temperature fields. Under the nonacceleration conditions, the momentum flux convergence is exactly counterbalanced by the Coriolis force of the mean meridional circulation [see Eq. (14.43)].

If we substitute the above $[v]$ and $[\omega]$ values into the continuity equation (14.41) we find after integration with respect to ϕ that

$$\frac{1}{f}\frac{\partial}{R\cos\phi\,\partial\phi}(R[u^*v^*]\cos^2\phi) - \frac{\partial}{\partial p}\left(R\cos\phi\,\frac{[v^*\theta^*]}{\partial\theta_s/\partial p}\right) = 0. \quad (14.45)$$

This is formally an expression for the divergence in the (ϕ, p) plane of a vector **F**. The components of this vector are

$$F^\phi = -R\cos\phi[u^*v^*] \quad (14.46a)$$

and

$$F^p = fR\cos\phi[v^*\theta^*](\partial\theta_s/\partial p)^{-1} \quad (14.46b)$$

which define the so-called Eliassen-Palm flux **F**. Its divergence is given by

$$\text{div }\mathbf{F} = \frac{1}{R\cos\phi}\frac{\partial}{\partial\phi}(F^\phi\cos\phi) + \frac{\partial}{\partial p}F^p.$$

Thus, expression (14.45) becomes $\text{div }\mathbf{F} = 0$. The vector **F** can be represented using arrows in the (ϕ, p) plane (see Sec. 14.5.6).

Let us consider now the more general case of steady motion with heat and friction included. We then find from the zonal momentum equation (14.39):

$$[v] = \frac{1}{fR\cos^2\phi}\frac{\partial[u^*v^*]\cos^2\phi}{\partial\phi} - \frac{1}{f}[F_\lambda] \quad (14.47a)$$

and from the thermodynamic equation (14.40):

$$[\omega] = -\frac{1}{(\partial\theta_s/\partial p)}\frac{\partial[v^*\theta^*]\cos\phi}{R\cos\phi\,\partial\phi} + \left(\frac{p_0}{p}\right)^\kappa\frac{[Q]}{c_p(\partial\theta_s/\partial p)}. \quad (14.47b)$$

When these expressions are inserted into the continuity equation we are led to the relation

$$\frac{\partial}{\partial p}\left\{\left(\frac{p_0}{p}\right)^\kappa\frac{[Q]}{c_p}\left(\frac{\partial\theta_s}{\partial p}\right)^{-1}\right\}$$

$$-\frac{\partial}{R^2\cos\phi\,\partial\phi}\left\{\frac{1}{f}[F_\lambda]R\cos\phi + \frac{1}{f}\text{div }\mathbf{F}\right\} = 0 \quad (14.48)$$

which combines the diabatic and friction forcings with the divergence of the E-P flux. According to Eqs. (14.47a) and (14.47b) the mean meridional circulation is forced not only by the mean heating and friction but also by the eddies. To show this more clearly, we shall now discuss the consequences of Eqs. (14.47a) and (14.47b) for the actual annual-mean conditions. Let us first consider Eq. (14.47). The earlier Fig. 11.7 of the observed flux of momentum shows that the transient eddy term dominates in the upper levels, whereas the friction term is more important in the lower levels. "Friction" may include surface drag, turbulent exchange of momentum, and possible drag due to cumulus convection (which may be of importance in the upper levels). The dominant convergence of momentum in midlatitudes causes an upper-level equatorward flow ($[v] < 0$ in the Northern Hemisphere; the upper branch of the Ferrel cell), whereas the observed divergence of momentum in tropical and subtropical latitudes induces upper-level poleward flux (the upper branch of the Hadley cell). On the other hand, at low levels and low latitudes where the easterlies prevail $[F_\lambda] > 0$ so that equatorward flow results (lower branch of the Hadley cell). In midlatitudes where $[F_\lambda] < 0$, poleward flow results (lower branch of the Ferrel cell).

Next let us consider Eq. (14.47b). The eddy convergence of heat is important in midlatitudes both at lower and upper levels as inferred from Fig. 13.5. It induces upward motions ($[\omega] < 0$, noting that $\partial\theta_s/\partial p < 0$). Divergence prevails mainly at subtropical latitudes generating subsiding motions ($[\omega] > 0$; the downward branches of the Hadley and Ferrel cells). The diabatic heating shown earlier in Fig. 13.2 generates strong upward motions ($[\omega] < 0$) in the ITCZ (ascending branch of Hadley cell) and somewhat weaker upward motions in the mid to high latitudes reinforcing the eddy effects of momentum (near the polar front). Diabatic cooling (due to radiation) is large at high latitudes and in the upper troposphere, thus inducing downward motions (downward branches of the polar and Ferrel cells).

When all these results are put together they lead to the three-cell structure shown before in Fig. 7.19. We may note that the midlatitude Ferrel cell is an indirect circulation with cold air rising and warm air sinking forced by the eddies.

Another, perhaps preferable, approach to understanding the net effects of the eddies on the mean meridional circulation is to study them through the divergence of the E-P flux in the combined equation (14.48) rather than through the separate eddy momentum and heat terms in Eqs. (14.47a) and (14.47b). This will be discussed further in the following sections.

14.5.4 Modified momentum and energy equations

In order to clarify the net effect of the disturbances on the mean flow, Andrews and McIntyre (1976) transformed the mean equations to bring out the E-P flux convergence in an explicit form by using a "residual mean circulation" defined by

$$\tilde{v} = [v] - \frac{\partial}{\partial p}\left(\frac{[v^*\theta^*]}{\partial\theta_s/\partial p}\right), \tag{14.49a}$$

$$\tilde{\omega} = [\omega] + \frac{1}{R\cos\phi}\frac{\partial}{\partial\phi}\left(\frac{[v^*\theta^*]\cos\phi}{\partial\theta_s/\partial p}\right). \tag{14.49b}$$

Formally, the $(\tilde{v}, \tilde{\omega})$ circulation is part of the mean meridional circulation which is not balanced by the convergence of the eddy enthalpy flux. Under these transformations the equation of continuity remains formally the same:

$$\frac{1}{R\cos\phi}\frac{\partial\tilde{v}\cos\phi}{\partial\phi} + \frac{\partial\tilde{\omega}}{\partial p} = 0. \tag{14.50}$$

The $[u]$ tendency equation (14.39) becomes

$$\frac{\partial[u]}{\partial t} = f\tilde{v} + [F_\lambda] + \frac{1}{R\cos\phi}\,\text{div}\,\mathbf{F} \tag{14.51}$$

and the $[\theta]$ tendency equation (14.40) becomes

$$\frac{\partial[\theta]}{\partial t} + \tilde{\omega}\frac{\partial\theta_s}{\partial p} = \left(\frac{p_0}{p}\right)^\kappa \frac{[Q]}{c_p}. \tag{14.52}$$

Thus, the net effect of the eddies in the forcing of $[u]$ and $[\theta]$ can be described by the divergence of the E-P flux. Within the quasi-geostrophic approximation div $\mathbf{F}$ represents the only internal forcing of the mean state by the disturbances, i.e., the eddy fluxes will force $[u]$ and $[\theta]$ only where div $\mathbf{F} \neq 0$.

14.5.5 Forcing of the mean meridional circulation

We can define a streamfunction ψ using Eq. (14.37) as was done in Sec. 7.4.3 such that

$$[v] = \frac{g}{2\pi R \cos \phi} \frac{\partial \psi}{\partial p}$$

and

$$[\omega] = -\frac{g}{2\pi R \cos \phi} \frac{\partial \psi}{R \partial \phi}.$$

In a similar way, we can define a streamfunction $\tilde{\psi}$ for $\tilde{v}$ and $\tilde{\omega}$ using Eq. (14.50):

$$\tilde{v} = \frac{g}{2\pi R \cos \phi} \frac{\partial \tilde{\psi}}{\partial p} \tag{14.53a}$$

and

$$\tilde{\omega} = -\frac{g}{2\pi R \cos \phi} \frac{\partial \tilde{\psi}}{R \partial \phi}. \tag{14.53b}$$

The integration of Eq. (14.53a) with the boundary condition $\tilde{\psi} = 0$ and $\tilde{\omega} = 0$ at the top of the atmosphere allows the evaluation of the streamfunction $\tilde{\psi}$:

$$\tilde{\psi} = 2\pi R \cos \phi \int_0^p \tilde{v} \, dp/g$$

$$= 2\pi R \cos \phi \int_0^p [v] \, dp/g$$

$$- \frac{2\pi R \cos \phi}{g} \frac{[v^*\theta^*]}{(\partial \theta_s/\partial p)}$$

or

$$\tilde{\psi} = \psi - \frac{2\pi R \cos \phi}{g} \frac{[v^*\theta^*]}{(\partial \theta_s/\partial p)}. \tag{14.54}$$

Using the thermal wind relationship (14.42) we can derive a diagnostic equation for $\tilde{\psi}$. In fact, if we differentiate the $\partial[u]/\partial t$ equation (14.51) with respect to p and the $\partial[\theta]/\partial t$ equation (14.52) with respect to ϕ, and if we subtract the two equations, we obtain a second-order differential equation in $\tilde{\psi}$:

$$\frac{f^2 g}{2\pi R \cos \phi} \frac{\partial^2 \tilde{\psi}}{\partial p^2} - \frac{g}{2\pi R \cos \phi \, \rho[\theta]} \frac{\partial}{R \partial \phi} \left(\frac{\partial \theta_s}{\partial p} \frac{\partial \tilde{\psi}}{R \partial \phi} \right)$$

$$= \frac{1}{\rho[T]} \frac{\partial}{R \partial \phi} \frac{[Q]}{c_p} - f \frac{\partial [F_\lambda]}{\partial p} - \frac{f}{R \cos \phi} \frac{\partial (\text{div } \mathbf{F})}{\partial p}. \tag{14.55}$$

This second-order partial differential equation in $\tilde{\psi}$ shows the interconnections between the various types of forcing phenomena and the resulting residual mean meridional circulation. The strength of the forcing depends on the eddies as well as the state of the mean flow.

An analogous equation can be obtained for ψ following the same methodology as before but using Eqs. (14.39) and 14.40):

$$\frac{f^2 g}{2\pi R \cos\phi} \frac{\partial^2 \psi}{\partial p^2} - \frac{g}{2\pi R \cos\phi \, \rho[\theta]} \frac{\partial}{R\partial\phi}\left(\frac{\partial[\theta]}{\partial p} \frac{\partial \psi}{R\partial\phi}\right)$$

$$= \frac{1}{\rho[T]} \frac{\partial}{R\partial\phi} \frac{[Q]}{c_p} - f \frac{\partial[F_\lambda]}{\partial p}$$

$$- \frac{1}{\rho[\theta]} \frac{\partial}{R\partial\phi}\left\{\frac{\partial}{R\cos\phi\,\partial\phi}([v^*\theta^*]\cos\phi)\right\} + \frac{f}{R\cos^2\phi} \frac{\partial^2}{\partial p \partial\phi}([u^*v^*]\cos^2\phi). \tag{14.56}$$

Equation (14.56) is similar to the equation given by Kuo (1956) and Pfeffer (1981). We should note that the last two eddy forcing terms on the right side of Eq. (14.56) can be nonzero even when div $\mathbf{F} = 0$.

14.5.6 Some examples of E-P flux diagrams

In order to obtain a graphical representation of the *E-P flux* $\mathbf{F}$ and its divergence in the (ϕ, p) plane, the procedures given by Edmon et al. (1980) are followed. Instead of the divergence of $\mathbf{F}$, the divergence weighted by the mass $[\Delta m = 2\pi R^2 \cos\phi \, d\phi \, (dp/g)]$ of an annular ring $d\phi \, dp$ is used:

$$\int \mathrm{div}\,\mathbf{F}\, dm = \frac{2\pi R^2}{g} \iint (\mathrm{div}\,\mathbf{F})\cos\phi \, d\phi \, dp = \iint \Delta d\phi \, dp, \tag{14.57}$$

where

$$\Delta = \frac{\partial}{\partial \phi}\left(\frac{2\pi R}{g}\cos\phi\, F^\phi\right) + \frac{\partial}{\partial p}\left(\frac{2\pi R^2}{g}\cos\phi\, F^p\right). \tag{14.58}$$

Thus, to have the appropriate vector representation of $\mathbf{F}$ in terms of the scale units of ϕ and p in the Cartesian plane (ϕ, p) the arguments of the divergence as given in expression (14.58) are used:

$$\widehat{F^\phi} = 2\pi R\, g^{-1}\cos\phi\, F^\phi = -\frac{2\pi R^2 \cos^2\phi}{g}[u^*v^*], \tag{14.59}$$

$$\widehat{F^p} = 2\pi R^2\, g^{-1}\cos\phi\, F^p = \frac{2\pi R^3 \cos^2\phi}{g} f[v^*\theta^*](\partial\theta/\partial p)^{-1}. \tag{14.60}$$

Cross sections showing the $\widehat{\mathbf{F}}$ vectors and the isolines of div $\widehat{\mathbf{F}}$ can then be constructed based on expressions (14.58), (14.59), and (14.60) using centered differences except at the bottom and top where one-sided difference schemes are used. As pointed out by Edmon et al. (1980), a pattern will look nondivergent if and only if div $\mathbf{F} = 0$.

Since we will consider seasonal mean conditions, we can separate the *E-P* fluxes into their transient and stationary components $\mathbf{F} = \mathbf{F}_{\mathrm{TE}} + \mathbf{F}_{\mathrm{SE}}$ with, e.g., $\mathbf{F}_{\mathrm{TE}} = (F^\phi_{\mathrm{TE}}, F^p_{\mathrm{TE}}) = \{-R\cos\phi\,[\overline{u'v'}],\, fR\cos\phi\,[\overline{v'\theta'}](\partial\theta_s/\partial p)^{-1}\}$. Thus, Fig. 14.9 shows cross sections of the *E-P* flux represented by arrows and contours of div $\widehat{\mathbf{F}}$ for both the transient and stationary eddy contributions for northern winter and summer conditions. The vertical component of $\widehat{\mathbf{F}}$ arises from the meridional eddy flux of heat and the horizontal component from the meridional eddy momentum flux. An upward pointing vector means that F^p is negative.

In general, maximum divergence is shown at low levels in mid and high latitudes. The vectors $\widehat{\mathbf{F}}$ point upward for the transient eddies during both winter and summer and for the stationary eddies during winter showing that the poleward heat flux is the

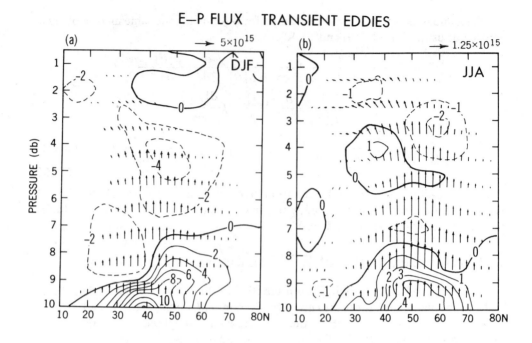
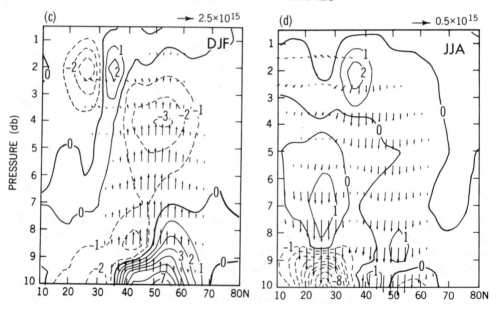

FIGURE 14.9 Cross sections of the Eliassen-Palm flux vectures $\hat{\mathbf{F}}$ plotted as arrows and of their divergence given by solid (positive) and dashed (negative) contours. Shown are the transient eddy (upper panel) and stationary eddy components (lower panel) of the *E-P fluxes* for mean northern winter and summer conditions for the period 1963–1973. Contour intervals are $2 \times 10^{15} \mathrm{m}^3$ for the transient eddy winter case and $1 \times 10^{15} \mathrm{m}^3$ for the other cases. The arrows are scaled differently in the various diagrams as indicated in the upper right-hand corner of each diagram. Each scale represents the value of the horizontal component $\hat{F}_\phi$ in m^3. The scale for the vertical component $\hat{F}_p$ is equal to the scale for $\hat{F}_\phi$ but multiplied by 62.2 kPa (1kPa = 10 mb), so that $\hat{F}_p$ is then in units of m^3 kPa.

main component of the *E-P flux* in the low and mid-troposphere and that baroclinic energy conversions dominate there. At upper levels, the arrows tilt towards the equator indicating important momentum convergence near the jet stream region.

The stationary eddy *E-P fluxes* are always smaller in magnitude than the corresponding transient fluxes. In winter the transient and stationary eddy patterns look quite similar, whereas they are very different in summer. In fact, the stationary eddy vectors tend to be directed downward in summer with highest values in the lower levels, indicating a weak equatorward heat flux.

14.6 OBSERVED ENERGY CYCLE IN THE OCEANS

14.6.1 Spatial distributions of the energy components

Based on Eqs. (14.17) and (14.18), it is possible to evaluate the available gravitational potential energy P as well as its mean zonal and stationary eddy components, P_M and P_{SE}. The data are not sufficient to give a reliable estimate of the transient eddy component P_{TE}. Thus, the mean annual values of P, P_M, and P_{SE} for the globe are shown in Table 14.2 as computed from the density variations at various levels using the vertical profile of $-\delta\tilde{\rho}/dz$ [see Fig. 8.16(a)] for the stability factor. The reference state was defined using data for the entire world ocean.

Because the seasonal variations tend to be small (much less than 10%) we show only the annual values in Fig. 14.10. The figure shows that equatorward of about 50° latitude the upper 500 m give the main contribution to the global integral of P_M. At high latitudes the deeper layers show large values but their contributions to the global integrals are not very important because of the small area involved. We see that with the exception of the intertropical regions the values of P_M are significantly larger than those of P_{SE}. The highest values of P_M are observed in mid to high latitudes with an apparent maximum in the Arctic region. However, we must note that the latitude scale in these figures is linear which, due to the sphericity of the earth, distorts the relative contribution of the various latitudes to the global integrals. Thus, the contributions of the polar regions are, in fact, small. The values of P_{SE} in Fig. 14.10 have a different latitudinal distribution from those of P_M with generally high values in the intertropical regions. These results are, of course, consistent with the geographical patterns of density shown before in Figs. 8.13(a) and 8.13(b).

As clearly shown in Fig. 14.11, the contributions to the available potential energy and its components decrease rapidly with depth in the first 100 to 200 m but decrease at a much slower rate below 200-m depth, actually showing a weak increase with depth below about 2000-m depth. The rate of decrease with depth is greater for P_M than P_{SE}. The integrated values of P_M and P_{SE} are given in Table 14.2 for the two hemispheres and the globe for the surface–1000-m and the surface–3000-m depth layers. It seems that the mean available potential energy P_M is somewhat larger in the Southern Hemisphere than in the Northern Hemisphere because of the larger north–south variance of $[\rho]$ in the Southern Hemisphere. However, for the stationary eddy component P_{SE}, the Southern Hemispheric values are smaller in view of the reduced east–west variance. This reduction is due to the greater uniformity and interconnec-

TABLE • 14.2. Estimates of the annual-mean available gravitational potential and kinetic energy components in the world ocean in terms of energy per unit area (10^5 J m^{-2}). Some energy ratios are also given (%). Inland seas, such as the Baltic Sea, Mediterranean Sea, Red Sea, Persian Gulf, and Hudson Bay have not been included.[a] The estimates of P, P_M, and P_{SE} were made for both the 0–1000- and 0–3000-m (in parentheses) layers (from Oort et al., 1989).

	NH	SH	Globe	Units
P_M	3.79	4.84	4.40	10^5 J m^{-2}
	(8.74)	(9.15)	(8.98)	
P_{SE}	1.38	1.11	1.22	10^5 J m^{-2}
	(3.89)	(2.48)	(3.06)	
P_{TE}[b]	0.66	0.32	0.46	10^5 J m^{-2}
P	5.83	6.27	6.08	10^5 J m^{-2}
	(13.29)	(11.95)	(12.50)	
K_M	0.004	0.007	0.006	10^5 J m^{-2}
K_{SE}	0.005	0.005	0.005	10^5 J m^{-2}
K_{TE}[b]	0.070	0.069	0.070	10^5 J m^{-2}
K	0.078	0.081	0.080	10^5 J m^{-2}
$K/(P+K)$	1.3	1.3	1.3	%
K_E/K	96	91	94	%
P_E/P	35	23	28	%

[a] To obtain hemispheric and global integrals the Northern Hemispheric, Southern Hemispheric, and global values should be multiplied by the appropriate ocean surface areas, namely 1.47, 2.05, and 3.52×10^{14} m^2, respectively.

[b] All time scales of variations less than one year (such as the annual cycle and mesoscale eddies) are included as transient eddies in the estimates of P_{TE} and K_{TE}.

tedness of the southern oceans compared to the northern oceans. The relative contribution of P_{SE} to the total available potential energy is about 20%.

Using a linear dependence of the density on temperature $\rho = \rho_0 (1 - \alpha T)$ with $\rho_0 = 1029$ kg m^{-3} and $\alpha = 0.000\,25$ °C^{-1}, we can obtain an approximate expression for P_{TE} that is more suitable for use with the observed temperature data:

$$P_{TE} = -\frac{1}{2} \int \frac{g \rho_0^2 \, \overline{T'^2} \alpha^2}{(\delta \tilde{\rho}/dz)} \, dV. \tag{14.61}$$

Using the annual-mean fields of $\overline{T'^2}$ at the surface presented before in Fig. 8.6(a) and vertical extrapolation, we can give a tentative global estimate for P_{TE} of 0.46×10^5 J m^{-2} (Oort et al., 1989).

We will now make some comparisons of the ocean values for the global available potential energy in the 0–1000-m depth layer with the corresponding annual-mean values for the atmosphere. We choose to use the 0–1000-m layer values rather than the values for the entire depth of the ocean because the direct interactions with the

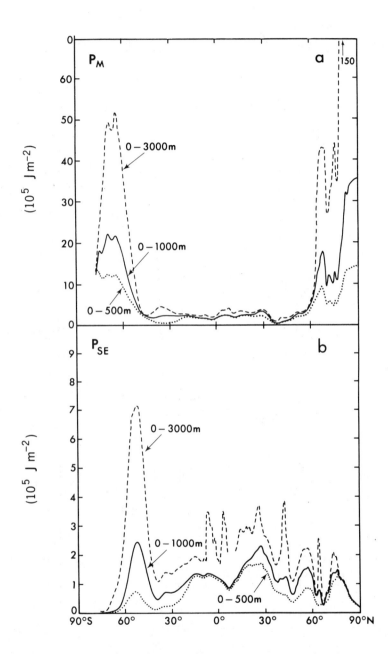

FIGURE 14.10. Meridional profiles of the contributions to the global integrals of the mean (a) and stationary eddy (b) available potential energy for annual-mean conditions for three layers 0–500, 0–1000, and 0–3000 m in the oceans in units of 10^5 J m^{-2} (after Oort *et al.*, 1989).

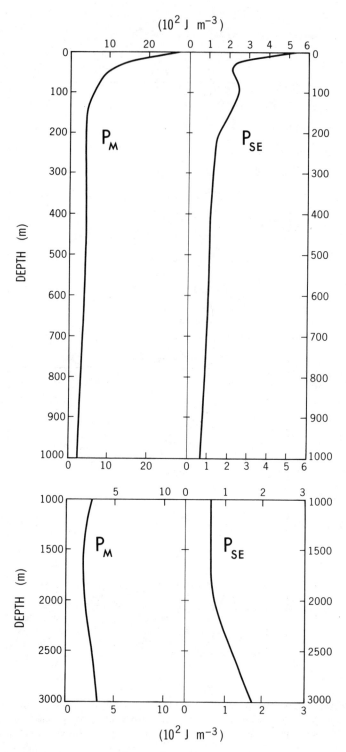

FIGURE 14.11. Vertical profiles of the contributions to the global integrals of the mean and stationary eddy available potential energy for annual-mean conditions in the oceans in units of 10^2 J m^{-3}. Note that the scales are different for the 0–1000 and 1000–3000 m layers.

atmosphere on decadal and shorter time scales are probably limited to the first 1000 m of the oceans:

$$\frac{P^O}{P^A} = \frac{6.08}{44.4} \times 0.69 = 0.09,$$

$$\frac{P^O_M}{P^A_M} = \frac{4.40}{33.3} \times 0.69 = 0.09,$$

$$\frac{P^O_{SE}}{P^A_{SE}} = \frac{1.22}{0.8} \times 0.69 = 1.0,$$

$$\frac{P^O_{TE}}{P^A_{TE}} = \frac{0.46}{10.3} \times 0.69 = 0.03,$$

where the multiplication factor 0.69 represents the fraction of the earth's surface taken up by the oceans. These ratios show that, in terms of available potential energy, the oceans are much less energetic ($\leqslant 10\%$) than the atmosphere, except for the stationary eddy component that is of relatively minor importance in both media in case of annual-mean statistics. The ratio of the transient eddy components shows how much more turbulent the atmosphere is compared with the oceans.

In order to get some insight regarding the differences in available potential energy for the atmosphere and oceans, we will compare the various factors that enter into expression (14.17) for P. For this purpose let us take for the atmospheric density variations a typical value of $\rho - \tilde{\rho} \sim 10^{-1}$ kg m^{-3} and for the mean stability throughout the depth of the atmosphere ($\sim 10\,000$ m) a value of $-\delta\tilde{\rho}/dz \sim 10^{-4}$ kg m^{-4}. For the corresponding oceanic parameters we will assume in the upper 200-m layer, where most of the activity occurs, a typical value of $\rho - \tilde{\rho} \sim 2$ kg m^{-3} and very small density variations below 200 m. For the mean stability we will assume a value of 0.8×10^{-2} kg m^{-4}.

Using these estimates we obtain for the ratio of the total available potential energy in the oceans and in the atmosphere a value of the order of 1/10 which agrees well with the ratio $P^O/P^A = 0.09$ found from actual observations. These ratios show that the two most important factors that explain the small ratio P^O/P^A are the larger stability in the first few 100 m's of the oceans and the smaller oceanic depth range in which the active processes take place. These two factors overcome the larger observed density variations in the oceans compared to those in the atmosphere.

Using ocean current data one can, in principle, also estimate the various modes of kinetic energy in the oceans using Eq. (14.19). However, no direct subsurface current observations are as yet available on a global scale so that one has to resort to a tentative vertical extrapolation scheme to obtain global kinetic energy values. The resulting values are shown in Table 14.2 but should be taken as tentative first estimates. Nevertheless, the values seem reasonable when compared with the global kinetic energy value of 0.05×10^5 J m^{-2} that would result with a uniform ocean velocity of 5 cm/s. The meridional profiles of the surface kinetic energy presented earlier in Fig. 8.22 showed that the kinetic energy is mainly contained in the transient eddies with maxima in the equatorial and subpolar regions.

Comparing the ratio of the kinetic and available potential energy in the oceans with the corresponding ratio in the atmosphere we find a much smaller value in the oceans:

$$(K/P)^O = 0.013 \quad \text{and} \quad (K/P)^A = 0.27.$$

These values imply that, in the hypothetical case that all external energy generating processes were cut off, the general circulation of the oceans would have a much larger relative source of energy than the atmospheric general circulation. Therefore, we would also expect the oceans to have a much longer intrinsic time scale than the atmosphere as we shall discuss further in the next section.

14.6.2 The energy cycle

A formal set of balance equations similar to Eqs. (14.23), (14.27), (14.30), and (14.33) for the atmosphere can be written to describe the energy cycle for the oceans. However, because of our very limited knowledge of the oceans, we will discuss here only the budgets of total available potential and total kinetic energy:

$$\frac{\partial P}{\partial t} = G(P) - C(P,K) - D(P), \tag{14.62}$$

$$\frac{\partial K}{\partial t} = G(K) + C(P,K) - D(K). \tag{14.63}$$

Since the dynamical coupling with the atmosphere is so important for the oceans we have written this effect explicitly as $G(K)$ rather than including it in the dissipation term $D(K)$ as was done for the atmosphere. The term $D(P)$ gives the rate of "dissipation" of available potential energy due to molecular and small-scale eddy mixing. Except in coarse-resolution general circulation models (see Chap. 17) $D(P)$ can be neglected as being very small.

For annual-mean conditions, the time rate of change terms on the left-hand side of Eqs. (14.62) and (14.63) are also very small so that the balance equations can be reduced to

$$0 = G(P) - C(P,K), \tag{14.64}$$

$$0 = G(K) + C(P,K) - D(K). \tag{14.65}$$

Let us first consider the rate of generation of available potential energy $G(P)$. This generation results from the processes that tend to increase the density of the surface waters in those regions where the density is high relative to the reference state, and to decrease the density of the surface waters where the density is relatively low. Thus, we can write

$$G(P) = -g \int \frac{(\partial \rho/\partial t)_{\text{sfc}}(\rho - \tilde{\rho})}{(\delta \tilde{\rho}/dz)} dV, \tag{14.66}$$

where

$$\left(\frac{\partial \rho}{\partial t}\right)_{\text{sfc}} = \left(\frac{\partial \rho}{\partial T}\right)\left(\frac{\partial T}{\partial t}\right)_{\text{sfc}} + \left(\frac{\partial \rho}{\partial S}\right)\left(\frac{\partial S}{\partial t}\right)_{\text{sfc}}$$

represents the tendency of the surface density to change due to the combined effects of (1) the net heating or cooling associated with the transfer of energy in the form of radiation, sensible heat, and latent heat across the atmosphere–ocean interface, and (2) the net freshening or saltening of the ocean by precipitation, evaporation, and runoff from land. These two types of air–sea interactions, together with advection effects by ocean currents, must be to a large extent responsible for generating the observed spatial and temporal variations in the surface density and the resulting baroclinicity in the oceans.

The observed surface density profile in Fig. 8.15 together with the annual profiles of the surface energy exchange in Fig. 13.24 and of $E - P$ in Fig. 12.16 can now be used to estimate $G(P_M)$ which must be the main term in $G(P)$. Thus, we find that the energy flux is the dominant contributor and that the net rate of generation of available potential energy due to both effects is on the order of 0.002 W m^{-2}.

The generation of kinetic energy in the oceans $G(K)$ results from the wind forcing through tangential stresses at the ocean surface (see Sec. 10.5):

$$G(K) = -\int \tau_0 \cdot \mathbf{v}_{oc} \, dA. \tag{14.67}$$

It is computed as the product of the atmospheric wind stress and the surface ocean velocity. This term represents only a small fraction of the total energy input by the winds because most ($\sim$99%) of the input is probably dissipated in the atmosphere–ocean boundary layer by turbulence, swell, waves, etc. Based on the data presented in Fig. 10.6, the wind energy input for the global ocean is found to be $G(K) = 0.007$ W m^{-2}.

The conversion term $C(P,K)$ that links the available potential and kinetic energy through the thermohaline circulations is given by

$$C(P,K) = -\int \rho g w \, dV. \tag{14.68}$$

It represents the rate at which work is done by the buoyancy forces in the oceans. No direct estimates of this conversion rate are available. However, we can estimate $C(P,K)$ from Eq. (14.64) to be equal to $G(P)$, i.e., about 0.002 W m^{-2}. Alternatively, we can use scaling arguments. For example, assuming in a 100-m thick layer a typical density variation of 1 kg m^{-3}, a typical vertical velocity fluctuation of 5×10^{-6} m s^{-1}, and a correlation of 0.4, we find for $C(P,K)$ the same value of 0.002 W m^{-2}.

Finally, for the rate of dissipation of kinetic energy $D(K)$ we obtain a residual value of 0.009 W m^{-2}. This dissipation must take place mainly through friction at the lateral boundaries and at the bottom of the oceans. The average dissipation time scale for the top 1000 m of the world ocean can now be calculated to be on the order of

$$\frac{(P+K)}{D(K)} = \frac{6.1 \times 10^5}{0.009} \text{ s}$$

$$\approx 780 \text{ days or about 2 years.}$$

Of course, the oceanic time scales must be shorter than two years in the surface layers, probably on the order of months, and much longer in the deep ocean, perhaps on the order of decades to centuries. In comparison we find for the atmosphere a time scale of only

$$\frac{(P+K)}{D(K)} = \frac{56.2 \times 10^5}{1.88} \text{ s}$$

$$= 35 \text{ days or about one month.}$$

The estimated values for the various terms in Eqs. (14.64) and (14.65) are shown together in a box diagram in Fig. 14.12. The energy amounts inside the boxes are taken from Table 14.2.

In comparing the energy cycles for the atmosphere in Fig. 14.8 and for the oceans in Fig. 14.12, we notice that the atmosphere is much more active in converting energy between its various forms. In fact, the generation of available potential energy $G(P)$,

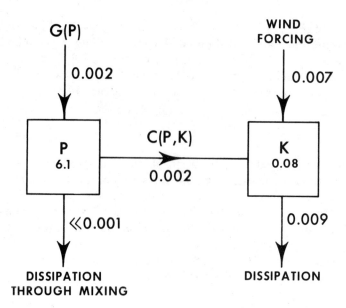

FIGURE 14.12. Schematic box diagram of the energy balance in the oceans. The energy amounts inside the boxes are in units of 10^5 J m^{-2}, and the generation, conversion, and dissipation rates in W m^{-2}.

the conversion from available potential to kinetic energy $C\,(P,K)$, and the dissipation of kinetic energy $D\,(K)$ are all on the order of 2 W m^{-2} in the atmosphere, whereas the corresponding values in the oceans appear to range between 0.002 and 0.009 W m^{-2}, i.e., two to three orders of magnitude smaller.

One of the central questions in oceanography concerns the relative importance of the surface winds [i.e., $G(K)$], or the thermohaline effects [i.e., $G(P)$ and $C\,(P,K)$] in maintaining the oceanic general circulation against dissipation through friction and turbulence. The flow diagram in Fig. 14.12, even with the uncertainties involved, suggests that the ocean currents are driven mainly by atmospheric forcing of the winds rather than by internal conversions in the oceans themselves.

CHAPTER 15

Entropy in the Climate System

15.1 INTRODUCTION

Most natural phenomena occurring in the climate system are characterized by great irreversibility and evolve in time with a marked increase in entropy (as we have seen in Sec. 2.4.1). For example, turbulent motions in the planetary boundary layer do not spontaneously develop into the large-scale organized flow of the general circulation. Neither can a cloud be reconstituted from the same water it lost previously through precipitation. Nor do rivers flow backwards from the sea to their headwaters. Ocean water does not decompose spontaneously into oxygen and hydrogen. None of these phenomena occur naturally in the climate system. Several papers dealing with the entropy budget of the atmosphere have been published before. Among them we can mention Wulf and Davis (1952), Lettau (1954), Dutton and Johnson (1967), Dutton (1976), Paltridge (1975), Fortak (1979), and Johnson (1989).

In this chapter, we will analyze some of the consequences of the second law of thermodynamics for the behavior of the climate system and, in particular, for atmospheric phenomena following the approach used by Peixoto *et al.* (1991).

The second law of thermodynamics implies that energy can only change from a higher to a lower level of availability, i.e., energy can only change from a more to a less usable or a less ordered form. The second law implies the existence of a function s, the entropy, that for an isolated system increases monotonically until it reaches its maximum value at the state of thermodynamic equilibrium:

$$\frac{ds}{dt} \geq 0.$$

An increase in entropy means a decrease in available energy and evolution toward a state of greater disorder.

The second law can also be extended to open systems that exchange energy and matter with their surroundings. We must then consider two components in the total entropy change dS. One component $d_e S$ represents the transfer of entropy across the boundaries of the open system and the other component $d_i S$ is the total entropy produced within the system. According to the second law, the rate of generation of entropy inside the system is always positive (see Fig. 15.1) so that (Prigogine, 1962)

$$\frac{dS}{dt} = \frac{d_e S}{dt} + \frac{d_i S}{dt} \tag{15.1}$$

and

$$\frac{d_i S}{dt} \geq 0. \tag{15.2}$$

Using this formulation the basic differences between reversible and irreversible processes become more clear. Only irreversible processes contribute to entropy production.

There are numerous irreversible processes in the climate system that lead to an increase of entropy. Among these are the absorption of solar and terrestrial radiation, melting of snow and ice, condensation, evaporation, erosion by the winds and running water, and the turbulent and molecular diffusion of gases, heat, and momentum.

How can we explain the high level of organization in certain atmospheric processes characterizing the weather, and the fact that these processes can evolve in an orderly way that, at first sight, would seem to lead to a decrease of entropy. For example, the zonal wind systems in the atmospheric general circulation are well-defined and highly organized; millions of tons of water evaporate each second from the earth's surface and are lifted up into the atmosphere against the force of gravity to feed the hydrological cycle; photosynthesis allows plants to grow year after year by absorbing carbon dioxide in the presence of sunlight. All these processes seem to proceed against the law of increase of entropy, and all occur due to the high quality (i.e., low entropy) of the incoming solar radiation and its systematic variation with latitude. It is, of course, the solar energy that heats the earth's surface and the atmosphere unevenly, thereby generating the global wind systems, producing the evaporation of water as one of the vital components of the hydrological cycle, and maintaining photosynthesis, among many other processes.

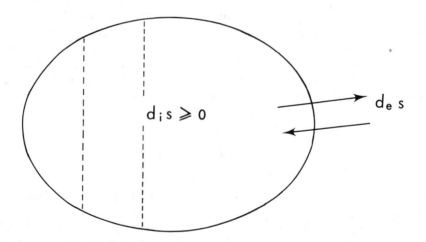

FIGURE 15.1. Schematic diagram of the climate system showing the transfer of entropy across the boundaries and the production of entropy inside the system.

Let us compare the quality of the short-wave solar radiation ($\lambda_{max} \approx 0.48\,\mu m$) with the quality of the long-wave radiation ($\lambda_{max} \approx 10\,\mu m$) emitted by the earth. According to the Planck equation, the energy of a photon ϵ is given by $\epsilon = h\nu$, and thus is inversely proportional to the wavelength ($\lambda = c/\nu$). Thus solar photons are richer in energy than terrestrial photons. In other words, the amount of entropy associated with the incoming solar radiation is much lower than the amount of entropy associated with the emitted terrestrial radiation, and the climate system receives high-quality "rich" energy and returns low-quality "impoverished" energy to space. Thus, solar radiation revitalizes the meteorological phenomena, feeds the hydrological cycle, and renovates the biosphere. If the earth were an isolated system there would be an unavoidable increase in entropy leading to a death-like uniformity of the planet earth. It is this capacity for permanent renovation that makes all natural phenomena possible in the climate system.

Most quantities in the physical world can increase or decrease with time, but entropy must always increase with time. The entropy can decrease locally during a given time interval, but only at the expense of a larger increase of entropy in the environment so that it results in a net increase in the global entropy. Entropy is "time's arrow," in the words of Eddington. It gives the direction in which time flows but it does not provide the rate at which time is increasing, so it cannot be used as a clock. Sometimes entropy increases more rapidly, and at other times more slowly; only rarely does it remain constant. The second law does not provide the speed of degradation.

Comparing the rate of degradation of energy by the different life forms on earth, we know that plant life can use the solar energy directly in photosynthesis to grow and reproduce itself, leading to a relatively slow rate of increase in entropy. Animal life, on the other hand, requires the consumption of plants or other animals to sustain itself, thereby consuming much more of the available solar energy than plants and accelerating the rate of increase of entropy and degradation of solar energy.

15.2 BALANCE EQUATION OF ENTROPY

The rate of change of entropy per unit mass is given by Eq. (3.44):

$$\frac{ds}{dt} = Q/T,$$

where Q is the rate of diabatic heating per unit mass. As we have seen before in Eq. (13.7), the diabatic heating includes radiational heating, the release of latent heat, and the heating by conduction and frictional dissipation:

$$Q = Q_h + Q_f = -\frac{1}{\rho}(\text{div}\,\mathbf{F}_{rad} + \text{div}\,\mathbf{J}_H^D)$$

$$-L(e-c) - \frac{1}{\rho}\tau{:}\text{grad}\,\mathbf{c}.$$

In the long run for the total atmosphere the entropy has to remain constant so that for the long-term mean (indicated by an overbar):

$$\int_V \overline{\rho \frac{ds}{dt}} \, dV = 0,$$

and

$$\int_V \overline{\rho \left(\frac{Q_h + Q_f}{T} \right)} \, dV = 0,$$

or, in terms of pressure, using the Poisson equation (3.47):

$$\int_V \overline{\rho \left(\frac{Q_h + Q_f}{p^\kappa} \right)} \, dV = 0,$$

where V is the volume of the global atmosphere.

Since T, p, and Q_f are always positive the integrals

$$\int_V \overline{\rho \frac{Q_f}{T}} \, dV$$

and

$$\int_V \overline{\rho \frac{Q_f}{p^\kappa}} \, dV$$

are also greater than zero, leading to

$$\int_V \overline{\rho \frac{Q_h}{T}} \, dV < 0 \qquad (15.3)$$

and

$$\int_V \overline{\rho \frac{Q_h}{p^\kappa}} \, dV < 0. \qquad (15.4)$$

Therefore, the Q_h and T or p fields have to be positively correlated, i.e., high values of Q_h must occur at high temperatures and high pressures and vice versa. This is what is actually observed in the atmosphere, namely heating occurs at low levels and cooling at high levels, together with heating in low latitudes and cooling in high latitudes (available potential energy is generated). In other words, the general circulation can be maintained against the disordering effects of friction only if we heat the warmer regions and cool the colder ones. A similar conclusion was reached in the previous chapter on available potential energy.

A balance equation of entropy can now be obtained by substituting the expression for diabatic heating (13.7) into Eq. (3.44):

$$\rho \frac{ds}{dt} = - \frac{\mathrm{div}\, \mathbf{F}_{\mathrm{rad}}}{T} - \frac{\mathrm{div}\, \mathbf{J}_H^D}{T} - L \frac{\rho(e-c)}{T} - \frac{\tau : \mathrm{grad}\, \mathbf{c}}{T} \qquad (15.5)$$

or

$$\rho \frac{ds}{dt} = - \mathrm{div}\, \frac{1}{T} \left(\mathbf{F}_{\mathrm{rad}} + \mathbf{J}_H^D \right) - \frac{1}{T^2} \left(\mathbf{F}_{\mathrm{rad}} + \mathbf{J}_H^D \right) \cdot \mathrm{grad}\, T$$

$$- L \frac{\rho(e-c)}{T} - \frac{\tau : \mathrm{grad}\, \mathbf{c}}{T}. \qquad (15.6)$$

Let us expand the total derivative on the left-hand side of Eqs. (15.5) and (15.6) using

$$\rho \frac{ds}{dt} = \frac{\partial \rho s}{\partial t} + \text{div}(\rho s \mathbf{c})$$

and let us integrate the two equations over the entire atmosphere. This leads to

$$\frac{\partial}{\partial t} \int_V \rho s \, dV = -\int_A \rho s \mathbf{c} \cdot \mathbf{n} \, dA - \int_V \frac{1}{T} \{\text{div } \mathbf{F}_{\text{rad}}$$

$$+ \text{div } \mathbf{J}_H^D + L \rho (e - c) + \tau : \text{grad } \mathbf{c}\} \, dV \quad (15.7)$$

and

$$\frac{\partial}{\partial t} \int_V \rho s \, dV = -\int_A \rho s \mathbf{c} \cdot \mathbf{n} \, dA - \int_A \frac{1}{T} \mathscr{F} \cdot \mathbf{n} \, dA - \int_V \frac{1}{T^2} (\mathscr{F} \cdot \text{grad } T) dV$$

$$- \int_V \frac{L}{T} \rho (e - c) dV - \int_V \frac{1}{T} (\tau : \text{grad } \mathbf{c}) dV, \quad (15.8)$$

where

$$\mathscr{F} = \mathbf{F}_{\text{rad}} + \mathbf{J}_H^D, \quad (15.9)$$

A is the area of the globe, and $\mathbf{n}$ is the unit vector directed outward at right angles to the boundaries of the atmosphere.

The first term on the right-hand side of Eq. (15.8) is associated with the net mass transport across the boundaries. This term vanishes at the top of the atmosphere since ρ decreases exponentially with height and s increases linearly; it also vanishes at the earth's surface where the normal component of the velocity $\mathbf{c} \cdot \mathbf{n}$ equals zero.

We may point out that Eq. (15.8) now has a form similar to that of Eq. (15.1), in which the transfer terms across the boundaries are separated from the generation terms within the system. Thus, the first two terms on the right-hand side of Eq. (15.8) represent the transport effect $d_e S/dt$ and the last three terms the generation effect $d_i S/dt$. The third and fifth terms have a bi-linear form in which one of the components of the product is a flux of a given quantity and the other is the gradient of the conjugate intensive state variable (de Groot and Mazur, 1984). For the climatic system these generation terms consist of the products of the rates at which the various irreversible processes take place (radiation, heat flow, diffusion, etc.) and the corresponding generalized force, e.g., the gradient of temperature. Equivalent versions of the preceding equations can be found in Batchelor (1957), Dutton (1976), and de Groot and Mazur (1984).

Equations (15.7) and (15.8) are only approximate equations since they do not take into proper account the entropy of the radiant energy fields. This constitutes a difficult problem to deal with outside equilibrium conditions because the nonequilibrium radiation entropy is not related in a simple way to the black body radiation entropy. However, some attempts have been made recently to evaluate the radiation entropy (see, e.g., Essex, 1984, and Lesins, 1990).

For an isolated system only the generation terms will remain, and their sum must always be positive. However, the individual generation terms may not be positive everywhere as possibly occurs in the case of radiation. Nevertheless, the latent heat release, frictional dissipation, and heat diffusion terms are always positive (Batchelor,

1967). This is easy to prove in the case of the heat diffusion term by assuming a type of Fickian law for $\mathbf{J}_H^D$,

$$\mathbf{J}_H^D = -K \operatorname{grad} T,$$

where K is the thermal conductivity. Its contribution to the entropy generation is then always positive since

$$-\int_V \frac{1}{T^2} \mathbf{J}_H^D \cdot \operatorname{grad} T \, dV = \int_V \frac{K}{T^2} (\operatorname{grad} T)^2 dV > 0.$$

It is convenient to rewrite the various boundary terms in Eq. (15.8) introducing an appropriate reference or equivalent temperature T^* for each component so that

$$\frac{1}{T^*} \equiv \frac{\int_A (1/T) \mathscr{F} \cdot \mathbf{n} \, dA}{\int_A \mathscr{F} \cdot \mathbf{n} \, dA}. \tag{15.10}$$

For example, for the radiation we find

$$-\int_A \frac{\mathbf{F}_{\text{rad}} \cdot \mathbf{n}}{T} dA = -\frac{1}{T^*_{\text{rad}}} \int_A \mathbf{F}_{\text{rad}} \cdot \mathbf{n} \, dA = G_{\text{rad}}/T^*_{\text{rad}}, \tag{15.11}$$

where G_{rad} is the net radiative flux across the boundaries into the atmosphere and T^*_{rad} the reference radiative temperature. For the other boundary terms we can obtain similar expressions so that Eq. (15.8) can be rewritten symbolically:

$$\frac{\partial S}{\partial t} = \frac{G_{\text{rad}}}{T^*_{\text{rad}}} + \frac{G_{\text{SH}}}{T^*_{\text{SH}}} + \sigma_{\text{rad}} + \sigma_{\text{LH}} + \sigma_{\text{SH}} + \sigma_{\text{dis}}. \tag{15.12}$$

In this equation, S is the total entropy of the atmosphere and $\sigma_{\text{rad}}, \sigma_{\text{LH}}, \sigma_{\text{SH}},$ and σ_{dis} denote the rate of generation of entropy by radiation, latent heat release, conduction of heat, and kinetic energy dissipation, respectively, which are all associated with irreversible processes in the atmosphere.

For the atmosphere as a whole, we will assume that the net rate at which latent heat is released can be approximated by the product of the observed precipitation rate at the earth's surface and the latent heat of condensation:

$$-L \int_0^\infty \rho(e-c) dz \approx \rho L P.$$

A large generation of entropy σ_{LH} must be connected with this phase transition. The mixing of tropical and polar air masses (with different temperatures and humidities) also contributes to the generation of entropy through the diffusion terms in Eq. (15.12).

For a sufficiently long interval of time we may assume that, on the average, the atmosphere is in a steady state. Therefore, averaging of the global entropy balance equation (15.12) over time leads to

$$\overline{\left(\frac{G_{\text{rad}}}{T^*_{\text{rad}}}\right)} + \overline{\left(\frac{G_{\text{SH}}}{T^*_{\text{SH}}}\right)} + \bar{\sigma}_{\text{rad}} + \bar{\sigma}_{\text{LH}} + \bar{\sigma}_{\text{SH}} + \bar{\sigma}_{\text{dis}} = 0. \tag{15.13}$$

Under steady-state conditions, the corresponding energy terms must also obey the energy balance equation

$$\bar{G}_{\text{rad}} + \bar{G}_{\text{LH}} + \bar{G}_{\text{SH}} = 0, \tag{15.14}$$

where $G_{\text{LH}} = -\int \mathbf{F}_{\text{LH}} \cdot \mathbf{n} \, dA$ represents the latent heat flux.

Let us now further discuss the reference temperature. Although this temperature has no direct physical interpretation, it gives some indication of the average temperature at which the processes occur. If we assume that the fluctuations of the reference temperatures and of G are small, i.e., $T'/\overline{T^*} \ll 1$ and $G'/G \ll 1$, then

$$\overline{G/T^*} = \overline{G}/\overline{T^*} - \overline{G'T'}/\overline{T^*}^2, \tag{15.15}$$

where $T' \equiv T^* - \overline{T^*}$ and $\overline{T^*}$ is the time average of the instantaneous reference temperature and not the reference temperature of the mean state, and similarly for G. The first term on the right-hand side of Eq. (15.15) represents the contribution of the time-mean flux of entropy and the second term the transient eddy contribution. Since $\overline{T^*}$ is very large compared to the fluctuations the last term in Eq. (15.15) can be neglected. Further, as preliminary calculations show, we can use the globally averaged mean temperature $\widetilde{T}$ instead of $\overline{T^*}$. Thus, Eq. (15.13) can be written as

$$\overline{G}_{\text{rad}}/\widetilde{T}_{\text{rad}} + \overline{G}_{\text{SH}}/\widetilde{T}_{\text{SH}} + \overline{\sigma}_{\text{rad}} + \overline{\sigma}_{\text{LH}} + \overline{\sigma}_{\text{SH}} + \overline{\sigma}_{\text{dis}} = 0, \tag{15.16}$$

which is used in the following calculations.

15.3 OBSERVED ENTROPY BUDGET OF THE ATMOSPHERE

15.3.1 Global entropy budget

To evaluate the entropy budget in the atmosphere we begin by estimating the first two terms in Eq. (15.16) that represent the fluxes of entropy across the top and bottom boundaries of the atmosphere. As shown in Eqs. (15.10) and (15.11), this requires computing both the mean energy fluxes G across the boundaries and the corresponding (energy-flux weighted) reference temperatures T^* for these boundaries.

To do the calculations properly, we would need to know the global distributions of both the temperature and radiation fluxes at the top of the atmosphere as well as the global distributions of the temperature, and radiation and sensible heat fluxes at the earth's surface. At the top of the atmosphere the radiation fields are reasonably well known from satellite observations [see Figs. 6.10(b) and 6.10(c)], but at the surface the flux fields are only poorly known (see Fig. 6.12 for radiation and Fig. 10.8 for sensible heat). The temperature distributions in the atmosphere are well known from the rawinsonde station analyses.

The annual-mean energy flux values at the top and bottom of the atmosphere given before in Figs. 6.3 and 14.1 (as percentages of the total incoming solar radiation) can be used to estimate the values of the entropy flux terms, assuming that the mean incident solar radiation is 340 W m^{-2} (solar constant of 1360 W m^{-2}; see Sec. 6.8.1). The corresponding reference temperatures could be estimated by taking into account the characteristic global distributions of the various fluxes and temperature fields, but as mentioned earlier, the averaged temperatures will be used instead. The results are presented in Fig. 15.2. The figure shows the chosen values of the mean temperatures, the energy fluxes, and the corresponding entropy fluxes at the top and bottom of the atmosphere. The mean temperatures for the downward solar radiation flux and the upward fluxes at the earth's surface were assumed to be 5760 (the

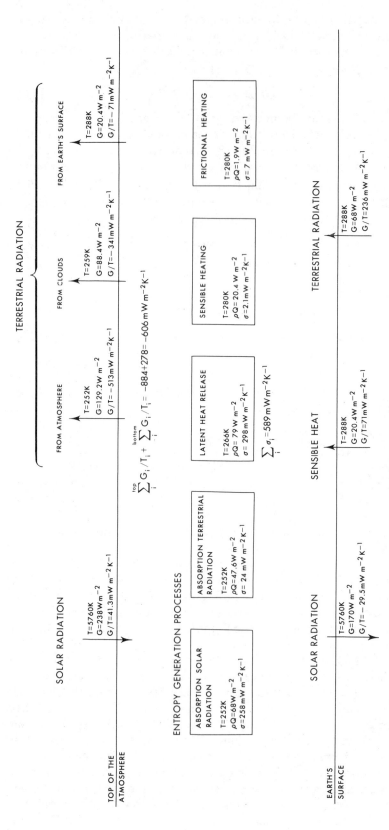

FIGURE 15.2. Entropy budget of the global atmosphere estimated for annual-mean conditions. Shown are the equivalent temperature T^* in K, the energy flux G in W m^{-2}, and the entropy flux G/T^* in units of mW m^{-2} K^{-1} for each energy component at the top of the atmosphere and at the earth's surface. A positive (negative) sign is used when the atmosphere gains (loses) entropy. The boxes in the middle of the figure contain for each component estimates of the atmospheric temperature at which the absorption takes place in K, the rate at which energy is absorbed ρQ in W m^{-2}, and the rate of entropy production σ in units of mW m^{-2} K^{-1} (from Peixoto et al., 1991).

effective temperature of the sun) and 288 K, respectively. We will use a positive sign for the entropy flux when it represents a gain of entropy and a negative sign in the case of a loss of entropy for the atmosphere.

The three terms in the upward flux of entropy at the top of the atmosphere are associated with the fluxes of infrared radiation emitted by the earth's surface (through the spectral window 8.5–11.0 μm), the atmosphere, and the clouds. For the emission of infrared radiation by the atmosphere a mean temperature of 252 K was used and for the emission by clouds a value of 259 K.

At the bottom of the atmosphere we estimated the values of the fluxes of entropy associated with the downward flux of solar radiation (-29.5 mW m^{-2} K^{-1}), the upward sensible heat flux (71 mW m^{-2} K^{-1}), and the emitted long-wave radiation (236 mW m^{-2} K^{-1}). As expected from the high quality of the incoming solar radiation the fluxes of entropy in the solar radiation are small both at the top and bottom of the atmosphere with values of 41.3 and -29.5 mW m^{-2} K^{-1}, respectively.

When we add all boundary fluxes together, the total flux of entropy at the top of the atmosphere is -884 mW m^{-2} K^{-1} and the total flux at the bottom of the atmosphere is 278 mW m^{-2} K^{-1}. Thus, according to Eq. (15.13) the mean rate of generation of entropy by internal processes in the atmosphere under steady conditions must be $\sigma = 884 - 278 = 606$ mW m^{-2} K^{-1}.

As we see from the values given in Fig. 15.2, the total amount of entropy exported by the climate system to space is $-71 - 513 - 341 = -925$ mW m^{-2} K^{-1}. This value is 22 times the amount of entropy imported by the incoming solar radiation at the top of the atmosphere (41.3 mW m^{-2} K^{-1}).

The largest fluxes of entropy are associated with the long-wave radiation fluxes (236 mW m^{-2} K^{-1} at the earth's surface and -925 mW m^{-2} K^{-1} at the top). The flux of entropy associated with turbulent and molecular diffusion processes is much smaller (71 mW m^{-2} K^{-1}).

Another approach is to directly compute the rate of production of entropy within the atmosphere by evaluating the individual contributions by the irreversible processes that occur inside the atmosphere. We will again make use of the energy flux values given in Figs. 6.3 and 14.1, and consider as an example the computation of the generation of entropy due to the absorption of solar radiation. This term can be evaluated assuming that the absorption occurs near the 500-mb level where the temperature is about 252 K, so that

$$\sigma_{\text{rad}}^{\text{SW}} = \frac{68}{252} - \frac{68}{5760} = 258 \text{ mW m}^{-2} \text{ K}^{-1}.$$

For the latent heat release we assume a net precipitation rate of 1 m/yr and a condensation temperature of 266 K so that

$$\sigma_{\text{LH}} = \frac{\rho L P}{T} = \frac{79}{266} = 298 \text{ mW m}^{-2} \text{ K}^{-1},$$

which represents the largest entropy generation term. For the sensible heat exchange processes we assumed an average temperature of 280 K as representative of the bottom layer of the atmosphere. The estimated temperatures, the rate of diabatic heating, and the rate of entropy generation for the various individual processes are shown in the rectangular boxes in the middle of Fig. 15.2.

Summing the entropy generation rates due to the various processes we find a value of 589 mW m^{-2} K^{-1} for the total mean rate of generation. This value can be compared with the previous value of 606 mW m^{-2} K^{-1} obtained from the global budget of the atmospheric energy. The difference is only 17 mW m^{-2} K^{-1} and can be attributed to (apart from errors) the approximations made and some neglected effects (e.g., mixing of air masses).

15.3.2 Regional entropy budgets

To obtain information about the distribution of the entropy sources and sinks in the atmosphere we present some results of the budgets on a regional scale. Due to the sphericity of the earth and the nonhomogeneous distribution of water and land over the globe there are large differences in insolation and in the entropy production both with respect to latitude and longitude. However, the differences in solar heating are, of course, most pronounced in the latitudinal direction and in particular between the equator and the poles. Thus, the entropy budgets for two limited regions will be compared, i.e., an equatorial zone bounded by latitudinal walls at 15 °N and 15 °S and a polar cap northward of 70 °N.

Conceptually the only difference between these regional budgets and the global budget comes from the existence of lateral boundaries. Thus, we must include the entropy fluxes across the lateral walls as new terms. These entropy fluxes are related mainly to the fluxes of potential energy and sensible heat since the radiative fluxes and the turbulent transfer of heat across the lateral boundaries are negligible.

The results for the horizontal and vertical fluxes across the boundaries are presented in Table 15.1 for the globe and the two regions. There is a lateral export of entropy away from the equatorial belt (-156 mW m^{-2} K^{-1}). On the other hand, we find a strong lateral inflow of entropy across 70 °N into the polar region (267 mW m^{-2} K^{-1}). The values at the top of the atmosphere show that the export of entropy to space is larger in the equatorial than in the polar atmosphere.

The sums of the boundary fluxes in the last column of Table 15.1 give a rough estimate of the net rate of generation of entropy inside each region. However, in view of their importance, the directly estimated rates of entropy generation by the various generating processes are shown in Table 15.2. The regional friction effects are not well known. However, we have assumed here that the regional values are the same as the global value of 7 mW m^{-2} K^{-1}. As expected, the main generation processes in the equatorial region are related to the absorption of solar radiation and the release of latent heat. The generation due to the absorption of long-wave terrestrial radiation and due to sensible heat are estimated to be relatively small. In the polar region the contributions of latent heat release and absorption of solar radiation are found to be about equally important. The total values in the last columns of Tables 15.1 and 15.2

TABLE • 15.1. Fluxes of entropy in mW m^{-2} K^{-1} across the boundaries of the global atmosphere, the equatorial zone (15 °N–15 °S), and the north polar cap (70 °N-NP).

	Lateral boundaries	Top	Bottom	Total
Globe	...	-884	278	-606
Equatorial zone	-156	-925	256	-825
Polar cap	267	-747	305	-175

TABLE • 15.2. Entropy generation terms in mW m^{-2} K^{-1} for the global atmosphere, the equatorial zone (15 °N–15 °S), and the north polar cap (70 °N-NP).

	Absorption of solar radiation	Absorption of terrestrial radiation	Latent heat release	Sensible heating	Frictional heating	Total
Globe	258	24	298	2	7	589
Equatorial zone	335	25	430	2	7	799
Polar cap	56	18	62	5	7	148

clearly show that the equatorial zone is by far more active than the polar region as far as the entropy generation is concerned.

The differences in the total rates of entropy generation as computed indirectly in Table 15.1 and directly in Table 15.2 are relatively small. They can be attributed to the various approximations in the formulation of the problem and in the actual calculations.

In summary, the values presented in Fig. 15.2 and Tables 15.1 and 15.2 show that the absorption of solar radiation and the release of latent heat are, by far, the largest sources of entropy in the atmosphere. The rate of production of entropy associated with the absorption of long-wave radiation is an order of magnitude smaller since the temperatures of infrared emission and absorption are not very different. Among the nonradiative processes, the water phase transitions dominate the entropy generation, especially at low latitudes.

CHAPTER 16

Interannual and Interdecadal Variability in the Climate System

16.1 INTRODUCTION

The meteorological, oceanic, and glacial records show considerable variability on all time scales. In general, there seems to be the tendency for a red spectrum. In other words, longer period phenomena tend to be associated with higher amplitudes of variability as was shown earlier by means of a schematic spectrum of the atmospheric temperature (see Fig. 2.7). We also found that no strict periodicities were in evidence in the temperature spectrum besides the diurnal and annual periods and their harmonics.

So far we have concentrated the discussions mainly on the long-term (multiyear) mean state of the climate system, i.e., on annual and seasonal mean values of the climatic components and on some integrated effects of the high-frequency variations. In these high-frequency variations we included in the atmospheric statistics the transient eddies associated with the day-to-day changing weather systems and in the oceanic statistics the mesoscale eddies that have longer time scales, typically ranging from weeks to months.

In this chapter, we will concentrate on some examples of fluctuations that are observed to occur in the climate system with time scales on the order of years to decades. We may note that the typical anomaly patterns that have been discovered in the atmosphere and oceans are clearly related to variations in the general circulations of the atmosphere and oceans, and in the interface conditions at the earth's surface, such as the sea surface temperature and the snow and ice cover (see, e.g., Namias, 1983). These variations on interannual and interdecadal time scales are of course of great importance for plant life, animal life, and especially for agriculture and other human affairs.

We will begin with the discussion of some well-defined interannual phenomena such as the quasibiennial cycle in the stratosphere, the El Niño-Southern oscillation (ENSO) in the tropics, and certain other regional teleconnection patterns that have been identified so far mainly in the atmosphere. Then we will proceed to a discussion of interdecadal fluctuations and trends as found in the concentration estimates of various atmospheric greenhouse gases and in the atmospheric and oceanic temperatures. Finally, we will discuss various statistics on certain special climatic events, such as droughts.

16.2 QUASIBIENNIAL OSCILLATION

16.2.1 Observed features

In addition to the expected annual and semiannual cycles prevalent in most of the atmosphere, we find in the tropical stratosphere a peculiar oscillation mainly in the zonal winds and temperature with an irregular period of generally slightly longer than two years, the so-called quasibiennial oscillation (QBO). The QBO signal in the zonal winds, as first described in detail by Reed *et al.* (1961) and Veryard and Ebdon (1961), is shown in Fig. 16.1 according to Naujokat (1986). It shows a pattern of alternating westerly and easterly winds over the equator. Since there is evidence of a high degree of zonal uniformity of the QBO signal around the equatorial belt the records of several equatorial stations were combined to produce a 32-year record from 1953 through 1984. We notice in Fig. 16.1 that the westerly and easterly winds alternate with an average period of about 27 months and that they reach extreme values on the order of -30 and 20 m s^{-1}. There appears to be a downward propagation of the QBO signal with a speed of about 1 km/month. The amplitude of the oscillation does not change much with height above 50 mb but decreases rapidly below the 50-mb level.

Most of the suggested explanations of the QBO phenomenon are given in terms of a zonally asymmetric wave forcing. Thus, the momentum source for the downward-propagating QBO is thought to be the absorption of upward-propagating equatorial waves from the troposphere as first discussed by Lindzen and Holton (1968) and later reviewed by Andrews *et al.* (1987) and Lindzen (1987). The wave absorption would take place at and below a critical level where the group velocity of the waves would go to zero. The rate of downward propagation and the amplitude of the QBO seem to be largely determined by the intensity and the phase speed of the upward-travelling waves, respectively (e.g., Lindzen, 1987). Thus, the QBO is probably a good example of a large-scale internal oscillation that results from wave-mean flow interactions in the atmosphere itself.

16.2.2 Possible solar-QBO-climate connections

Although there has been an almost continuous search for solar-weather relationships since systematic weather observations began, no convincing connections with tropospheric weather and climate have been established up until perhaps very recently. The most important problem has been and remains the lack of a physical mechanism by which an energetically weak signal of solar variability can influence conditions in the lower atmosphere. Such a mechanism would almost certainly require a special triggering effect, such as the formation of additional condensation nuclei to stimulate the release of latent heat or, e.g., the mechanism proposed by Schuurmans (1969) to account for the statistical relationships he reported between solar flares and the tropospheric circulation. In his theory, the solar flare produced particles interact

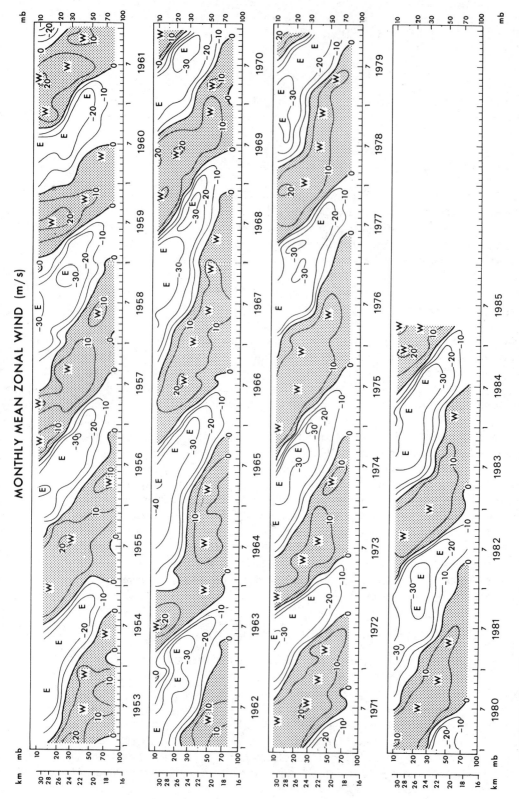

FIGURE 16.1. Time-height section of the monthly-mean zonal wind component near the equator in m s^{-1}, showing the QBO signal of alternating westerly and easterly winds. The section is based on data from Canton Island (3 °S, 172 °W; 1953–1967), Gan (1 °S, 73 °E; 1967–1975), and Singapore (1 °N, 104 °E; 1976–1985) (after Naujokat, 1986).

with some of the atmospheric greenhouse gases (e.g., H_2O and O_3) in the upper troposphere and lower stratosphere leading first to a change in the radiation balance and then to anomalies in the pressure patterns and in the atmospheric circulation.

Perhaps the most convincing statistical evidence of a solar-weather relationship is found in the recent work of Labitzke and van Loon (1988) and van Loon and Labitzke (1988). As a new link they claim that the quasibiennial oscillation may be an important factor in modulating possible solar effects on the atmosphere. For example, Fig. 16.2 (top portion) from their work shows, besides a 10.7-cm solar flux record, a record of the winter temperatures at 30 mb during the last three 11-year solar cycles. Obviously the correlation with the solar or sunspot activity is very low and insignificant. However, when the winters are stratified according to whether the QBO is in the westerly or easterly phase (see middle and lower parts of Fig. 16.2), the 30-mb temperature data are found to be strongly correlated with the sunspot activity. These results are suggestive of a true relationship.

Even at the earth's surface, the correlations between solar activity and sea level pressure or surface temperature as shown in Fig. 16.3 are unusually high and appear to explain an important fraction of the total interannual variability in the winter circulation. The typical circulation anomalies connected with the surface pressure pattern in Fig. 16.3(a) are consistent with the surface temperature anomalies shown in Fig. 16.3(b). For example, the figures show the tendency for (geostrophic) winds from the south and warmer temperatures (warm air advection) over the west coast of North America, and the tendency for (geostrophic) winds from the north and colder temperatures (cold air advection) over the southeastern U.S.

If confirmed by independent data samples this solar-QBO-weather relationship will be an important, new, and so far unexplained factor in the climate.

16.3 ENSO PHENOMENON

The air-sea interactions that occur on monthly and seasonal time scales are generally connected with short-term variability in the atmosphere or with the seasonal forcing of the atmosphere-ocean system ($\mathscr{A} \cup \mathscr{O}$). On the interannual time scale there are no large external forcings of this system so that the variations must arise from internal interactions with many positive and negative feedbacks. The most spectacular example of an internal variation is the ENSO phenomenon that may be regarded as a free oscillation of the ocean-atmosphere system. It is the only true global-scale oscillation that has been identified so far.

As the name suggests ENSO consists of two components. The first (mainly oceanic) component, El Niño, has historically been associated with a weak, warm current appearing along the coast of Ecuador and Peru annually around Christmas time (therefore the Spanish name El Niño for the Christ child), replacing the usual cold waters of the Peru current. However, more recently the name El Niño tends to be used for a much larger scale phenomenon that occurs not annually but every three to seven years in which the normally cold waters over the entire eastern equatorial Pacific Ocean show a dramatic warming of several °C. Also, very large anomalies in the oceanic and atmospheric circulations and in the global weather are associated with these changes in the equatorial sea surface temperatures.

The second (mainly atmospheric) component of ENSO, the Southern Oscillation, was first named and described by Walker (1924) and further documented by, e.g.,

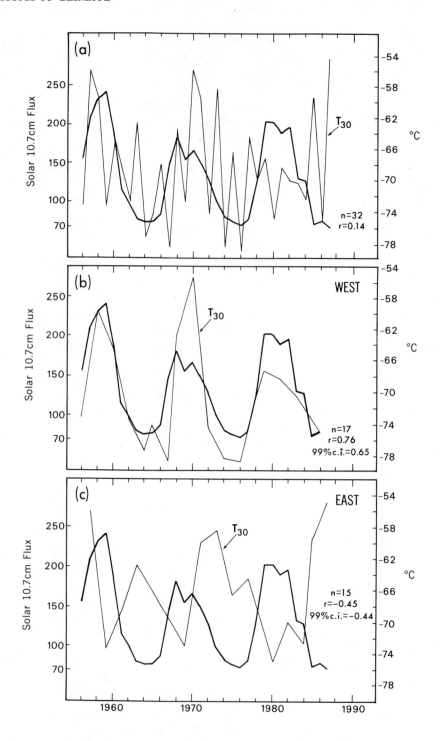

FIGURE 16.2. Time series of the 10.7-cm solar flux (thick solid lines) in units of 10^{-22} W m^{-2} Hz^{-1}. Thin solid lines show the mean 30-mb temperature in °C at the North Pole for all 32 winters from 1956 through 1987 (a), for 17 winters in the west phase ($\bar{u} > 0$) of the QBO (b), and for 15 winters in the east phase ($\bar{u} < 0$) of the QBO (c). Winter conditions are taken as the average of the months of January and February. The number of winters n, the correlation coefficient r between the solar flux and the 30-mb winter temperature, and the 99% confidence level are shown on the right in each figure (after Labitzke and van Loon, 1988).

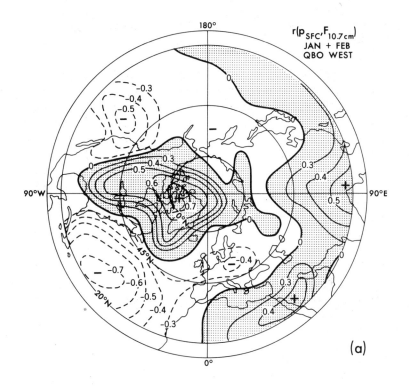

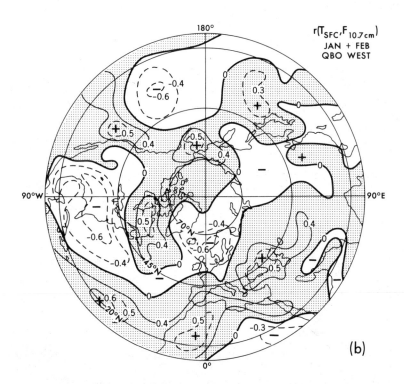

FIGURE 16.3. Horizontal distributions of the correlation coefficients between the 10.7-cm solar flux and the sea level pressure (a), and between the 10.7-cm solar flux and the surface air temperature (b) for 19 winters from the period 1953–1986 in the west phase ($\bar{u} > 0$) of the QBO. Areas of positive correlation are shaded (after van Loon and Labitzke, 1988).

Walker and Bliss (1932, 1937) and Berlage (1966). This oscillation is associated with large east–west shifts of mass in the tropical atmosphere between the Indian and West Pacific Oceans and the East Pacific Ocean.

Bjerknes (1969) was one of the first to show the possible connections between the El Niño and Southern Oscillation phenomena and to point out that they may be considered two aspects of one global-scale oscillation in the combined ocean–atmosphere system.

Thus, it is to some extent arbitrary whether we begin the discussion of the ENSO phenomenon in the oceanic or in the atmospheric branch. It is well known that the atmosphere influences the oceans mainly through anomalies in the stress exerted by the surface winds, whereas the ocean in turn influences the atmosphere mainly through anomalies in the sea surface temperatures and in the associated upward fluxes of sensible and latent heat (see Fig. 16.4). Although certain time-lag relationships and typical sequences of events show up in the various atmospheric and oceanic quantities during ENSO, it has proven very difficult to unravel why and where a particular ENSO event starts and why it ends.

Superposed on the zonal-mean Hadley circulations discussed in previous chapters there are also important east–west circulations in the equatorial atmosphere. A schematic of the so-called Walker circulation over the equatorial Pacific Ocean is shown for normal, non-ENSO conditions in Fig. 16.5. The figure shows an intense broad region of rising motions over the warm waters of the Indian Archipelago and strong sinking motions over the cold waters in the eastern equatorial Pacific. There are two other regions of upward motion over eastern equatorial Africa and the Amazon area. The associated sinking motions are found over the slightly cooler western Indian Ocean and the cold waters of the eastern equatorial Atlantic, respectively.

During ENSO conditions [Fig. 16.6(a)], the waters in the central and eastern equatorial Pacific warm up, the western waters slightly cool down, and the convection is enhanced over the central and eastern equatorial Pacific Ocean but is reduced over the Indonesian area. During anti-ENSO conditions [Fig. 16.6(b)], the waters in the central and eastern equatorial Pacific are colder than normal and convection is re-

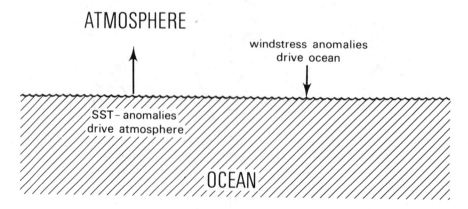

FIGURE 16.4. Schematic diagram showing the circular path of the principal atmosphere–ocean interactions. Sea surface temperature (SST) anomalies tend to lead to anomalies in the atmospheric general circulation and in the surface winds, while the resulting anomalies in the surface windstress lead to anomalies in the ocean circulation and eventually to SST anomalies, thereby closing the loop.

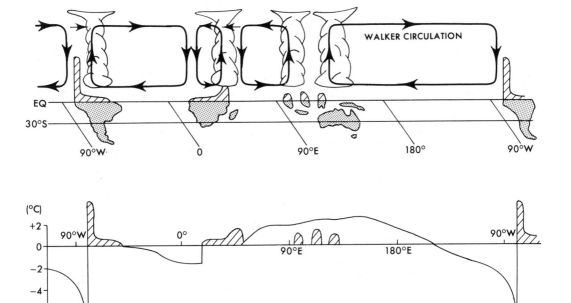

FIGURE 16.5. Schematic diagram of the normal Walker circulation along the equator during non-ENSO conditions. Rising air and heavy rains tend to occur over Indonesia and the western Pacific, southeast Africa, and the Amazon area in South America, while sinking air and desert conditions prevail over the eastern equatorial Pacific and southwest Africa (see also Fig. 7.24). The strongest branch of the Walker circulation over the Pacific is related to the very warm SST in the west Pacific where the air is rising, and the cool SST in the east Pacific where the air is sinking. The SST departures from the zonal-mean along the equator are shown in the lower part of the figure (after Wyrtki, 1982).

duced over these regions, whereas the convection over the Indonesian area is enhanced. The so-called normal conditions represent an average of both ENSO and anti-ENSO episodes but best resemble the weak anti-ENSO (cold) conditions.

The wind forcing and the response in the ocean for normal, ENSO, and anti-ENSO conditions are shown schematically in Fig. 16.7. Under normal conditions [Fig. 16.7(a)] the westward trade winds maintain a difference in sea level of about 40 cm between the east and west coasts as well as a strong slope in the thermocline inside the ocean. When for some reason the westward atmospheric pressure gradient decreases and the trade winds over the central equatorial Pacific weaken [see Figs. 16.6(a) and 16.7(b)] the east–west slopes in sea level and in the thermocline decrease, warm waters flood over or replace the cold waters of the eastern equatorial Pacific, and an ENSO event has started. If, on the other hand, the trade winds strengthen [see Figs. 16.6(b) and 16.7(c)] the east–west slopes increase and the east–west oceanic temperature gradients intensify, leading to anti-El Niño, that Philander (1989) called "La Niña" conditions.

In the early stages of an ENSO episode large shifts in atmospheric mass occur so that the surface pressure decreases over almost the entire eastern Pacific and increases over the western Pacific and Indian Oceans. This global facet of the Southern Oscillation is shown in Fig. 16.8(a) in the form of an analyzed map of the correlation coefficients between the annual-mean sea-level pressures over the globe and the surface pressure in Darwin, Australia (12 °S, 131 °E).

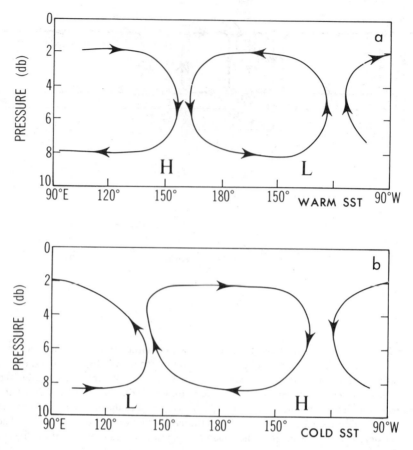

FIGURE 16.6. Schematic diagram of the departures from normal (as shown in Fig. 16.5) for the circulation in a vertical plane along the equator during (a) warm ENSO conditions and (b) cold anti-ENSO conditions. The weakening of the normal trade winds over the central equatorial Pacific in (a) and the strengthening of the tradewinds in (b) are important factors for the response in the oceans (adapted from Julian and Chervin, 1978).

The worldwide sea surface temperature changes connected with the changes in surface pressure at Darwin are shown in Fig. 16.8(b). The map shows very high positive correlations (up to 0.90) with the Darwin pressure not only over the central and eastern tropical Pacific, but also over most of the tropical Indian and Atlantic Oceans. The actual temperature differences between ENSO conditions (relatively high pressure at Darwin) and anti-ENSO conditions (relatively low pressure at Darwin) range from an average value of about 0.5 °C over the tropical oceans to extreme values of about 2 °C in the eastern equatorial Pacific (see, e.g., Fig. 4.1). Over the Indonesian Archipelago and in midlatitudes there is a tendency for slightly cooler (warmer) temperatures during ENSO (anti-ENSO) conditions.

The traditional index of the Southern Oscillation is given by the difference in surface pressure between two stations at or near the maximum and minimum values on the correlation map [Fig. 16.8(a)], i.e., between Darwin and Easter Island (27 °S, 109 °W). For example, when this index is low negative a strong ENSO event is in progress. A time series of this index for each month of the period January 1949 to December 1988 is shown in Fig. 16.9(a). The generally accepted ENSO events are marked by arrows below the figure (see Rasmusson and Carpenter, 1982). We may

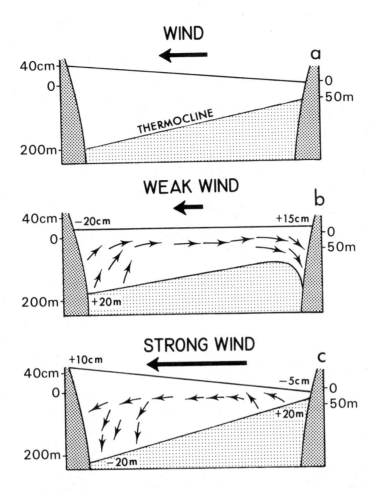

FIGURE 16.7. Response of the thermal structure of the equatorial Pacific to changes in the surface winds. (a) Under normal easterly trade wind conditions sea level rises to the west and the thermocline deepens. (b) When the winds relax, water sloshes east, which leads to a rise in sea level and a deepening of the thermocline near the South American coast. In the western Pacific, sea level drops and the thermocline rises (El Niño conditions). (c) The normal situation is amplified during strong trade winds [anti-El Niño or La Niña conditions (after Wyrtki, 1982)].

note that the unsmoothed curve for the Southern oscillation index shows a high-frequency component that is probably associated with transient 30–60 day oscillations that are very common in the tropics but will not be discussed further here (for more information see, e.g., Madden and Julian, 1971; Knutson and Weickmann, 1987; and Lau and Chan, 1986).

For comparison, the record of the monthly wind stress over the central equatorial Pacific is shown in Fig. 16.9(b) and the record of the average SST anomalies in the eastern equatorial Pacific in Fig. 16.9(c), both in the form of unsmoothed and smoothed curves. The unsmoothed curve in Fig. 16.9(c) is almost identical to the time series of the first EOF mode in the Pacific Ocean shown before in Fig. 4.1(b). This close correspondence clearly demonstrates the dominance of the ENSO phenomenon in the sea surface temperature records.

A detailed comparison of the three sets of graphs in Fig. 16.9 confirms what was shown schematically in Figs. 16.6 and 16.7. Thus, ENSO events are characterized by

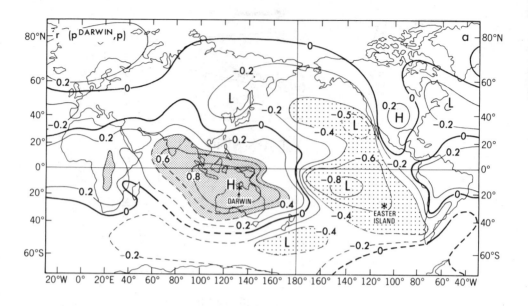
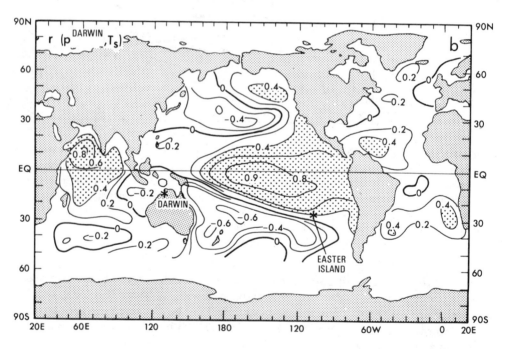

FIGURE 16.8. (a) Horizontal distribution of the correlation coefficient between annual-mean sea level pressure anomalies over the globe and the corresponding pressure anomalies in Darwin, Australia (12 °S, 131 °E) as a measure of the Southern Oscillation. The map shows that global shifts of atmospheric mass take place during ENSO episodes (adapted from Trenberth and Shea, 1987). (b) Horizontal distribution of the correlation coefficient between annual-mean sea level pressure anomalies in Darwin and sea surface temperature anomalies over the globe for the 30-year period 1950–1979. Areas with anomalies greater than 0.4 are stippled.

a low negative Southern Oscillation Index (i.e., reduced westward pressure gradient over the equatorial Pacific), weaker than normal trade winds over the central Pacific, and warmer than normal sea surface temperatures in the eastern equatorial Pacific. On the other hand, anti-ENSO (La Niña) events are characterized by a high positive Southern Oscillation Index (i.e., an increased westward pressure gradient over the equatorial Pacific), stronger trade winds over the central Pacific, and cooler sea surface temperatures in the eastern equatorial Pacific. To quantify these possible links, the actual correlation coefficients between the various sets of curves in Fig. 16.9 are given in Table 16.1. We find that the smoothed wind stress values (τ_x^{CEP}) show an extreme correlation value of -0.65 when τ_x^{CEP} lags the Southern Oscillation Index (Δp) by two months, and that the smoothed sea surface temperature anomalies (T_s^{EEP}) show an extreme correlation value of -0.83 when T_s^{EEP} lags Δp by 4.5 months. It seems from these observations that variations in the pressure gradient are first followed by changes in the wind stress, and later by changes in temperature, as was suggested before in Fig. 16.7.

As can be seen in Fig. 16.9, ENSO events occur at variable intervals ranging from about two to seven years. The average interval is on the order of 40 months. We should note that in 1982/1983 an exceptionally strong ENSO event was observed. As presented clearly in Fig. 16.9, the years 1982–1983 show the most negative pressure anomalies and the highest SST anomalies observed during the past 40 years. During early 1988 and 1989, we have been experiencing a cold phase or anti-ENSO conditions.

Perhaps the most important aspect of an ENSO event is the change in the precipitation patterns over the globe (as shown schematically in Fig. 16.10). The figure shows a composite map of the regions of abnormally wet (dense shading) and abnormally dry conditions (light shading) associated with a typical ENSO event. Within each region the approximate period of extreme conditions was determined during the 24-month period starting with the July month preceding the event, designated by Jul($-$), continuing through the June month following the event, designated by Jun($+$). The index 0 behind a month refers to the year of El Niño as defined by Rasmusson and Carpenter (1982). Thus, for example, April(0) reflects conditions near or just before the time of maximum SST anomalies along the Ecuador–Peru coast. During the early months of year "0" "desert-like" regions over the central and eastern equatorial Pacific Ocean and over the coastal regions of Ecuador and Peru may receive torrential rains. We should stress that there are often important differences in the character and evolution of individual ENSO events, and that the "average" conditions represent only an idealized picture of the real conditions.

The global precipitation map in Fig. 16.10 shows strong increases in precipitation over the central Pacific, in the narrow coastal zone of Ecuador and Peru, over a region just south of India, and over eastern equatorial Africa. However overall, the areas of reduced rainfall seem to dominate over the areas of increased rainfall. Thus, relative drought conditions are found in the tropics over the western tropical Pacific Ocean, Indonesia, Australia, India, southeastern Africa, and northeastern South America. Recent research is being conducted on the lag relations of deficient summer monsoon rains over India (e.g., Rasmusson and Carpenter, 1983) and of droughts over northeast Brazil ("Nordeste"; e.g., Hastenrath and Heller, 1977) with ENSO events and on the possibility of predicting future droughts over these regions.

In midlatitudes Fig. 16.10 shows a weak tendency for regional increases in rainfall over North America and southeast South America following ENSO. However, these correlations as well as other correlations for different climatic variables are relatively weak.

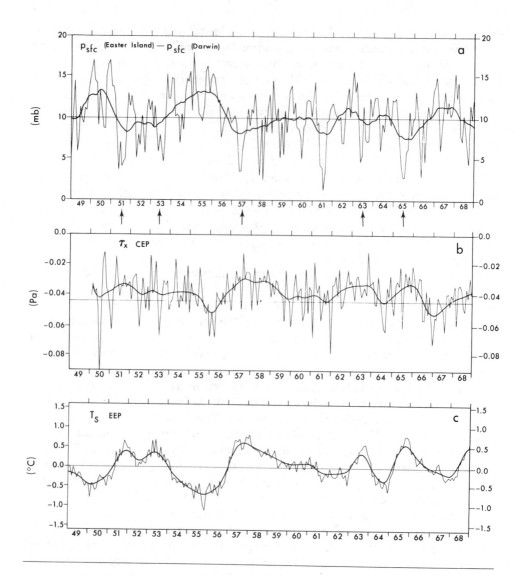

Some other aspects of the ENSO phenomenon are given in the form of zonal-mean differences between warm ENSO and cold anti-ENSO events in Fig. 16.11. Besides a weakening and a zonal shift of the normal east–west Walker oscillation over the equator (shown before in Figs. 16.5 and 16.6), the north–south Hadley circulations become stronger. This is shown in Fig. 16.11(b) by the enhanced inflow of mass near the surface into the ITCZ, the resulting enhanced rising motions between about 10 °S and 10 °N, and the enhanced sinking motions in the subtropics of each hemisphere [Fig. 16.11(c)]. As can be understood from the conservation of absolute angular momentum the increased Hadley circulations also lead to a strengthening of the eastward flow in the subtropical jets by a few m s^{-1} as shown in Fig. 16.11(d). As pointed out by Bjerknes (1969) the bulk of the changes in the zonal-mean values are due to the more local changes over the eastern tropical Pacific.

The temperatures in the free atmosphere are also profoundly affected by events in the equatorial Pacific Ocean. A zonal-mean cross section of the warm–cold temperature differences in Fig. 16.12 shows values on the order of + 0.5 to + 1 °C over the

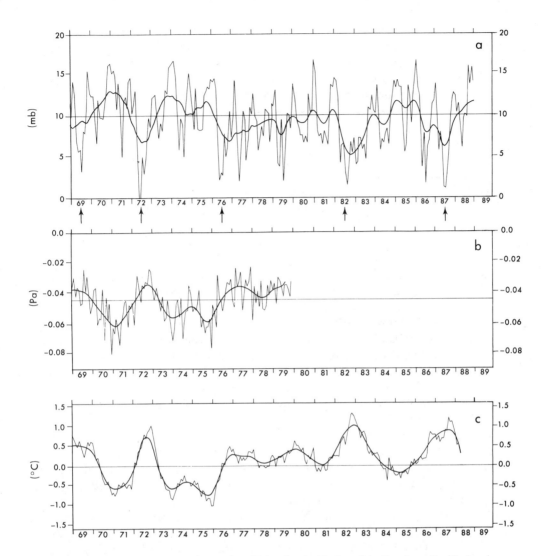

FIGURE 16.9. Monthly-mean time series of (a) a classical index of the Southern Oscillation, i.e., the pressure difference between Easter Island and Darwin in mb (after Wyrtki, 1982, updated with data from Ropelewski, 1989, personal communication). (b) The zonal wind stress over the central equatorial Pacific region τ_x^{CEP} (8 °S–4 °N, 160 °E–130 °W) in units of Pa (1 Pa = 10 dyn cm^{-2}), and (c) the sea surface temperature anomalies in the eastern equatorial Pacific region T_s^{EEP} (20 °S–20 °N, 180–80 °W) in °C. The most important El Niño events (i.e., when warm water covers the entire eastern equatorial Pacific Ocean) according to Rasmusson and Carpenter (1981, updated) are indicated by arrows at the bottom of (a). The thick solid lines indicate a 12-month running mean in (a) and 15-month weighted means in (b) and (c) (obtained by using a Gaussian-type filter with weights 0.012, 0.025, 0.040, 0.061, 0.083, 0.101, 0.117, and 0.122 at the central point).

entire tropical troposphere and values on the order of -0.5 °C or less in midlatitudes. So apparently convection and upper-level release of latent heat must be very effective in distributing the heat vertically in the tropics. The colder temperatures in the lower stratosphere (at the top of the figure) must be due to overshooting of the convective clouds into the more stable stratosphere. When we compare the time series of the SST anomalies in the eastern equatorial Pacific with the time series of the

TABLE • 16.1. Maximum correlation coefficients between the Southern Oscillation Index (Δp) and two other ENSO indicators with the corresponding lag values (in months) for the unsmoothed and smoothed curves shown in Fig. 16.9. A positive value of the lag indicates that the "other" ENSO indicator follows Δp.

	Unsmoothed		Smoothed	
	r	lag	r	lag
$r(\Delta p, \tau_x^{CEP})$	-0.30	1.5	-0.65	2
$r(\Delta p, T_s^{EEP})$	-0.57	4	-0.83	4.5

atmospheric temperature averaged vertically and horizontally over the entire Northern Hemisphere as presented in Fig. 16.13, we find that the curves are highly correlated. The actual maximum correlation coefficients are $r = 0.67$ for the unsmoothed curves when the atmosphere lags the ocean by about three months, and $r = 0.82$ for the smoothed curves when the atmosphere lags the ocean by about four months (see also Pan and Oort, 1983, where based on a shorter record a somewhat longer lag of about six months was found). Since correlations with the mean Southern Hemisphere temperatures (not shown) are very similar and also very high, it is clear that a large part of the observed variability in the global atmosphere must be connected with the ENSO events.

Based on the results presented above, we may conclude that an increased knowledge and understanding of the ENSO phenomenon will likely lead to considerable skill in forecasting climatic anomalies in the tropics more than six months ahead, but that there would be only limited improvement in the prediction in mid and high latitudes. Because of its obvious socio-economic importance, an intense effort is presently being made to study the ENSO phenomenon. It is clear that an explanation of what happens during ENSO must include a detailed understanding of the complex feedback processes between the oceans and the atmosphere. In this highly interactive system it may prove very difficult to determine cause and effect.

16.4 REGIONAL TELECONNECTIONS

Several other more regional oscillations have been discovered besides ENSO that are perhaps less spectacular than the global ENSO phenomenon but are found to be of considerable significance for describing regional climate anomalies (e.g., Bjerknes, 1964). These regional oscillations may also be regarded as free internal oscillations in the atmosphere–ocean system.

Some attractive methods for determining regional oscillations and associated teleconnection patterns in the atmosphere are to use empirical orthogonal function (EOF) analyses (see Sec. 4.3 and Appendix B) or to compute correlation maps. Two examples of such correlation maps for the North Atlantic Oscillation (NAO) are given in Fig. 16.14 taken from Wallace and Gutzler (1981). The first example shows the field of the correlation coefficient between the sea-level pressure at the point (65 °N, 20 °W) near the semipermanent Icelandic low and the sea-level pressure at all

VARIABILITY IN THE CLIMATE SYSTEM 427

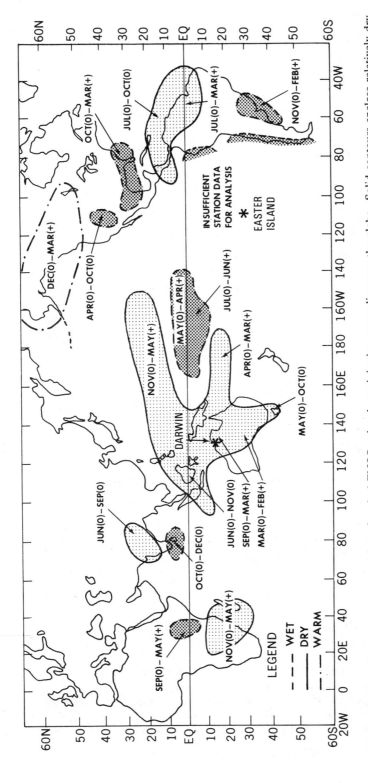

FIGURE 16.10. Schematic representation of typical ENSO-related precipitation anomalies over the globe. Solid contours enclose relatively dry regions (light shading) and dashed contours enclose relatively wet regions (heavier shading). The approximate period of extreme conditions relative to the typical El Niño (0) year is also shown for the various regions (adapted from Ropelewski and Halpert, 1987).

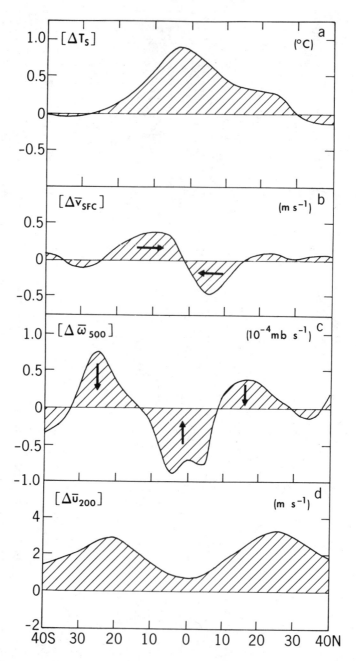

FIGURE 16.11. Meridional profiles from 40 °S to 40 °N of the zonal-mean difference between warm (El Niño) and cold (La Niña) events for the sea surface temperature (a), the northward component of the surface wind (b), the vertical wind component at 500 mb (a value of $\omega = -10^{-4}$ mb s^{-1} is equivalent with an upward velocity $w \approx 0.14$ cm s^{-1}), and the zonal wind component at 200 mb (d) during northern winter (adapted from Pan and Oort, 1983).

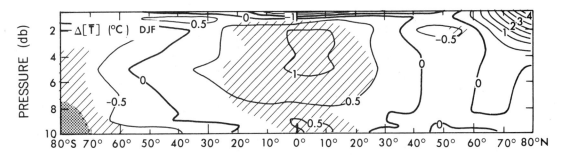

FIGURE 16.12. Zonal-mean cross section of the average temperature difference in the atmosphere for the cases that the sea surface temperatures in the eastern equatorial Pacific were relatively warm (El Niño) and for those cases that they were relatively cold for northern winter (from Pan and Oort, 1983).

grid points north of 20 °N. Besides the expected decrease of the correlation with increasing distance from the base point we find high negative correlations in an area to the southeast of the Azores, where one often finds the semipermanent Azores high-pressure system. Of course, the negative correlation means that an abnormally intense Icelandic low tends to go together with an abnormally strong Azores high, and that, conversely, a weak Icelandic low tends to go together with a weak Azores high. The first situation implies a strong north–south pressure gradient associated with strong westerly winds over the central North Atlantic Ocean, abnormally strong advection of cold polar air masses from the north over western Greenland, and abnormally strong advection of warm tropical air masses from the south over northwestern Europe. The second situation of a weaker Icelandic low and a weaker Azores high would imply weaker westerly winds over the central North Atlantic Ocean, weaker than normal flow from the north over Western Greenland, and weaker than normal flow from the south over northwestern Europe.

These facts are shown more clearly in Fig. 16.14(b) which gives the correlation between the Iceland–Azores pressure difference and the mean temperature in the lower troposphere. As expected, the correlation is strongly positive over the northwestern Atlantic where an intense Icelandic low would mean abnormally cold conditions and a weak Icelandic low abnormally warm conditions. Over northwestern Europe the situation is reversed and the correlation is negative, indicating relatively warm conditions when the Icelandic low is intense and relatively cold conditions when it is weak. Thus, we frequently find a "see-saw" pattern between the winter temperatures over western Greenland and northwestern Europe (see van Loon and Rogers, 1978).

A synopsis of the so-called centers of action in the Northern Hemisphere is given schematically in Fig. 16.15 [taken again from Wallace and Gutzler (1981)]. The two composite maps in Fig. 16.15 are based on the calculations of one-point correlation patterns for each grid point in the Northern Hemisphere in case of the surface pressure (a) and the midtropospheric geopotential height (b) fields. More specifically, the composite maps were obtained by plotting at the location of each grid point the value of the strongest negative correlation found in the one-point correlation map for that particular point. For example, for the point (65 °W, 20 °W) we find from its one-point correlation map in Fig. 16.14(a) a value of -0.76. This value was then plotted at (65 °N, 20 °W) in Fig. 16.15(a). However, for convenience, the sign was left out and the values were multiplied by 100 in Fig. 16.15.

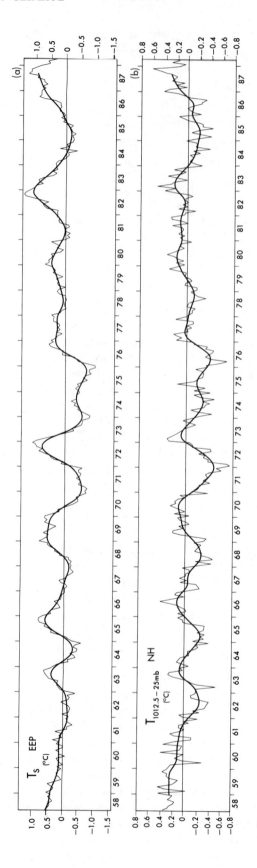

FIGURE 16.13. Time series of monthly-mean sea-surface temperature anomalies in the eastern equatorial Pacific (a) and monthly-mean atmospheric temperature anomalies averaged over the entire Northern Hemispheric mass between the surface and about 25-km height (b). Smoothed curves show 15-month Gaussian-type filtered values (adapted from Pan and Oort, 1983, updated).

VARIABILITY IN THE CLIMATE SYSTEM 431

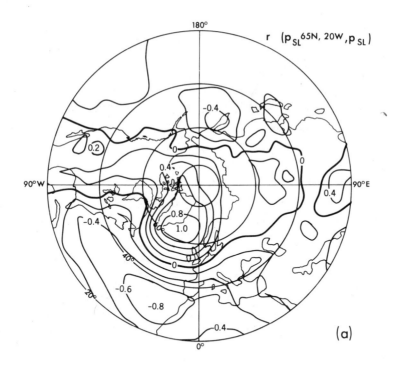

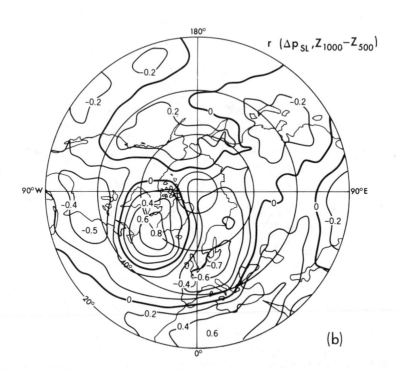

FIGURE 16.14. The North Atlantic Oscillation: (a) Map of the correlation coefficient between the sea-level pressure at (65 °N, 20 °W) and sea-level pressure at all grid points north of 20 °N. (b) Map of the correlation coefficient of the pressure difference, (65 °N, 20 °W) minus (30 °N, 20 °W), and the 1000–500-mb thickness at all grid points north of 20 °N. The 1000–500-mb thickness is a measure of the mean temperature in the lower troposphere. Contour interval is 0.2 (after Wallace and Gutzler, 1981).

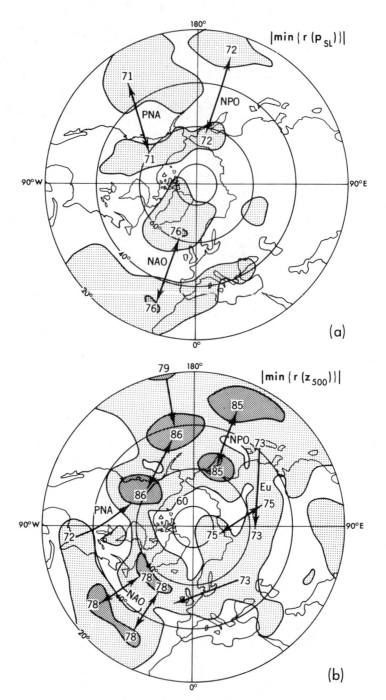

FIGURE 16.15. Spatial distribution of the strongest negative correlation r_{ij} at each grid point as taken from the corresponding one-point correlation map for the same grid point for (a) sea-level pressure and (b) 500-mb height based on 45 winter months (1963–1977). Negative signs have been omitted and the correlation coefficients are multiplied by 100. Regions where $r_{ij} > -0.60$ are unshaded; $-0.60 \geqslant r_{ij} > -0.75$ stippled lightly; $r_{ij} \leqslant -0.75$ stippled heavily. Arrows connect centers of strongest teleconnectivity with the grid point which exhibits strongest negative correlation on their respective one-point correlation maps. The most dominant patterns on the maps are the Pacific—North American Oscillation (PNA), the North Pacific Oscillation (NPO), the North Atlantic Oscillation (NAO), and the Eurasian Oscillations (EU) (after Wallace and Gutzler, 1981).

At the surface the most dominant patterns in Fig. 16.15(a) can be identified as the North Atlantic Oscillation, the Pacific—North American Oscillation (PNA), and the North Pacific Oscillation (NPO). In the midtroposphere we find several oscillation patterns over the Eurasian continent (EU) in addition to the previous ones. Although other oscillations, such as a zonally symmetric see-saw in the sea-level pressures at middle and polar latitudes (Lorenz, 1951), have been described in the earlier meteorological literature, the patterns described so far appear to be the most important ones in the Northern Hemisphere winter circulation. Some of these oscillations tend to occur throughout the observational record while others occur only infrequently. The two more stable patterns appear to be the North Atlantic Oscillation and the Pacific—North American Oscillation. The last one is sometimes found to be associated with SST anomalies in the equatorial Pacific (ENSO; Horel and Wallace, 1981) and at other times with midlatitude SST anomalies (e.g., Wallace and Jiang, 1987). In any case, we can conclude that all oscillation patterns are associated with important regional anomalies in both temperature and precipitation. Furthermore, the more persistent the anomalies in surface pressure are, the stronger the advection and the greater the anomalies in temperature and precipitation tend to be. Of course, these atmospheric anomalies must, in general, be connected with (i.e., either the result or the cause of) anomalies in the large-scale general circulation of the atmosphere and possibly the oceans. They may also be connected with anomalies in the boundary conditions at the earth's surface, such as sea surface temperature and snow and ice cover (see, e.g., the extensive investigations summarized in Namias, 1983). All these oscillations seem to be "typical" modes of the atmospheric system and must affect the regional predictability of weather and climate anomalies. These topics are actively pursued in present climate research.

16.5 INTERDECADAL FLUCTUATIONS AND TRENDS

16.5.1 Anthropogenic influences

Human activities can interfere in the climate through changes in the composition and structure of the atmosphere, through the release of heat into the atmosphere, and through changes in the albedo and other properties of the earth's surface thus changing the radiation balance. Man is inadvertently altering the composition of the atmosphere in a continuous manner because he changes the concentration of some of the atmospheric constituents already in existence, and because he introduces new substances into the atmosphere that under normal natural conditions would not exist. We may mention the emission of gases, smoke, and particulate matter from the industrial complexes, from urban centers (automobiles, etc.), and from agricultural practices such as the burning of trees, deforestation, and the massive use of fertilizers, leading to the pollution of the atmosphere and waters. With deforestation and overgrazing the soil is eroded and prevailing winds may remove large quantities of top soil which later will be spread out into the atmosphere. These processes will produce substantial changes in the surface albedo affecting then the radiation and moisture balances near the surface and eventually leading to desertification.

The gases and particles thrown into the atmosphere may affect the climate because: (a) they alter the radiation balance of both the local and the global atmosphere, leading to modifications in the thermal and dynamical structure of the atmosphere, (b) they interfere in the photochemical equilibrium of the stratosphere and especially in the balance of ozone (freons), and (c) they can form acids in either the gas or liquid phase as fog or cloud droplets which will lead to acid precipitation.

Man-made aerosols may also interfere with the radiation balance through reflection, scattering, and absorption of the short-wave solar radiation leading likely to a slight, net cooling of the atmosphere (Bryson, 1972).

16.5.2 Atmospheric gases

16.5.2.1 Carbon dioxide

Among the gases released into the atmosphere by human activities, one of the most important is CO_2. Carbon dioxide is well-mixed in the atmosphere with an almost uniform mixing ratio of, at present, about 350 parts per million by volume (ppmv). It is a small but essential component of the atmosphere because along with water vapor and ozone it plays a very important role in heating the atmosphere. As mentioned in Chap. 6, CO_2 molecules are transparent for short-wave solar radiation but are strong absorbers for infrared radiation emitted by the earth's surface. Through absorption CO_2 prevents some of the radiation emitted by the surface from being lost into space. This mechanism is called the CO_2–greenhouse effect (see Sec. 2.5.3). The more CO_2 there is in the atmosphere the more radiation will be absorbed, leading to a warming effect in the lower atmosphere. Thus, the increase in CO_2 and other trace gases may have profound and long-term effects on the climate.

The carbon in the climatic system is contained in four major reservoirs: the atmosphere, biosphere, oceans, and lithosphere including the fossil fuels. The transfers of carbon between these reservoirs are complex and not well known. They lead to the carbon cycle that links all parts of the climate system together with many interactions and feedback processes, some positive and some negative. The carbon cycle has been the subject of much recent research (e.g., Keeling *et al.*, 1982). Carbon in the atmosphere mainly exists in the form of CO_2. The amounts of carbon in the atmosphere and the terrestrial vegetation are about the same, but are very small compared to the amounts in the earth's crust and oceans (e.g., Emanuel *et al.*, 1985). Model computations give an average atmospheric residence time for CO_2 connected with its recycling on the order of five to seven years (National Academy of Sciences, 1979), which is short compared to the residence times in the other reservoirs. Thus, due to its large mobility, the atmosphere plays an important role in the carbon cycle.

The amount of CO_2 has increased by about 25% since the beginning of the industrial revolution ($\approx$ 1850) when its mixing ratio was on the order of 280 ppmv (Neftel *et al.*, 1985). Figure 16.16 gives the time series of CO_2 mixing ratio measured since 1958 at the high-altitude observatory of the isolated Mauna Loa mountain of Hawaii (19.5 °N, 155.6 °W; elevation 3397 m above sea level). It shows the general behavior of the evolution of CO_2 in the atmosphere. We note that the rate of increase (i.e., the slope of the smoothed curve) has been on the average about 1 ppmv/year but that it has increased to 1.5 ppmv/year in recent years, partly in response to the rates at which fossil fuels have been and are being burned. Besides this anthropogenic source there are also important exchanges of CO_2 with the oceans and biosphere. The interactions with the biosphere occur through photosynthesis and oxidation processes. The net effect of deforestation, such as the reduction of tropical forests, the extensive use of fertilizers, and the general decay of organic matter may constitute a small source of atmospheric CO_2 compared to the fossil fuel input but it is still thought to be sizeable (perhaps ~ 20%). On the other hand, the idea promoted by, e.g., Dyson (1977), that plants and trees would grow better in a richer CO_2 environment ("CO_2 fertilization") and thereby would absorb some of the atmospheric CO_2, cannot be discounted.

The quasi-sinusoidal annual variation in Fig. 16.16 with a range of about 6 ppmv at Mauna Loa is associated with the annual cycle in photosynthesis. The phase of this regular breathing of the global biosphere is dominated by the Northern Hemisphere vegetation cycle, with consumption of CO_2 during spring and summer leading to a minimum concentration at Mauna Loa in fall, and with release of CO_2 into the atmosphere during the late fall and winter leading to a maximum concentration at Mauna Loa in spring (see, e.g., Tucker et al., 1986). Although the general trend in the annual-mean CO_2 concentration at various geographical locations is found to be the same as the trend at Mauna Loa, the amplitude of the annual cycle is much larger over the high-latitude continents in the Northern Hemisphere and almost zero at high southern latitudes over Antarctica (e.g., Keeling et al., 1989).

Considering the Mauna Loa record in Fig. 16.16 in more detail one also finds significant interannual variations that seem to be related with the ENSO cycle (e.g., Bacastow, 1976). According to Keeling and Revelle (1985) the observed increase of atmospheric CO_2 just following an ENSO event cannot be due to warming of the

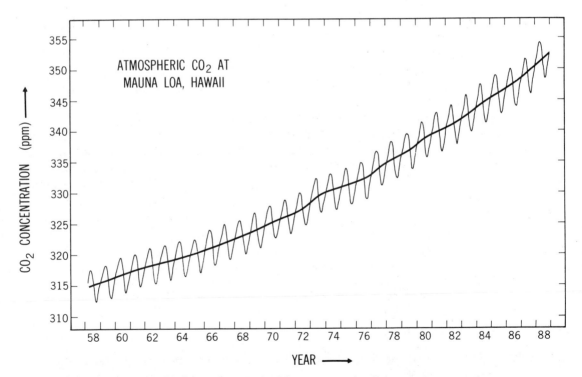

FIGURE 16.16. Time series of the atmospheric CO_2 concentration in ppmv as measured at Mauna Loa Observatory, Hawaii (19.5 °N, 155 °W) between 1958 and 1988 after Keeling et al. (1989). The thick solid curve is a spline fit to the monthly data showing the long-term trend. The trend at Mauna Loa is generally accepted to be representative for the trend in global mean conditions.

equatorial Pacific Ocean since the associated decrease of equatorial upwelling would actually cut off the normal local source of CO_2 from the subsurface, CO_2-rich equatorial ocean waters. Thus, a decrease in the intake of atmospheric CO_2 in other ocean areas, a decrease in photosynthesis or an increased decomposition of vegetation must overcompensate for this effect. In fact, Keeling and Revelle's (1985) studies of the $^{13}C/^{12}C$ isotope ratio suggest that the withering of tropical land vegetation and fires associated with drought conditions may lead to a release of CO_2 from the decay and burning of wood and plant material during ENSO episodes (see Fig. 16.10). This may be a dominant factor in the observed increase of atmospheric CO_2 immediately following ENSO conditions.

There is a gross imbalance between the estimated anthropogenic input of CO_2 and the observed change in atmospheric concentration. In fact, the atmosphere appears to retain only about half of the amount that is produced by the burning of fuels. It is now generally accepted that the oceans constitute the principal absorber for atmospheric CO_2 and that approximately one-half of the newly released CO_2 is removed from the atmosphere by absorption at the ocean surface (e.g., Broecker et al., 1979). However, the detailed mechanisms are not yet well understood.

Accepting that the burning of fossil fuels will continue to be the dominant source of CO_2 for the atmosphere and that the atmosphere will continue to retain at most one-half of the CO_2 produced, we can predict the atmospheric CO_2 amounts for various scenarios of the future consumption of fossils fuels. If we maintain the present rate of fuel consumption, the mixing ratio of CO_2 will double with respect to the preindustrial value by the year 2200. However, if instead the actual energy use were to grow at a rate of 2.5% per year, doubling of CO_2 would occur by the middle of the next century (National Academy of Sciences, 1979). With the doubling of CO_2 a sizeable change in the global-mean surface temperature of about 3 ± 1.5 °C would be expected (e.g., Manabe, 1983). The possible CO_2 heating effect was already noted by Arrhenius in the late 19th century and later by Callender (1938). Other radiatively active trace gases (CFC's, CH_4, and N_2O) are also being introduced into the atmosphere by man. Their combined effect may, in the near future, lead to a possible doubling of the greenhouse effect due to CO_2 alone (e.g., Ramanathan et al., 1985).

There is little empirical evidence to allow us to predict with any confidence the effects of increasing CO_2 on climate. Therefore, we have to rely on mathematical simulations using a variety of climate models (see, e.g., Schlesinger and Mitchell, 1987; Hansen et al., 1988; and Bryan et al., 1988) as will be discussed further in the next chapter.

16.5.2.2 Ozone

Ozone (O_3) is found in minute quantities throughout the atmosphere with the largest concentrations (number of molecules per unit volume) in the lower stratosphere between 12 and 30 km and a maximum at about 25-km height. If all ozone in an atmospheric column were brought to the mean temperature and pressure at sea level (STP), its global-mean thickness would be about 0.30 cm. Around 90% of the ozone is found in the stratosphere and the remaining 10% in the troposphere.

Although present in only small quantities, ozone plays a critical role in the biosphere by absorbing the solar ultraviolet radiation that would otherwise reach the earth's surface. The resulting heating of the upper atmosphere leads to a temperature increase with height in the stratosphere and a maximum value of about 270 K near the stratopause (see Fig. 2.3). This heating provides a major source of energy to drive the circulation of the upper stratosphere and mesosphere.

In the troposphere, ozone is produced mainly in urban air through oxidation of unburned hydrocarbons with nitrogen oxide as a catalyst, although globally substantial ozone amounts result from the downward transport of ozone-rich stratospheric air. Due to its oxidizing properties, tropospheric ozone plays an important role in the generation of photochemical smog and in the formation of acid rain from sulfur dioxide (SO_2) and nitrogen dioxide (NO_2). Because ozone absorbs thermal infrared radiation at wavelengths near 9.6 μm (see Fig. 6.2), it also acts as a greenhouse gas in the troposphere.

In the stratosphere, ozone results from the photodissociation of molecular oxygen by solar ultraviolet radiation (see Sec. 6.3.4) and the subsequent recombination reaction between atomic and molecular oxygen in the presence of another molecule (M):

$$O_2 + h\nu \to O + O,$$
$$O + O_2 + M \to O_3 + M.$$

In the balance, ozone is destroyed in the stratosphere by catalytic reactions with nitrogen, hydrogen, and chlorine radicals (i.e., molecules with an odd number of electrons), each playing an important role at different altitudes. For example, nitric oxide (NO) can react with ozone to form nitrogen dioxide and molecular oxygen. Nitrogen dioxide may then react with atomic oxygen to reform nitric oxide and molecular oxygen (Crutzen, 1970; Johnston, 1971):

$$NO + O_3 \to NO_2 + O_2,$$
$$NO_2 + O \to NO + O_2.$$

The nitric oxide released in the second reaction is then ready to start destroying ozone again (catalytic reaction). Thus, a small amount of nitric oxide can destroy a large amount of ozone.

Based on the photochemical theory of the formation of ozone we would expect that its maximum mixing ratio would occur at about 30-km height in the middle stratosphere over the tropical regions, where the solar radiation is most intense, and that minimum mixing ratios would be found over the polar regions. However, considering the total amount of ozone in a vertical column of air the almost opposite situation is observed with maximum amounts occurring at high latitudes. Thus, atmospheric transports by mean and eddy (wave) motions (mainly along isentropic surfaces) must play a central role in redistributing the ozone with latitude (see, e.g., Newell, 1963). Mainly due to dynamic effects, the high-latitude ozone mixing ratios undergo an annual cycle with highest values in the late winter and early spring.

Since the early 1970s there has been a growing concern that human activities may upset the natural ozone balance and reduce its concentration in the stratosphere. These activities include, among others:

(a) the emission into the atmosphere of industrial chlorofluorocarbons (CFC-11 and CFC-12) which are used in refrigerants, foams, spray propellants, etc.;

(b) the generation of nitric oxide and nitrogen dioxide by thermal decomposition of air (78% N_2 and 21% O_2). This decomposition is associated with the high temperatures that occur in internal combustion engines and may be a potential threat to the O_3 layer in the case of supersonic and future hypersonic stratospheric flights;

(c) the intensive use of nitrate fertilizers results in an increase of the production of nitrous oxide (N_2O) which is then released into the atmosphere and broken apart in the stratosphere to generate nitric oxide.

On the basis of early model results it was expected for some time that ozone concentrations in the atmosphere would decrease at moderate rates with the increased use of the man-made chemicals mentioned above. However, in 1985 a completely unexpected strong decrease in ozone was discovered over Antarctica. Changes of about 50% in total ozone during the last decade were found to occur in the lower stratosphere, but only during the spring season (Farman *et al.*, 1985). Such a strong seasonal depletion, as shown in Fig. 16.17, was not anticipated by the models and came as a complete surprise (Solomon, 1988). The severe spring depletion of ozone over Antarctica came to be known as the "ozone hole."

Farman *et al.* (1985) have suggested that the ozone hole is connected with the chemistry of chlorofluorocarbon. The chlorofluorocarbons are relatively inert species in the lower atmosphere and are transported upward without being destroyed. These gases enter the stratosphere mainly in the tropics through large-scale upward motions and by deep convection. When the chlorofluorocarbons reach the middle stratosphere they decompose into free chlorine atoms (Cl) and other components under the influence of solar ultraviolet radiation. The free chlorine atoms can then react with ozone leading to ozone destruction:

$$Cl + O_3 \rightarrow ClO + O_2,$$
$$ClO + O \rightarrow Cl + O_2.$$

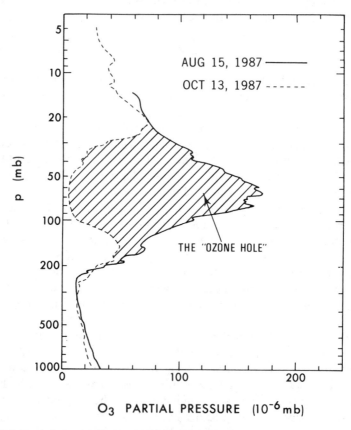

FIGURE 16.17. Observed vertical profiles of ozone concentration (expressed as a partial pressure in 10^{-6} mb) at Halley Bay (76 °S, 27 °W) before and after the development of the Antarctic ozone hole in the spring of 1987 between about 200 and 30 mb (or about 12- and 22-km height). These measurements by balloon-borne ozonesondes at Halley Bay are fairly representative for the recent large-scale depletion of ozone within the polar vortex during the southern spring (from Gardiner, 1989).

This chemical mechanism may be very efficient in the depletion of ozone in the middle stratosphere (e.g., Molina and Rowland, 1974). It is worth noting that the lifetime of the CFC's in the atmosphere is on the order of 100 years. Thus, even if the production of CFC's were stopped immediately their effects on ozone would still be present for several decades.

Farman *et al.* (1985) also pointed out that the cold conditions, prevailing in the upper atmosphere over Antarctica during winter and spring, must be important contributing factors in the depletion of ozone. However, it soon became clear that ozone destruction by chlorine (as proposed by Molina and Rowland, 1974) could not explain by itself the intensity, height, seasonal nature, and regional occurrence of the ozone hole.

The peculiar situation in the Antarctic stratosphere is apparently related to the combination of the chlorine chemistry, the very low temperatures (below 190 K) observed in the large circumpolar vortex during winter, and the reduction of planetary wave activity in the Southern Hemisphere compared to the Northern Hemisphere. The strong zonal circulation isolates the stratospheric air inside the vortex from the outside air so that the chlorine chemistry can occur in near isolation and not be impeded by the formation of stable reservoir species involving odd nitrogen, as in a "contained laboratory reaction vessel." Thus, the ozone hole can spread over a large portion of the vortex. An important further element in this process seems to be the presence of polar stratospheric clouds (PSC's). The existence of ice crystals in these clouds is important not only because nitric acid can be absorbed by the crystals but also because they favor heterogeneous chemical reactions (i.e., reactions in which both solid and gas phases are involved). These heterogeneous reactions can free the chlorine atoms from the stable reservoir species of chlorine (e.g., chlorine nitrate $ClONO_2$) to serve as a catalyst in the photochemical destruction of ozone (Toon *et al.*, 1986).

Some recent satellite observations seem to indicate that ozone depletion may also be occurring over the Arctic regions and, although the ozone destruction occurs much more slowly outside the polar regions, it may even be taking place on a global scale (Stolarski *et al.*, 1991).

16.5.3 Surface temperatures

Returning to a more traditional meteorological quantity (the temperature), we will first discuss the observational evidence for its long-term variability near the earth's surface. Surface temperature records have been available since about the 1860s. A recently analyzed 130-year time series of land-based annual-mean surface air temperatures is shown in Fig. 16.18. It shows in the Northern Hemisphere a general heating until about 1940, a cooling between about 1940 and 1970, and a warming since about 1970 whereas in the Southern Hemisphere a steady warming is observed throughout the entire record. The decrease in Northern Hemisphere surface temperature after about 1940 may be just a natural fluctuation but it is still intriguing since it took place in spite of the atmospheric heating which one would expect due to the (largely man-made) increase in atmospheric CO_2 shown in Fig. 16.16. Nevertheless, the recent warming trend in the 1970s and especially in the 1980s is consistent with the heating associated with the increases in CO_2 and other trace gases. It is good to keep in mind, however, that there are many unanswered questions as to the importance of other effects such as changes in solar input, volcanic activity, and aerosols. Ultimately we will have to include all these effects in our

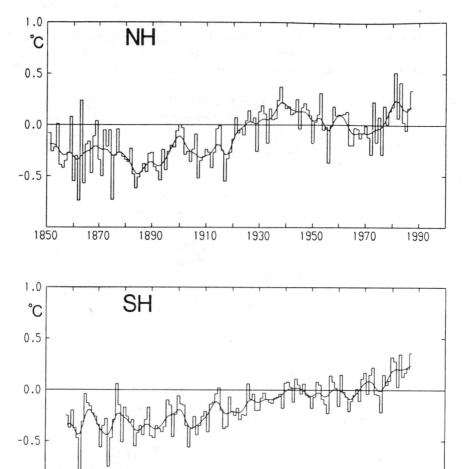

FIGURE 16.18. Time series of the annual-mean surface air temperature (in °C) averaged over the Northern (a) and Southern (b) Hemispheres using land-based historical records from the early 1850s through 1987. Both series are expressed as departures from the reference period 1951–70. Smooth curves show ten-year Gaussian filtered values (from Jones, 1988).

climate models in addition to purely internal factors connected with atmosphere–ocean–cryosphere–lithosphere interactions as was done in a preliminary fashion by, e.g., Schneider and Mass (1975).

A further issue regarding the early historical records arises when we compare the hemispheric curves of surface temperature anomalies in Fig. 16.18 based on land and island data with the comparable curves in Fig. 16.19 based on historical ship data. After about 1900 the land and ocean curves compare quite well, but before 1900 there is a marked discrepancy with very low temperatures in the land records and relatively high temperatures in the ship records. Although attempts have been made (Jones *et al.*, 1986) the differences in the early records have not been satisfactorily explained so far, casting some doubt on the notion of a clear CO_2-related warming in the historical records. On the other hand, we should not discard the possibility that land and ocean may respond differently to the same climatic forcing.

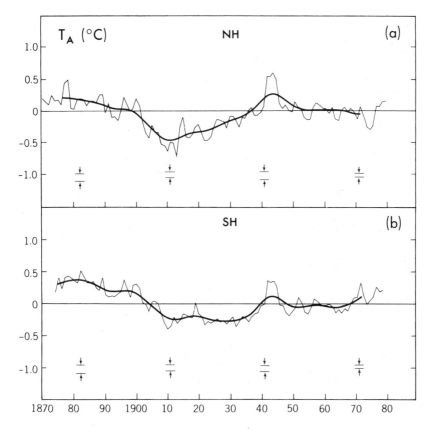

FIGURE 16.19. Time series of the annual-mean surface air temperature anomalies (in °C) averaged over the Northern (a) and Southern (b) Hemisphere oceans between 1870 and 1979. The dominant trends are shown by decadally smoothed curves. The range of the 95% confidence limits for the decadal curves is shown by a few horizontal line segments below the curves (from Oort et al., 1987).

16.5.4 Upper air temperatures

The historical records for the upper air are of a much shorter duration than at the earth's surface, and have only become available since the gradual development of the rawinsonde network following World War II. Nevertheless, these records are of great value since they give a much more representative picture of the state of the atmosphere than the surface records.

Of primary interest is of course the global mean temperature, as a measure of the heat content of the atmosphere. Some of the records for the 30-year period May 1958–April 1988 are shown in Fig. 16.20. The figure shows the time series of the mean temperature for the Northern and Southern Hemispheres averaged vertically between the surface and 25 mb. The temperature variations in the two hemispheres are highly correlated with a maximum correlation coefficient of $r = 0.50$ at zero lag for the unsmoothed series and $r = 0.68$ for the smoothed series. As noticed before in Fig. 16.13, the interannual variability in the global atmosphere is dominated by the tropical ENSO signal with a period between about three and seven years. There is no evidence of a trend in the hemispheric mean temperatures. However, if we separate the contributions from the troposphere and lower stratosphere, as is done in Fig.

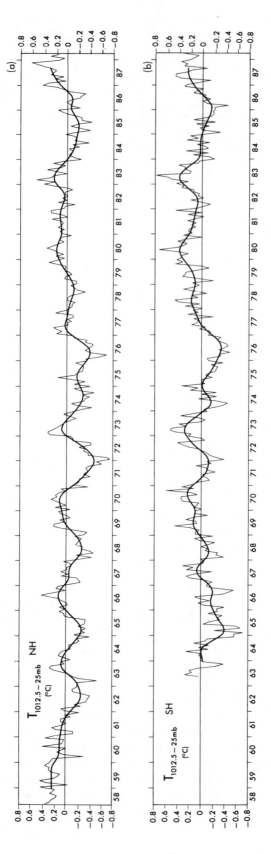

FIGURE 16.20. Time series of monthly-mean atmospheric temperature anomalies in °C averaged over the Northern (a) and Southern (b) Hemispheric mass between the surface and about 25-km height for the period May 1958–April 1988. No Southern Hemispheric values are presented for the 1958–1963 period because of sparse data coverage. The anomalies are taken with respect to the 1963–1973 mean conditions. The smooth curves show 15-month Gaussian-type filtered values.

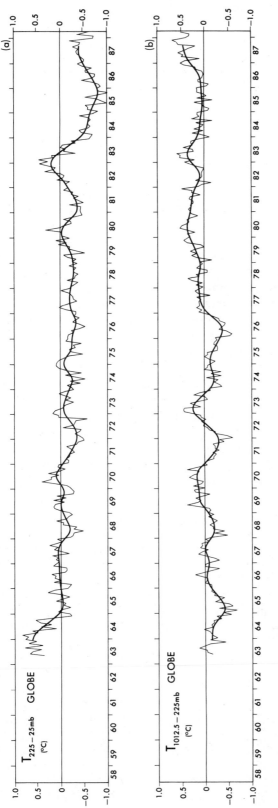

FIGURE 16.21. Time series of monthly-mean global air temperature anomalies (in °C) between May 1963 and April 1988 for (a) the lower stratosphere between 225 and 25 mb and (b) the troposphere between the surface and 225 mb. The anomalies are taken with respect to the 1963–1973 mean conditions. The smooth curves show the 15-month Gaussian-type filtered values. The peaks and valleys in the tropospheric curve largely reflect the ENSO (warm SST) and anti-ENSO (cool SST) conditions, respectively (see also Fig. 16.13). Two of the main peaks in the lower stratospheric curve appear to be caused by major volcanic eruptions (i.e., Agung, February 1963 and El Chichon, March 1982). The curves further show some weak evidence for a heating of the troposphere and a cooling in the lower stratosphere during the 1963–1988 period that is not inconsistent with the predicted greenhouse effect.

16.21, there seems to be weak evidence for a general heating in the troposphere and a general cooling in the lower stratosphere during the last 25 years, roughly as one would expect on the basis of the increase in CO_2 in the air (see Fig. 17.8; Angell, 1988; Karoly, 1987 and 1989; Kuo *et al.*, 1990; Houghton *et al.*, 1990; Oort and Liu, 1991). There is no significant correlation between the temperature variations in the troposphere and lower stratosphere.

16.6 SOME SPECIAL CLIMATIC PHENOMENA

If the anomalous atmospheric circulation patterns mentioned earlier in Sec. 16.4 would persist in time they would lead, through the anomalous advection of cold or warm air masses, to large regional departures in temperature and precipitation from normal conditions. Such departures might occur in the form of cold waves, heat waves, droughts and floods that would exert large stresses on the biosphere and, in particular, on man's living conditions. Moreover, in agriculture the present crops have been developed to make optimum use of the "normal" weather and climate conditions. This practice provides much less flexibility and adaptability during times of abnormal weather and climate conditions, and may thus lead to more wide-spread crop failures.

An example of fluctuations in a climatic index that is particularly important for North America is given in Fig. 16.22. It shows a time series between 1881 and 1985 of annual-mean precipitation anomalies for the Great Plains region. Inspection of Fig. 16.22 shows clear evidence of the well-known drought periods of the early 1930s and 1950s. We should note that not only precipitation but also evaporation and depletion of soil moisture are important factors in drought. An important nonmeteorological factor in drought is the question of supply and demand since the definition of drought generally includes, either explicitly or implicitly, a statement regarding the lack of sufficient water to meet human requirements.

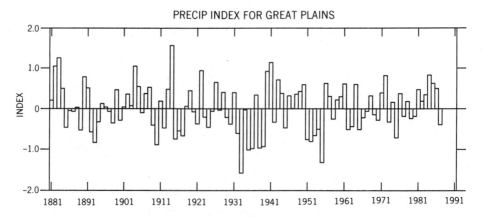

FIGURE 16.22. Time series of annual-mean precipitation anomalies averaged over the Great Plains of the U.S.A. for the period 1881 through 1988 in units of standard deviations. The annual-mean values were computed for the "water" year, i.e., October–September. The first point is for October 1881–September 1882 (personal communication, C. F. Ropelewski, Climate Analysis Center/NMC, Washington, D.C., 1989).

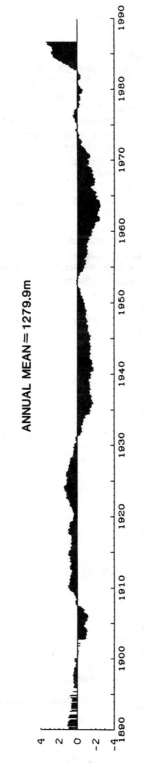

FIGURE 16.23. Time series of lake-level anomalies for the Great Salt Lake in the U.S.A. for the period 1890–1986 (in m). Note the very high lake levels in the mid-1980s (from Cayan et al., 1989).

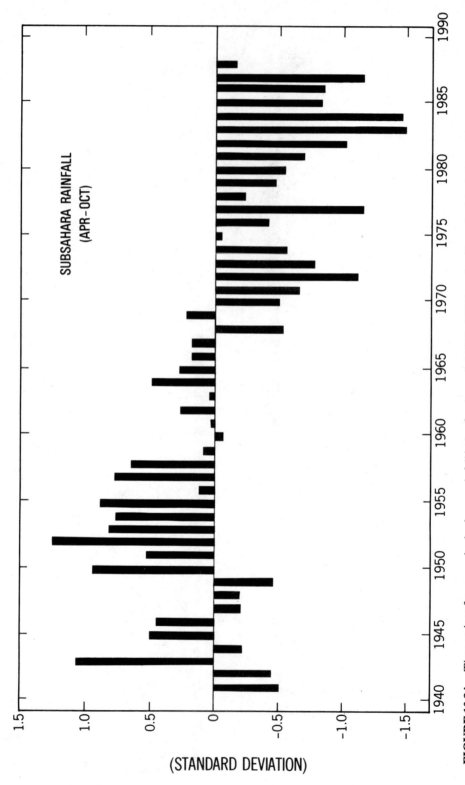

FIGURE 16.24. Time series of an annual subsaharan rainfall index for the period 1941–1988 in units of standard deviations. The index was computed by averaging the rainfall departures from normal for 20 stations in the West African Sahel (after Lamb and Peppler, 1991).

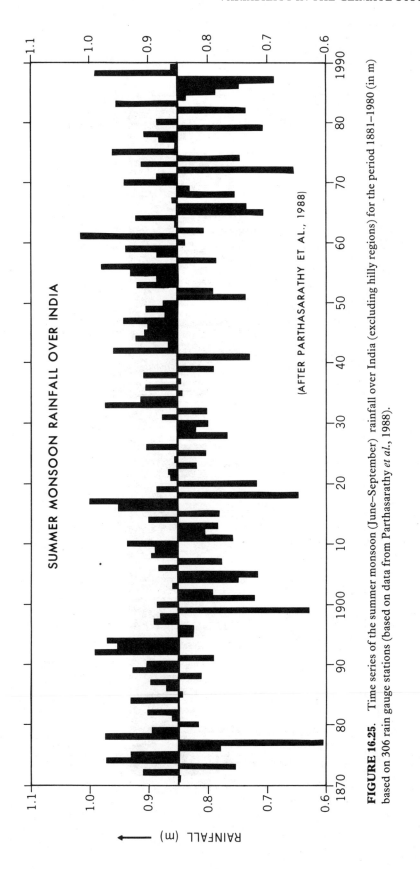

FIGURE 16.25. Time series of the summer monsoon (June–September) rainfall over India (excluding hilly regions) for the period 1881–1980 (in m) based on 306 rain gauge stations (based on data from Parthasarathy et al., 1988).

448 PHYSICS OF CLIMATE

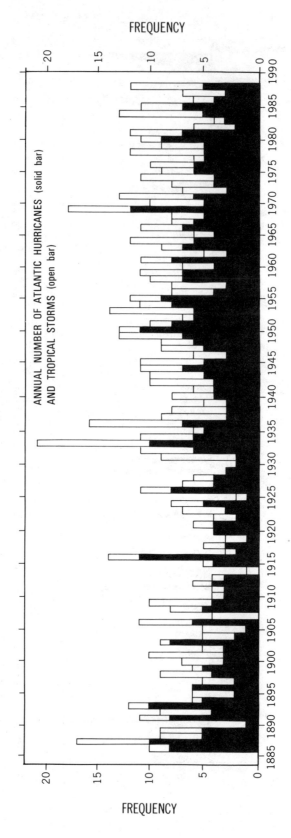

FIGURE 16.26. Time series of the annual number of Atlantic tropical cyclones reaching at least tropical storm strength (open bar) and those reaching hurricane strength (solid bar) for 1886–1988. The average numbers of tropical storms and hurricanes per year are 8.4 and 4.9, respectively (adapted from Neumann *et al.*, 1981, updated).

Another interesting climatic indicator is found in lake levels. As an example, Fig. 16.23 shows the large historical excursions of the level of the Great Salt Lake with very low levels during the 1930s, 1940s, late 1950s and 1960s, and extremely high levels in recent times during the early 1980s. The lake forms an integrating factor of the precipitation and river flow anomalies in its catchment basin and shows therefore fairly long-period variations (a "reddening" of the spectrum). A nonclimatic factor that has recently become important is the extensive irrigational use of water before it reaches the lakes, reducing some of the usefulness of lake levels as a purely climatic indicator unless one can make a correction for this effect.

As a third example, a rainfall index for the Subsaharan region in Africa is shown in Fig. 16.24 from Lamb and Peppler (1991). It shows the dramatic decrease in rainfall since the late 1960s and the continued severe drought conditions from the early 1970s up to 1987. Such a prolonged drought period is unique in the historical record available since about 1900, and there are speculations that desertification by man may be a contributing factor to the present drought. However, there is no generally accepted explanation of why this prolonged drought is occurring and no known way of predicting when such a drought will end.

As a fourth example, the record of the monsoon rainfall over India is shown in Fig. 16.25. On the interannual time scale there is a significant correlation with the ENSO cycle, in the sense that "bad" summer monsoons (insufficient rain) tend to follow the peak of an ENSO event and "good" summer monsoons (sufficient rain) tend to occur during an anti-ENSO episode. There is no clear evidence of long-term trends, although one can recognize some episodes with ample rains, e.g., during the 1920s through 1950s and some episodes with insufficient rains during the 1900s, 1910s, 1960s, and 1980s that are, of course, of crucial importance for the Indian people and their economy.

Due to the potential destructive power of tropical storms it seems appropriate to show here a plot of the frequency of tropical storms and hurricanes, as another example of an important regional phenomenon. From the climate point of view these storms that occur mainly in the early fall provide another mechanism to transport heat and water vapor from the tropics to midlatitudes. It is well-known that these tropical vortices develop over the warm waters of the tropical oceans (e.g., in the case of the North Atlantic often near the Cape Verde Islands). The large generation of available potential energy associated with the release of latent heat provides the source for the kinetic energy that keeps the circulation in the storms alive. A plot of the annual number of North Atlantic storms since 1885 is shown in Fig. 16.26. It is evident that there are certain years and even groups of years when the storms are much more frequent than normal.

CHAPTER 17

Mathematical Simulation of Climate

17.1 INTRODUCTION

As we have discussed in Chaps. 2 and 16, the climate of the earth has undergone large changes in the past as shown from geological evidence, proxy data, and, more recently, historical records. The climate is changing now and will change in the future. These changes have a profound influence on life on earth as well as on the environment.

If we had a general theory of climate and climatic change (our ultimate goal; see Fig. 1.1), it would be possible to explain the causes of climatic variability, to get a better understanding of the observed changes, and finally to predict their occurrence in the future. Unfortunately, as mentioned in the Introduction, there is not at present such a comprehensive theory and, which is still worse, it is not to be expected in the near future. Furthermore, there exist no physical models which can simulate the complex behavior of the climate system in a laboratory environment in an adequate way. For example, the nonlinear interactions between the various subsystems are impossible to reproduce, even partially, in any laboratory experiment. The laboratory analogs (dishpan or annulus experiments) obtained so far can only yield some general dynamical characteristics of the rotating ocean–atmosphere system in a very preliminary and coarse way.

On the other hand, we know that the external factors (solar radiation, absorbing gases in the atmosphere, ice cover, etc.) and the thermo-hydrodynamical quantities that characterize the climate (temperature, density, velocity, moisture content, salinity, etc.) are all interrelated through a set of physical laws expressed by the various equations, as discussed and analyzed in previous chapters. These equations are based on the general principles of conservation of mass, momentum, and energy. Taken together with the physical and chemical laws that govern the composition of the components of the climate system, they constitute the fundamental theoretical basis for studying the climate and its evolution. These three-dimensional equations are nonlinear and each variable is related to the others. Any change in one of the variables may induce variations in the others which, in turn, will generate a feedback on the original variable.

The set of coupled partial differential equations can, in principle, be solved subject to the solar radiation input and other specified boundary and initial conditions that define an instantaneous state of the climate system. This mathematically closed set of equations constitutes a well-posed mathematical problem and forms the core of all mathematical model simulations of climate.

As mentioned several times before, the climatic processes are very intricate and some of them are still poorly understood. They involve various feedback mechanisms whose effects are not always known. Further, the equations expressing the governing physical laws are, mathematically speaking, very complex. Also the boundary conditions are not always sufficiently defined because they are based on observational data which are sometimes incomplete or not accurate enough.

All these difficulties create the unavoidable need for introducing simplifications and specializations in studying the climate. Such simplifications and specializations can be accomplished through different methods of averaging and parameterizing, resulting in the great variety of models that are in use today. By and large, the simplifications fall into the two broad areas of physics and mathematics. The first area of simplifications deals with the selection of the natural processes that will be represented in the model, whereas the second area deals with the resolution of the model equations in space and time.

Due to practical limitations and our limited understanding of the many scales of the natural phenomena that occur in the climate system, it is not possible to duplicate all processes in a single general model. Thus, in a given model some processes are selected and treated in great detail, whereas the others are ignored or treated in a less elaborate way. In principle, the resolution in time and space is dictated by the type of study but, in practice, it is also limited by the available computational resources. In general, the finer the resolution the more reliable the results.

At this point we must emphasize the decisive role played by high-speed computers in climate modeling and simulations. Without the growth in computational power and the consequent reduction of costs, most of the progress in climate modeling could not have happened (see Fig. 17.1). Furthermore, the easy accessibility to computing facilities has strongly contributed to spread interest in performing climate studies. It is the use of both mathematics and high-speed computers that makes it possible to simulate various physical processes occurring in the climate system, allowing a better understanding of the working of the atmosphere, hydrosphere, and cryosphere.

It is important to note that different classes of models require different kinds of validation, i.e., different levels of comparison between the results of the theoretical model and the corresponding results obtained from the real atmosphere.

In summary, mathematical models provide a new way to not only understand the climate behavior but to also explore the possibility of predicting future climatic developments. Furthermore, the mathematical simulations provide us with a powerful tool to conduct "experiments" for a great variety of prescribed climate scenarios.

17.2 MATHEMATICAL AND PHYSICAL STRUCTURE OF CLIMATE MODELS

17.2.1 Basic parameters of a climate model

We will now consider in a more precise manner what constitutes a physical–mathematical model of climate. We assume that climate represents a statistical, qua-

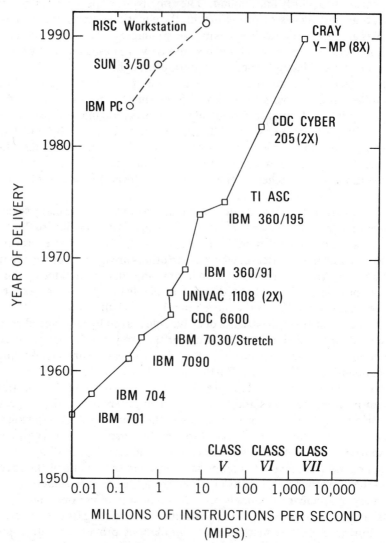

FIGURE 17.1. Example of the growth in computational power (in MIPS) during the last three decades at the Geophysical Fluid Dynamics Laboratory/NOAA, one of the pioneering institutions in the numerical simulation and diagnosis of climate. Shown is the succession of main-line computers culminating in the recent acquisition of a Class VII (8-processor) Cray Y-MP computer, as well as the new line of desktop computers introduced in the 1980s. We may note that many of the analyses of both the real and model atmosphere and oceans described in this book were obtained using the computer facilities at GFDL (courtesy B. Ross, J. Smagorinsky, and J. Welsh, GFDL).

si-steady state of the atmosphere that depends only on the boundary conditions. A climate model is a set of specialized thermo-hydrodynamic equations with usually prescribed boundary and initial conditions, certain given values of the physical constants, and specified schemes of parameterization of the subgrid-scale fluxes of water vapor, momentum, and energy. The physical constants include planetary data, such as the radius of the earth, the acceleration of gravity, the angular velocity of rotation, and internal constants, such as total mass, chemical composition of the air and oceans, specific heats, latent heats of phase transitions of water, and radiative transfer parameters. The boundary conditions include the solar energy input as influenced by the orbital parameters (ellipticity, obliquity, and longitude of perihelion), surface topography, surface roughness, heat capacity of the underlying surface, albedo, soil moisture capacity, vegetation, etc.

As we have seen in Chap. 2, in order to understand the climate system as a whole we must also consider the behavior of each of its individual components and the strong interactions between these components. However, in practice, it seems impossible to do this because of the wide range of time and space scales involved. For example, the time scales for the atmosphere were found to be 10^{-3} to 10^{-1} years, for the oceans 10^{-1} to 10^3 years, and for the cryosphere 10^0 to 10^5 years. As an intermediate solution, we can often take the atmosphere as the main system with the continental surface topography, the surface roughness, and the sea surface temperatures as boundary conditions, and only study the thermo-hydrodynamic equations for the atmosphere.

17.2.2 Model equations

A general view of the major components of the mathematical simulation of climate in the atmosphere is given in Fig. 17.2, showing the basic conservation equations and the external factors, namely radiation and other interactive processes, such as the transfers of momentum, heat, and water substance across the earth's surface. Using modeler's language, we would say that Fig. 17.2 portrays the three main components in a climate model, i.e., the "dynamics," "physics," and "other factors."

The dynamics deals with the numerical schemes of the equations of motion and the large-scale transports of mass, water vapor, and energy. Although the dynamics are, of course, a part of physics, it makes some sense here to distinguish between them since, historically, the climate models were an extension of the early models developed for numerical weather prediction that were based mainly on the equations of motion. In applying the numerical methods to the study of climate, it became obvious that the boundary conditions and more elaborate physical processes had to be included (as it is now also current practice in numerical weather prediction). This explains the apparent inconsistency of the modeler's language between dynamics and physics.

The physics comprise specialized schemes by which (1) the input of solar radiation and the emission and absorption of terrestrial radiation are treated within the climate system, (2) the water in the atmosphere is dealt with including cloud formation, cloud distribution, precipitation, and the release of latent heat, and (3) the thermodynamics and the balance of energy are taken into account.

The other factors include surface processes which link the atmosphere with the oceans, and with the continental surfaces through the transfers of mass, momentum, sensible heat, and latent heat; the land–sea–ice contrasts and the resulting variations in the albedo; the topography; the soil moisture and vegetation cover, etc. In addition, we must consider the nature of the various parameterizations used (see Sec. 17.2.4).

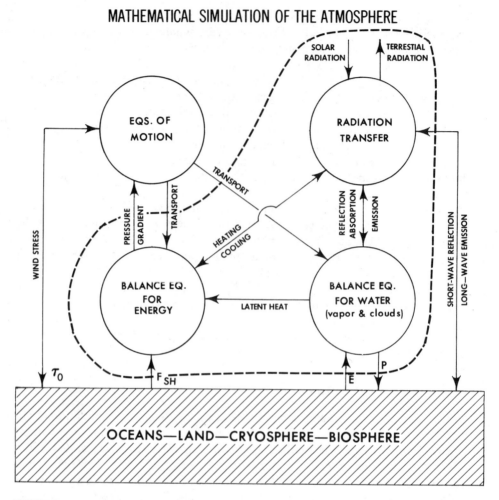

FIGURE 17.2. Major components of the mathematical simulation of climate (the equation of state has been left out from the diagram). The domain within the dashed curve corresponds to the physics, and the outside domain to the dynamics and other factors mentioned in the text.

The governing equations of the atmospheric models are given essentially by the equations of motion, the first law of thermodynamics, the balance equation for water vapor, and the equation of state. They form a closed system of equations with specified boundary conditions. Similar equations apply for the oceans except that a balance equation for salinity rather than for water vapor is used and the equation of state is more complex than in the atmospheric case. At this point, it seems convenient to summarize the equations as they are commonly used in climate modeling, which are simplified versions of the equations presented in Chap. 3. Using the same notations and symbols, we obtain then

(a) the horizontal equations of motion

$$\frac{du}{dt} = \left(f + u\frac{\tan \phi}{R}\right)v - \frac{1}{\rho}\frac{\partial p}{R \cos \phi \partial \lambda} + F_\lambda, \qquad (17.1)$$

$$\frac{dv}{dt} = -\left(f + u\frac{\tan \phi}{R}\right)u - \frac{1}{\rho}\frac{\partial p}{R \partial \phi} + F_\phi, \qquad (17.2)$$

where

$$\frac{d}{dt} = \frac{\partial}{\partial t} + u\frac{\partial}{R\cos\phi\,\partial\lambda} + v\frac{\partial}{R\partial\phi} + w\frac{\partial}{\partial z};$$

(b) the hydrostatic law

$$\frac{\partial p}{\partial z} = -\rho g; \tag{17.3}$$

(c) the continuity equation

$$\frac{\partial \rho}{\partial t} = -\frac{1}{R\cos\phi}\left[\frac{\partial}{\partial\lambda}(\rho u) + \frac{\partial}{\partial\phi}(\rho v\cos\phi)\right]$$
$$- \frac{\partial}{\partial z}(\rho w); \tag{17.4}$$

(d) the first law of thermodynamics (balance equation for energy)

$$c_p \frac{d\ln T}{dt} - R_d \frac{d\ln p}{dt} = \frac{Q}{T}, \tag{17.5}$$

where [see Eq. (13.7)]

$$Q = -\alpha \operatorname{div} \mathbf{F}_{\text{rad}} - L(e-c) - \alpha \operatorname{div} \mathbf{J}_H^D$$
$$- \alpha \tau : \operatorname{grad} \mathbf{c}; \tag{17.6}$$

(e) the equation of state for air

$$p = \rho R_d T(1 + 0.61q); \tag{17.7}$$

and (f) the balance equation for water vapor

$$\frac{dq}{dt} = (e - c) + D. \tag{17.8}$$

The basic predicted variables are the wind components u and v, temperature T, specific humidity q, and, in most formulations, the surface pressure p_0. The prediction equation for surface pressure is obtained from the continuity equation (17.4) by vertical integration using the hydrostatic assumption.

The computation of the net radiative heating term in Eq. (17.6) requires, as shown before in Eq. (6.38), information on the fluxes of solar and terrestrial radiation expressed in terms of the vertical distributions of temperature and the radiatively active atmospheric constituents including clouds. This is given basically by Eqs. (6.16) or (6.18) for the solar radiation and Eqs. (6.34a) and (6.34b) for the terrestrial radiation.

In constructing a numerical model of the atmosphere, one of the first decisions to make is the choice of the vertical coordinate. Several systems have been used besides the z system, such as the p system, the θ system (potential temperature), and the σ system where $\sigma = p/p_0$ and p_0 is the surface pressure (Phillips, 1957).

The most common system presently used in numerical modeling is the σ system. We may note that $\sigma = 0$ at the top of the atmosphere, where $p = 0$, and that $\sigma = 1$ at the earth's surface, where $p = p_0$. With this system the problem of having a vertical coordinate system that intersects the mountains is reduced, because the $\sigma = 1$ surface coincides, even over mountains, with the earth's surface. Furthermore, with this system it is easier to incorporate the vertical exchange processes in the planetary boundary layer.

It is obvious that the σ-coordinate system requires some formal changes in the equations. For example, the pressure gradient force is written in the form

$$-\alpha \operatorname{grad}_z p = -\operatorname{grad}_\sigma \phi - \sigma \alpha \operatorname{grad} p_0, \qquad (17.9)$$

where ϕ is the geopotential. Over steep terrain, the two terms on the right-hand side of Eq. (17.9) have almost the same large value but the opposite sign so that the finite difference formulation of these terms may lead to large errors in the pressure gradient force. The operator d/dt becomes in the σ system

$$\frac{d}{dt} = \frac{\partial}{\partial t} + \mathbf{v}\cdot\operatorname{grad}_\sigma + \dot{\sigma}\,\frac{\partial}{\partial \sigma},$$

where

$$\dot{\sigma} = \frac{d\sigma}{dt}. \qquad (17.10)$$

Analogous considerations can be made for an ocean model with some adjustments concerning the boundary conditions (such as the surface wind stress and surface temperature), internal constants, and specified parameterization schemes.

17.2.3 Necessity of using numerical integrations

The climate equations are highly nonlinear in nature and no general analytical methods are available to solve them. It suffices to say that, e.g.,

$$\frac{du}{dt} = \frac{\partial u}{\partial t} + u\,\frac{\partial u}{\partial x} + v\,\frac{\partial u}{\partial y} + w\,\frac{\partial u}{\partial z},$$

where, of course, the last three terms are nonlinear. We then have to rely on numerical analyses to provide the desired solutions.

Thus, the partial differential equations have to be either replaced by equivalent finite difference equations or solved by spectral techniques. Even so, the primitive equations contain all scales of motion from sound and gravity waves up to planetary waves. Some of these scales of motion do not influence the weather processes significantly and may produce a noise level during integration that obscures the real meteorological signal. This appears to be the case for sound and some of the gravity waves. These must then be filtered. The filtering may be accomplished by using the hydrostatic equilibrium hypothesis and, in some models, by using the quasigeostrophic approximation in middle and high latitudes (see discussions in Secs. 3.2.2 and 3.2.3).

The utilization of numerical methods may give rise to other problems of a mathematical nature, such as the convergence of the solutions and their stability. Basically in the finite difference techniques, we are forced to substitute a discretum x_i, y_j, z_k, t_l for a continuum x, y, z, t, where $i = 1, 2, 3,..., I$, $j = 1, 2,..., J$, $k = 1, 2,..., K$, and $l = 1, 2,..., L$. The computational resolution is given by the increments $\Delta x, \Delta y, \Delta z$, and Δt and must be fixed in agreement with the nature of the problem studied.

In the case of the atmosphere and oceans, the number of levels in the vertical K is usually on the order of 10 to 20, with variable spacing ranging from a few 100 m's in the surface boundary layer to a few km's in the free atmosphere. Although the separation between gridpoints in the horizontal, e.g., 200 to 300 km, far exceeds the separation in the vertical, the number of gridpoints in either the east–west or north–south direction needed to cover the globe is an order of magnitude greater than in the vertical.

The total number of horizontal gridpoints is given by A/Δ^2 where A is the area of the globe and Δ the average horizontal grid distance. If n is the number of variables necessary to define the state of the atmosphere at each point, then at a given time the total number of variables is $n \times K \times A/\Delta^2$. This last number represents the number of degrees of freedom for the system we are dealing with.

The increment of time Δt cannot be chosen arbitrarily. Its choice depends on the numerical time-stepping schemes used. For example, in the centered difference scheme, it must obey the Courant–Friedrichs–Lewy condition in order to secure the stability and convergence of the integration

$$c \, \Delta t / \Delta x \leqslant 1, \tag{17.11}$$

where c denotes the speed of the fastest mode allowed. Strictly speaking, this condition can be applied only to linear systems. However, it also gives some idea of the necessary limitations for the time step in nonlinear systems. Implicit methods to solve the equations do not require a pre-established time step but their calculation is more complex (Richtmeyer and Morton, 1967).

Among several possible schemes (Arakawa, 1988) we will follow the usual approach of approximating the spatial derivatives of a variable ψ at a given point (i,j,k) by a centered difference between the values of ψ at the neighboring gridpoints of the discretum:

$$\frac{\partial \psi}{\partial x} = \frac{\psi_{i+1,j,k} - \psi_{i-1,j,k}}{2\Delta x},$$

$$\frac{\partial \psi}{\partial y} = \frac{\psi_{i,j+1,k} - \psi_{i,j-1,k}}{2\Delta y}, \tag{17.12}$$

$$\frac{\partial \psi}{\partial z} = \frac{\psi_{i,j,k+1} - \psi_{i,j,k-1}}{2\Delta z}.$$

The centered difference scheme has to be modified at the boundaries using one-sided differences, forward or backward.

Using the prediction equations, the time rate of change at each grid point for a given instant τ, $(\partial \psi / \partial t)_\tau$, can be computed from the instantaneous distribution of the variables appearing on the right-hand side of the equations. So that at each point (i,j,k) using forward differencing:

$$\psi^{\tau+1} = \psi^{\tau} + \Delta t \left(\frac{\partial \psi}{\partial t} \right)_\tau \tag{17.13a}$$

or centered differencing:

$$\psi^{\tau+1} = \psi^{\tau-1} + 2\Delta t \left(\frac{\partial \psi}{\partial t} \right)_\tau. \tag{17.13b}$$

The last scheme is known as the leap-frog method.

By repeated application of the leap-frog method the state of the atmosphere can be evaluated, in principle, through any desired period of time. In practice, however, this particular system of finite difference equations is computationally unstable. For numerical weather prediction purposes and short-range forecasts typical values are $\Delta x \simeq 100$ km, $\Delta y \simeq 100$ km, $\Delta t \simeq 5$ min, whereas Δz ranges from about 50 m in the boundary layer to several kilometers in the free atmosphere. However, for climate studies the typical values are usually larger for the sake of computational economy (see Fig. 17.3).

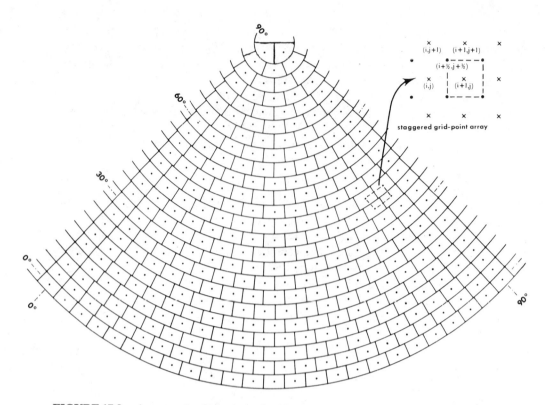

FIGURE 17.3. An example of a horizontal grid system (the so-called Kurihara grid) with 24 grid points between the pole and equator, used in general circulation model experiments (only 1/8 of globe is shown). A staggered grid-point array for performing the needed finite difference computations is shown as an insert in the top right-hand corner. Other systems in use are regular latitude–longitude grids and various spherical harmonic systems (after Kurihara and Holloway, **1967**).

When we consider the number of degrees of freedom of the system, the number of operations required at each grid point, and the number of time steps to cover the forecast or simulation period, it is obvious that the integration requires very powerful computers to perform the calculations (see Fig. 17.1; Smagorinsky, 1974).

17.2.4 Parameterizations

As we have seen, the phenomena occurring in the climate system span a very wide range of both time and space scales. Indeed, whereas the general circulation features extend for 1000 km or more in space and last for days to weeks, the boundary layer turbulent flows persist only for a few minutes with dimensions of the order of centimeters to meters. Nevertheless, the turbulent motions are of paramount importance in transferring mass, momentum, and energy across the air–sea and air–land interface and the boundary layers. As shown in Chap. 10, these small-scale processes can be treated statistically and their effects can be related to average conditions over much longer periods of time and larger space scales using the so-called gradient-flux and bulk aerodynamic methods of approach.

These techniques have been extended to express the effects of all unresolved scales in terms of the larger scale phenomena that are directly evaluated in the model. This technique is called parametric representation or parameterization, and is used extensively in numerical modeling. The parameterization plays a very important role mainly when we deal with the physics and other factors of a model.

To solve the model equations (17.1)–(17.8) we have to treat the system at a discrete number of instants in time and at a discrete number of points in space, substituting the continuum by a spatial grid of isolated points. The dimensions of the basic grid determine the minimum scale of phenomena that can be resolved by the equations. Scales smaller than the basic grid are known as subgrid scales. For example, clouds, sharp fronts, gravity waves, and turbulence are usually not resolved explicitly and therefore represent subgrid scale phenomena. However, they are thought to be very important in the various budgets and especially in the vertical transports of momentum, heat, and water vapor. Thus, their effects must be included in some way in the models. For example, cumulus convection will influence the general circulation through the release of latent heat, the vertical fluxes of heat, moisture, and momentum, and the interactions with radiation that accompany it. The drag resulting from the breaking of orographically induced gravity waves has also been parameterized recently because without it the model tropospheric winds tend to be too strong (Boer et al., 1984; Palmer et al., 1986). Through parameterization techniques it is possible to describe the effects of all subgrid-scale eddies on the large-scale phenomena in both the horizontal and vertical directions using appropriate diffusion schemes. However, there is no guarantee that this approach is the correct way of incorporating subgrid scale processes in the general circulation. The spatial resolution, the time step of integration, the parameterization techniques, and the physical and chemical constants are all very important factors defining the model.

We must be careful not to use parameterizations indiscriminately as a mathematical device just to obtain a closure condition. Whenever possible, parameterizations should be based primarily on physical arguments rather than on mathematical expediency. Also they should be quantitatively validated against observations.

The surface processes are usually parameterized based on the bulk or drag formulations given in Chap. 10. For example, for momentum (10.30) we have

$$\tau_0 = -\rho C_D |\mathbf{v}(z)| \mathbf{v}(z),$$

for heat [Eq. (10.34)]

$$F_{\text{SH}} = -\rho c_p C_H |\mathbf{v}(z)| [T(z) - T(0)],$$

and for water vapor [Eq. (10.38)]

$$E = -\rho C_W |\mathbf{v}(z)| [q(z) - q(0)],$$

where some uncertainty is involved because of the dependence of the drag coefficients on static stability, clouds, etc.

In the boundary layer, the vertical eddy fluxes are frequently parameterized using the gradient-flux approach [see, e.g., Eq. (10.20) for heat and Eq. (10.21) for momentum].

The parameterization of cumulus convection has so far been performed using many different schemes. At present, the three main types are (a) adjustment schemes (Manabe et al., 1965), (b) moisture convergence schemes (Kuo, 1965, 1974), and (c) the Arakawa–Schubert (1974) scheme.

The simplest scheme is Manabe's adjustment scheme. It assumes that in a moist convectively unstable atmosphere buoyant convection prevails at the subgrid scales thereby removing large-scale super moist adiabatic instabilities. Because of the efficient vertical mixing and the continuous formation and evaporation of water droplets, the equivalent potential temperature in the convective layer is assumed to be uniform in the vertical.

The Kuo scheme assumes that in a moist, convectively unstable atmosphere penetrative convection will develop, leading to the formation of cumulus clouds. This occurs in such a way that only a fraction of the total water vapor that converges in the boundary layer is condensed above the lifting condensation level and precipitates out, while the remainder stays in the atmosphere, increasing its moisture content.

The Arakawa–Schubert scheme is a more complex adjustment scheme incorporating the effects of an ensemble of cumulus clouds and their interactions with the environment. In this last scheme, the cloud formation is assumed to be the net product of the rising of buoyant air, the entrainment of environmental air into the clouds, the detrainment from the clouds, and the subsidence outside the clouds.

17.2.5 Nature of the mathematical solutions

The simplifications imposed on the equations and on their transcription into numerical form, truncation errors, and the errors in the observations all lead to errors in the initial and boundary conditions. Because of the nonlinear and unstable characteristics of the atmosphere, an initial error can grow at such a high rate that it can mask the real meteorological signal, thus limiting the predictability by the model. The rate of error growth depends on the scales of motion considered. The smaller the scale, the shorter its predictability in the model. As discussed before, for synoptic-scale weather patterns the limit of predictability is estimated to be on the order of two to three weeks (see Fig. 2.5), which means that no individual, instantaneous state of the atmosphere can be predicted beyond about three weeks in spite of further perfections in the equations and in the accuracy of the initial data.

The predictive skill actually attained using numerical weather prediction models is, of course, much shorter than two or three weeks. Examples of the skills are shown in Fig. 17.4 for the mid-1960s and the late 1980s. As a measure of the skill, the correlation coefficients $r(t)$ between the observed and predicted time changes of the Northern Hemisphere 500-mb height field from its initial value were chosen. The correlation coefficients were computed using the expression

$$r(t) = \frac{\{\Delta z^p(t)\Delta z(t)\}}{\sqrt{\{\Delta z^p(t)^2\}}\sqrt{\{\Delta z(t)^2\}}}, \quad (17.14)$$

where $\Delta z_{ij}(t) = z_{ij}(t) - \bar{z}_{ij}$ and $\Delta z^p_{ij}(t) = z^p_{ij}(t) - \bar{z}_{ij}$ are the observed and predicted geopotential height departures from the climatological normal $\bar{z}_{ij}$ at t days at grid point (i,j), respectively, and the bracelets

$$\{\Delta z\} \equiv \sum_{i,j} \Delta z_{ij} / N$$

indicate an average over all grid points between 20 °N and the north pole (N = total number of grid points). The persistence correlation defined by

$$r^p(t) = E\left(\frac{\{\Delta z(0)\Delta z(t)\}}{\sqrt{\{\Delta z(0)^2\}}\sqrt{\{\Delta z(t)^2\}}}\right) \quad (17.15)$$

can be used as a measure of the no-skill case. On the right-hand side of Eq. (17.15) the expected value (E) for an ensemble of forecasts is given. This value is approximately equal to the averaged correlation for the various forecasts made [i.e., 12 winter cases in Fig. 17.4(a) and 90 winter cases in Fig. 17.4(b)]. For further information on skill scores and their potential deficiencies, see Brier and Roger (1951).

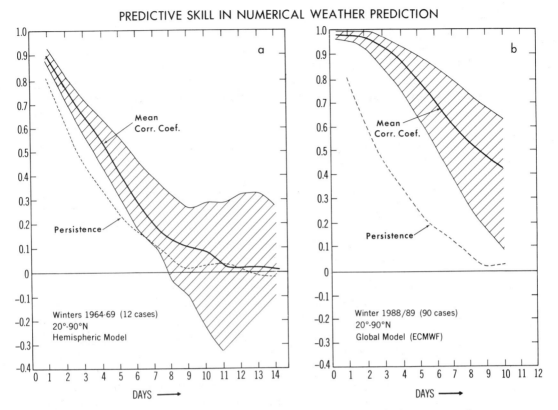

FIGURE 17.4. The predictive skill in numerical weather prediction and its improvement from the mid-1960s (a) to the late 1980s (b). Shown is the decrease with time of the correlation coefficient between predicted and observed 500-mb geopotential height anomalies for 12 two-week predictions in (a) and 90 ten-day predictions in (b). The correlation coefficients were computed using all grid-point values north of 20 °N. The thick solid lines show the mean correlation values averaged over all individual cases, whereas the shaded areas give a measure of the range of the individual cases (i.e., about 5% of the predictions lie below the shaded area and 5% above it). For comparison, the persistence curve (dashed) indicates the no-skill forecast. The growth in (useful) predictive skill from less than one week in the mid-1960s to about ten days in the 1980s is evident. The data in (a) are from Miyakoda et al. (1972) and in (b) from unpublished ECMWF statistics (courtesy L. Bengtsson).

Comparing the two sets of graphs in Fig. 17.4 that are fairly typical for the predictive skill in the mid-1960s and late 1980s, we see a remarkable growth during the last two decades. This growth is due largely to improvements in the input data, initialization procedures, and prediction models. For example, the change from an hemispheric model in Fig. 17.4(a) to a global model in Fig. 17.4(b) is essential for forecasts beyond about five days since the influence of the tropics and Southern Hemisphere will begin to affect the Northern Hemispheric circulation in midlatitudes after a few days.

The growth in (useful) predictive skill from less than one week in the mid-1960s to about ten days in the 1980s is very clear. Further, growth up to possibly two to three weeks may still be expected, especially in blocking situations.

We may note that a blocking condition is usually formed by a persistent high-pressure system (anticyclone) near the western sides of the continents in middle latitudes. With this type of situation the trajectories of low-level disturbances and of the upper air jets are split into two branches as if they were blocked by the high-pressure system. This situation may persist for several weeks.

There are, of course, some intrinsic limitations in applying a finite-difference scheme repetitively in order to simulate the long-term climate. However, in studying climate we are not interested in an individual instantaneous state *per se* but only in an ensemble of states. This is very similar to the situation that occurs in statistical mechanics. Therefore, we may instead study the statistical moments of various orders for the atmospheric variables as well as their evolution in time when the external boundary conditions are given. These conditions determine the type of circulation regime and, in principle, the statistical behavior of the atmosphere. For example, in spite of difficulties in predicting the exact location and trajectory of a particular depression or other perturbation it is possible to obtain a statistical distribution function for the positions and trajectories of the entire class of perturbations.

In an early conference to study the potential of climate forecasting, von Neumann (1960) distinguishes three classes of predictions. The first class includes short-range predictions of day-to-day weather events, the second class medium-to-long range weather or climate predictions from weeks to months and possibly years, and finally the third class, the climate equilibrium (steady state) simulation. As envisioned by von Neumann, much progress has been made since the 1950s in the first and third class of problems. However, in the second class of problems progress has been more limited, because both initial conditions and imposed boundary conditions are equally important to obtain the final solution.

More recently Lorenz (1975) gave a slightly different version considering two types of climate predictability, i.e., the predictability of the first and second kind. The first kind is defined as the prediction of the statistical properties of the climate system with a given initial state as if we were dealing with a long-range forecast problem. This type corresponds to a variant of the second class of problems in von Neumann's classification.

The predictability of the second kind is defined as the determination of the response of the climate system to changes in the external forcings. These predictions are not concerned directly with the chronological evolution of the climate state but rather with the long-term average of the statistical properties of climate. This corresponds to the third class of von Neumann.

As we have discussed in Chap. 2, the important question of uniqueness of the statistical equilibrium solution has also been dealt with by Lorenz (1969). He concluded that for certain hydrodynamical systems more than one equilibrium state may exist that will satisfy the same governing equations and same boundary conditions. This point is very important, and raises the problem of climate determinism.

Most modeling studies of climate assume that the climate is transitive. They are thus concerned with the steady-state or equilibrium solutions under given boundary conditions and with some prescribed internal variables and an adequate set of empirical constants. The time derivatives of the various quantities are supposed to be negligible compared to the other terms in the climate equations since no systematic trends with time have been observed. The quasi-equilibrium solutions enable us to compare the results of the simulation with the observed mean global climate. Furthermore, there is the possibility of testing the sensitivity of the climate models. A sensitivity analysis may consist in studying the response of the model to changes in the individual parameters, the model constants, or the parameterizations on which the equilibrium solution depends. Of course, in conducting a sensitivity experiment for a particular parameter, the values of all remaining parameters must be kept constant. A sensitivity analysis of, e.g., the parameterization constants is important to test the validity of a model, since it may reveal which parameters have the most pronounced influence on the solution and hence must be determined with greatest accuracy.

17.3 HIERARCHY OF CLIMATE MODELS

We will now describe some ways to classify the various types of mathematical models of climate that are presently being used. Thus, we will follow a kind of hierarchy in terms of the number of space dimensions included and the degree of model complexity emphasizing the physical foundations. For general reviews, see Schneider and Dickinson (1974), Smagorinsky (1974), Saltzman (1978), Ramanathan and Coakley (1978), North et al. (1981), and Schlesinger (1988).

A useful systematization of climate models can be obtained by taking into account the modes of averaging in space. Let us consider first the three-dimensional climate fields, denoted by 3-D(λ,ϕ,z). They can be averaged in the vertical leading to two-dimensional [2-D(λ,ϕ)] models, or in the zonal direction leading to two-dimensional [2-D(ϕ,z)] models which, of course, are axially symmetric models. If these two-dimensional models are now averaged in one of the other dimensions, we obtain one-dimensional [1-D(λ), 1-D(ϕ), or 1-D(z)] models. The integration of these last models with respect to the remaining dimension will lead to global mean or zero-dimensional [0-D] models.

As an example of a 0-D model, we can mention the simplest model of climate, which considers only one variable, the temperature. As we discussed in Sec. 6.8.1, this temperature may be regarded as the long-term average of the global-mean temperature that results from the balance between the absorbed solar and the emitted terrestrial radiation. Thus, the computed value of the radiative equilibrium temperature $T_e = 255$ K would be equal to the surface temperature if the atmosphere were transparent for thermal radiation. However, the actually observed value of the mean surface temperature $T_{\text{sfc}} = 288$ K is very different from the value of T_e so that the increment of 33 K must be attributed to the atmospheric greenhouse effect.

The more averaging operations are performed the lower the model resolution needed but the more parameterizations are required. As we will see, the 3-D general circulation models (GCM's) have the highest resolution but require the least amount of parameterization. In the GCM's, such phenomena as synoptic weather systems are explicitly taken into account. On the other hand, the global-mean models demand the most elaborate parameterization schemes.

Perhaps more suggestive than this type of classification based on geometrical considerations is the systematization of climate models according to their contents, i.e., the dynamics, physics, and other factors. The most frequently used type of model is the deterministic model. In this group there are two main subgroups: the explicit dynamical models and the statistical dynamical models. Both subgroups have several variants as we will see.

In general, the explicit dynamical models are three-dimensional models [3-D(λ,ϕ,z)] based on the synoptic equations [Eqs. (17.1)–(17.8)]. The evolution of the large-scale transient eddy circulations is followed on a day-to-day basis similarly as in numerical weather forecasting. The final climate statistics are evaluated from the daily simulations in the same way as is done usually in climatological studies based on the actual observations (see Fig. 5.9).

These models can further be subdivided into two types:

(1) GCM's that require a long-term integration of the three-dimensional equations on a synoptic basis. The equations can be solved using a three-dimensional network of grid points in the geometrical space or using a complete spherical harmonic description at various levels in the wave number domain. The grid-point GCM's generally require a high resolution in space and an integration with relatively short time steps. In view of their importance and their wide use in applications we will mainly discuss the GCM's in the later sections.

(2) Truncated wave models that include only a limited number of wave components. They give a low resolution of the physical phenomena involved and require that the eddies can be described by a small number of spectral components (Lorenz, 1960).

The statistical dynamical models (SDM's) are based on time-averaged equations in which the effects of large-scale transient eddies are parameterized in terms of the mean quantities. The main models in this group are the energy balance models (EBM's), the radiative—convective models (RCM's), and the two-dimensional models. These models are useful for the practical reason that they consume much less computer time than full general circulation models, and give the possibility for integrating the model even over geological times (e.g., Suarez and Held, 1979; Berger et al., 1990).

Saltzman (1978) has suggested that the real atmosphere may operate in a statistically deterministic way described by a set of simplified climate equations. These parameterized equations would directly govern the atmospheric climatic statistics such as monthly or longer term averages, variances, and higher moments of the probability distributions of climatic variables.

For general surveys of the SDM's, see Saltzman (1978) and MacCracken and Ghan (1987).

17.4 GENERAL CIRCULATION MODELS

17.4.1 Some general features

General circulation models are time-dependent hemispheric or global models. They simulate explicitly, with as much fidelity as possible, the various processes occurring in the atmosphere and oceans. They all use the three components of the models mentioned before in Sec. 17.2.1, i.e., the dynamics, physics, and a large number of other factors as shown in Fig. 17.2 and in Eqs. (17.1)–(17.8).

These models have a huge number of degrees of freedom because the equations must be solved at all grid points corresponding with the spatial resolution chosen. Of course, these climate models are resolution-bound in the sense that all processes occurring at smaller scales cannot be represented and resolved by the model grid. The subgrid scale phenomena have to be parameterized, whenever and wherever possible, in terms of the resolved macroscale variables. This parameterization offers only a partial resolution of the problem. In fact, when one attempts to adjust the parameterization to better fit the observed subgrid scale processes for one parameter this sometimes leads to a worse simulation for some other parameters of the global system.

In view of their high level of simulation capacity, the GCM's constitute a powerful tool for climate research studies since these models have the potential to simulate the climate faithfully under a great variety of boundary conditions. However, it is imperative that the model results be compared with the large-scale statistics, such as the mean fields, their variances and covariances, obtained from the real world as described in earlier chapters (6–15). Such comparisons are critical for a validation of any GCM.

17.4.2 The development of general circulation modeling

The first general circulation model of the atmosphere was developed by Phillips (1956). It was a quasigeostrophic, two-layer, hemispheric model which included friction and nonadiabatic heating effects in a simple manner. The model, although very simple, could develop a zonal flow with eddies in middle latitudes generated through the baroclinic instability mechanism. This experiment demonstrated the feasibility of simulating the global atmospheric circulation numerically.

An important next step in the development of the GCM's was accomplished in a study by Smagorinsky (1963). In a two-layer model he solved the primitive equations numerically for an extended period of 60 days which clearly showed the development of baroclinic perturbations in middle latitudes. The motions were confined to a zone bound by the equator and 64° latitude so that long planetary waves could develop. The kinematics of the motions on the sphere were taken into consideration using a Mercator projection.

Later many other GCM's were developed incorporating new features or using different types of parameterization, such as Mintz and Arakawa (Mintz, 1965), Smagorinsky *et al.* (1965), Manabe *et al.* (1965), Leith (1965), Kasahara and Washington (1967), etc. These were followed by even more sophisticated models (see surveys by Schneider and Dickinson, 1974; Smagorinsky, 1974).

With the improvement of computing facilities it was possible to increase the spatial and therefore the temporal resolution allowing a more detailed simulation of the atmospheric flow (Manabe *et al.*, 1970). It also became possible to improve the modeling of the vertical structure of the atmosphere by, e.g., including an atmospheric boundary layer with several levels below 1-km height in order to better link the free atmosphere with the earth's surface through a turbulent boundary layer.

The simulation of the general circulation in the world ocean was attempted first by Bryan and Cox (1967) using the basic thermo-hydrodynamic equations for the oceans with the hydrostatic assumption. The resolutions in the horizontal and vertical directions were higher than those used in atmospheric models. To minimize the effects of the unresolved eddies an unrealistically high viscosity had to be used to establish a steady circulation. In fact, the small dimensions of the observed eddies in the ocean (a few hundred kilometers rather than thousands of kilometers in the atmosphere), would require a grid distance on the order of tens of kilometers which has been attempted only recently by Semtner and Chervin (1988) in a global integration. Thus, in general, the present ocean models are not yet as far advanced as the atmospheric models due to, among other factors, the wider range of time scales (see Fig. 2.5).

As other important component models of the full climate system we should mention sea–ice models (e.g., Washington and Parkinson, 1986; Hibler, 1988), land–surface models (e.g., Sud *et al.*, 1988), and biosphere models (e.g., Dickinson, 1984; Sellers *et al.*, 1986). Such models are being developed now and should eventually be included in the coupled ocean–atmosphere models to be discussed next.

17.4.3 Coupled ocean–atmosphere models

Since the atmosphere and oceans are strongly connected through the exchanges of momentum, heat, and matter, it is obvious that a coupled atmosphere–ocean model is very desirable. However, the integration of a fully coupled model including the atmosphere, oceans, land, and cryosphere with such different internal time scales poses almost insurmountable difficulties in reaching a final solution, even if all interacting processes were completely understood. The reason is that the model should ideally be run at time steps applicable to the fastest component of the system, i.e., the atmosphere. As a consequence of these difficulties, three main directions of approach have been used so far. The first one is to drastically simplify the oceanic component of the coupled model while explicitly resolving the atmospheric processes. The second direction is to simplify the atmospheric forcing and to explicitly resolve the oceanic processes. Finally, the third direction is to attempt to resolve explicitly both the atmospheric and oceanic processes as best as we can with our present knowledge and the available computer resources.

17.4.3.1 Simplified ocean processes

In the context of the coupled ocean–atmosphere models, the simplest ocean model is probably the swamp model. In this model, the oceanic role is only to act as an unlimited source of water vapor for the atmosphere without storing or transporting heat and other quantities.

A second step is to use an ocean surface mixed layer coupled to an explicit atmospheric GCM, where the ocean has the capacity to both store heat and to supply water vapor to the atmosphere.

The next step before using a full ocean GCM would be to include part of the ocean dynamics, by resolving in a few layers some of the large-scale oceanic processes but parameterizing all the remaining processes.

17.4.3.2 Simplified atmospheric processes, stochastic models

One of the most drastic ways to simplify the atmospheric component of a coupled model is to simulate the atmospheric effects on the ocean by using a stochastic forcing. In introducing this type of model, Hasselmann (1976) argued that because of the large differences in internal time scales, a perturbation in the atmosphere might lead to very different response times in the ocean.

Hasselmann's (1976) basic idea is that a long-period signal in the ocean may be regarded as a superposition of a large number of random, short-period signals. An analogy can be made with statistical mechanics where certain macroscopic variables, such as temperature and pressure, can be determined by the statistics of the microscopic, molecular motions. The evolution of the slowly changing variables can then be expressed in terms of the faster variables using stochastic differential equations, such as the Fokker–Planck equation.

In the ocean–atmosphere system the white noise spectrum of the atmosphere will tend to produce an oceanic response with a red spectrum in which most of the variance is confined to the long-period fluctuations (Hasselmann, 1976; Sutera, 1981; and Hasselmann, 1988). According to stochastic theory, the random input by the atmosphere has to be bounded by the variance of the atmospheric fluctuations as obtained from observations.

From the oceanic perspective its long-period changes will generate new boundary conditions, i.e., new sea surface temperature and sea–ice distributions, for the atmosphere with the possibility to force the atmosphere into new quasi-equilibrium states.

17.4.3.3 Fully coupled models

In the first global coupled ocean–atmosphere GCM constructed by Manabe (1969) and Bryan (1969), a special technique for integration was used to circumvent the difficulty of having two interacting subsystems with different time scales. They started with a horizontally uniform but vertically stratified ocean temperature distribution, and an isothermal atmosphere at rest. After a sufficiently long interval of time (on the order of one week to avoid unrealistic transient ocean circulations) a mean wind stress and mean fluxes of heat and moisture obtained from the atmospheric model integration were used as new boundary conditions at the ocean surface for further integration of the ocean model. Then, after a certain interval of time, a new sea surface temperature distribution generated by integration of the ocean GCM was applied as a new boundary condition to the atmosphere, and so on. Basically this technique constitutes an alternating, repetitive process. The results for the zonal-mean temperature are synthesized in Fig. 17.5 from a later paper which includes the seasonal variations. It is clear that the model reproduces the gross features of the observed temperature distribution.

The coupled models are still at a fairly primitive stage of development (Hasselmann, 1988) but they will probably play a more and more important role in the near future to provide acceptable climate predictions.

17.5 STATISTICAL DYNAMICAL MODELS

17.5.1 Energy balance models

The EBM's are one-dimensional $[1\text{-}D(\phi)]$ models based on the first law of thermodynamics. They are sometimes completed with parameterized heat fluxes. Energy balance models predict the temperature changes at the earth's surface under the condition that the net radiation flux is zero. The value of the solar constant is prescribed and the albedo of the earth's surface is usually parameterized in terms of the surface temperature. These models do not explicitly include any dynamics in the atmosphere or ocean. However, they sometimes include a statistical–dynamical approach. Thus, the EBM's give steady-state or equilibrium solutions of the energy equation (13.21) when the boundary conditions and certain internal parameters are prescribed.

The basic equation used in these models is a simplified version of Eq. (13.21) that can be written as

$$\rho c \frac{\partial T}{\partial t} = -\rho \operatorname{div} \mathbf{J}_E - \operatorname{div} \mathbf{F}_{\text{rad}} \qquad (17.16)$$

or in an even more simplified form as

$$\rho c \frac{\partial T(\phi)}{\partial t} = F_{\text{rad}}^{\downarrow} - F_{\text{rad}}^{\uparrow} + \tau(\phi), \qquad (17.17)$$

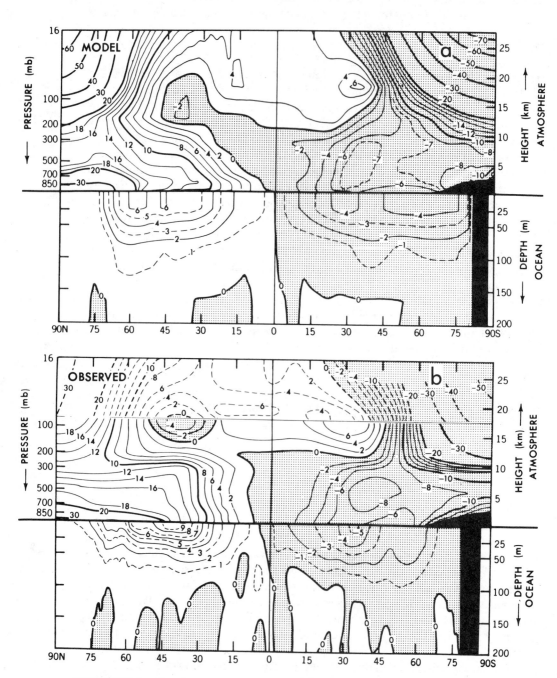

FIGURE 17.5. Zonal-mean cross sections of the seasonal range in temperature (August–February) in the atmosphere–ocean system for both simulated (top diagram) and observed (bottom diagram) conditions after Manabe *et al.* (1979). The units are °C.

where c is the heat capacity and $\tau(\phi)$ is the horizontal convergence of energy into the latitudinal belt. For the steady-state solution the left-hand side of Eq. (17.17) is equal to zero.

The models require parameterization of the various terms. For the downward solar radiation one uses $F_{SW}^{\downarrow}(1 - A_{sfc})$, where the surface albedo is parameterized mainly in terms of the surface temperature (see Sec. 6.7). The infrared radiation is also parameterized in terms of the surface temperature using a semi-empirical relation. The horizontal flux divergence term $\tau(\phi)$ has been parameterized in various ways. For example, Budyko (1969) simply used a Newtonian law of cooling in which

$$\tau(\phi) = -\beta[T_{sfc}(\phi) - \{T_{sfc}\}], \qquad (17.18)$$

where β is an empirical coefficient and $\{T_{sfc}\}$ is the global-mean surface temperature.

Sellers (1969) identified three separate components for the meridional flux of energy, i.e., the sensible heat transport by the atmosphere, the sensible heat transport by the oceans, and the transport of latent heat by the atmosphere. These fluxes were parameterized using the gradient-flux approach for the eddies, while retaining the mean meridional circulation terms.

With the simple zonally averaged models mentioned above we may compute the equilibrium temperature as a function of latitude. Simple as they may seem, these models can give valuable information about the sensitivity of the climatic system to changes in planetary albedo or in solar radiation input. However, they seem to be more sensitive to perturbations than the real atmosphere and are now mainly of academic interest.

For example, a decrease of the solar constant by as little as 2% (Budyko, 1969; Sellers, 1969) might lead to glaciation of the earth, whereas an increase of less than 1% might lead to melting of the polar ice caps. These conclusions came from the 1-D(ϕ) models with a simple positive ice-albedo feedback mechanism. Furthermore, the one-dimensional models allowed Budyko and Sellers to estimate the influence of a change in radiation on the temperature in each latitudinal zone after adding the flux convergence of heat due to atmospheric motions. With this simple choice of parameters the final latitudinal distribution of temperature was found to be very similar to the observed zonal-mean distribution. Budyko (1969) also determined the extent of the polar ice caps, assuming that the ice boundaries coincided roughly with the $-10\ °C$ isotherm.

Schneider and Gal-Chen (1973) were among the first to demonstrate that, in these models, more than one solution (e.g., the present climate and a completely ice-covered earth) can be obtained with the present value of the solar constant so that the climate system, at least according to these simple models, appears to be intransitive.

A general review of the energy-balance climate models was published by North *et al.* (1981) and North (1988), covering this topic in great detail.

17.5.2 Radiative—convective models

The RCM's are one-dimensional [1-D(z)] models based on the thermodynamic energy equation (17.5) and a convective adjustment scheme. They were designed to evaluate the vertical profile of temperature in the atmosphere taking into consideration the heating and cooling effects by solar and infrared radiation at various levels of the atmosphere.

The evaluation of the net radiative fluxes requires a radiative transfer model and a prescribed vertical distribution of carbon dioxide, ozone, and, in some models, water vapor and clouds. The atmosphere is divided vertically into layers. The radiative equilibrium temperature for each layer may then be obtained by integrating the thermodynamic energy equation in time from an initial temperature profile until equilibrium is reached. However, this purely radiative thermodynamic model gives a superadiabatic vertical temperature gradient with temperatures that are too cold in the upper troposphere and too warm in the lower troposphere compared to observations. This stratification is unstable leading to overturnings associated with convection.

To obtain a more realistic profile of the temperature, the thermodynamic energy equation must include some extra terms related to the nonradiative transfer of energy from the earth's surface to the atmosphere, as well as to the convective redistribution of energy within the atmosphere. These two processes have been parameterized in various ways. A very simple scheme to compensate for these two effects was devised in a pioneering study by Manabe and Strickler (1964) (see also Sec. 17.2.4). They introduced a convective adjustment in which the lapse rate is not permitted to exceed a mean tropospheric lapse rate of 6.5 K km^{-1} during the integration of the model. This adjustment is equivalent to an upward transport of heat by convective motions. It leads to a thermal equilibrium rather than a radiative equilibrium and to a more realistic vertical-mean lapse rate. Manabe and Stricker (1964) further assumed a fixed vertical distribution of specific humidity with height. Later this hypothesis was modified by Manabe and Wetherald (1967) to a more realistic one, in which the relative humidity $U(z)$ was assumed to vary as a linear function of height. In the last model, the moisture distribution could be evaluated from the relation

$$q = \frac{0.622 U\, e_s(T)}{p - 0.378 U\, e_s(T)} \tag{17.19}$$

since $q = 0.622\, e/(p - 0.378\, e)$ and $U = e/e_s$. The saturation vapor pressure e_s is only a function of temperature and can be obtained from the Clausius–Clapeyron equation (3.65).

Manabe and Strickler's (1964) one-dimensional radiative-convective model can be considered to represent a globally averaged model. Starting with isothermal, initial conditions the model developed within one year a vertical temperature structure with a troposphere and a stratosphere separated by a well-defined tropopause. The temperature profile obtained was very close to the mean observed profile in the atmosphere (see Fig. 17.6). For a review of the 1-D(z) radiative—convective models see Ramanathan and Coakley (1978).

17.5.3 Two-dimensional statistical dynamical models

The 2-D SDM's may include the thermodynamic energy equation, the equations for the mean zonal and meridional motions, and parameterizations of the transports of heat and momentum. The time-averaged equations are written in such a way that all source terms and the eddy transports are put on the right-hand side of the equations. The time-mean state of the atmosphere in these statistical–dynamical models may be then regarded as a regime forced not only by external processes, such as radiation, orography, surface friction, and evaporation, but also by the large-scale transient eddy transports of heat and momentum. In the geometrical scheme, the two-dimensional SDM's can be subdivided into the zonally symmetric 2-D(ϕ,z) models and the vertical-mean 2-D(λ,ϕ) models.

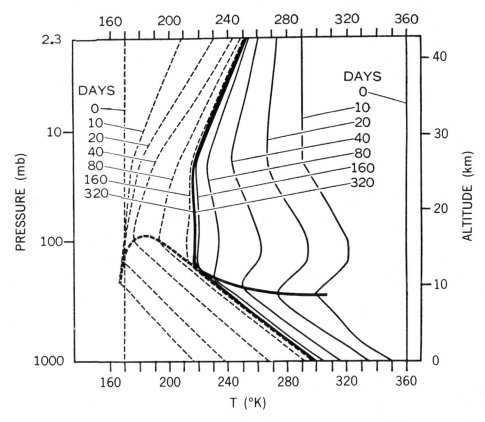

FIGURE 17.6. Approach to the state of thermal equilibrium from a warm (solid lines) and a cold (dashed lines) isothermal atmosphere using a 1-D(z) model (adapted from Manabe and Strickler, 1964).

17.5.3.1 Zonally symmetric SDM's

The 2-D(ϕ,z) models are designed to simulate the zonal-mean structure of the atmosphere and, in particular, the mean meridional circulations and their influence on the temperature field. In these models longitudinal variations are not resolved. The equations used are often very simplified and reduced versions of Eqs. (3.81)–(3.83), bringing out the eddy contributions which must be parameterized. Thus, the zonal equation of motion (3.82a) is frequently reduced to

$$\frac{\partial [u]}{\partial t} = f[v] - \frac{\partial [u^*v^*]}{R\partial \phi} + [F_\lambda], \tag{17.20}$$

the meridional equation of motion (3.82b) to

$$f[u] + R_d[T]\frac{\partial (\ln[p])}{R\partial \phi} = 0, \tag{17.21}$$

the equation of state (3.59) to

$$[p] = R_d[\rho][T], \tag{17.22}$$

the hydrostatic equilibrium condition (3.13) to

$$\frac{\partial (\ln[p])}{\partial z} = -g/(R_d[T]), \tag{17.23}$$

the equation of continuity (3.81) to

$$\frac{\partial [\rho][v]}{R\partial \phi} + \frac{\partial [\rho][w]}{\partial z} = 0, \tag{17.24}$$

and the thermodynamic equation (3.83) to

$$\frac{\partial [T]}{\partial t} = -\frac{\partial [v^*T^*]}{R\partial \phi} - \frac{\partial [w^*T^*]}{\partial z}$$

$$- [w](\gamma_d - [\gamma]) + \frac{[Q]}{c_p}. \tag{17.25}$$

These axially symmetric models are able to simulate rather realistically the mean characteristics of the atmosphere. However, both the horizontal and vertical transports must be parameterized very carefully in these models.

The 2-D(ϕ,z) models have been used, among others, by Saltzman (1968), Green (1970), Kurihara (1970), Stone (1974), and more recently by MacCracken and Ghan (1987). They parameterized the transient and stationary eddies in terms of the zonal-mean parameters usually based on baroclinic instability theory. Frequently the gradient flux approach has been used to parameterize the mean meridional flux of enthalpy due to transient eddies $[\overline{v'T'}]$, so that

$$[\overline{v'T'}] = -K_A \frac{\partial [\overline{T}]}{R\partial \phi}, \tag{17.26}$$

where K_A is an empirical transport coefficient. The justification for such an approach is that the meridional eddy flux of enthalpy is due to the existence of baroclinic waves that, in turn, are driven by the meridional gradient of the mean zonal temperature. However, we must keep in mind that there exists another important component of the eddy flux $[\overline{v}^*\overline{T}^*]$ that is associated with the large quasi-stationary waves in the general circulation and is forced by topography and land–sea temperature contrast. Thus, we cannot expect a very good agreement between the observed fluxes and those obtained from parameterizations based on baroclinic processes alone.

As for the transient eddy momentum flux $[\overline{u'v'}]$, the diffusive model leads to the expression

$$[\overline{u'v'}] = -K \frac{\partial [\overline{u}]}{R\partial \phi}, \tag{17.27}$$

which is inadequate because observations show that momentum can be transported both up and down (see Sec. 11.2.3) the meridional gradient of the mean zonal motion $[\overline{u}]$. This is why recent parameterizations are often reformulated in terms of the gradient of potential vorticity (Green, 1970).

17.5.3.2 Vertical-mean SDM's

The 2-D(λ,ϕ) statistical–dynamical models are designed to study the longitudinal variations of climate in the vertical mean or at the earth's surface and to take land–sea contrasts into account. Indeed, in the real world the observed variations from land to sea can be almost as pronounced as the latitudinal variations. This can be seen clearly in the earlier figures for the observed net radiation (Fig. 6.12), temperature (Fig. 7.4), and humidity (Fig. 12.3) distributions at the earth's surface. The main difficulty in this group of SDM's lies again in the problem of formulating physically sound parameterizations for the large-scale eddy processes (Saltzman, 1978). In one such approach pioneered by Adem (1964) it is assumed that the latitudinal and longitudinal heat transports may be parameterized in terms of a local horizontal "Austausch" coefficient (Clapp, 1970) in a way analogous to Eq. (17.26).

17.6 USES AND APPLICATIONS OF MODELS

17.6.1 Some general remarks

The mathematical simulation of climate has reached a level of development that cannot be regarded anymore as a purely academic exercise. However, the model simulations must be validated through a systematic comparison of their output with the observed values as presented earlier in Chaps. 6–16. Thus, it seems essential that the models not only simulate the mean observed features (see Chaps. 6–10), but also that they can reproduce the observed transient behavior of the system as it affects the transports of angular momentum, water vapor, and energy in the atmosphere and oceans (see Chaps. 11–15). Further, for climate change studies the models should exhibit interannual and interdecadal variations of the same nature and magnitude as are observed (see Chap. 16). Only after a certain degree of confidence in the models' ability to represent the real climate system has been obtained, should they be used for applications. In particular, the temptation to make use of the models' predictions before they are fully tested must be avoided.

After proper validation, the models can be used as a means to obtain predictions of the response of the climate to external forcings (predictions of the second kind; see Sec. 17.2.5). Such experiments may also be used to study feedback processes and to identify the principal elements involved. It is clear that the models constitute a powerful new tool to conduct indoor experiments of the natural climatic system. For example, through simulation it is possible to evaluate the sensitivity of the components of the climatic system to variations in solar radiation, surface topography, and composition of the atmosphere. In fact, the simulation is the only way of getting answers to some important questions that now start to emerge. How would the temperature in the atmosphere respond to an increase in carbon dioxide? What would be the consequences if the solar constant were subject to variations? What would be the critical decrease in the solar constant to produce a new glaciation of the earth? How was the climate during the last glaciation? What is the influence of a change in the atmospheric heat balance on the intensity of the hydrological cycle? What are the effects of changing the surface albedo through overgrazing or deforestation? What are the impacts of other human activities, such as industrialization, changes in agricultural practices and urbanization, on the climatic system?

Furthermore, simulation allows us to better understand the mechanisms involved in the nonlinear feedback processes in the climatic system. Thus, one can test different working hypotheses or ideas regarding, e.g., the influence of clouds, aerosols, dust, CO_2, and ozone on climate, and the movement of tracer elements in the atmosphere and oceans. Tentative answers to some of the previous questions are starting to appear. For a review of the many types of climate sensitivity experiments performed so far we may refer to, e.g., GARP (1979) and, in the case of CO_2, to Schlesinger and Mitchell (1987).

17.6.2 Data assimilation and network testing

Since the early 1960s extensive work has been done at the major numerical weather prediction centers of the various countries to generate the proper initial conditions for regional, hemispheric, or global weather predictions. Much experience has been gained on how to error check and assimilate the great variety of available atmospheric data, such as upper-air rawinsonde reports, surface station data over land, surface ship data, aircraft reports and satellite-derived temperatures and cloud winds, to name only a few, into a coherent four-dimensional picture of the state of the atmosphere.

Often a 6- or 12-hr forecast has been used as a first guess for the initial conditions for the NWP models, which are then updated with the actual data received through the Global Telecommunication System (GTS) based on observations taken during the most recent few hours. The numerical model is an ideal framework to incorporate this deluge of information and to arrive at a physically consistent description of the present state. Therefore, it is not surprising that these daily analyses of the present state of the atmosphere are finding an increasing use in research on weather and climate (see Sec. 5.4). In fact, they form an attractive alternative to the mathematical interpolation techniques used traditionally in climate research (see Fig. 5.9). However, we should mention that in this book we have chosen to use the traditional approach of analysis because of several limitations in the model-analyzed data sets produced so far. The principal limitations are (1) the strong dependence of many of the statistics on the particular model and model parameterizations used, (2) our inability to decide and agree on what parameterizations are the best, (3) the frequent changes (updates) in the operational NWP models that make it almost impossible to separate real from model-induced climatic changes, and (4) the incompleteness of the present NWP records.

However, with the recent advances in model simulation it may become worthwhile in the near future to go back in time and to reanalyze the historical data using a "frozen" version of the "best" presently available GCM at the highest feasible resolution. In spite of some unavoidable flaws, like the particular parameterizations used, such a reanalysis project could provide a coherent self-consistent and physically based description of the evolution of the global atmosphere since the early 1950s. The reanalysis should lead to a more realistic description of the climate and its evolution than is possible using the traditional methods.

So far we have mentioned only the numerical weather prediction models, but recently numerical ocean models have also been used to assimilate the much sparser ocean data available and to produce a physically consistent representation of the state of the oceans (e.g., Derber and Rosati, 1989; Leetmaa and Ji, 1989). One known fact that, to some extent, may alleviate the data deficiencies in the oceans is that the ocean currents are driven mainly by the stresses exerted by the surface winds (see Secs. 10.5 and 14.6), and that these surface stresses are routinely available from the operational NWP models. Therefore, a coupled ocean–atmosphere model approach may be a great asset in analyzing the real-time oceanic state and in generating the needed oceanic data sets for climate research.

Other useful applications of models are to use them for the design and testing of observational networks. For example, NWP models have been employed in the planning of special regional networks during the First GARP Global Weather Experiment (see Sec. 5.1.4). Another application is to test the influence of the spatial gaps in the rawinsonde network on climate statistics through a GCM simulation (Oort, 1978).

17.6.3 Modeling of the hydrological cycle

Due to its importance for hydrology, we will describe an atmospheric GCM with a hydrological cycle developed by Manabe and Holloway (1975), using the so-called bucket model of Budyko (1956). The GCM includes continents with mountains and soil that has a uniform "field capacity" of 15-cm water. In other words, the maximum amount of water the soil can retain without generating runoff is 15 cm. In the case of the oceans they assumed an infinite field capacity but no redistribution by ocean currents, i.e., a swamp-type ocean model. The model basically computes the fields of temperature, wind, and water vapor for the atmosphere, and derives the rate of precipitation (rain or snow) minus evaporation from the convergence of moisture transport in the atmosphere. The evaporation is inferred through a simple scheme which assumes that evaporation is a function of soil moisture and potential evaporation (see Sec. 10.7). The rates of change of soil moisture and snow depth are then computed keeping track of the water and heat budgets of the ground. Runoff occurs when the accumulated soil moisture exceeds the prescribed field capacity of 15 cm as if the bucket were full. The excess water is assumed to flow directly into the surrounding seas without further infiltration. The total precipitation and other hydrological parameters compare well with those observed in the real atmosphere. For example, the model depicts the equatorial rain zone and the major subtropical deserts (see Fig. 17.7). It is interesting to note that the climate obtained (Manabe et al., 1979) fits well with the widely used climate classification by Köppen (1931).

Several other aspects of the land boundary conditions have also been considered in GCM's in order to understand the role of land processes in the general circulation. We may mention the effects of changes in the evaporation and soil moisture characteristics of the ground and of variations in the surface roughness on the intensity of the hydrological cycle (Sud et al., 1988).

17.6.4 Modeling of the ENSO phenomena

The modeling of air–sea interactions has been developed to a sufficiently high level to permit the study of important phenomena, such as the El Niño—Southern Oscillation. Thus, a variety of atmospheric, oceanic, and coupled ocean–atmosphere GCM's have been run in recent attempts to simulate the ENSO phenomena. For example, a series of numerical experiments was conducted at GFDL by Lau (1981 and 1985). He demonstrated that an atmospheric model could duplicate an ENSO-like oscillation when the sea surface temperature (SST) conditions were prescribed to change with time according to the observations, whereas the ENSO signal was absent when climatological mean SST conditions were prescribed. It is interesting to note that recent runs with simple coupled models exhibit quite regular ENSO oscillations but that runs using more complex coupled models undergo irregular ENSO events better resembling the real world phenomena (e.g., Cane and Zebiak, 1985; Schopf and Suarez, 1988; Philander et al., 1989).

Therefore, it seems that the modeling experiments of the ENSO phenomena may be used profitably as a way to measure our progress in improving the ocean–atmosphere coupled models.

17.6.5 Modeling of the CO_2 effects

Among the gases released into the atmosphere by human activities, the most important component is probably carbon dioxide. There is little empirical evidence

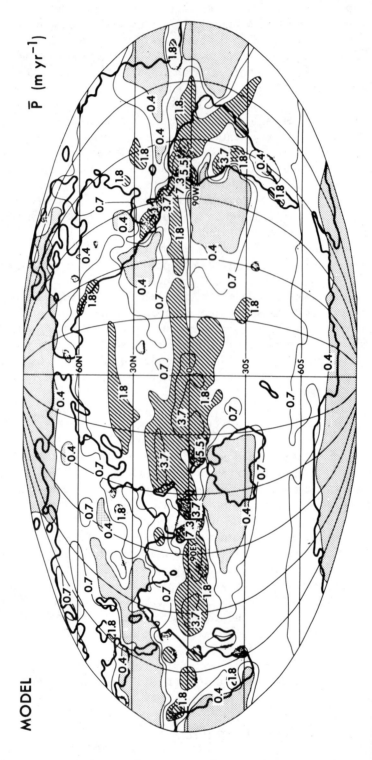

FIGURE 17.7. Global distribution of the annual-mean rate of precipitation (m yr^{-1}) simulated by a general circulation model after Manabe and Holloway (1975). [Compare with observed precipitation rates in Fig. 7.24(a).] Stippled areas indicate $\bar{P} \leq 0.4$ m yr^{-1} and shaded areas $\bar{P} \geq 1.8$ m yr^{-1}.

to allow us to predict with some confidence the effects of the increasing CO_2 and other trace gases on climate. Therefore, we have to rely on mathematical simulations using a variety of climate models. Using a one-dimensional radiative-convective equilibrium model Manabe and Wetherald (1967) computed the vertical distribution of the global-mean temperature for a CO_2 concentration of 300 ppm, as well as twice and half this value (see Fig. 17.8). They computed a change in surface temperature of slightly more than 2 °C when the relative humidity was kept constant. Strong cooling was found in the stratosphere when the CO_2 concentration was increased.

Later many investigators (see review paper by Schlesinger and Mitchell, 1987) have used GCM's to calculate the changes in the three-dimensional structure of the atmosphere resulting from a doubling of the CO_2 concentration. In practically all experiments, a significant general warming of the troposphere due to the doubling of CO_2 and the corresponding increased greenhouse effect was computed with an average global rise in surface temperature of about 3 °C. The largest warmings tend to be found in the polar regions as a result of the positive temperature-snow albedo feedback and the high static stability at those latitudes, which then limits the convection and confines the temperature changes to the lowest levels.

The resulting climatic changes would markedly affect the hydrological cycle increasing both the overall evaporation and precipitation, and consequently the runoff. Furthermore, selective warming of the polar regions could lead to the melting of snow and ice, especially in the Antarctic, which might conceivably cause a 5-m rise in sea level during the next century with significant flooding of coastal regions (Hansen et al., 1981). In fact, the rise in sea level might become the main signature of the CO_2 effect.

The agricultural productivity depends critically not only on the incoming solar radiation (through photosynthesis) but also on the temperature, precipitation, cloudiness, and, of course, the soil fertility. Thus, any climatic change, such as the change implied by a CO_2 increase, could affect this productivity. The changes in CO_2 would further influence other segments of the biosphere, such as the forests and fisheries. The fisheries would be affected mainly through changes in the upwelling of nutrient waters.

We should stress the general premise in all these studies that all factors besides the CO_2 concentration are assumed to remain the same. For example, the influence of possible changes in cloud cover is difficult to evaluate, and has generally not been included in the models in an adequate way, in spite of the fact that clouds could significantly alter the temperature response in the atmosphere. In a global assessment of the CO_2 problem the National Academy of Sciences (1979, 1982) concluded that if carbon dioxide continues to increase, there is no reason to doubt that climatic changes will result and no reason to believe that these changes will be negligible.

The models are also useful to study the global distributions and transports of other trace constituents in the atmosphere and oceans, such as ozone, N_2O, and tritium, introducing some chemistry in the models (see Sec. 16.5.2; Mahlman et al., 1986).

17.6.6 Effects of mountains and the simulation of an ice-age climate

The effects of mountains on the general circulation of the atmosphere were studied by, e.g., Kasahara et al. (1973) and Manabe and Terpstra (1974) through a set of numerical experiments. The simulated circulation in a model with an actual topo-

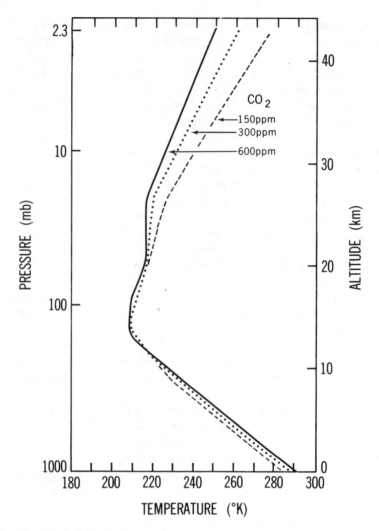

FIGURE 17.8. Vertical distribution of temperature simulated in a 1-D(z) radiative equilibrium model for CO_2 concentrations of 300 (the atmospheric value estimated for the year 1860), 600, and 150 ppm (adapted from Manabe and Wetherald, 1967).

graphy was compared with the corresponding circulation in a model without mountains, keeping the land–sea distribution the same. The results show that the mountains are important in explaining many of the quasistationary features of the general circulation, like the cyclogenesis at the leeside of the mountains and the distribution of precipitation. This type of experiment has been conducted by several investigators to study different aspects of the physiography of the continents, e.g., the importance of mountains in the development of deserts (e.g., Ruddiman and Kutzbach, 1989; Manabe and Broccoli, 1990).

Using various types of models there have been some attempts to model the ice-age climate. These models use boundary conditions at the earth's surface (extent of land and sea ice, sea surface temperature, and surface albedo) prescribed according to paleoclimatological evidence, and evaluate the atmospheric structure as an equilibrium solution. Among them we refer to Saltzman and Vernekar (1975) who used a statistical–dynamical model, Gates (1976) who used a two-layer GCM, and Williams *et al.* (1974), Manabe and Broccoli (1985), and Kutzbach *et al.* (1989) who used a multilayer GCM.

17.6.7 Sensitivity to changes in astronomical parameters

The effects of changing the solar constant on climate were already mentioned in Sec. 17.5 when we discussed the works of Budyko (1969) and Sellers (1969). Later using a GCM the same problem was attacked by Wetherald and Manabe (1975). Their results show a significant warming of the lower troposphere at high latitudes resulting in a reduction of the meridional temperature gradient, and of the sensible heat transport due to baroclinic waves in middle latitudes; they also show a decrease in the oceanic heat transport. The computed increase in temperature with increasing solar constant is in qualitative agreement with the results of Budyko and Sellers. Another interesting result from the GCM is the large sensitivity of the intensity of the hydrological cycle to small changes in the solar constant.

By altering the rotation rate, obliquity and diurnal period, earth-like model atmospheres have also been shown to produce a wide range of circulation forms resembling those observed on the other planets (Williams and Holloway, 1982).

APPENDIX A
Analysis in Terms of Fourier Components

A1. INTRODUCTION

The basic fields of the atmosphere that we need to analyze are the wind components (u,v,ω), temperature (T), pressure (p), geopotential height (z), and specific humidity (q). The boundary is defined by the earth's topography, or by the surface pressure (p_0).

All these quantities vary in space and time. They are single-valued and piecewise differentiable functions of the form $f(\lambda,\phi,p,t)$. For a fixed isobaric level at a given instant, the function along a latitude circle becomes an exclusive function of longitude λ defined in the interval $[0,2\pi]$:

$$f(\lambda,\phi,p,t)_{\phi,p,t} \equiv f(\lambda).$$

Thus, in general terms, we are concerned with the transformation of a function defined in the λ space $[0,2\pi]$ into a series of functions defined in the wave number domain, through a Fourier transform $\mathscr{F}(\ )$. Symbolically we can write:

$$\mathscr{F}[f(\lambda)] = F(k) \quad (k=0,1,2,3,...),$$

where k is the wave number. The wave number is defined here as the number of waves along the latitude circle $k = 2\pi R \cos \phi / L$, where L is the wavelength. The functions $f(\lambda)$ and $F(k)$ that we will consider are listed in Table A.1.

We are also interested in certain derived quadratic fields, such as the kinetic energy $K = (u^2 + v^2)/2$, the available potential energy $P = c_p \Gamma T^2/2$, and the transports of angular momentum $\mathsf{v}u$, enthalpy $c_p \mathsf{v}T$, moisture $\mathsf{v}q$, and geopotential energy $\mathsf{v}gz$. These quantities can be reduced to the product of two functions $f(\lambda)g(\lambda)$ defined in the same interval $[0,2\pi]$, or to $f(\lambda)f(\lambda)$, where $g(\lambda) \equiv f(\lambda)$.

From a formal point of view we have two main problems: (1) to determine the Fourier transform of a well-behaved function $f(\lambda)$, defined in an interval $[0,2\pi]$, and (2) to derive the Fourier transform of the product of two functions $f(\lambda)g(\lambda)$, both defined in the same interval $[0,2\pi]$ and with cyclic continuity.

In the following section, we will discuss these two problems in general terms without specifying the meteorological nature of the variables. Later, we will provide various applications of this method of analysis to meteorological fields.

TABLE• A.1. Symbols used for the Fourier pairs of the horizontal and vertical wind components u, v, and ω, the temperature T, the geopotential height z, the specific humidity q, and the diabatic heating rate h.

$f(\lambda)$	u	v	ω	T	z	q	h
$F(k)$	U	V	Ω	B	Z	Q	H
$F_1(k) + iF_2(k)$	$U_1 + iU_2$	$V_1 + iV_2$	$\Omega_1 + i\Omega_2$	$B_1 + iB_2$	$Z_1 + iZ_2$	$Q_1 + iQ_2$	$H_1 + iH_2$

A2. THE FOURIER SPECTRUM

Any real function which satisfies the Dirichlet conditions in the interval $[0, 2\pi]$ can be expanded in a Fourier series:

$$f(\lambda) = \sum_{k=-\infty}^{\infty} F(k) e^{-ik\lambda}, \tag{A1}$$

where the complex coefficients $F(k)$ are given by

$$F(k) = \frac{1}{2\pi} \int_0^{2\pi} f(\lambda) e^{ik\lambda} \, d\lambda \quad (k = 0, 1, 2, \ldots). \tag{A2}$$

As we are concerned with the Fourier representation of functions along a given latitude circle, λ denotes the longitude and k is the number of waves around the latitude circle.

Equations (A1) and (A2) constitute a Fourier transform pair. $F(k)$, which is the representation of $f(\lambda)$ in the wave number domain, is called the complex spectral function of $f(\lambda)$. Symbolically, we have, for Eq. (A2),

$$F(k) \equiv \mathscr{F}\{f(\lambda)\}$$

meaning that $F(k)$ is the Fourier transform of $f(\lambda)$. Expression (A2) can be written in the form

$$F(k) = \frac{1}{2\pi} \int_0^{2\pi} f(\lambda) \cos k\lambda \, d\lambda + \frac{i}{2\pi} \int_0^{2\pi} f(\lambda) \sin k\lambda \, d\lambda$$

$$= F_1(k) + iF_2(k), \tag{A3}$$

where $F_1(k)$ and $F_2(k)$ are the real and imaginary parts of $F(k)$. It is easy to verify that

$$F_1(-k) = F_1(k),$$
$$F_2(-k) = -F_2(k), \tag{A4}$$

i.e., the real part is an even function of k and the imaginary part an odd function of k. Furthermore,

$$F(-k) = F_1(k) - iF_2(k) = \widehat{F}(k) \tag{A5}$$

if we denote the complex conjugate of $F(k)$ by $\widehat{F}(k)$.

The spectral function $F(k)$ can also be written in the exponential form

$$F(k) = |F(k)| \, e^{ik\epsilon_k}, \tag{A6}$$

where the amplitude $|F(k)|$ and the phase angle ϵ_k are given in terms of the real and imaginary parts by the expression

$$|F(k)| = \{F_1^2(k) + F_2^2(k)\}^{1/2} \tag{A7}$$

and

$$\epsilon_k = \frac{1}{k}\tan^{-1}\left\{\frac{F_2(k)}{F_1(k)}\right\} \tag{A8}$$

in terms of wave number.

Inserting Eq. (A6) into Eq. (A1) we get a new form for the Fourier expansion of $f(\lambda)$ in the wave domain:

$$f(\lambda) = \sum_{k=-\infty}^{\infty} |F(k)|\, e^{-ik(\lambda - \epsilon_k)} \tag{A9}$$

$$= F(0) + 2\sum_{k=1}^{\infty} |F(k)|\cos k(\lambda - \epsilon_k). \tag{A10}$$

The functions $|F(k)|$ and ϵ_k are called the amplitude spectrum and phase spectrum.

Although the function $f(\lambda)$ is real it should be noted that the coefficients $F(k)$ are complex. The representation for $f(\lambda)$, as expressed by Eq. (A1), is known as the complex Fourier series. Expression (A9) is another form of representation for $f(\lambda)$, which explicitly contains both the amplitude and the phase angle. It may readily be shown that Eq. (A1) is equivalent to Eq. (A10) in terms of cosines and sines. In fact, from Eq. (A1) we may represent $f(\lambda)$ as

$$f(\lambda) = F_1(0) + \sum_{k=1}^{\infty} \{F(k)e^{-ik\lambda} + \hat{F}(k)e^{ik\lambda}\} \tag{A11}$$

or

$$f(\lambda) = F_1(0) + 2\sum_{k=1}^{\infty} \{F_1(k)\cos k\lambda + F_2(k)\sin k\lambda\} \tag{A12}$$

since $F_2(0) = 0$.

Now if one writes

$$a_k = 2F_1(k),$$
$$b_k = 2F_2(k), \tag{A13}$$

the previous expression reduces to the familiar form of the Fourier series in terms of sine and cosine components:

$$f(\lambda) = \frac{a_0}{2} + \sum_{k=1}^{\infty} \{a_k \cos k\lambda + b_k \sin k\lambda\}$$

$$= \sum_{k=0}^{\infty} c_k \cos k(\lambda - \theta_k), \tag{A14}$$

where

$$c_0 = \frac{a_0}{2},$$
$$c_k = (a_k^2 + b_k^2)^{1/2} \quad \text{for} \quad k = 1, ..., \infty,$$

and

$$\theta_k = \frac{1}{k}\tan^{-1}(b_k/a_k).$$

Comparing expressions (A10) and (A14) we see that the amplitude c_k is equal to twice the amplitude of the complex component $|F(k)|$, and that the phase angle θ_k is the same as the phase angle ϵ_k for the complex component $|F(k)|$.

The complex form of the Fourier series is often more convenient to use due to its relative compactness and the greater ease with which the exponential function may be manipulated with respect to algebraic, differential, and integral operations. The harmonic phase angles are implicitly contained in the complex form and hence need not be written in an explicit form during the operations. For example, the derivative of $f(\lambda)$ may be written using Eq. (A1) as

$$\frac{\partial f}{\partial \lambda} = -\sum_{k=-\infty}^{\infty} ikF(k)e^{-ik\lambda} \tag{A15}$$

or

$$\mathscr{F}\left[\frac{\partial f}{\partial \lambda}\right] = -ikF(k). \tag{A16}$$

Similarly, the transform of the integral of $f(\lambda)$ is $-F(k)/ik$.

For a different variable, ϕ, we have

$$\frac{\partial f}{\partial \phi} = \sum_{k=-\infty}^{\infty} \frac{\partial F(k)}{\partial \phi} e^{-ik\lambda} \tag{A17}$$

so that

$$\mathscr{F}\left(\frac{\partial f}{\partial \phi}\right) = \frac{\partial F(k)}{\partial \phi}. \tag{A18}$$

Thus, the derivative of $f(\lambda)$ in the λ space and the derivative of $F(k)$ in the k domain constitute a Fourier transform pair.

Similarly as the function $f(\lambda)$ may be represented graphically in the space domain, the equivalent function $F(k)$ may be plotted in the wave number domain. Because the function $F(k)$ is in general complex, it is necessary to make two graphs in order to represent it completely. The quantities most often used are the magnitude $|F(k)|$ and the phase angle ϵ_k, although the real and imaginary parts of $F(k)$ could also be used. Since the wave number has only discrete values, the amplitude and the phase spectra are not continuous functions but are line spectra.

A3. MULTIPLICATION AND PARSEVAL THEOREMS

When a function $p(\lambda)$ is a product of two component functions:

$$p(\lambda) = f(\lambda) g(\lambda) \tag{A19}$$

we want to know how its Fourier transform can be expressed in terms of the individual transforms of the component $f(\lambda)$ and $g(\lambda)$. The components are assumed to be piecewise continuously differentiable in the interval $[0, 2\pi]$. From Eqs. (A1) and (A2) we then have

$$\mathscr{F}\{p(\lambda)\} \equiv \frac{1}{2\pi} \int_0^{2\pi} f(\lambda) g(\lambda) e^{ik\lambda} d\lambda$$

$$= \frac{1}{2\pi} \int_0^{2\pi} f(\lambda) \left\{\sum_{j=-\infty}^{\infty} G(j) e^{-ij\lambda}\right\} e^{ik\lambda} d\lambda.$$

If $g(\lambda)$ is uniformly convergent in $[0, 2\pi]$, the order of summation and integration can be interchanged, so that

$$\mathscr{F}\{p(\lambda)\} = \sum_{j=-\infty}^{\infty} \left\{ G(j) \frac{1}{2\pi} \int_0^{2\pi} f(\lambda) e^{i(k-j)\lambda} \, d\lambda \right\} \tag{A20a}$$

or

$$\mathscr{F}\{p(\lambda)\} \equiv P(k) = \sum_{j=-\infty}^{\infty} G(j) F(k-j) \tag{A20b}$$

which is the desired result. This is known as the "multiplication theorem." Thus, the spectral function of order k of the product of two functions involves more than the disturbances or waves of order k of both functions. It combines the disturbance of wave number k of one function with the disturbances of all wave numbers j of the other function. It represents a nonlinear wave–wave interaction process. Expression (A20) remains valid if F and G are interchanged.

For many problems arising in spectral analysis it is important to introduce the following integral relationship:

$$p(\lambda) = \sum_{\zeta=-\infty}^{\infty} f(\lambda - \zeta) g(\zeta) \, d\zeta \tag{A21a}$$

which is known as the convolution of $f(\lambda)$ and $g(\lambda)$, and is represented by

$$f(\lambda) * g(\lambda). \tag{A21b}$$

Convolution may be regarded either as expressing a relationship or as an interaction between the two functions, $f(\lambda)$ and $g(\lambda)$, that generates a new function $p(\lambda)$. In the convolution, no distinction is made between the role of $f(\lambda)$ and the role of $g(\lambda)$ so that

$$f(\lambda) * g(\lambda) = g(\lambda) * f(\lambda). \tag{A22}$$

Using the same approach for Eq. (A21a) as that used to derive Eq. (A20b), we see that convolution in the wave number domain corresponds to multiplication in the space domain. Similarly convolution in the space domain corresponds to multiplication in the wave number domain.

If we set k equal to zero in expression (A20b) we obtain the generalized Parseval theorem:

$$P(0) = \frac{1}{2\pi} \int_0^{2\pi} f(\lambda) g(\lambda) \, d\lambda$$

$$= \sum_{j=-\infty}^{\infty} G(j) F(-j)$$

$$= \sum_{j=-\infty}^{\infty} G(j) \hat{F}(j)$$

$$= \sum_{j=-\infty}^{\infty} \Phi_{fg}(j), \tag{A23}$$

where

$$\Phi_{fg}(j) = \hat{F}(j) G(j) \tag{A24}$$

is called the covariance spectrum (co-spectrum) of $f(\lambda)$ and $g(\lambda)$. In establishing expression (A23), we made use of the fact that the complex conjugate $\widehat{F}(j)$ is equal to $F(-j)$. It is easy to see that

$$\Phi_{fg}(j) = \widehat{\Phi}_{gf}(j). \tag{A25}$$

Expression (A23) may be rewritten in an alternative form:

$$P(0) = F(0) G(0) + \sum_{j=1}^{\infty} \{G(j) F(-j) + G(-j) F(j)\} \tag{A26}$$

or using

$$F(j) = |F(j)| \exp\{ij\theta_f(j)\} \text{ and } G(j) = |G(j)| \exp\{ij\theta_g(j)\}, \tag{A27}$$

expression (A26) takes the form

$$P(0) = F(0) G(0) + 2 \sum_{j=1}^{\infty} |F(j)| |G(j)| \cos j [\theta_f(j) - \theta_g(j)]. \tag{A28}$$

The interesting result is that the average value of the product $p(\lambda) = f(\lambda)g(\lambda)$ depends only upon the products of the amplitudes of the spectral components of the same order. None of the cross product terms arising from the product of the spectra of different orders of $f(\lambda)$ and $g(\lambda)$ contribute to the average value of this product. The phase angle of the spectrum is the difference between the phase angles of the spectral components of the same wave number j, so that the greatest value of $P(0)$ is obtained when $F(j)$ and $G(j)$ are in phase.

When $f(\lambda) = g(\lambda)$ expression (A23) reduces to

$$\frac{1}{2\pi} \int_0^{2\pi} f^2(\lambda) \, d\lambda = \sum_{j=-\infty}^{\infty} F(j) F(-j)$$

$$= \sum_{j=-\infty}^{\infty} F(j) \widehat{F}(j)$$

$$= \sum_{j=-\infty}^{\infty} |F(j)|^2$$

$$= \sum_{j=-\infty}^{\infty} \Phi_{ff}(j), \tag{A29}$$

which is another version of Parseval's theorem. The quantity $\Phi_{ff}(j) = |F(j)|^2$ is the variance spectrum (power spectrum) of $f(\lambda)$.

Expression (A29) states that the square of $f(\lambda)$ integrated over its range in the λ domain equals the sum of the squares of the corresponding spectral functions $F(j)$ over all wave numbers, except for a proportionality constant.

Noting that the average value of $f(\lambda)g(\lambda)$ in $[0, 2\pi]$ is

$$[fg] = \frac{1}{2\pi} \int_0^{2\pi} f(\lambda)g(\lambda) d\lambda = [f][g] + [f^*g^*], \tag{A30}$$

where the asterisk denotes, as before in the previous chapters, a deviation from the zonal mean, we conclude that the covariance $[f^*g^*]$ is given by

$$[f^*g^*] = \sum_{\substack{j=-\infty \\ j \neq 0}}^{\infty} \Phi_{fg}(j). \tag{A31}$$

Similarly

$$[f^{*2}] = \sum_{\substack{j=-\infty \\ j \neq 0}}^{\infty} \Phi_{ff}(j). \tag{A32}$$

Expressions (A31) and (A32) show how the covariance and variance can be expanded in the wave number domain. The corresponding spectral components are the components of the cospectrum and power spectrum. This result constitutes a special case of the famous Wiener theorem.

A4. SPECTRAL FUNCTIONS OF THE METEOROLOGICAL VARIABLES AND EQUATIONS

A4.1 Linear quantities

Let us consider some applications of the previous discussions to various meteorological quantities. As we have seen before, the zonal-mean value of a meteorological quantity $f(\lambda)$ is defined by

$$[f(\lambda)] = \frac{1}{2\pi} \int_0^{2\pi} f(\lambda) \, d\lambda = F(0).$$

Since

$$f(\lambda) = [f(\lambda)] + f^*(\lambda),$$

we conclude from Eqs. (A10) and (A12) that

$$f^*(\lambda) = 2 \sum_{k=1}^{\infty} |F(k)| \cos k(\lambda - \epsilon_k)$$

$$= 2 \sum_{k=1}^{\infty} \{F_1(k) \cos k\lambda + F_2(k) \sin k\lambda\}. \quad (A33)$$

The last expression shows that a zonal anomaly results from the superposition of various waves. Each of these waves may be regarded as an individual disturbance with a space scale:

$$L(k) = (2\pi R \cos \phi)/k \sim k^{-1} \quad (A34)$$

and with an amplitude $|F(k)|$. The phase angle ϵ_k determines the geographical distribution of the wave, since it indicates the longitude of the first maximum. For example, for the eastward component of the wind u (using the notation given in Table A.1) we can write for the spectral function $U(k)$:

$$U(k) = \frac{1}{2\pi} \int_0^{2\pi} u(\lambda) e^{ik\lambda} \, d\lambda = U_1(k) + iU_2(k), \quad (A35)$$

where $U_1(k)$ and $U_2(k)$ are the real and imaginary parts of the complex spectrum $U(k)$. The zonal perturbation of u according to Eq. (A33) is given by

$$u^*(\lambda) = 2 \sum_{k=1}^{\infty} |U(k)| \cos k[\lambda - \epsilon_u(k)] \quad (A36)$$

$$= 2 \sum_{k=1}^{\infty} \{U_1(k) \cos k\lambda + U_2(k) \sin k\lambda\}. \quad (A37)$$

The amplitude and the phase angle are given explicitly by the relation

$$|U(k)| = \{U_1^2(k) + U_2^2(k)\}^{1/2} \tag{A38}$$

and

$$\epsilon_u(k) = \frac{1}{k}\tan^{-1}\left\{\frac{U_2(k)}{U_1(k)}\right\}. \tag{A39}$$

For the northward component of the wind $v(\lambda)$ we can write similar expressions as Eqs. (A35)–(A39).

For the temperature $T(\lambda)$ we have

$$T(\lambda) = \sum_{k=-\infty}^{\infty} B(k)\, e^{-ik\lambda}$$

$$= B(0) + 2\sum_{k=1}^{\infty} |B(k)|\cos k[\lambda - \epsilon_T(k)]$$

$$= [T(\lambda)] + T^*(\lambda) \tag{A40}$$

and

$$B(k) = \frac{1}{2\pi}\int_0^{2\pi} T(\lambda)e^{ik\lambda}\, d\lambda$$

$$= B_1(k) + iB_2(k). \tag{A41}$$

The amplitude and phase are

$$|B(k)| = \{B_1^2(k) + B_2^2(k)\}^{1/2} \tag{A42}$$

and

$$\epsilon_T(k) = \frac{1}{k}\tan^{-1}\left\{\frac{B_2(k)}{B_1(k)}\right\}. \tag{A43}$$

For the specific humidity $q(\lambda)$, the equivalent expressions are

$$q(\lambda) = [q(\lambda)] + q^*(\lambda)$$

$$= Q(0) + 2\sum_{k=1}^{\infty} |Q(k)|\cos k[\lambda - \epsilon_q(k)], \tag{A44}$$

$$Q(k) = Q_1(k) + iQ_2(k), \tag{A45}$$

and

$$\epsilon_q(k) = \frac{1}{k}\tan^{-1}\left\{\frac{Q_2(k)}{Q_1(k)}\right\}. \tag{A46}$$

A4.2 Quadratic quantities

Let us now consider the horizontal kinetic energy per unit mass $K(\lambda,\phi,p,t)$. At a given pressure level and at a time t we may write, using Parseval's theorem (A29),

$$[K] = \frac{1}{2\pi}\int_0^{2\pi}(u^2 + v^2)\, d\lambda$$

$$= ([u]^2 + [v]^2)/2 + \sum_{k=1}^{\infty}\{\Phi_{uu}(k) + \Phi_{vv}(k)\}, \tag{A47}$$

where the sum of the spectral components, $\Phi_{uu}(k) + \Phi_{vv}(k)$, represents the kinetic energy at wave number k, i.e., the contribution of the k wave to the eddy kinetic energy.

Of course, $([u]^2 + [v]^2)/2$ is the zero-order component of

$$\Phi_{uu}(k) + \Phi_{vv}(k) \equiv |U(k)|^2 + |V(k)|^2. \tag{A48}$$

Furthermore

$$|U(k)|^2 = U_1^2(k) + U_2^2(k) \tag{A49}$$

and

$$|V(k)|^2 = V_1^2(k) + V_2^2(k). \tag{A50}$$

The variance spectrum of the temperature is related to the available potential energy. For example, at a latitude ϕ the contribution to the total available potential energy $P = P_M + P_E$ per unit mass at a given instant is

$$P = \frac{c_p}{2} \Gamma [T]^{*2} + \frac{c_p}{2} \Gamma [T^{*2}]. \tag{A51}$$

In the above expressions, Γ is a measure of the static stability parameter and $[T]^*$ is the departure of the zonal-mean temperature from its global-mean value (see Chap. 14). Using Parseval's theorem (A29) we may write for the total available potential energy in the wave domain

$$P = \frac{c_p}{2} \Gamma \sum_{k=-\infty}^{\infty} \Phi_{TT}(k), \tag{A52}$$

where $\Phi_{TT}(k) = B_1^2(k) + B_2^2(k)$ is the power spectrum of the temperature. It includes the contributions by the zonal mean ($k=0$) and the eddies of wave number k.

The northward eddy transport of relative zonal angular momentum across latitude ϕ, per unit pressure difference and per unit time [see Eq. (11.13)] is given by

$$\mathcal{J} = \frac{2\pi R^2 \cos^2 \phi}{g} [u^* v^*]. \tag{A53}$$

According to the generalized Parseval's theorem (A23) or (A26), expression (A53) can be written in terms of the spectral functions as follows:

$$\mathcal{J} = \frac{2\pi R^2 \cos^2 \phi}{g} \sum_{\substack{k=-\infty \\ k \neq 0}}^{\infty} \Phi_{uv}(k)$$

$$= \frac{2\pi R^2 \cos^2 \phi}{g} \sum_{k=1}^{\infty} \{U(k)V(-k) + U(-k)V(k)\}$$

$$= \frac{2\pi R^2 \cos^2 \phi}{g} \sum_{k=1}^{\infty} \{U_1(k)V_1(k) + U_2(k)V_2(k)\}. \tag{A54}$$

The evaluation of these spectra from atmospheric data along a latitude circle will show the relative contributions of the various scales of disturbances in the horizontal transport of angular momentum (Saltzman and Peixoto, 1957). Based on Eq. (A28) we see that the transport of angular momentum by a given wave depends on the degree to which disturbances of that wave number in the v field are in phase with disturbances of the same wave number in the u field, as well as on the amplitude of the waves.

Similarly, for the meridional eddy transport of sensible heat across a latitude ϕ the spectral function is given by the co-spectrum Φ_{vT}:

$$\begin{aligned}
\mathcal{J}_H &\equiv \frac{2\pi R \cos\phi}{g} c_p [v^*T^*] \\
&= \frac{2\pi R \cos\phi}{g} c_p \sum_{\substack{k=-\infty \\ k\neq 0}}^{\infty} \Phi_{vT}(k) \\
&= \frac{2\pi R \cos\Phi}{g} c_p \sum_{k=1}^{\infty} \{V(k)B(-k) + V(-k)B(k)\} \\
&= \frac{2\pi R \cos\phi}{g} c_p \sum_{k=1}^{\infty} \{V_1(k)B_1(k) + V_2(k)B_2(k)\}.
\end{aligned}$$
(A55)

Finally, the meridional eddy transport of water vapor across a latitude ϕ is given by

$$\begin{aligned}
\mathcal{J}_q &= \frac{2\pi R \cos\phi}{g} [v^*q^*] \\
&= \frac{2\pi R \cos\phi}{g} \sum_{\substack{k=-\infty \\ k\neq 0}}^{\infty} \Phi_{vq}(k) \\
&= \frac{2\pi R \cos\phi}{g} \sum_{k=1}^{\infty} \{V_1(k)Q_1(k) + V_2(k)Q_2(k)\}.
\end{aligned}$$
(A56)

A4.3 Meteorological equations

The basic meteorological equations in the space domain can also be transformed into equivalent equations in the wave number domain by using the relations presented in the previous sections (Saltzman, 1957; van Mieghem, 1961; Saltzman and Teweles, 1964; and Wiin Nielsen et al., 1963). These transformations can be made by applying the Fourier transform $\mathcal{F}$ (i.e., multiplying the equations by $e^{ik\lambda}/2\pi$ and integrating over λ).

For example, the geostrophic wind equations (3.16) in the wave number domain take the form

$$\begin{cases} U(k) = -\dfrac{g}{fR} \dfrac{\partial Z(k)}{\partial \phi} \\ V(k) = \dfrac{gki}{fR \cos\phi} Z(k) \end{cases}$$
(A57)

and the hydrostatic equation (3.13) becomes

$$\frac{\partial Z(k)}{\partial p} = -\frac{R_d}{gp} B(k),$$
(A58)

where R_d is the dry gas constant.

Let us consider next the thermodynamic energy equation (3.58) in spherical coordinates:

$$\frac{\partial T}{\partial t} = -\left\{\frac{u}{R\cos\phi}\frac{\partial T}{\partial \lambda} + \frac{v}{R}\frac{\partial T}{\partial \phi} + \omega\frac{\partial T}{\partial p}\right\}$$
$$+ \frac{R_d}{c_p p}\omega T + \frac{h}{c_p}. \tag{A59}$$

The symbol h is used here for the diabatic heating rate instead of Q to avoid confusion with the specific humidity. Using the above techniques and applying the multiplication theorem (A20b), the k-order spectral function of Eq. (A59) takes the form

$$\frac{\partial}{\partial t}B(k) = -\sum_{\substack{j=-\infty \\ j\neq k}}^{\infty}\left\{\frac{-ij}{R\cos\phi}B(j)U(k-j)\right.$$
$$+ \frac{1}{R}\frac{\partial B(j)}{\partial \phi}V(k-j) + \frac{\partial B(j)}{\partial p}\Omega(k-j)$$
$$\left. + \frac{R_d}{c_p p}B(j)\Omega(k-j)\right\} + \frac{H(k)}{c_p}. \tag{A60}$$

We have seen in the previous section that the kinetic and available potential energy can also be written in the wave number domain as a sum of contributions from various spatial perturbations. As was done above for the thermodynamic heat equation, we can derive spectral equations for the rate of change of the total kinetic energy and available potential energy for wave number zero, $K(0)$ and $P(0)$, as well as for the different nonzero wave numbers (eddy components), $K(k)$ and $P(k)$.

As expected from the nature of the multiplication theorem, wave–wave interactions between different wave numbers will appear in these equations. They represent the nonlinear interactions between eddies of different scales. Thus, the wave–wave terms represent the rate of transfer of energy to a perturbation with a given wave number from all other wave number perturbations.

As a relatively simple example, we give here the gain in the zonal-mean available potential energy resulting from interactions between all eddies and the zonal-mean temperature. This gain was discussed earlier in Chap. 14 and was given basically by the integrand of $C(P_M, P_E)$ in Eq. (14.32):

$$S(0) = -c_p\Gamma[v^*T^*]\frac{\partial[T]}{R\partial\phi} - c_p\Gamma[\omega^*T^*]\frac{\partial}{\partial p}\{[T] - \tilde{T}\}. \tag{A61}$$

In the spectral domain, this expression can be transformed by applying the various techniques developed above (spectral derivatives, multiplication theorem, etc.) into the form (see, e.g., Saltzman, 1970)

$$S(0) = -c_p\Gamma\sum_{\substack{j=-\infty \\ j\neq 0}}^{\infty}\left\{B(j)V(-j)\frac{\partial B(0)}{R\partial\phi} + B(j)\Omega(-j)\frac{\partial B(0)}{\partial p}\right\}. \tag{A62}$$

APPENDIX B

Analysis in Terms of Empirical Orthogonal Functions (EOF's)

B1. THE PROBLEM

Let us consider a real-valued geophysical field $f(x,t)$ defined simultaneously at M positions denoted as x with N observations at times t. In other words, we are dealing with an ensemble of N instantaneous samples (maps) of a scalar field $f(x,t)$ defined at M stations.

Alternatively we can assume that each sample n constitutes a map with M elements that can be organized in an $M \times 1$ array of data represented by a column vector $\mathbf{f}_n$:

$$(\mathbf{f}_n) \equiv \begin{bmatrix} f_{1n} \\ f_{2n} \\ \cdot \\ \cdot \\ \cdot \\ f_{Mn} \end{bmatrix}, \tag{B1}$$

where $n = 1,...,N$. When we consider all N maps together, we obtain an array of N column vectors forming an $M \times N$ rectangular matrix, $\mathbf{F}$, with M rows (or M time series at each station) and N columns (maps). The matrix element f_{mn} represents the observation made at station m at time n. Instead of the data matrix, we will use another matrix whose elements are functions of deviations from their mean values, namely the covariance matrix $\mathbf{R}$, as we will see next.

The set of the N vectors can be represented in an M-dimensional linear vector space spanned by an arbitrary unit basis $\{\mathbf{u}_1, \mathbf{u}_2,..., \mathbf{u}_M\}$. The N data vectors are directed from the origin to a point in the M space. If there is some correlation between the N vectors we expect that the distribution of their extremities will be organized in clusters or along some preferred directions (see Fig. B.1).

The problem that we want to solve is to find an orthogonal basis in the vector space $\{\mathbf{e}_1, \mathbf{e}_2,..., \mathbf{e}_M\}$ instead of the original basis such that each vector $\mathbf{e}_m$ best represents the cluster of the original data vectors $\mathbf{f}_n$ with $n = 1,...,N$. This is equivalent to

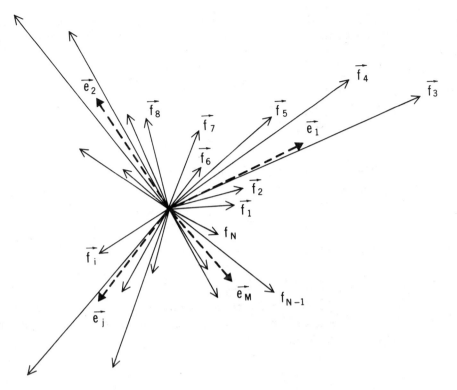

FIGURE B.1. Example of a possible configuration of the station data vectors $\mathbf{f}_n$ (the subscript $n = 1,..., N$ denotes a particular instant of time) and the empirical orthogonal vectors $\mathbf{e}_m$ ($m = 1,..., M$ with, in general, $M \ll N$).

finding a set of M vectors, $\mathbf{e}_m$, whose orientation is such that the sum of the squares of the projections of all the N observation vectors $\mathbf{f}_n$ onto each $\mathbf{e}_m$ is maximized sequentially. We assume that the vectors of the set $\{\mathbf{e}\}$ are mutually orthonormal so that by definition of the inner product:

$$(\mathbf{e}_m \cdot \mathbf{e}_j) = \mathbf{e}_m^T \mathbf{e}_j = \begin{cases} 1 & \text{for } m = j \\ 0 & \text{for } m \neq j \end{cases},$$

where $\mathbf{e}_m^T$ is the transpose vector of $\mathbf{e}_m$. The set of vectors $\{\mathbf{e}\}$ are called the empirical orthogonal functions (EOF's).

B2. SOLUTION OF THE PROBLEM

In mathematical terms, we want to maximize the expression

$$\frac{1}{N} \sum_{n=1}^{N} [\mathbf{f}_n \cdot \mathbf{e}_m]^2 \tag{B2}$$

for $m = 1, 2, ..., M$ subject to the conditions

$$\mathbf{e}_m^T \mathbf{e}_m = 1 \quad \text{and} \quad \mathbf{e}_m^T \mathbf{e}_j = 0 \quad \text{for all } j \neq m. \tag{B3}$$

Expression (B2) can be written in matrix form. Considering the definition of the inner product and noting that the transpose of the product of two matrices is the product of the transposes in reverse order, we have

$$\frac{1}{N}\sum_{n=1}^{N}[\mathbf{f}_n\cdot\mathbf{e}_m]^2 = \frac{1}{N}[\mathbf{e}_m^T\mathbf{F}\ \mathbf{F}^T\mathbf{e}_m] = \mathbf{e}_m^T\mathbf{R}\ \mathbf{e}_m, \tag{B4}$$

where $\mathbf{R}$ is the matrix defined by

$$\mathbf{R} = \frac{1}{N}\mathbf{F}\ \mathbf{F}^T.$$

This is a $M \times M$ real symmetric matrix and is the covariance matrix of the data. Its elements are given by

$$r_{mj} = \frac{1}{N}\sum_{n=1}^{N}f_{mn}f_{jn}$$

and the diagonal elements are the variances

$$r_{mm} = \left(\sum_{n=1}^{N}f_{mn}f_{mn}\right)/N.$$

The maximizing of $[\mathbf{e}_m^T\mathbf{R}\ \mathbf{e}_m]$ subject to the conditions (B3) constitutes a variational problem leading to an eigenvalue or characteristic value problem. Thus, we obtain an equation of the form

$$\mathbf{R}\mathbf{e}_m = \lambda_m\mathbf{e}_m$$

or

$$(\mathbf{R} - \lambda I)\mathbf{e}_m = 0, \tag{B5}$$

where the vector $\mathbf{e}_m$ is the characteristic vector associated with the characteristic value λ_m (introduced as a Lagrange multiplier) of the matrix $\mathbf{R}$, and I is the unit matrix of order M. The matrix $\mathbf{L} = \lambda I$ is a diagonal matrix with the eigenvalues λ_m as diagonal elements.

Equation (B5) leads to an homogeneous system of M linear equations of M unknowns. This homogeneous system possesses nontrivial solutions if and only if the determinant of the coefficients of the matrix $\mathbf{R} - \lambda I$ vanishes, i.e., if

$$|\mathbf{R} - \lambda I| \equiv \begin{vmatrix} r_{11} - \lambda & r_{12} & \cdots & r_{1M} \\ r_{21} & r_{22} - \lambda & \cdots & r_{2M} \\ \cdots & \cdots & \cdots & \cdots \\ r_{M1} & r_{M2} & \cdots & r_{MM} - \lambda \end{vmatrix} = 0. \tag{B6}$$

This leads to an algebraic equation of degree M in λ, known as the characteristic equation of $\mathbf{R}$. The M solutions $\lambda_1, \lambda_2, ..., \lambda_M$ are real and positive, because $\mathbf{R}$ is symmetric and positive definite. Since $\mathbf{R}$ is symmetric, its trace (the sum of the diagonal elements of the matrix) is invariant under a basis transformation and, thus, is equal to the sum of the eigenvalues

$$\sum_{m=1}^{M} r_{mm} = \sum_{m=1}^{M} \lambda_m.$$

This means that each eigenvalue λ_m explains a fraction of the total explained variance, i.e.,

$$\lambda_m \bigg/ \sum_{i=1}^{M} \lambda_i. \tag{B7}$$

For each value λ_m ($m = 1,...,M$), Eq. (B6) leads to a vector solution $\mathbf{e}_m$ which is the eigenvector associated with λ_m. Any two eigenvectors associated with different eigenvalues are orthonormal, so that the inverse of the matrix $\mathbf{E}$, consisting of the M eigenvectors, equals its transpose. The eigenvalues in the $\mathbf{L}$ matrix are arranged in decreasing order of magnitude so that $\lambda_1 \geq \lambda_2 \geq \cdots \geq \lambda_M$. Thus, the first mode $\mathbf{e}_1$ associated with λ_1 (the gravest mode) explains the largest fraction of the total variance of the data.

The set of the M independent and mutually orthogonal eigenvectors, each scaled to have length 1, can be taken as an ordered orthonormal basis $\{\mathbf{e}_1, \mathbf{e}_2, ..., \mathbf{e}_M\}$ in the M-vector space. This gives the optimum representation of the set of the observation data vectors, $\mathbf{f}_n$.

Thus, any observation vector $\mathbf{f}_n$ can be expressed as a linear combination of the M eigenvectors, $\mathbf{e}_m$:

$$\mathbf{f}_n = \sum_{m=1}^{M} c_{mn} \mathbf{e}_m, \tag{B8}$$

where the coefficients c_{mn} are the components or the projections of $\mathbf{f}_n$ on $\mathbf{e}_m$ ($m = 1,...,M$) so that

$$c_{mn} = \mathbf{e}_m^T \mathbf{f}_n. \tag{B9}$$

These coefficients represent the weight of a certain mode $\mathbf{e}_m$ in describing the observations $\mathbf{f}_n$. Furthermore, the coefficients c_{mn} are the elements of an $M \times N$ matrix $\mathbf{C}$, such that

$$\mathbf{C} = \mathbf{E}^T \mathbf{F} \tag{B10}$$

with

$$\mathbf{F} = \mathbf{E} \mathbf{C}. \tag{B11}$$

Because the N observations refer to different times, the elements of a row vector $[c_{m1}, c_{m2}, ..., c_{mN}]$ give the values of the coefficients associated with a given eigenvector $\mathbf{e}_m$. It is important to note that the row vectors $\mathbf{c}_m$ are also mutually orthogonal.

Noting that

$$\sum_{n=1}^{N} \mathbf{f}_n \cdot \mathbf{f}_n = \sum_{n=1}^{N} \left(\sum_{m=1}^{M} c_{mn} \mathbf{e}_m \right) \left(\sum_{j=1}^{M} c_{jn} \mathbf{e}_j \right)$$

$$= \sum_{m=1}^{M} \sum_{n=1}^{N} c_{mn}^2$$

and

$$\sum_{n=1}^{N} \mathbf{f}_n \cdot \mathbf{f}_n = N \sum_{m=1}^{M} \lambda_m,$$

we find

$$N^{-1} \sum_{n=1}^{N} c_{mn}^2 = \lambda_m.$$

Thus, the eigenvalues are the mean-square values of the expansion coefficients of the various modes. The magnitude of the correlations is described completely by the set (λ_m) of the eigenvalues.

Equations (B8) and (B9) are formally similar to the corresponding expressions in the Fourier analysis, where the $\mathbf{e}_m$ are like sine or cosine functions and the c_{mn} are equivalent to the coefficients of a Fourier expansion.

An example of an EOF analysis was given before in Fig. 4.1 and discussed in Chap. 4.

B3. VARIABILITY OF TWO-DIMENSIONAL VECTOR FIELDS

The EOF treatment of the variability of two-dimensional vector fields (e.g., the horizontal wind, water vapor transport, etc.) can be achieved by introduction of a complex representation of the vector fields (e.g., Salstein et al., 1983; Richman, 1986).

Let us consider a two-dimensional vector $\mathbf{V} = V_\lambda \mathbf{i}_\lambda + V_\phi \mathbf{i}_\phi$, where $\mathbf{i}_\lambda$ and $\mathbf{i}_\phi$ are the unit vectors in the zonal and meridional directions, respectively. We can obtain a representation in the complex domain using $V = V_\lambda + iV_\phi$, where $i = (-1)^{1/2}$.

The corresponding $M \times N$ rectangular data matrix consists now of complex elements. From this matrix a complex covariance matrix is derived by multiplying the complex data matrix by its complex conjugate transpose. The covariance matrix $\mathbf{A}$ is hermitian or self-adjoint ($\mathbf{A} = \mathbf{A}^+$ where $\mathbf{A}^+$ is formed by transposing the complex conjugate of $\mathbf{A}$). If a matrix is hermitian its eigenvalues are real. The technique followed is the same as in the real case, so that we can write

$$f^c_{mn} = \sum_{k=1}^{M} c^c_{kn} e^c_{mk},$$

where f^c_{mn} is the departure of the complex value of $\mathbf{V}$ from its long-term mean at the nth time and at the mth station. The eigenvectors and the corresponding time series are complex numbers.

The complex coefficient c^c_{kn} of the time series can also be expressed in terms of its magnitude (d) and its phase (θ), so that

$$c^c = d\, e^{i\theta}.$$

Analogous to the scalar case where the sign of each eigenvector is arbitrary the complex eigenvector contains an arbitrary rotation. Thus, the entire complex field of a derived eigenvector could be multiplied by $e^{i\alpha}$ and it would still remain an eigenvector solution of the covariance matrix. Its corresponding time series, however, would have to be rotated by the angle $-\alpha$, so that the product of the eigenvector and its time series remains invariant. To make the physical interpretation easier it is useful to shift the phase of each complex eigenvector so that the mean-square value of the imaginary part is minimized. Under these conditions the phase of each eigenvector assumes the optimum orientation, i.e., in the direction of the axis along which most of the variance occurs. Then the phase angle α for such a shift would be

$$\alpha = \frac{1}{2} \tan^{-1} \frac{\overline{2\,\mathrm{Re}(V_n) \times \mathrm{Im}(V_n)}}{\overline{\mathrm{Im}(V_n)^2} - \overline{\mathrm{Re}(V_n)^2}},$$

where Re and Im denote the real and the imaginary parts, respectively, of the vectors V_n that form the time series projections of the derived eigenvector solutions onto the data. The overbar indicates the average over each time interval of the time series.

References

Abramopoulos, F., C. Rosenzweig, and B. Choudhury, 1988: "Improved ground hydrology calculations for global climate models (GCMs): Soil water movement and evapotranspiration." J. Climate **1**, 921–941.

Adem, J., 1964: "On the physical basis for the numerical prediction of monthly and seasonal temperatures in the troposphere–ocean–continent system." Mon. Weather Rev. **92**, 91–103.

Andrews, D. G. and M. E. McIntyre, 1976: "Planetary waves in horizontal and vertical shear: the generalized Eliassen-Palm relation and the mean zonal acceleration." J. Atmos. Sci. **33**, 2031–2048.

Andrews, D. G., J. R. Holton, and C. B. Leovy, 1987: *Middle Atmosphere Dynamics*. Academic, New York, 489 pp.

Angell, J. K., 1988: "Variations and trends in tropospheric and stratospheric global temperatures, 1958-87." J. Climate **1**, 1296–1313.

Arakawa, A., 1988: "Finite difference methods in climate modeling." In *Physically based Modelling and Simulation of Climate and Climatic Change* (M. E. Schlesinger, editor), Kluwer Academic, Dordrecht, pp. 79–168.

Arakawa, A. and W. H. Schubert, 1974: "Interaction of a cumulus cloud ensemble with the large-scale environment: Part I." J. Atmos. Sci. **31**, 674–701.

Arpe, K., A. Hollingsworth, M. S. Tracton, A. C. Lorenc, S. Uppala, and P. Kallberg, 1985: "The response of numerical weather prediction systems to FGGE level IIb data. Part II: Forecast verifications and implications for predictability." Q. J. R. Meteor. Soc. **111**, 67–101.

Arya, S. P., 1988: *Introduction to Micrometeorology*. Academic, New York, 307 pp.

Bacastow, R. B., 1976: "Modulation of atmospheric carbon dioxide by the Southern Oscillation." Nature **261**, 116–118.

Barnes, R. T. H., R. Hide, A. A. White, and C. A. Wilson, 1983: "Atmospheric angular momentum fluctuations, length of day changes, and polar motion." Proc. R. Soc. London Ser. A **387**, 31–73.

Barry, R. G., 1985: "The cryosphere and climate change." In *Detecting the Climatic Effects of Increasing Carbon Dioxide* (M. C. MacCracken and F. M. Luther, editors), DOE/ER-0235, pp. 109–141.

Batchelor, G. K., 1967: *An Introduction to Fluid Dynamics*. Cambridge University, New York, 615 pp. (reprinted 1977).

Baumgartner, A. and E. Reichel, 1975: *The World Water Balance*. Elsevier, Amsterdam, 179 pp.

Bengtsson, L., M. Kanamitsu, P. Kallberg, and S. Uppala, 1982: "FGGE four-dimensional data assimilation at ECMWF." Bull. Am. Meteor. Soc. **63**, 29–43.

Bentley, C. R., 1984: "Some aspects of the cryosphere and its role in climatic change." In *Climate Processes and Climate Sensitivity* (J. E. Hansen and T. Takahashi, editors), American Geophysical Union, Washington, D.C., pp. 207–220.

Berger, A. L., 1978: "Long-term variations of daily insolation and quaternary climatic changes." J. Atmos. Sci. **35**, 2362–2367.

Berger, A. L., H. Galleé, Th. Fichefet, I. Marsiat, and Ch. Tricot, 1990: "Testing the astronomical theory with a coupled climate-ice sheet model." Global Planet. Change **89**, 135–141.

Bergman, K. H., A. Hecht, and S. H. Schneider, 1981: "Climate models." Physics Today **34**, 44–51.

Berlage, H. P., 1966: "The Southern Oscillation and world weather." Royal Netherlands Meteor. Inst., Mededelingen en Verhandelingen No. 88, De Bilt, The Netherlands, 152 pp.

Berlyand, T. G. and L. A. Strokina, 1980: "Global distribution of total cloudiness." Gidrometeoizdat, Leningrad, 71 pp.

Bjerknes, J., 1964: Atlantic air–sea interaction." Adv. Geophys. **10**, 1–81.

Bjerknes, J., 1969: "Atmospheric teleconnections from the Equatorial Pacific." Mon. Weather Rev. **97**, 163–172.

Boer, G. J., N. A. McFarlane, R. Laprise, J. D. Henderson, and J.-P. Blanchet, 1984: "The Canadian Climate Center spectral atmospheric general circulation model." Atmosphere–Ocean **22**, 397–429.

Bourke, W., 1988: "Spectral methods in global climate and weather prediction models." In *Physically based Modelling and Simulation of Climate and Climatic Change* (M. E. Schlesinger, editor), Kluwer Academic, Dordrecht, pp. 169–220.

Bowen, I. S., 1926: "The ratio of heat losses by conduction and by evaporation from any water surface." Phys. Rev. **27**, 779–787.

Bowman, K. P., 1985: "Sensitivity of an energy balance climate model with predicted snowfall rates." Tellus A **37**, 233–248.

Bracewell, R. N., 1978: *The Fourier Transform and its Applications*, 2nd ed. McGraw–Hill, New York, 444 pp.

Brier, G. W. and R. A. Roger, 1951: "Verification of weather forecasts." *Compendium of Meteorology*, American Meteorological Society, Boston, MA, pp. 841–848.

Broecker, W. S., T. Takahashi, H. J. Simpson, and T.-H. Peng, 1979: "Fate of fossil fuel carbon dioxide and the global carbon budget." Science **206**, 409–418.

Bromwich, D. H., 1988: "Snowfall in high southern latitudes." Rev. Geophys. **26**, 149–168.

Bryan, F., 1986: "High latitude salinity effects and interhemispheric thermohaline circulations." Nature **323**, 301–304.

Bryan, K., 1969: "Climate and the ocean circulation: III. The ocean model." Mon. Weather Rev. **97**, 806–827.

Bryan, K., 1982: "Poleward heat transport by the ocean: Observations and models." Annu. Rev. Earth Planet. Sci. **10**, 15–38.

Bryan, K. and M. D. Cox, 1967: "A numerical investigation of the oceanic general circulation." Tellus **19**, 54–80.

Bryan, K. and L. J. Lewis, 1979: "A water mass model of the world ocean." J. Geophys. Res. **84**, 2503–2517.

Bryan, K., S. Manabe, and M. J. Spelman, 1988: "Interhemispheric asymmetry in the transient response of a coupled ocean–atmosphere model to CO_2 forcing." J. Phys. Oceanogr. **18**, 851–867.

Bryson, R. A., 1972: "Climate modification by air pollution." In *The Environmental Future* (N. Poluin, editor) MacMillan, London, pp. 134–174.

Budyko, M. I., 1956: "The heat balance of the earth's surface (in Russian)." Gidrometeorologicheshoe Isdatel'stvo, Leningrad, 255 pp. (Translated from Russian by N. A. Stepanova, Office of Technical Services, U. S. Dept. of Commerce, Washington, D. C., 1958.)

Budyko, M. I., 1963: "Atlas of the heat balance of the earth (in Russian)." Globnaia Geofiz. Observ., Moscow, 69 pp.

Budyko, M. I., 1969: "The effect of solar radiation variations on the climate of the earth." Tellus **21**, 611–619.

Budyko, M. I., 1982: *The Earth's Climate: Past and Future*. Academic, New York, 307 pp.

Budyko, M. I., 1986: *The Evolution of the Biosphere*. Reidel, Dordrecht, 423 pp.

Callender, G. S., 1938: "The artificial production of carbon dioxide and its influence on temperature." Q. J. R. Meteor. Soc. **64**, 223–240.

Campbell, G. G. and T. H. Vonder Haar, 1980: "Climatology of radiation budget measurements from satellites." Atm. Sci. Paper No. 323, Dept. Atmos. Sci., Colorado State University, 74 pp.

Campbell, G. G. and T. H. Vonder Haar, 1983: "Latitude average radiation budget over land and ocean." Colorado State University, Fort Collins, CO (unpublished manuscript).

Cane, M. A. and S. E. Zebiak, 1985: "A theory for El Niño and Southern Oscillation." Science **228**, 1085–1087.

Carissimo, B. C., A. H. Oort, and T. H. Vonder Haar, 1985: "Estimating the meridional energy transports in the atmosphere and oceans." J. Phys. Oceanogr. **15**, 82–91.

Cayan, D. R., J. V. Gardner, J. M. Landwehr, J. Namias, and D. H. Peterson, 1989: In *Aspects of Climate Variability in the Pacific and the Western Americas* (D. H. Peterson, editor), Introduction, xiii–xvi. Geophys. Mon. **55**, American Geophysical Union, Washington, D.C.

Cess, R. D., G. L. Potter, J. P. Blanchet, G. J. Boer, S. J. Ghan, J. T. Kiehl, H. LeTreut, Z.-X. Li, X.-Z Liang, J. F. B. Mitchell, J.-J. Morcrette, D. A. Randall, M. R. Riches, E. Roeckner, U. Schlese, A. Slingo, K. E. Taylor, W. M. Washington, R. T. Wetherald, and I. Yagai, 1989: "Interpretation of cloud-climate feedback as produced by 14 atmospheric general circulation models." Science **245**, 513–516.

Charney, J. G., 1947: "The dynamics of long waves in a baroclinic westerly current." J. Meteor. **4**, 135–162.

Charney, J. G., 1948: "On the scale of atmospheric motions." Geophys. Publ. **17**, 17 pp.

Charney, J. G., and P. G. Drazin, 1961: "Propagation of planetary-scale disturbances from the lower into the upper atmosphere." J. Geophys. Res. **66**, 83–109.

Charney, J. G., W. J. Quirk, S. H. Chow, and J. Kornfield, 1977: "A comparative study of the effects of albedo change on drought in semiarid regions." J. Atmos. Sci. **34**, 1366–1385.

Christodoulidis, D. C., D. E. Smith, R. G. Williamson, and S. M. Klosko, 1988: "Observed tidal breaking in the earth/moon/sun system." J. Geophys. Res. **93**, 6216–6236.

Clapp, P. F., 1970: "Parameterization of macroscale transient heat transport for use in a mean-motion model of the general circulation." J. Appl. Meteor. **9**, 554–563.

Cressman, G. P., 1959: "An operational objective analysis system." Mon. Weather Rev. **87**, 367–374.

Crowley, T. J., 1983: "The geological record of climatic change." Rev. Geophys. Space Phys. **21**, 828–877.

Crowley, T. J. and G. R. North, 1991: *Paleoclimatology*. Oxford University, Oxford, 352 pp.

Crutzen, P. J., 1970: "The influence of nitrogen oxides on the atmospheric ozone content." Q. J. R. Meteor. Soc. **96**, 320–325.

de Groot, S. R. and P. Mazur, 1984: *Non-Equilibrium Thermodynamics*. Dover, New York, 510 pp.

Derber, J. and A. Rosati, 1989: "A global oceanic data assimilation scheme." J. Phys. Ocean **19**, 1333–1347.

Dickinson, R. E., 1984: "Modeling evapotranspiration for three-dimensional global climate models." Climate Processes and Climate Sensitivity **5**, 58–72.

Dobson, F. W., F. P. Bretherton, D. M. Burridge, J. Crease, E. B. Kraus, and T. H. Vonder Haar, 1982: "The 'CAGE' experiment: A feasibility study." World Climate Program, **22**, World Meteorological Organization, Geneva.

Dopplick, T. G., 1979: "Radiative heating of the global atmosphere: Corrigendum." J. Atmos. Sci. **36**, 1812–1817.

Dutton, J. A., 1976. *The Ceaseless Wind: An Introduction to the Theory of Atmospheric Motion*, McGraw Hill, New York, 556 pp. [reprinted and revised, Dover, New York, 1986, 617 pp.]

Dutton, J. A. and D. R. Johnson, 1967: "The theory of available potential energy and a variational approach to atmospheric energetics." Adv. Geophys. **12**, 333–436.

Dyson, F. J., 1977: "Can we control the carbon dioxide in the atmosphere?" Energy **2**, 287–291.

Eady, E. T., 1949: "Long waves and cyclone waves." Tellus **1**, 33–52.

Edmon, H. J., Jr., B. J. Hoskins, and M. E. McIntyre, 1980: "Eliassen-Palm cross sections for the troposphere." J. Atmos. Sci. **37**, 2600–2616.

Eliassen, A., and E. Palm, 1961: "On the transfer of energy in stationary mountain waves." Geopfysiske Publikasjoner **22**, 1–23.

Ellis, J. S., T. H. Vonder Haar, S. Levitus, and A. H. Oort, 1978: "The annual variation in the global heat balance of the earth." J. Geophys. Res. **83**, 1958–1962.

Emanuel, W. R., I. Y.-S. Fung, G. G. Killough, B. Moore III, and T.-H. Peng, 1985: "Modeling of the global carbon cycle and changes in the atmospheric carbon dioxide levels." U.S. Dept. of Energy, DOE/ER-0239, pp. 141–173.

Essex, C., 1984: Radiation and the irreversible thermodynamics of climate. J. Atmos. Sci. **41**, 1985–1991.

Farman, J. C., B. G. Gardiner, and J. D. Shanklin, 1985: "Large losses of total ozone in Antarctica reveal seasonal Cl_x/NO_x interaction." Nature **315**, 207–210.

Fleming, R. J., T. M. Kaneshige, and W. E. McGovern, 1979a: "The Global Weather Experiment: I. The observational phase through the first special observing period." Bull. Am. Meteor. Soc. **60**, 649–659.

Fleming, R. J., T. M. Kaneshige, W. E. McGovern, and T. E. Bryan, 1979b: "The Global Weather Experiment: II. The second special observing period." Bull. Am. Meteor. Soc. **60**, 1316–1322.

Fofonoff, N. P., 1962: "Physical properties of sea water," *The Sea* (M. N. Hill, editor), Wiley, New York, Vol. 1, pp. 3–30.

Fortak, H. G., 1979: "Entropy and climate." In *Man's Impact on Climate* (W. Bach, J. Pankrath, and W. Kellogg, editors), Elsevier, Amsterdam, pp. 1–14.

Gardiner, B. G., 1989: "The Antarctic ozone hole." Weather **44**, 291–298.

GARP, 1979: *Report of the JOC study conference on climate models: Performance, intercomparison, and sensitivity studies* (W. L. Gates, editor), GARP Publ. No. 22 (2 volumes), 606 pp.

Gast, P. R., 1965: "Solar electromagnetic radiation." In *Handbook of Geophysics and Space Environments*, Air Force Cambridge Research Laboratory, U.S. Air Force, pp. 16.1–16.9.

Gates, W. L., 1976: "The numerical simulation of ice-age climate with a global general circulation model." J. Atmos. Sci. **33**, 1844–1873.

Goody, R. M., 1964: *Atmospheric Radiation: I. Theoretical Basis*. Clarendon, Oxford, 436 pp.

Green, J. S. A., 1970: "Transfer properties of the large-scale eddies and the general circulation of the atmosphere." Q. J. R. Meteor. Soc. **96**, 157–185.

Guillemin, E. A., 1948: *The Mathematics of Circuit Analysis: Extensions to the Mathematical Training of Electrical Engineers*. MIT, Cambridge, MA, 590 pp.

Hall, M. M. and H. L. Bryden, 1982: "Direct estimates and mechanisms of ocean heat transport." Deep-Sea Res. **29**, 339–359.

Hansen, J., D. Johnson, A. Lacis, S. Lebedeff, P. Lee, D. Rind, and G. Russell, 1981: "Climate impact of increasing atmospheric carbon dioxide." Science **213**, 957–966.

Hansen, J., I. Fung, A. Lacis, D. Rind, S. Lebedeff, R. Ruedy, G. Russell, and P. Stone, 1988: "Global climate changes as forecast by Goddard Institute for Space Studies three-dimensional model." J. Geophys. Res. **93**, 9341–9364.

Hantel, M., 1974: "On the display of the atmospheric circulation with stream functions." Mon. Weather Rev. **102**, 649–661.

Hantel, M. and J. M. Hacker, 1978: "On the vertical eddy transports in the northern atmosphere: II. Vertical eddy momentum transport for summer and winter." J. Geophys. Res. **83**, 1305–1318.

Hardy, D. M. and J. J. Walton, 1978: "Principal components analysis of vector wind measurements." J. Appl. Meteorol. **17**, 1153–1162.

Harris, R. G., A. Thomasell, Jr., and J. G. Welsh, 1966: "Studies of techniques for the analysis and prediction of temperature in the ocean, Part III: Automated analysis and prediction." Interim Report, prepared by Travelers Research Center, Inc., for U.S. Naval Oceanographic Office, Contract No. N62306-1675, 97 pp.

Hasselmann, K., 1976: "Stochastic climate models, Part 1. Theory." Tellus **28**, 473–485.

Hasselmann, K., 1988: "Some problems in the numerical simulation of climate variability using high-resolution coupled models." In *Physically based Modelling and Simulation of Climate and Climatic Change* (M. E. Schlesinger, editor), Kluwer Academic, Dordrecht, pp. 583–614.

Hastenrath, S. 1982: "On meridional heat transports in the World Ocean." J. Phys. Oceanogr. **12**, 922–927.

Hastenrath, S. and L. Heller, 1977: "Dynamics of climatic hazards in northeast Brazil." Q. J. R. Meteor. Soc. **103**, 77–92.

Hayashi, Y., 1982: "Space-time spectral analysis and its applications to atmospheric waves." J. Met. Soc. Jpn. **60**, 156–171.

Heckley, W. A., 1985: "Systematic errors of the ECMWF operational forecasting model in tropical regions." Q. J. R. Meteor. Soc. **111**, 709–738.

Hellerman, S., 1967: "An updated estimate of the wind stress on the World Ocean." Mon. Weather Rev. **95**, 607–626 (see also corrections **96**, 63–74).

Hellerman, S., 1981: "The net meridional flux of water by the oceans from evaporation and precipitation estimates." GFDL/NOAA, Princeton, NJ (unpublished manuscript), 13 pp.

Hellerman, S. and M. Rosenstein, 1983: "Normal monthly wind stress over the World Ocean with error estimates." J. Phys. Oceanogr. **13**, 1093–1104.

Hibler, W. D. III, 1988: "Modelling sea ice thermodynamics and dynamics in climate studies." In *"Physically based Modelling and Simulation of Climate and Climatic Change* (M. E. Schlesinger, editor), Kluwer Academic, Dordrecht, pp. 509–563.

Hide, R., N. T. Birch, L. V. Morrison, D. J. Shea, and A. A. White, 1980: "Atmospheric angular momentum fluctuations and changes in the length of the day." Nature **286**, 114–117.

Holopainen, E. O., 1982: "Long-term budget of zonal momentum in the free atmosphere over Europe in winter." Q. J. R. Meteor. Soc. **108**, 95–102.

Holopainen, E. O. and A. H. Oort, 1981: "Mean surface stress curl over the oceans as determined from the vorticity budget of the atmosphere." J. Atmos. Sci. **38**, 262–280.

Holton, J. R., 1972: *An Introduction to Dynamic Meteorology.* Academic, New York, 319 pp.

Horel, J. D. and J. M. Wallace, 1981: "Planetary-scale atmospheric phenomena associated with the Southern Oscillation." Mon. Weather Rev. **109**, 813–829.

Hoskins, B. J. and D. J. Karoly, 1981: "The steady, linear response of a spherical atmosphere to thermal and orographic forcing." J. Atmos. Sci. **38**, 1179–1196.

Hoskins, B. J., I. N. James, and G. H. White, 1983: "The shape, propagation, and mean-flow interaction of large-scale weather systems." J. Atmos. Sci. **40**, 1595–1612.

Houghton, H. G., 1985: *Physical Meteorology.* MIT, Cambridge, MA, 442 pp.

Houghton, J. T., G. J. Jenkins, and J. J. Ephraums (editors): "Climate Change; The IPCC Scientific Assessment." Cambridge University Press, 365 pp.

Howard, J. N., D. L. Burch, and D. Williams, 1955: "Near-infrared transmission through synthetic atmospheres." Geophys. Res. Papers No. 40, Geophys. Res. Dir., Air Force Cambridge Research Center, 244 pp.

Imbrie, J. and K. P. Imbrie, 1979: *Ice Ages, Solving the Mystery.* Enslow, Hillside, NJ, 224 pp.

Iqbal, M., 1983: *An Introduction to Solar Radiation.* Academic, New York, 390 pp.

Jacobs, S. S. and R. G. Fairbanks, 1985: "Origin and evolution of water masses near the Antarctic continental margin: Evidence from $H_2^{18}O/H_2^{16}O$ ratios in sea water." In *Oceanology of the Antarctic Continental Shelf* (S. S. Jacobs, editor), American Geophysical Union, Washington, D.C., pp. 59–85.

Jaeger, L., 1976: "Monthly precipitation maps for the entire earth (in German)." Ber. Deutschen Wetterdienstes **18**, 38 pp.

Johnson, D. R., 1989: "The forcing and maintenance of global monsoonal circulations: An isentropic analysis." Adv. Geophys. **31**, 43–316.

Johnston, H. S., 1971: "Reduction of stratospheric ozone by nitrogen oxide catalysts from supersonic transport exhaust." Science **173**, 517–522.

Jones, P. D., 1988: "Hemispheric surface air temperature variations: Recent trends and an update to 1987." J. Climate **1**, 654–660.

Jones, P. D., T. M. L. Wigley, and P. B. Wright, 1986: "Global temperature variations between 1861 and 1984." Nature **322**, 430–434.

Julian, P. R. and R. M. Chervin, 1978: "A study of the Southern Oscillation and Walker circulation phenomenon." Mon. Weather Rev. **106**, 1433–1451.

Kaplan, L. D., 1959: "Inference of atmospheric structure from remote radiation measurements." J. Opt. Soc. Am. **39**, 1004–1007.

Karoly, D. J., 1987: "Southern Hemisphere temperature trends: A possible greenhouse gas effect?" Geophys. Res. Lett. **14**, 1139–1141.

Karoly, D. J., 1989: "Northern Hemisphere temperature trends: A possible greenhouse gas effect?" Geophys. Res. Lett. **16**, 465–468.

Karoly, D. J. and A. H. Oort, 1987: "A comparison with Southern Hemisphere circulation statistics based on GFDL and Australian analyses." Mon. Weather Rev. **115**, 2033–2059.

Kasahara, A. and W. M. Washington, 1967: "NCAR global general circulation model of the atmosphere." Mon. Weather Rev. **95**, 389–402.

Kasahara, A., T. Sasamori, and W. M. Washington, 1973: "Simulation experiments with a 12-layer stratospheric global circulation model. I. Dynamical effect of the Earth's orography and thermal influence of continentality." J. Atmos. Sci. **30**, 1229–1251.

Keeling, C. D., R. B. Bacastow, and T. P. Whorf, 1982: *Carbon Dioxide Review: 1982* (W. C. Clark, editor), Oxford University, Oxford, pp. 377–385.

Keeling, C. D. and R. Revelle, 1985: "Effects of El Niño/Southern Oscillation on the atmospheric content of carbon dioxide." Meteoritics **20**, 437–451.

Keeling, C. D., R. B. Bacastow, A. F. Carter, S. C. Piper, T. P. Whorf, M. Heimann, W. G. Mook, and H. Roeloffzen, 1989: "A three-dimensional model of atmospheric CO_2 transport based on observed winds: I. Analysis of observational data." In *Aspects of Climate Variability in the Pacific and the Western Americas* (D. H. Peterson, editor), Geophysical Monograph 55, American Geophysical Union, Washington, D.C., pp. 165–236.

Knutson, T. R. and K. M. Weickmann, 1987: "30–60 day atmospheric oscillations: Composite life cycles of convection anomalies." Mon. Weather Rev. **115**, 1407–1436.

Köppen, W., 1931: *The Climates of the Earth*. (in German) de Gruyter, Berlin, 388 pp.

Kubota, S., 1954: "Numerical prediction by the quasi-double Fourier Series." Papers Meteor. Geophys. **5**, Meteor. Soc. Japan, Tokyo, 144–152.

Kukla, G. J. and H. J. Kukla, 1974: "Increased surface albedo in the Northern Hemisphere." Science **183**, 709–714.

Kuo, C., C. Lindberg, and D. J. Thomson, 1990: "Coherence established between atmospheric carbon dioxide and global temperature." Nature **343**, 709–714.

Kuo, H. L., 1956: "Forced and free meridional circulations in the atmosphere." J. Meteor. **13**, 561–568.

Kuo, H. L., 1965: "On formation and intensification of tropical cyclones through latent heat release by cumulus convection." J. Atmos. Sci. **22**, 40–63.

Kuo, H. L., 1974: "Further studies of the parameterization of the influence of cumulus convection on large-scale flow." J. Atmos. Sci. **31**, 1232–1240.

Kurihara, Y., 1970: "A statistical–dynamical model of the general circulation of the atmosphere." J. Atmos. Sci. **27**, 847–870.

Kurihara, Y. and J. L. Holloway, Jr., 1967: "Numerical integration of a nine-level global primitive equation model formulated by the box method." Mon. Weather Rev. **95**, 509–530.

Kutzbach, J. E., P. J. Guetter, W. F. Ruddiman, and W. L. Prell, 1989: "Sensitivity of climate to Late Cenozoic uplift in Southern Asia and the American West: Numerical experiments." J. Geophys. Res. **94**, 18,393–18,407.

Labitzke, K. and H. van Loon, 1988: "Associations between the 11-year solar cycle, the QBO, and the atmosphere. Part I: The troposphere and stratosphere in the northern hemisphere in winter." J. Atmos. Terr. Phys. **50**, 197–206.

Lamb, H. H., 1972: *Climate: Present, Past and Future*. Methuen, London, Vol. 1, 613 pp.

Lamb, H. H., 1977: *Climate: Present, Past and Future. Vol. 2. Climatic History and the Future*. Methuen, London, 835 pp.

Lamb, P. J. and R. A. Peppler, 1991: West Africa. In *Teleconnections Linking Worldwide Climate Anomalies: Scientific Basis and Societal Impact* (M. H. Glantz, R. W. Katz, and N. Nicholls, editors), Cambridge University, New York.

Large, W. G. and S. Pond, 1981: "Open ocean momentum flux measurements in moderate to strong winds." J. Phys. Oceanogr. **11**, 324–336.

Large, W. G. and S. Pond, 1982: "Sensible and latent heat flux measurements over the ocean." J. Phys. Ocean **12**, 464–482.

Lau, K. M. and P. H. Chan, 1986: "The 40–50 day oscillation and the El Nino/Southern Oscillation: A new perspective." Bull. Am. Meteor. Soc. **67**, 533–534.

Lau, N.-C. 1981: "A diagnostic study of recurrent meteorological anomalies appearing in a 15-year simulation with a GFDL general circulation model." Mon. Weather Rev. **109**, 2287–2311.

Lau, N.-C., 1985: "Modeling the seasonal dependence of the atmospheric response to observed El Niños in 1962–76." Mon. Weather Rev. **113**, 1970–1996.

Lau, N.-C. and A. H. Oort, 1981: "A comparative study of observed Northern Hemisphere circulation statistics based on GFDL and NMC analyses. Part I: The time-mean fields." Mon. Weather Rev. **109**, 1380–1403.

Lau, N.-C. and A. H. Oort, 1982: "A comparative study of observed Northern Hemisphere circulation statistics based on GFDL and NMC analyses. Part II: Transient eddy statistics and the energy cycle." Mon. Weather Rev. **110**, 889–906.

Leetmaa, A. and M. Ji, 1989: "Operational hindcasting of the tropical Pacific." Dyn. Atmos. Oceans **13**, 465–490.

Leith, C. E., 1965: "Numerical simulation of the earth's atmosphere." Methods Comput. Phys. **4**, 1–28.

Leith, C. E., 1978: "Predictability of climate." Nature **276**, 352–355.

Lesins, G. B., 1990: On the relationship between radiative entropy and temperature distributions. J. Atmos. Sci. **47**, 795–803.

Lettau, H., 1954: "A study of the mass, momentum, and energy budget of the atmosphere." Arch. Meteor., Geophys. Bioklimatologie 133–157.

Levitus, S., 1982: Climatological Atlas of the World Ocean. NOAA Professional Paper No. 13, U.S. Government Printing Office, Washington, D.C., 163 pp.

Levitus, S., 1984: "Annual cycle of temperature and heat storage in the world ocean." J. Phys. Oceanogr. **14**, 727–746.

Levitus, S. and A. H. Oort, 1977: "Global analysis of oceanographic data." Bull. Am. Meteor. Soc. **58**, 1270–1284.

Lindzen, R. S., 1987: "On the development of the theory of the QBO." Bull. Am. Meteor. Soc. **68**, 329–337.

Lindzen, R. S. and J. R. Holton, 1968: "A theory of the quasibiennial oscillation." J. Atmos. Sci. **25**, 1095–1107.

Liou, K.-N., 1980: *An Introduction to Atmospheric Radiation.* Academic, New York, 392 pp.

List, R. J. (editor), 1951: *Meteorological Table,* 6th ed., Smithsonian Institute, Washington, D.C., 527 pp.

Lockwood, J. G., 1979: *Causes of Climate.* Wiley, New York, 260 pp.

Lorenz, E. N., 1951: "Seasonal and irregular variations of the Northern Hemisphere sea-level pressure profile." J. Meteor. **8**, 52–59.

Lorenz, E. N., 1955: "Available potential energy and the maintenance of the general circulation." Tellus **7**, 157–167.

Lorenz, E. N., 1956: "Empirical orthogonal functions and statistical weather prediction." MIT, Dept. of Meteorology, Science Report 1, 49 pp.

Lorenz, E. N., 1960: "Maximum simplification of the dynamic equations." Tellus **12**, 243–254.

Lorenz, E. N., 1963: "Deterministic nonperiodic flow." J. Atmos. Sci. **20**, 130–141.

Lorenz, E. N., 1967: "The Nature and Theory of the General Circulation of the Atmosphere." WMO Publication, 218, World Meteorological Organization, Geneva, Switzerland, 161 pp.

Lorenz, E. N., 1969: "The predictability of a flow which possesses many scales of motion." Tellus **21**, 289–307.

Lorenz, E. N., 1975: "Climate predictability." In *The Physical Basis of Climate and Climate Modelling.* World Meteorological Organization, Geneva, Switzerland, GARP, **16**, 132–136.

L'vovitch, M. I., 1979: "World water resources and their future" (translated from 1974 Russian text by Amer. Geophys. Union). American Geophysical Union, Washington, D.C., 415 pp.

L'vovitch, M. I. and S. P. Ovtchinnikov, 1964: "River drainage. Physical-geographical atlas of the world." Academy of Sciences, USSR and Central Administration of Geodesy and Cartography of USSR, Moscow, 208 pp.

MacCracken, M. C. and S. Ghan, 1987: "Design and use of zonally averaged climate models." UCRL-94338, University of California, Livermore, CA, 44 pp.

Madden, R. and P. R. Julian, 1971: "Detection of a 40–50 day oscillation in the zonal wind in the tropical Pacific." J. Atmos. Sci. **28**, 702–708.

Mahlman, J. D., H. Levy II, and W. J. Moxim, 1986: "Three-dimensional simulations of stratospheric N_2O: Predictions for other trace constituents." J. Geophys. Res. **91**, 2687–2707.

Mak, M.-K., 1969: "Laterally driven stochastic motions in the tropics." J. Atmos. Sci. **26**, 41–64.

Manabe, S., 1969: "Climate and the ocean circulation: 2. The atmospheric circulation and the effect of heat transfer by ocean currents." Mon. Weather Rev. **97**, 775–805.

Manabe, S., 1983: "Carbon dioxide and climatic change." Adv. Geophys. **25**, 39–82.

Manabe, S. and A. J. Broccoli, 1985: "The influence of continental ice sheets on the climate of an ice age." J. Geophys. Res. **90**, 2167–2190.

Manabe, S. and A. J. Broccoli, 1990: "Mountains and arid climates in middle latitudes." Science **247**, 192–195.

Manabe, S. and J. L. Holloway, Jr., 1975: "The seasonal variation of the hydrological cycle as simulated by a global model of the atmosphere." J. Geophys. Res. **80**, 1617–1649.

Manabe, S. and R. F. Strickler, 1964: "Thermal equilibrium of the atmosphere with a convective adjustment." J. Atmos. Sci. **21**, 361–385.

Manabe, S. and T. Terpstra, 1974: "The effects of mountains on the general circulation of the atmosphere as identified by numerical experiments." J. Atmos. Sci. **31**, 3–42.

Manabe, S. and R. T. Wetherald, 1967: "Thermal equilibrium of the atmosphere with a given distribution of relative humidity." J. Atmos. Sci. **24**, 241–259.

Manabe, S., K. Bryan, and M. J. Spelman, 1979: "A global ocean–atmosphere climate model with seasonal variation for future studies of climate sensitivity." Dyn. Atm. Oceans **3**, 393–426.

Manabe, S., J. L. Holloway, Jr., and H. M. Stone, 1970: "Tropical circulation in a time integration of a global model of the atmosphere." J. Atmos. Sci. **27**, 580–613.

Manabe, S., J. Smagorinsky, and R. F. Strickler, 1965: "Simulated climatology of a general circulation model with a hydrological cycle." Mon. Weather Rev. **93**, 769–798.

Margules, M., 1903: "Ueber die Energie der Stürme." Jahrb. Zentralanst. Meteor. 1–26. (Translated by C. Abbe, 1910: The Mechanics of the Earth's Atmosphere. Smithsonian Misc. Collect. **51**, Washington, D.C., 553–595.)

Mass, C. F. and D. A. Portman, 1989: "Major volcanic eruptions and climate: A critical evaluation." J. Climate **2**, 566–593.

Masuda, K., 1988: "Meridional heat transport by the atmosphere and the ocean: Analysis of FGGE data." Tellus A **40**, 285–302.

McCarthy, D. D. and A. K. Babcock, 1986: "The length of day since 1656." Phys. Earth Planet. Interface **44**, 281–292.

Meehl, G. A., 1980: "Observed world ocean seasonal surface currents on a 5° grid." NCAR Technical Note, TN/IA-159, 23 pp.

Milankovitch, M., 1941: "History of radiation on the earth and its use for the problem of the ice ages" (in German). K. Serb. Akad. Beogr. Spec. Publ. 132 (Translated by the Israel Program for Scientific Translations, Jerusalem, 1969), 633 pp.

Miller, J. R., G. L. Russell, and L.-C. Tsang, 1983: "Annual oceanic heat transports computed from an atmospheric model." Dyn. Atm. Oceans **7**, 95–109.

Mintz, Y., 1965: "Very long-term global integration of the primitive equations of atmospheric motion: (an experiment in climate simulation)." WMO Tech. Notes **66**, 119–143 (and Am. Met. Soc. Meteor. Monograph **8**, 20–36, 1968).

Mitchell, J. M., 1976: "An overview of climatic variability and its causal mechanisms." Quaternary Res. **6**, 481–493.

Miyakoda, K., G. D. Hembree, R. F. Strickler, and I. Shulman, 1972: "Cumulative results of extended forecast experiments. I. Model performance for winter cases." Mon. Weather Rev. **100**, 836–855.

Molina, M. and F. S. Rowland, 1974: Stratospheric sink for chlorofluoromethanes: Chlorine atom-catalyzed destruction of ozone." Nature **249**, 810–812.

Molteni, F. and S. Tibaldi, 1985: "Climatology and systematic error of rainfall forecasts at ECMWF." ECMWF Tech. Report 51, Shinfield Park, Reading, Berkshire, England, 88 pp.

Munk, W. H., 1966: "Abyssal recipes." Deep-Sea Res. **13**, 707–730.

Mysak, L. A., 1986: "El Niño, interannual variability, and fisheries in the Northeast Pacific Ocean." Can. J. Fish. Aquat. Sci. **43**, 464–497.

Nakamura, N. and A. H. Oort, 1988: "Atmospheric heat budgets of the polar regions." J. Geophys. Res. **93**, 9510–9524.

Namias, J., 1983: "Short period climatic variations." *Collected Works of J. Namias, 1975 Through 1982, Vol. III.* University of California, San Diego, 393 pp.

National Academy of Sciences, 1975: "Understanding Climatic Change: A Program for Action." Natl. Acad. Sci., Washington, D.C., 239 pp.

National Academy of Sciences, Climate Research Board, 1979: "Carbon Dioxide and Climate: A Scientific Assessment." Natl. Acad. Sci., Washington, D.C., 22 pp.

National Academy of Sciences, Climate Research Committee, 1982: "Carbon Dioxide and Climate: A Second Assessment." Natl. Acad. Sci., Washington, D.C., 72 pp.

National Climatic Data Center, 1987: Monthly Climatic Data for the World, Vol. 40, Nos. 1 and 7. NOAA, NCDC, Asheville, NC.

Naujokat, B., 1986: "An update of the observed quasibiennial oscillation of the stratospheric winds over the tropics." J. Atmos. Sci. **43**, 1873–1877.

Neftel, A., E. Moor, H. Oeschger, and B. Stauffer, 1985: "Evidence from polar ice cores for the increase in atmospheric CO_2 in the last two centuries." Nature **315**, 45–47.

Neumann, C. J., G. W. Cry, E. L. Caso, and B. R. Jarvinen, 1981: "Tropical cyclones of the North Atlantic Ocean, 1871-1980." U.S. Dept. of Commerce, Natl. Climatic Center, Asheville, N.C., U.S. Government Printing Office, Washington, D.C., 174 pp.

Neumann, G. and W. J. Pierson, Jr., 1966: *Principles of Physical Oceanography.* Prentice-Hall, Englewood Cliffs, NJ, 545 pp.

Newell, R. E., 1963: "Transfer through the tropopause and within the stratosphere." Q. J. R. Meteor. Soc. **89**, 167–204.

Newell, R. E., D. G. Vincent, T. G. Dopplick, D. Ferruza, and J. W. Kidson, 1970: "The energy balance of the global atmosphere." In *The Global Circulation of the Atmosphere* (G. A. Corby, editor), Royal Meteorological Society, London, pp. 42–90.

Newell, R. E., J. W. Kidson, D. G. Vincent, and G. J. Boer, 1972: *The General Circulation of the Tropical Atmosphere, Vol. 1.* MIT, Cambridge, MA, 258 pp.

Newell, R. E., J. W. Kidson, D. G. Vincent, and G. J. Boer, 1974: *The General Circulation of the Tropical Atmosphere, Vol. 2.* MIT, Cambridge, MA, 371 pp.

Newton, C. W., 1971a: "Global angular momentum balance: Earth torques and atmospheric fluxes." J. Atmos. Sci. **28**, 1329–1341.

Newton, C. W., 1971b: "Mountain torques in the global angular momentum balance." J. Atmos. Sci. **28**, 623–628.

North, G. R., 1988: "Lessons from energy balance models." In *Physically based Modelling and Simulation of Climate and Climatic Change* (M. E. Schlesinger, editor), Kluwer Academic, Dordrecht, pp. 627–651.

North, G. R., R. F. Cahalan, and J. A. Coakley, Jr., 1981: "Energy-balance climate models." Rev. Geophys. Space Phys. **19**, 91–121.

Oerlemans, J. and C. J. van der Veen, 1984: *Ice Sheets and Climate.* Reidel, Dordrecht, 217 pp.

Oke, T. R., 1987: *Boundary Layer Climates,* 2nd ed. Halsted, New York, 435 pp.

Oort, A. H., 1978: "Adequacy of the rawinsonde network for global circulation studies tested through numerical model output." Mon. Weather Rev. **106**, 174–195.

Oort, A. H., 1983: "Global atmospheric circulation statistics, 1958–1973." NOAA Professional Paper No. 14, U.S. Government Printing Office, Washington, D.C., 180 pp. + 47 microfiches.

Oort, A. H., 1985: "Balance conditions in the earth's climate system." Adv. Geophys. A **28**, 75–98.

Oort, A. H., 1989: "Angular momentum cycle in the atmosphere–ocean–solid earth system." Bull. Am. Meteor. Soc. **70**, 1231–1242.

Oort, A. H., and H. Liu, 1991: "Upper air temperature trends over the globe, 1958–1989." J. Climate (in press).

Oort, A. H. and J. P. Peixoto, 1983: "Global angular momentum and energy balance requirements from observations." Adv. Geophys. **25**, 355–490.

Oort, A. H. and T. H. Vonder Haar, 1976: "On the observed annual cycle in the ocean–atmosphere heat balance over the Northern Hemisphere." J. Phys. Oceanogr. **6**, 781–800.

Oort, A. H., Y.-H. Pan, R. W. Reynolds, and C. F. Ropelewski, 1987: "Historical trends in the surface temperature over the oceans based on the COADS." Climate Dyn. **2**, 29–38.

Oort, A. H., S. C. Ascher, S. Levitus, and J. P. Peixoto, 1989: "New estimates of the available potential energy in the world ocean." J. Geophys. Res. **94**, 3187–3200.

Overland, J. E. and R. W. Preisendorfer, 1982: "A significance test for principal components applied to a cyclone climatology." Mon. Weather Rev. **110**, 1–4.

Palmén, E. and C. W. Newton, 1969: *Atmospheric Circulation Systems*. Academic, New York, 606 pp.

Palmer, T. N., G. J. Shutts, and R. Swinbank, 1986: "Alleviation of a systematic westerly bias in general circulation and numerical weather prediction models through an orographic gravity wave drag parameterization." Q. J. R. Meteor. Soc. **112**, 1001–1039.

Paltridge, G. W., 1975: "Global dynamics and climate—A system of minimum entropy exchange." Quart. J. R. Met. Soc. **101**, 475–484.

Pan, Y.-H. and A. H. Oort, 1983: "Global climate variations connected with sea surface temperature anomalies in the eastern equatorial Pacific Ocean for the 1958–1973 period." Mon. Weather Rev. **111**, 1244–1258.

Parthasarathy, B., A. A. Munot, and D. R. Kothawale, 1988: "Regression model for estimation of Indian food grain production from summer monsoon rainfall." Agr. Forest Meteor. **42**, 167–182.

Peixoto, J. P., 1960: "Hemispheric temperature conditions during the year 1950." MIT, Dept. of Meteorology, Planet. Circ. Proj. Sci. Rept. **4**, 211 pp.

Peixoto, J. P., 1973: "Atmospheric vapor flux computations for hydrological purposes." WMO Publ. 357, Geneva, Switzerland, 83 pp.

Peixoto, J. P., 1974: "Enthalpy distribution in the atmosphere over the Southern Hemisphere." Rev. Ital. Geofis. **23**, 223–242.

Peixoto, J. P. and M. A. Kettani, 1973: "The control of the water cycle." Sci. Am. **228**, 46–61.

Peixoto, J. P. and A. H. Oort, 1974: "The annual distribution of atmospheric energy on a planetary scale." J. Geophys. Res., **79**, 2149–2159.

Peixoto, J. P. and A. H. Oort, 1983: "The atmospheric branch of the hydrological cycle and climate." In *Variations of the Global Water Budget*. Reidel, London, pp. 5–65.

Peixoto, J. P. and A. H. Oort, 1984: "Physics of climate." Rev. Mod. Phys. **56**, 365–429.

Peixoto, J. P., A. H. Oort, M. de Almeida, and A. Tomé, 1991: "Entropy budget of the atmosphere." J. Geophys. Res. **96**, 10981–10988.

Penman, H. L., 1948: "Natural evaporation from open water, bare soil, and grass." Proc. R. Soc. London Ser. A **193**, 120–145.

Pfeffer, R. L., 1981: "Wave-mean flow interactions in the atmosphere." J. Atmos. Sci. **38**, 1340–1359.

Philander, S. G. H., 1989: *El Niño, La Niña, and the Southern Oscillation*. Academic, New York, 293 pp.

Philander, S. G. H., N.-C. Lau, R. C. Pacanowski, and M. J. Nath, 1989: "Two different simulations of the Southern Oscillation and El Niño with coupled ocean–atmosphere general circulation models." Philos. Trans. R. Soc. London Ser. A **329**, 167–178.

Phillips, N. A., 1956: "The general circulation of the atmosphere: A numerical experiment." Q. J. R. Meteor. Soc. **82**, 123–164.

Phillips, N. A., 1957: "A coordinate system having some special advantages for numerical forecasting." J. Meteor. **14**, 184–185.

Pickard, G. L. and W. J. Emery, 1982: *Descriptive Physical Oceanography*, 4th ed. Pergamon, New York, 249 pp.

Piola, A. R., H. A. Figueroa, and A. A. Bianchi, 1987: "Some aspects of the surface circulation south of 20 °S revealed by First GARP Global Experiment drifters." J. Geophys. Res. **92**, 5101–5114.

Priestley, C. H. B., 1951: "A survey of the stress between the ocean and atmosphere." Aust. J. Sci. Res. A **4**, 315–328.

Prigogine, I., 1962: *Introduction to Nonequilibrium Thermodynamics*. Wiley–Interscience, New York, 119 pp.

Radok, U., T. J. Brown, I. N. Smith, W. F. Budd, and D. Jenssen, 1986: "On the surging potential of polar ice streams: Part IV. Antarctic ice accumulation basins and their main discharge regions." CIRES, University of Colorado, Boulder, Report DOE/ER/60197-5, Addendum 5 pp.

Ramanathan, V., R. D. Cess, E. F. Harrison, P. Minnis, B. R. Barkstrom, E. Ahmad, and D. Hartmann, 1989: "Cloud-radiative forcing and climate: Results from the Earth Radiation Budget Experiment." Science **243**, 57–63.

Ramanathan, V. and J. A. Coakley, Jr., 1978: "Climate modeling through radiative–convective models." Rev. Geophys. Space Phys. **16**, 465–489.

Ramanathan, V., R. J. Cicerone, H. B. Singh, and J. T. Kiehl, 1985: "Trace gas trends and their potential role in climate change." J. Geophys. Res. **90**, 5547–5566.

Rasmusson, E. M. and T. H. Carpenter, 1982: "Variations in tropical sea surface temperature and surface wind fields associated with the Southern Oscillation/El Niño. Mon. Weather Rev. **110**, 354–384.

Rasmusson, E. M. and T. H. Carpenter, 1983: "The relationship between eastern Equatorial Pacific sea surface temperatures and rainfall over India and Sri Lanka." Mon. Weather Rev. **111**, 517–528.

Reed, R. J., W. J. Campbell, L. A. Rasmussen, and D. G. Rogers, 1961: "Evidence of a downward propagating, annual wind reversal in the equatorial stratosphere." J. Geophys. Res. **66**, 813–818.

Reid, R. O., B. A. Elliott, and D. B. Olson, 1981: "Available potential energy: A clarification." J. Phys. Oceanogr. **11**, 15–29.

Reif, F., 1965: *Fundamentals of Statistical and Thermal Physics*. McGraw–Hill, New York, 651 pp.

Richman, M. B., 1986: "Rotation of principal components." J. Climate **6**, 293–335.

Richtmyer, R. D. and K. W. Morton, 1967: *Difference Methods for Initial-Value Problems*, 2nd ed. Interscience, New York, 405 pp.

Ropelewski, C. F., 1989: "Monitoring large-scale cryosphere/atmosphere interactions." Adv. Space Res. **9**, 213–218.

Ropelewski, C. F. and M. S. Halpert, 1987: "Global and regional scale precipitation patterns associated with the El Niño/Southern Oscillation." Mon. Weather Rev. **115**, 1606–1626.

Rosen, R. D. and D. A. Salstein, 1980: "A comparison between circulation statistics computed from conventional data and NMC Hough analyses." Mon. Weather Rev. **108**, 1226–1247.

Rosen, R. D. and D. A. Salstein, 1983: "Variations in atmospheric angular momentum on global and regional scales, and the length of day." J. Geophys. Res. **88**, 5451–5470.

Rosen, R. D., D. A. Salstein, and J. P. Peixoto, 1979: "Variability in the annual fields of large-scale atmospheric water vapor transport." Mon. Weather Rev. **107**, 26–37.

Rosen, R. D., D. A. Salstein, J. P. Peixoto, A. H. Oort, and N.-C. Lau, 1985: "Circulation statistics derived from level III-b and station-based analyses during FGGE." Mon. Weather Rev. **113**, 65–88.

Rosen, R. D., D. A. Salstein, A. J. Miller, and K. Arpe, 1987: "Accuracy of atmospheric angular momentum estimates from operational analyses." Mon. Weather Rev. **115**, 1627–1639.

Rosen, R. D., D. A. Salstein, and T. M. Wood, 1990: "Discrepancies in the earth–atmosphere angular momentum budget." J. Geophys. Res. **95**, 265–279.

Ruddiman, W. F. and J. E. Kutzbach, 1989: "Forcing of late cenozoic Northern Hemisphere climate by plateau uplift in Southern Asia and the American West." J. Geophys. Res. **94**, 18,409–18,427.

Russell, G. L., J. R. Miller, and L.-C. Tsang, 1985: "Seasonal oceanic heat transports computed from an atmospheric model." Dyn. Atm. Oceans **9**, 253–271.

Salstein, D. A. and R. D. Rosen, 1986: "Earth rotation as a proxy for interannual variability in atmospheric circulation, 1860–present." J. Climate Appl. Meteor. **25**, 1870–1877.

Salstein, D. A., R. D. Rosen, and J. P. Peixoto, 1983: "Modes of variability in annual hemispheric water vapor and transport fields." J. Atmos. Sci. **40**, 788–803.

Saltzman, B., 1957: "Equations governing the energetics of the larger scales of atmospheric turbulence in the domain of wave number." J. Meteor. **14**, 513–523.

Saltzman, B., 1968: "Steady-state solutions for axially symmetric climatic variables." Pure Appl. Geophys. **69**, 237–259.

Saltzman, B., 1970: "Large-scale atmospheric energetics in the wave number domain." Rev. Geophys. Space Phys. **8**, 289–302.

Saltzman, B., 1977: "Global mass and energy requirements for glacial oscillations and their implications for mean ocean temperature oscillations." Tellus **29**, 205–212.

Saltzman, B., 1978: "A survey of statistical–dynamical models of the terrestrial climate." Adv. Geophys. **20**, 183–304.

Saltzman, B., 1983: "Climate systems analysis." Adv. Geophys. **25**, 173–233.

Saltzman, B. and J. P. Peixoto, 1957: "Harmonic analysis of the mean northern hemisphere wind field for the year 1950." Q. J. R. Meteor. Soc. **83**, 360–364.

Saltzman, B. and S. Teweles, 1964: "Further statistics on the exchange of kinetic energy between harmonic components of the atmospheric flow." Tellus **16**, 432–435.

Saltzman, B. and A. D. Vernekar, 1975: "A solution for the northern hemisphere climatic zonation during a glacial maximum." Quaternary Res. **5**, 307–320.

Schlesinger, M. E. and J. F. B. Mitchell, 1987: "Climate model simulations of the equilibrium climatic response to increased carbon dioxide." Rev. Geophys. **25**, 760–798.

Schlesinger, M. E. (editor) 1988: *Physically based Modelling and Simulation of Climate and Climatic Change.* NATO ASI Series, Series C: Math. and Phys. Sciences. Kluwer Academic, Dordrecht, Vol. 243, 1084 pp. (two parts).

Schneider, S. H. and R. E. Dickinson, 1974: "Climate modeling." Rev. Geophys. Space Phys. **12**, 447–493.

Schneider, S. H. and T. Gal-Chen, 1973: "Numerical experiments in climate." J. Geophys. Res. **78**, 6182–6194.

Schneider, S. H. and C. Mass, 1975: "Volcanic dust, sunspots, and temperature trends." Science **190**, 741–746.

Schopf, P. S. and M. J. Suarez, 1988: "Vacillations in a coupled ocean–atmosphere model." J. Atmos. Sci. **45**, 549–566.

Schuurmans, C. J. E., 1969: "The influence of solar flares on the tropospheric circulation." KNMI Mededelingen en Verhandelingen **92**, 123 pp., De Bilt, The Netherlands.

Sellers, W. D., 1965: *Physical Climatology.* University of Chicago, Chicago, 272 pp.

Sellers, W. D., 1969: "A global climate model based on the energy balance of the earth–atmosphere system." J. Appl. Meteor. **8**, 396–400.

Sellers, P. J., Y. Mintz, Y. C. Sud, and A. Dalcher, 1986: "A simple biosphere model (SiB) for use within general circulation models." J. Atmos. Sci. **43**, 505–531.

Semtner, A. J. and R. M. Chervin, 1988: "A simulation of the global ocean circulation with resolved eddies." J. Geophys. Res. **93**, 15,502–15,522.

Smagorinsky, J., 1963: "General circulation experiments with the primitive equations. 1. The basic experiment." Mon. Weather Rev. **91**, 99–164.

Smagorinsky, J., 1974: "Global atmospheric modeling and the numerical simulation of climate." In *Weather and Climate Modification* (W. N. Hess editor), Wiley, New York, pp. 633–686.

Smagorinsky, J., 1983: "The beginnings of numerical weather prediction and general circulation modeling: Early recollections." Adv. Geophys. **25**, 3–37.

Smagorinsky, J., S. Manabe, and J. L. Holloway, 1965: "Numerical results from a nine-level general circulation model of the atmosphere." Mon. Weather Rev. **93**, 727–768.

Smith, M. L. and F. A. Dahlen, 1981: "The period and Q of the Chandler wobble." Geophys. J. R. Astron. Soc. **64**, 223–281.

Smith, R., 1987: *Electronics, Circuits and Devices*, 3rd ed. Wiley, New York, 476 pp.

Solomon, S., 1988: "The mystery of the Antarctic ozone "hole". " Rev. Geophys. **26**, 131–148.

Sommerfeld, A. J., 1953: *Partial Differential Equations in Physics* (translated from German). Academic, New York, 335 pp.

Starr, V. P., 1948: "An essay on the general circulation of the earth's atmosphere." J. Meteor. **5**, 39–43.

Starr, V. P., 1951: "Applications of energy principles to the general circulation." In *Compendium of Meteorology*. American Meteorological Society, Boston, MA, pp. 568–574.

Starr, V. P., 1953: "Note concerning, the nature of the large-scale eddies in the atmosphere." Tellus **5**, 494–498.

Starr, V. P., 1968: *Physics of Negative Viscosity Phenomena*. McGraw–Hill, New York, 256 pp.

Starr, V. P. and J. P. Peixoto, 1958: "On the global balance of water vapor and the hydrology of deserts." Tellus **10**, 189–194.

Starr, V. P. and J. P. Peixoto, 1971: "Pole-to-pole eddy transport of water vapor in the atmosphere during the IGY." Arch. Met. Geophys. Biokl. A **20**, 85–114.

Starr, V. P. and J. M. Wallace, 1964: "Mechanics of eddy processes in the tropical troposphere." Pure Appl. Geophys. **58**, 138–144.

Starr, V. P. and R. M. White, 1954: "Balance requirements of the general circulation." Air Force Cambridge Research Center, Geophys. Res. Papers, **35**, 57 pp.

Starr, V. P., J. P. Peixoto, and N. E. Gaut, 1970a: "Momentum and zonal kinetic energy balance of the atmosphere from five years of hemispheric data." Tellus **22**, 251–274.

Starr, V. P., J. P. Peixoto, and G. E. Sims, 1970b: "A method for the study of the zonal kinetic energy balance in the atmosphere." Pure Appl. Geophys. **80**, 346–358.

Stephens, G. L., G. G. Campbell, and T. H. Vonder Haar, 1981: "Earth radiation budgets." J. Geophys. Res. **86**, 9739–9760.

Stidd, C. K., 1975: "Meridional profiles of ship drift components." J. Geophys. Res. **80**, 1679–1682.

Stolarski, R. S., P. Bloomfield, R. D. McPeters, and J. R. Herman, 1991: "Total ozone trends deduced from Nimbus 7 TOMS data." Geophys. Res. Lett. **18**, 1015–1018.

Stommel, H., 1960: *The Gulf Stream: A Physical and Dynamical Description*. University of California, Berkeley and Los Angeles, CA, 197 pp.

Stone, P. H., 1974: "The meridional variation of the eddy heat fluxes by baroclinic waves and their parameterization." J. Atmos. Sci. **31**, 444–456.

Suarez, M. J. and I. M. Held, 1979: "The sensitivity of an energy balance climate model to variations in the orbital parameters." J. Geophys. Res. **84**, 4825–4836.

Sud, Y. C., J. Shukla, and Y. Mintz, 1988: "Influence of land surface roughness on atmospheric circulation and precipitation: A sensitivity study with a general circulation model." J. Appl. Meteor. **27**, 1036–1054.

Sutera, A., 1981: "On stochastic perturbation and long-term climate behavior." Q. J. R. Meteor. Soc. **107**, 137–151.

Swanson, G. S. and K. E. Trenberth, 1981: "Trends in the Southern Hemisphere tropospheric circulation." Mon. Weather Rev. **109**, 1879–1889.

Tisza, L., 1966: *Generalized Thermodynamics*. MIT, Cambridge, MA, 384 pp.

Tolmazin, D., 1985: *Elements of Dynamic Oceanography*. Allen and Unwin, Winchester, MA, 181 pp.

Toon, O. B., P. Hamill, R. P. Turco, and J. Pinto, 1986: "Condensation of HNO_3 and HCL in winter polar stratospheres." Geophys. Res. Lett. **13**, 1284–1287.

Trenberth, K. E., 1979: "Mean annual poleward energy transports by the oceans in the Southern Hemisphere." Dyn. Atmos. Oceans **4**, 57–64.

Trenberth, K. E., 1981: "Seasonal variations in global sea level pressure and the total mass of the atmosphere." J. Geophys. Res. **86**, 5238–5246.

Trenberth, K. E. and D. J. Shea, 1987: "On the evolution of the Southern Oscillation." Mon. Weather Rev. **115**, 3078–3096.

Trenberth, K. E., J. R. Christy, and J. G. Olson, 1987: "Global atmospheric mass, surface pressure, and water vapor variations. J. Geophys. Res. **92**, 14,815–14,826.

Trewartha, 1968: *An Introduction to Climate* (4th ed.). McGraw–Hill, New York, 408 pp.

Tucker, C. J., I. Y. Fung, C. D. Keeling, and R. M. Gammon, 1986: "Relationship between atmospheric CO_2 variations and a satellite-derived vegetation index." Nature **319**, 195–199.

Ulrych, T. J. and T. N. Bishop, 1975: "Maximum entropy spectral analysis and autoregressive decomposition." Rev. Geophys. Space Phys. **13**, 183–200.

Untersteiner, N., 1984: "The cryosphere." In *The Global Climate* (J. T. Houghton, editor), Cambridge University, New York, pp. 121–140.

U.S. Standard Atmosphere, 1976: NOAA, NASA, USAF, Washington, D.C., 227 pp.

van de Hulst, H. C., 1957: *Light Scattering by Small Particles*. Wiley, New York, 470 pp.

van Loon, H. and K. Labitzke, 1988: "Association between the 11-year solar cycle, the QBO, and the atmosphere. Part II: Surface and 700 mb in the Northern Hemisphere in winter." J. Climate **1**, 905–920.

van Loon, H. and J. C. Rogers, 1978: "The seesaw in winter temperature between Greenland and Northern Europe. Part I: General description." Mon. Weather Rev. **106**, 296–310.

van Loon, H., J. J. Taljaard, R. L. Jenne, and H. L. Crutcher, 1971: "Climate of the upper air, Southern Hemisphere: 2. Zonal geostrophic winds." NCAR TN/STR-57, NAVAIR 50-1C-56, 40 pp.

van Mieghem, J., 1961: "Zonal harmonic analysis of the Northern Hemisphere geostrophic wind field." UGGI, Monograph **8**, 57 pp.

Veryard, R. G. and R. A. Ebdon, 1961: "Fluctuations in tropical stratospheric winds." Met. Mag. **90**, 125–143.

Vinnichenko, N. K., 1970: "The kinetic energy spectrum in the free atmosphere—1 second to 5 years." Tellus **22**, 158–166.

Vonder Haar, T. H. and A. H. Oort, 1973: "New estimates of annual poleward energy transport by Northern Hemisphere oceans." J. Phys. Oceanogr. **3**, 169–172.

Von Neuman, J., 1960: "Some remarks on the problem of forecasting climatic fluctuations." In *Dynamics of Climate* (R. L. Preffer, editor), Pergamon, New York, pp. 9–11.

Vowinckel, E. and S. Orvig, 1970: "The climate of the North Polar Basin." *World Survey of Climatology*. Elsevier, Amsterdam, Vol. 14, pp. 129–252.

Walker, G. T., 1924: "Correlation in seasonal variations of weather, IX: A further study of world weather." Mem. Indian Meteor. Dept. **24**, 275–332.

Walker, G. T. and E. W. Bliss, 1932: "World Weather V." Mem. R. Meteor. Soc. **4**, 53–84.

Walker, G. T. and E. W. Bliss, 1937: "World Weather VI." Mem. R. Meteor. Soc. **4**, 119–139.

Wallace, J. M. and D. S. Gutzler, 1981: "Teleconnections in the geopotential height field during the Northern Hemisphere winter." Mon. Weather Rev. **109**, 784–812.

Wallace, J. M. and P. V. Hobbs, 1977: *Atmospheric Science: An Introductory Survey*. Academic, New York, 467 pp.

Wallace, J. M. and Q. Jiang, 1987: "On the observed structures of the interannual variability of the atmosphere/ocean climate system." In *Atmospheric and Oceanic Variability* (H. Cattle, editor), Royal Meteorological Society, Bracknell, England, pp. 17–43.

Wallace, J. M., S. Tibaldi, and A. J. Simmons, 1983: "Reduction of systematic forecast errors in the ECMWF model through the introduction of an envelope orography." Q. J. R. Meteor. Soc. **109**, 683–717.

Washington, W. M. and C. L. Parkinson, 1986: *An Introduction to Three-dimensional Climate Modeling*. University Science, Mill Valley, CA, 422 pp.

Weare, B. C., A. R. Navato, and R. E. Newell, 1976: "Empirical orthogonal analysis of Pacific sea surface temperatures." J. Phys. Ocean **6**, 671–678.

Weare, B. C. and A. Soong, 1990: "Adjusted NOAA outgoing long-wave and net solar irradiances." Q. J. R. Meteor. Soc. **116**, 205–219.

Wetherald, R. T. and S. Manabe, 1975: "The effects of changing the solar constant on the climate of a general circulation model." J. Atmos. Sci. **32**, 2044–2059.

White, R. M., 1949: "The role of the mountains in the angular momentum balance of the atmosphere." J. Meteor. **6**, 353–355.

Wiin-Nielsen, A., J. A. Brown, and M. Drake, 1963: "On atmospheric energy conversions between the zonal flow and the eddies." Tellus **15**, 261–279.

Williams, G. P. and J. L. Holloway, Jr., 1982: "The range and unity of planetary circulations." Nature **297**, 295–299.

Williams, J., R. G. Barry, and W. M. Washington, 1974: "Simulation of the atmospheric circulation using the NCAR global circulation model with ice age boundary conditions." J. Appl. Meteor. **13**, 305–317.

WMO, 1973: "One hundred years of international cooperation in meteorology (1873–1973). An historical review." World Meteorological Organization, Geneva, Switzerland, Vol. 345, 53 pp.

Wulf, O. R. and L. Davis, Jr., 1952: "On the efficiency of the engine driving the atmospheric circulation." J. Meteor. **9**, 79–82.

Wyrtki, K., 1982: "The Southern Oscillation, ocean–atmosphere interaction, and El Niño." Marine Technol. Soc. J. **16**, 3–10.

Name Index

A

Abramopoulos, F., 236
Adem, J., 473
Andrews, D. G., 389, 413
Angell, J. K., 444
Arakawa, A., 457, 459, 460, 465
Arpe, K., 90
Arrhenius, S. A., 436
Arya, S. P., 220, 228

B

Babcock, A. K., 248
Bacastow, R. B., 435
Barnes, R. T. H., 247
Barry, R. G., 211, 213
Batchelor, G. K., 405
Baumgartner, A., 169, 170, 171, 172, 271, 298
Bengtsson, L., 89, 461
Bentley, C. R., 211
Berger, A. L., 99, 464
Bergman, K. H., 20
Berlage, H. P., 418
Berlyand, T. G., 173, 174
Bishop, T. N., 67
Bjerknes, J., 50, 58, 368, 418, 424, 426
Bliss, E. W., 418
Boer, G. J., 459
Bourke, W., 66
Bowen, I. S., 218, 219, 237, 239
Bowman, K. P., 127
Bracewell, R. N., 67
Brier, G. W., 460
Broccoli, A. J., 359, 478, 479
Broecker, W. S., 436
Bromwich, D. H., 305
Bryan, F., 210
Bryan, K., 203, 204, 436, 465, 467
Bryden, H. L., 343
Bryson, R. A., 434
Budyko, M. I., 7, 125, 126, 233, 235, 238, 351, 469, 475, 479

C

Callender, G. S., 436
Campbell, G. G., 121, 123, 125, 126, 127, 128, 130
Cane, M. A., 475
Carissimo, B. C., 345, 347
Carpenter, T. H., 420, 423, 425
Cayan, D. R., 445
Cess, R. D., 30
Chan, P. H., 421
Charney, J. G., 30, 37, 337, 387
Chervin, R. M., 420, 465
Christodoulidis, D. C., 247
Clapp, P. F., 473
Coakley, J. A. Jr., 463, 470
Courant, R., 457
Cox, M. D., 465
Cressman, G. P., 84, 87
Crowley, T. J., 7, 31
Crutzen, P. J., 437

D

Dahlen, F. A., 243
Davis, L. Jr., 401
de Groot, S. R., 405
Derber, J., 474
Dickinson, R. E., 463, 465
Dobson, F. W., 343, 344
Dopplick, T. G., 116
Drazin, P. G., 387
Dutton, J. A., 401, 405
Dyson, F. J., 435

E

Eady, E. T., 337
Ebdon, R. A., 413
Edmon, H. J., Jr., 386, 391
Eliassen, A., 386, 388, 392
Ellis, J. S., 120, 348, 351
Emanuel, W. R., 434
Emery, W. J., 203
Essex, C., 405

F

Fairbanks, R. G., 306
Farman, J. C., 438, 439
Fels, S. B., 93
Fleming, R. J., 80
Fofonoff, N. P., 55
Fortak, H. G., 401
Friedrichs, K. O., 457

G

Gal-Chen, T., 469
Gardiner, B. G., 438
Gast, P. R., 92
Gates, W. L., 478
Ghan, S., 464, 472
Goody, R. M., 93, 109
Green, J. S. A., 472
Guillemin, E. A., 67
Gutzler, D. S., 66, 426, 429, 431, 432

H

Hacker, J. M., 259
Hall, M. M., 343
Halpert, M. S., 427
Hansen, J., 436, 477
Hantel, M., 259
Hardy, D. M., 69
Harris, R. G., 84
Hasselmann, K., 466, 467
Hastenrath, S., 343, 345, 423
Hayashi, Y., 65
Heckley, W. A., 90
Held, I. M., 464
Heller, L., 423
Hellerman, S., 230, 261, 262, 300
Hibler, W. D. III, 465
Hide, R., 247
Hobbs, P. V., 15, 53, 100
Holloway, J. L. Jr., 160, 458, 475, 476, 479
Holopainen, E. O., 260
Holton, J. R., 35, 413
Horel, J. D., 433
Hoskins, B. J., 66
Houghton, H. G., 78, 98, 101, 102, 103, 104, 107, 113, 240
Houghton, J. T., 444
Howard, J. N., 93

I

Imbrie, J. (and K. P.), 99
Iqbal, M., 99, 100

J

Jacobs, S. S., 306
Jaeger, L., 167, 168
Ji, M., 474
Jiang, Q., 433
Johnson, D. R., 401
Johnston, H. S., 437
Jones, P. D., 440
Julian, P. R., 420, 421

K

Kaplan, L. D., 78
Karoly, D. J., 66, 90, 444
Kasahara, A., 465, 477
Keeling, C. D., 434, 435, 436
Kettani, M. A., 271
Knutson, T. R., 421
Koppen, W., 475
Kubota, S., 66
Kukla, G. J., 214, 348
Kuo, C., 444
Kuo, H. L., 391, 459, 460
Kurihara, Y., 458, 472
Kutzbach, J. E., 478, 479

L

Labitzke, K., 415, 416, 417
Lamb, H. H., 7
Lamb, P. J., 446, 449
Lambert, 97
Large, W. G., 228, 233
Lau, K. M., 421
Lau, N.-C., 89, 90, 475
Leetmaa, A., 474
Leith, C. E., 21, 465
Lesins, G. B., 405
Lettau, H., 401
Levitus, S., 55, 76, 77, 83, 87, 88, 179, 182, 183, 186, 188, 191, 192, 195, 200, 203, 353, 354
Lewis, L. J., 203, 204
Lewy, H., 457
Lindzen, R. S., 413
Liou, K.-N., 109
List, R. J., 100
Liu, H., 144
Lockwood, J. G., 306
Lorenz, E. N., 1, 7, 22, 68, 90, 260, 369, 433, 462, 464
Lorentz, 108, 109
L'vovitch, M. I., 273

M

MacCracken, M. C., 464, 472
Madden, R., 421
Mahlman, J. D., 477
Mak, M.-K., 327
Manabe, S., 115, 348, 359, 436, 459, 465, 467, 468, 470, 471, 475, 476, 477, 478, 479
Margules, M., 367
Mass, C. F., 18, 440
Masuda, K., 343
Mazur, P., 405
McCarthy, D. D., 248
McIntyre, M. E., 389

Meehl, G. A., 196
Milankovitch, M., 24, 99
Miller, J. R., 343
Mintz, Y., 465
Mitchell, J. M., 24, 25, 29, 436, 473, 477
Miyakoda, K., 90, 461
Molina, M., 439
Molteni, F., 90
Morton, K. W., 84, 457
Munk, W. H., 221
Mysak, L. A., 180

N

Nakamura, N., 127, 357, 358, 359, 360, 362, 363
Namias, J., 412, 433
Naujokat, B., 413, 414
Neftel, A., 434
Neumann, C. J., 448
Neumann, G., 56
Newell, R. E., 247, 320, 321, 437
Newton, C. W., 260, 263, 338
North, G. R., 7, 463, 469

O

Oerlemans, J., 211
Oke, T. R., 221
Oort, A. H., 81, 85, 89, 90, 127, 129, 133, 135, 136, 139, 156, 165, 171, 200, 244, 249, 256, 258, 259, 260, 261, 262, 263, 264, 265, 266, 267, 268, 269, 274, 287, 288, 292, 294, 296, 297, 299, 301, 312, 322, 323, 326, 328, 329, 330, 331, 332, 333, 334, 336, 337, 339, 340, 342, 352, 354, 357, 358, 359, 360, 362, 363, 371, 375, 377, 380, 381, 382, 383, 385, 394, 395, 426, 428, 429, 430, 441, 444, 474
Oppenheimer, M. J., 435
Orvig, S., 306
Overland, J. E., 69
Ovtchinnikov, S P., 273

P

Palm, E., 386, 387, 388, 392
Palmen, E., 338
Palmer, T. N., 90, 459
Paltridge, G. W., 401
Pan, Y.-H., 426, 428, 429, 430
Parkinson, C. L., 212, 465
Parthasarathy, B., 447
Peixoto, J. P., 129, 156, 165, 171, 244, 256, 258, 259, 261, 262, 263, 264, 271, 274, 275, 277, 287, 288, 292, 294, 296, 297, 299, 301, 312, 322, 323, 324, 326, 328, 329, 330, 331, 332, 333, 334, 336, 337, 339, 352, 354, 375, 377, 380, 381, 382, 383, 385, 401, 408, 489
Penman, H. L., 238, 239
Peppler, R. A., 446, 449
Pfeffer, R. L., 391
Philander, S. G. H., 419, 475
Phillips, N. A., 455, 465
Pickard, G. L., 203
Pierson, W. J. Jr., 56
Piola, A. R., 203, 206
Pond, S., 228, 233
Portman, D. A., 18,
Preisendorfer, R. W., 69
Priestly, C. H. B., 261
Prigogine, I., 402

R

Radok, U., 306
Ramanathan, V., 30, 116, 436, 463, 470
Rasmusson, E. M., 420, 423, 425
Reed, R. J., 413
Reichel, E., 169, 170, 171, 172, 271, 298
Reid, R. O., 371
Reif, F., 13
Revelle, R., 435, 436
Richman, M. B., 69, 496
Richtmeyer, R. D., 84, 457
Roger, R. A., 460
Rogers, J. C., 429
Ropelewski, C. F., 213, 425, 427, 444
Rosati, A., 474
Rosen, R. D., 81, 84, 90, 244, 245, 246, 247, 248
Rosenstein, M., 230, 262
Ross, B. B., 452
Rowland, F. S., 439
Ruddiman, W. F., 478
Russell, G. L., 343

S

Salstein, D. A., 69, 90, 244, 245, 248, 496
Saltzman, B., 21, 22, 65, 66, 130, 463, 464, 472, 478, 489, 490, 491
Schlesinger, M. E., 29, 436, 463, 473, 477
Schneider, S. H., 440, 463, 465, 469
Schopf, P. S., 475
Schubert, W. H., 459, 460
Schuurmans, C. J. E., 413
Schwarzkopf, M. D., 93
Sellers, W. D., 171, 172, 235, 465, 469, 479
Semtner, A. J., 465
Shea, D. J., 422
Smagorinsky, J., 6, 452, 458, 463, 465
Smith, M. L., 243
Smith, R., 26
Solomon, S., 438
Sommerfeld, A. J., 67
Soong, A., 120
Starr, V. P., 7, 32, 61, 81, 247, 255, 256, 258, 265, 277, 296, 327, 338, 384
Stephens, G. L., 79, 84
Stidd, C. K., 196
Stolarski, R. S., 439
Stommel, H., 199
Stone, P. H., 472

Strickler, R. F., 115, 470, 471
Strokina, L. A., 173, 174
Suarez, M. J., 464, 475
Sud, Y. C., 465, 475
Sutera, A., 466
Swanson, G. S., 156

T

Terpstra, T., 477
Teweles, S., 490
Tibaldi, S., 90
Tisza, L., 9
Tolmazin, D., 177
Toon, O. B., 439
Trenberth, K. E., 135, 156, 342, 422
Trewartha, 6
Tucker, C. J., 435

U

Ulrych, T. J., 67
Untersteiner, N., 208, 209, 212, 215, 361

V

van de Hulst, H. C., 103
van der Veen, C. J., 211
van Loon, H., 156, 415, 416, 417, 429
van Mieghem, J., 490

Vernekar, A. D., 478
Veryard, R. G., 413
Vinnichenko, N. K., 16
Voigt, 109
Von Neuman, J., 462
Vonder Haar, T. H., 121, 123, 125, 126, 127, 128, 129, 130, 342
Vowinckel, E., 306

W

Wallace, J. M., 15, 53, 66, 90, 100, 327, 426, 429, 431, 432, 433
Walton, J. J., 69
Washington, W. M., 212, 465
Weare, B. C., 68, 69, 120
Weickmann, K. M., 421
Welsh, J. G., 84, 452
Wetherald, R. T., 470, 477, 478, 479
White, R. M., 61, 260
Wiin-Nielsen, A., 490
Williams, G. P., 160, 479
Williams, J., 478
Wulf, O. R., 401
Wyrtki, K., 419, 421, 425

Z

Zebiak, S. E., 475

Subject Index

A

Absorption,
  bands, 106–109
  solar radiation, 101,102
  spectrum of atmosphere, 92, 93
Absorptivity,
  definition, 96
Acid rain, 434, 437
Aerosols, 18, 101, 434
  absorption solar energy, 102
  formation, 240
Ageostrophic,
  current component, 41
  effects, 149, 318
  wind component, 39
Air-sea interactions, 418
    (see also Exchange at earth's surface)
Aitken nuclei, 101, 240
Albedo,
  definition, 96
  earth's surface, 103, 104, 116
  feedback, 29, 30, 210
  global, 118–122, 128, 129
Aliasing, 67
Analysis methods,
  objective, 84–90
Angular momentum,
  atmospheric estimates, 244, 246
  balance equation, 247–254
  definition (Ω, relative), 242
  ocean estimates, 244, 246
  solid earth estimates, 243, 246
  surface exchange, 261
  transport density vector, 252–255
  transport, Fourier components, 489
Antarctic,
  angular momentum balance, 257
  carbon dioxide, 435
  circumpolar current, 181, 184, 198, 199, 206, 343
  convergence zone, 190
  heat balance, 126, 233, 332, 353–364

  ice sheet, 17, 127, 210, 211, 215, 273, 302, 477
  ozone, 438
  sea ice, 207, 212–214, 306
  surface pressure, 198
  temperature, 137, 180, 212, 379
  water budget, 172, 174, 285, 302–307
Anthropogenic influences, 433–439
    (see also Greenhouse effect)
Arctic, 210, 211, 302
  cloudiness, 175
  heat balance, 126, 233, 353–364, 393
  ice, 212, 213, 273
  salinity ocean, 189, 190
  temperature, 180, 181, 379
  water balance, 172, 174, 302–307
Aridity index (E/P), 171, 172
Astronomical parameters modeling, 479
Astronomical unit (AU), 98, 119
Autobarotropic conditions, 57
Availability, 367, 401
  atmosphere definition, 367, 368
  ocean definition, 371
Available potential energy atmosphere,
  balance equations, 377–379
  definition, 368–370
  Fourier components, 489, 491
  observed cycle, 379–385
    (see also Energy cycle)
Available potential energy oceans,
  definition, 370–373
  observed energy cycle, 393–400

B

Baroclinic,
  component, 64
  conditions, 45, 57, 143, 148, 367
  instability, 24, 337, 383, 465, 472
Baroclinicity vector, 57, 156
Barotropic,
  component, 64
  conditions, 57, 148, 367
Bathythermograph, expendable (XBT), 75, 78

513

Beer-Bougert-Lambert law, 97
Bjerknes theorem, 50, 58, 368
Black body radiation, 95, 96
Blocking condition of weather systems, 461, 462
Boundary layer,
   exchange processes, 217–240
   planetary, 217, 222–224
Bowen constant, 237
Bowen ratio, 218, 219, 237, 238
Box diagram,
   atmosphere energetics (Lorenz), 378, 382, 383
   ocean energetics, 399, 400
Brunt-Väisälä frequency,
   atmosphere, 49, 141, 142
   ocean, 56, 194, 197, 372 (see also Stability)
Bulk aerodynamic method,
   heat, 233, 361, 459
   momentum, 228, 261, 459
   water vapor, 237, 459

C

Carbon dioxide,
   absorption bands, 107, 108
   feedback, 30, 118
   increase impact, 6, 434–436, 439–444
   modeling, 475, 477, 478
   radiative changes, 115, 116
Carnot cycle, 367, 385
Centers of action atmosphere, 429
Circulation,
   atmosphere observed, 149–162
   mean meridional circulation, 156–160, 387–390
   ocean observed, 176, 177, 196, 198–203
   ocean western intensification, 199, 203
Clausius-Clapeyron equation, 53, 239, 470
Climate,
   classification, 6, 8
   definition, 1, 9
   equations, 312–319
   models hierarchy, 463, 464 (see also Models)
   system definition, 19, 20
Climatic Optimum, 24
Cloudiness,
   feedback, 29, 30
   observed, 173–175
Clouds,
   effects on radiation balance, 103, 114, 115, 121
   parameterization, 459, 460
Computational,
   resolution, 456
   stability, 457
Computer resources and requirements, 451, 452, 458
Continuity equation of mass
   (see Equation of continuity)

Convective adjustment, 459, 470
Coriolis,
   acceleration (force), 35, 36, 38, 39, 50, 149, 213, 223, 242, 387
   parameter, 35, 37, 44, 45, 180
Courant-Friedrichs-Lewy stability condition, 457
Cretaceous, 31
Cryosphere,
   extent, 207–210
   general, 17, 18, 207–211
   observed energy budget, 353–364
   observed water budget, 302–307
Cumulus parameterization, 459, 460
Curtis-Godson approximation, 110

D

Dalton,
   evaporation law, 238
   pressure law, 52
Darcy law, 226
Data,
   analysis methods, 84–88
   assimilation in NWP models, 89, 474
   observational networks, 70–81, 88–90
   processing techniques, 81–84
"Dead" state atmosphere, 367, 368, 403
Deforestation, 433, 435
Density oceans,
   definition potential, 55
   global distribution observed, 190, 193, 194
   vertical structure observed, 194–196
Desert, 167–170
Desertification, 30, 433, 449
Diabatic heating (see Heating)
Difference schemes numerical integration, 457, 458
Diffusion,
   heat, 225, 233, 406, 472
   momentum, 225–228, 472
   water vapor, 236, 237, 275
Dirichlet condition, 482
Dissipation of kinetic energy, 224, 399
   (see also Friction)
Doppler broadening spectral lines, 108, 109
Drought conditions, 167, 168
   ENSO related, 423, 427
   Great Plains, 444, 445, 449
   Indian monsoon, 447, 449
   Subsahara, 446, 449

E

Eccentricity (see Orbital parameters)
Eddington time's arrow, 403
Eddy-mean flow interactions
   (see Wave-mean flow interactions)
Efficiency atmospheric heat engine, 367, 385

SUBJECT INDEX

Ekman,
  equations, 41
  layer, 42, 223
  pumping, 42, 179, 184, 199, 228
  transport, 179–181, 203, 228–230
Eliassen-Palm flux (E-P flux), 386–393
El Niño defined, 415
El Niño/Southern Oscillation (ENSO)
  phenomenon, 68, 69, 415–429
  CO_2 variations, 435, 436
  effect on atmospheric temperature, 441–443
  modeling, 475
Elsasser band model, 109
Emissivity, 105, 114
Empirical orthogonal functions (EOF;
  eigenvector analysis), 67–69 421, 426
  theory, 492–496
Energy atmosphere,
  balance equations, 310–312
  basic forms, 308
Energy balance surface,
  equations, 116, 117, 218–220, 237, 354
    (see also Radiation balance)
Energy cycle atmosphere,
  formulation, 310–319, 377–379
  Fourier components, 491
  observed, 382–385
Energy cycle oceans, 398–400
Energy, kinetic atmosphere, 162, 308
  balance equations, 375–377
  Fourier components, 488, 489
  observed, 162–165
Energy, kinetic oceans,
  generation, 231, 232
  observed, 204–206
Energy storage, 346
  atmosphere observed, 323, 349–352
  cryosphere, 213–215, 346, 348
  oceans observed, 339–341, 348–351, 354, 362
Energy transfer,
  atmosphere-ocean, 231, 232, 352, 361
Energy transports atmosphere,
  formulation, 313–317, 322, 324, 353
  observed, 324–336, 338, 339, 347, 360, 363
Energy transports ocean, 341–347, 356, 363
Ensemble,
  average, 12, 13, 21
  forecasts, 460
  samples (maps), 492
Enthalpy (see Heat)
Entropy, 47, 367, 371, 401–403
  balance equation, 403–405
  budget, 407–411
Equations,
  climate model, 453–456
  continuity, 32–34, 58, 59, 386, 455, 472
  energy balance, 310–312 (see also Energy)
  motion in atmosphere, 34–40, 58, 59, 386, 387, 454, 471
  motion in ocean, 40–42
  state, 51–55, 58, 59, 455, 471
  thermodynamic energy, 46, 51, 58, 59, 310, 311, 313, 386, 387, 455, 467
  vorticity (see Vorticity equation)
  water vapor, 58, 59, 455
  zonal mean form, 158, 386, 387
ERBE satellite data, 79, 120
Ergodic assumption, 13, 21
European Centre for Medium Range Weather Forecasts (ECMWF), 90, 461
Evaporation,
  formulation, 236–239, 276
  observed, 169–172, 174, 234, 235, 271
  power, 238
Evapotranspiration, 234–236, 239
Exchange at earth's surface,
  momentum, 226–231, 418
  sensible heat, 232, 233, 235, 418
  water vapor (latent heat), 233–239, 418

F

Factors,
  external, 8
  internal, 8
Feedback, 12
  cryosphere, 207, 210
  processes, 26–31
Ferrel cell, 39, 157, 158, 160, 264, 265, 291, 383, 388, 389
Fickian law for diffusion, 219, 226, 236, 406
Field capacity,
  ground, 235
  bucket model, 475
Filtering basic equations, 37–41, 456
First GARP Global Experiment (FGGE), 80
First law of thermodynamics, 46, 47
Flux gradient relationships,
  heat, 225, 233
  momentum, 225, 226
  water vapor, 236, 237
Fokker-Planck equation, 466
Forcing,
  eddies-mean flow, 385–393
  external, 19–21
Fourier,
  analyses, 66, 69, 481–487
  law for fluxes, 226
  spectrum, 482, 483, 484, 495
  transform, 65, 481, 482, 484, 490
Friction,
  eddy, 36
  force, 34, 36, 37, 40
  stress, 36, 38
  torque, 247, 250, 252, 260–263
  velocity, 227

Frictional dissipation of energy, 224
  atmospheric energy cycle, 376–378, 383–385
  ocean energy cycle, 398–400

G

Gain in feedbacks, 26–29
Gaussian type filter, 425, 430, 440, 442, 443
General circulation models (GCM), 463–466
  (see also Models)
Geopotential,
  defined, 35
  height structure observed, 144–149
Geostrophic,
  balance, 38, 39, 41
  ocean current, 41, 199–203
  wind formulation, 38, 39
  wind, Fourier components, 490
  wind, observations, 144, 149, 151, 152
Geothermal heating, 119
Gibbs,
  equation, 371
  phase rule, 52, 54
  phenomenon, 67
Glaciers, 211, 212
Global Atmospheric Research Program
  (GARP), 80, 473, 474
Global Energy and Water Cycle Experiment
  (GEWEX), 80
Global Telecommunication System (GTS),
  80, 474
Global Weather Experiment (FGGE), 80
Goody band model, 109
Gravitational potential, 31
Gravity waves, 6, 36, 459
  (see also Parameterization)
Greenhouse effect, 29, 30
  CO_2 and other trace gases, 115–118, 434–436,
    444, 477, 478
  water vapor, 30, 115
Gulf Stream, 40, 42, 83, 170, 176, 179, 184, 231,
  233, 275, 332, 336
  influence on climate Western Europe,
    176–179
  sensible, latent heat fluxes, 170, 233, 275,
    332, 336
  wind generation, 231

H

Hadley cell, 39, 157, 158, 160, 259, 261, 263, 264,
  287, 291, 293, 327, 330, 338, 383, 388, 389,
  418, 424
Halocline, 190, 192
Heat engine,
  entropy, 401–411
  ocean-atmosphere, 365–400
Heat storage (see Energy storage)
Heat transport,
  atmosphere, 129, 130, 324–327, 345

Fourier components, 490
  oceans, 129, 130, 341–347
Heating, diabatic of the atmosphere, 114–116,
  388, 389
  defined, 47, 310, 403
  observed distribution, 320, 321
  role in energy cycle, 377–379, 382–385
Hour angle, 99
Humidity,
  modeling, 470
  observed distribution, 279–284
  relative, definition, 53
  specific, definition, 52
Hurricane statistics, 448, 449
Hydrological cycle,
  atmospheric branch, 270, 271
  balance equation for water vapor, 274–276
  classic equation of hydrology, 273, 274
  modeling, 475
  terrestrial branch, 270, 271
    (see also Water vapor)
Hydrostatic equilibrium, 33, 38, 40, 41, 471, 490

I

Ice,
  flow of sea ice, 303, 304, 306, 361
  sea ice, 208, 209, 212–215, 361
  sheets, 17, 207–211
Ice age,
  Little Ice Age, 24
  modeling, 478, 479
Intensification (western) of ocean currents,
  199, 200, 203
International,
  Geophysical Year (IGY), 80
  Geosphere-Biosphere Program (IGBP), 81
  Meteorological Organization (IMO), 22, 78
  Polar Year, 80
Intertropical Convergence Zone (ITCZ),
  definition, 133

J

Jupiter circulation, 7, 160
  modeling, 479

K

von Kármán constant, 227
Kinetic energy,
  spectrum, 15, 16 (see also Energy)
Kirchhoff's law, 96, 97, 105, 106, 110
Kuroshio, 40, 42, 83, 170, 179, 184, 231, 233,
  275, 332, 336

L

La Niña, 419, 423, 428
  (see also El Niño/Southern Oscillation)
Lake level, Great Lake, 445, 449

Lagrange multiplier, 494
Lagrangian,
  form of equations, 58, 59
Lapse rate,
  dry adiabatic, 48, 49, 51, 141, 370
  moist adiabatic, 54, 141, 142
    (see also Stability static)
Latent heat, 308
  release, equations, 310–319
  release, observed, 321
  surface flux, 218, 219, 238, 239 (see also Evaporation, Heating diabatic, Energy)
Leap-frog scheme, 457
Length of day, 99
  estimates, 245–248
Lithosphere,
  description, 18
  continental torque, 266–269
  exchange angular momentum with ocean and atmosphere, 247, 249
Logarithmic wind profile, 227
Lorentz line shape, 108–110

M

Mass atmosphere,
  balance, 131, 132, 136, 137
  conservation equation, 32–34
  distribution, 133–137
  flow near surface, 133, 134
Mean meridional circulation, 62, 63, 387–390
  angular momentum transport observed, 255–259, 264, 265
  energy transport observed, 332–334
  mass transport observed, 156–160
  water vapor transport observed, 290–293
    (see also Ferrel cell, Hadley cell)
Mesoscale eddies ocean, 184, 186, 196, 206
Mie theory of scattering, 103
Mixed layer,
  atmosphere, 222, 223
  ocean modeling, 466
  ocean, 181, 194, 195
Mixing-length theory, 227
Mixing ratio water vapor (see Humidity)
Models,
  basic components, 454
  basic parameters, 451, 453
  classification (hierarchy), 463
  coupled ocean-atmosphere, 466–468
  energy balance (EBM), 467–470
  equations, 453–456
  general circulation (GCM), 464, 465
  history, 465
  radiative-convective (RCM), 115, 116, 464, 469, 470
  resolution, 456–458, 465
  statistical dynamical (SDM), 464, 470–473
  stochastic, 466, 467 (see also Numerical weather prediction)
Moment of inertia,
  atmosphere, 244
  earth, 243
Monitoring of climate, 2–4
Monsoon precipitation India, 447, 449
Mountain torque,
  formulation, 250–252
  observed, 261, 263 (see also Continental torque, Angular momentum)

N

National Meteorological Center (NMC),
  analyses, 89, 90, 244, 245
  grid, 85
  standard atmosphere, 75, 143, 145, 148, 322
Networks (observational),
  atmosphere, 70–75
  oceans, 72, 75–78
  satellite, 77, 78
  testing using models, 474
Newton,
  law of cooling, 361
  law of viscosity, 36, 222
  second law of motion, 32, 34, 469
Nimbus satellite data, 79, 120
Numerical weather prediction,
  analyses, 88–90
  models, 80, 453, 457, 474
Nyquist frequency, 67

O

Obliquity of ecliptic (see Orbital parameters)
Optical depth (thickness), 97
Orbital parameters,
  axial precession, 24, 98
  eccentricity, 24, 98
  modeling, 479
  obliquity, 24, 98
Oscillations,
  Eurasian (EU), 432, 433
  North Atlantic (NAO), 426, 429, 431, 432
  North Pacific (NPO), 432, 433
  Pacific-North American (PNA), 432, 433
Ozone,
  absorption bands, 107, 108
  absorption solar energy, 101, 102
  formation and depletion processes, 436–439
  hole, 438, 439
  pollution, 434
  radiative changes, 115, 116

P

Parameterizations,
  climate models, 458–460, 464, 469–472
  definition, 458

Parseval theorem, 485, 486, 488, 489
Pauses,
   tropopause, stratopause, mesopause
      definitions, 14, 15
Penman formula (evaporation), 238, 239
Permafrost, 208, 209, 215
Photosynthesis, 402, 403
Planck,
   constant, 105
   law, 95, 96, 97, 112, 113, 403
Pleistocene, 17, 31, 99, 215
Poisson's equations,
   for adiabatic process, 47, 404
   partial differential equation
      (used in objective analyses), 84
Polar regions,
   data distribution atmosphere, 74
   energy budget, 353–364
   hydrology, 302–307
      (see also Antarctic and Arctic)
Pollution, 433 (see also Carbon dioxide, Ozone)
Polyn'ya, 361
Potential,
   gravitational, 34
   temperature definition, 47
   temperature distribution, 140, 141
Prandtl,
   layer, 223
   mixing length model, 227, 233
Precipitable water,
   balance equation, 275
   definition, 274
   observed distribution, 278, 280–282
Precipitation,
   acid, 434
   observed, 165–168, 171, 172, 174, 271
Predictability,
   climate, 462
   ENSO, 426
   limits, 22, 23, 460–462
Prediction, weather, 22
Predictive skill in numerical weather prediction,
   460–462 (see also Models)
Pressure,
   coordinate system, 33
   distribution, 131–137
Pressure broadening spectral lines, 108
Prévost principle, 91, 104
Process, thermodynamic, 9, 10 (see also System)
Properties, extensive and intensive, 11
Pycnocline, 194, 195

Q

Quasibiennial oscillation (QBO), 413–415
   solar-QBO connection, 413, 415–417

R

Radiation balance,
   atmosphere, 114–116
   global, 94, 95
   polar cap, 357–360
   surface, 116, 117
   top of the atmosphere, 118–130
      (see also Energy balance)
Radiation flux,
   surface, 125, 126, 218, 219
   top of the atmosphere, 121–130
Radiative,
   equilibrium temperature of the earth,
      118, 463, 470 (see also Models)
   heating of the atmosphere, 114–116, 319, 321
   transfer equation, 110–113
Radiosonde (rawinsonde) network, 73, 74
Raoult's law, 180
Rayleigh scattering, 102, 103
Reflectivity definition, 96 (see also Albedo)
Relaxation,
   method to solve Poisson's equation, 84
   time, 14, 17–20 (see also Time scale)
   response time, 14, 17–20 (see also Time scale)
Resolution, time and space, 61–69
   models, 456–458, 465
Reynolds,
   expansion, 66
   number defined, 222, 223
   stress, 386
Richardson number for stability, 224, 225, 237
River runoff (see Runoff)
Rossby number, 39, 40, 387
Roughness length, 227, 228
Runoff,
   aerial, 297–299
   ratio, 171, 172
   river, 171, 172, 174, 271, 273, 293, 300,
      306, 307
   river influence on salinity ocean, 187–189
   subterranean, 273
   surface, 171, 172, 174, 215

S

Sahel drought region, 2, 168, 446
Salinity,
   global distribution surface, 187–189, 296
   vertical structure, 190–192
Satellite radiation data, 76, 78, 79, 83, 84, 118,
   120–130, 358, 359
Scale analysis, 37–41, 156, 456
Scattering,
   Mie, 103
   Rayleigh, 102, 103
   solar radiation, 102, 103
Schwarzchild equation in radiative transfer,
   110, 111

Sea,
  ice cover, 208, 209, 212–215, 306
  level changes, 17, 477
  surface temperatures, 176, 178–181, 184–187
Ship drift data, 196, 198–203
Sigma coordinate system, 455, 456
Sigma-t, 55
  observed structure, 194, 195
Snow,
  cover, 208, 209, 214, 215
  cover feedback, 29, 30, 207, 210
  properties, 17
Soil,
  albedo, 104
  heating characteristics, 221
  moisture, 273, 274, 475
Solar,
  climate connections, 413, 415–417
  constant, defined, 98, 118
  constant variations modeling, 469, 479
  irradiance, 91–94, 98–100
  spectrum, 92, 93
Solenoid,
  isosteric-isobaric, 56, 57
  term in vorticity equation, 45
  tube, 50, 156
Somali Current, 179, 196, 228, 232
Sound,
  speed, 55, 309
  wave filtering, 37, 456
Southern Oscillation, 62, 415, 418, 475
  (see also El Niño/Southern Oscillation)
  Index (SOI), 420, 423–426
Specific humidity definition, 52, 274
Spectral,
  analysis, 65, 66, 481–491
  models, 464
  window for terrestrial radiation, 93, 116
Spectrum,
  atmospheric kinetic energy, 15, 16, 466
  atmospheric temperature, 24, 25, 412
  solar energy, 92, 93, 98
Spherical harmonic expansion
  of meteorological fields, 66
Stability, static atmosphere,
  available potential energy, 370
  definition, 48–51
  moist, 54, 141, 142
  in terms of Richardson number, 224, 225
  (see also Hydrostatic equilibrium)
Stability, static in ocean,
  available potential energy, 372
  defined, 56
  observed distribution, 194, 195, 197
Standard atmosphere NMC, 75, 148
State, reference,
  availability, 367–372
  climate, 21–23
  entropy, 401–403

Stefan-Boltzmann,
  constant, 218
  law, 95, 96, 104, 117, 118, 123
Stokes,
  stream function, 158
  theorem, 43
Storage of energy,
  atmosphere, 320, 322, 323, 348–352, 357
  climatic system, 346, 355
  cryosphere, 207–209, 211–215, 348, 361–364
  oceans, 339–341, 348–351, 353, 354, 362
Storage of heat,
  atmosphere, 126, 221
  ground, equations, 219–222
  oceans, 126, 221 (see also Storage of energy)
Storm, tropical, 448, 449
Stream function,
  angular momentum,
    defined, 258
    observed cross sections, 259, 264
  atmospheric mass,
    defined, 158, 390, 391
    observed, 159, 160
  energy,
    defined, 338
    observed, 339
  oceanic mass, 204
  water vapor,
    defined, 300
    observed, 301, 302
Stress (see Wind stress)
  tensor defined, 36
Subgrid-scale processes,
  defined, 459
  parameterization, 458–460, 464
Surface drag (see Wind stress)
Surface exchange processes, 216–240
  angular momentum, 259–263
    (see also Torque)
  energy, 217–222, 231, 232, 352, 366
  latent heat, 218, 219
  momentum, 222, 223, 226–231
  radiation, 103, 104, 116, 117, 125, 126, 219,
  sensible heat, 218–222, 232, 233, 235, 361
  water, 165–170, 233, 239
Sverdrup transport, 42, 180
Systems, thermodynamic,
  external, internal, closed, open, cyclic, etc.,
    8–13, 27–29
  transitive, intransitive, almost intransitive,
    22, 23 (see also Predictability)

T

Tchebycheff polynomials, 69
Teleconnection patterns, 426, 427, 429, 431–433
  (see also ENSO)
Temperature atmosphere,
  global distribution surface, 137–139, 234

potential, 47, 140, 141
  equivalent, 54, 141, 142
  variability, 142–144
  vertical structure, 139–141
  virtual, 52, 132, 148
Temperature oceans,
  global distribution surface, 176, 178–181
  potential, defined, 55
  variability, 184–187
  vertical structure, 181–184
Thermal,
  conductivity, diffusivity, 219–222, 361
  wind, 58, 156, 387
Thermocline, permanent, 181–184
Thermohaline circulation, 181, 203, 204, 399
Tidal friction moon, 247
Tilted trough model atmosphere, 253, 255
Time scales, 19, 20, 37, 412, 453, 458
  atmospheric dissipation, 399
  droughts, 444–447, 449
  greenhouse warming, 434–436, 440, 444, 477
  numerical modeling, 458, 466, 467
  oceanic dissipation, 399
  sea ice, 212
  water in climatic system, 273, 284
    (see also Relaxation time)
Time step in numerical modeling, 457
Tomography acoustic, 2
Torque,
  continental, 266–269
  friction, 247, 250, 252, 260, 263, 268
  mountain, 250–252, 260, 263, 268
  pressure, 247, 250
Trace gases,
  heating effect, 436
  ozone destruction, 437–439
  radiative effects, 116, 118
Transfer function in feedbacks, 26–29
Transmissivity, 96, 97
  atmospheric layer, 111–113
  diffuse, 113
  function, 109
Tropical Ocean Global Atmosphere (TOGA) Experiment, 80
Truncation problem, 67
Turbidity air (see Aerosols)
Turbulence generation and maintenance, 222–225 (see also Surface exchange processes)
Twisting term in vorticity equation, 46

U

Upwelling, 179–181, 228
U.S. Standard Atmosphere, 15

V

Variations,
  forced, 12, 22–26
  free, 12, 22–26
Venus runaway greenhouse effect, 31
Viscosity, dynamic coefficient, 36
  eddy coefficient, 36, 226
  kinematic coefficient, 36, 222–224
  molecular, 36
  negative, 265, 376, 384, 472
Voigt line shape, 109
Volcanic eruptions, 18, 272
  Agung, 443
  El Chichón, 78, 443
  Mount St. Helens, 78 (see also Aerosols)
Vorticity,
  absolute, planetary, relative, 43, 44
  balance oceans, 199–203
  definition, 42, 43
  equation, 45, 46, 50
  potential, definition, 50
  stretching or tipping term, 41

W

Walker circulation, 418–420, 424
Wave-mean flow interactions,
  global energy cycle, 376–378, 382–385
  maintenance and forcing zonal-mean state, 385–393
  spectral domain, 491
Water vapor,
  absorption bands, 106, 107
  balance equation, 274–276
  divergence, 293, 295–297
  greenhouse effect, 30, 115
  Fourier components, 490
  meridional transport, 287–290
  modes of transport, 277, 285, 286, 289–291
  radiative effects, 115
  stream function, 300–302
  vertical transport, 291, 293
  zonal transport, 285–288
Webber formulation, 44
Wien's displacement law, 96, 104
Wiener theorem, 487
Wind,
  drag, 228–230
  forcing oceans, 399, 400
  meridional component, 133, 134, 151, 152, 156, 157, 202
  stress, 41, 228–230
  thermal, equation, 58, 156, 387
  vertical component, 157, 158
  zonal component,
    surface, 133, 134, 201
    upper air, 151–155
World Climate Program, 80
World Meteorological Organization (WMO), 78, 80
World Ocean Circulation Experiment (WOCE), 80
World Weather Watch, 80